RENEWALS 458-4574

DATE DUE

GAYLORD			PRINTED IN U.S.A.

Quantitative Fund Management

CHAPMAN & HALL/CRC
Financial Mathematics Series

Aims and scope:
The field of financial mathematics forms an ever-expanding slice of the financial sector. This series aims to capture new developments and summarize what is known over the whole spectrum of this field. It will include a broad range of textbooks, reference works and handbooks that are meant to appeal to both academics and practitioners. The inclusion of numerical code and concrete real-world examples is highly encouraged.

Series Editors

M.A.H. Dempster
Centre for Financial Research
Judge Business School
University of Cambridge

Dilip B. Madan
Robert H. Smith School of Business
University of Maryland

Rama Cont
Center for Financial Engineering
Columbia University
New York

Published Titles

American-Style Derivatives; Valuation and Computation, *Jerome Detemple*

Credit Risk: Models, Derivatives, and Management, *Niklas Wagner*

Engineering BGM, *Alan Brace*

Financial Modelling with Jump Processes, *Rama Cont and Peter Tankov*

An Introduction to Credit Risk Modeling, *Christian Bluhm, Ludger Overbeck, and Christoph Wagner*

Introduction to Stochastic Calculus Applied to Finance, Second Edition, *Damien Lamberton and Bernard Lapeyre*

Numerical Methods for Finance, *John A. D. Appleby, David C. Edelman, and John J. H. Miller*

Portfolio Optimization and Performance Analysis, *Jean-Luc Prigent*

Quantitative Fund Management, *M. A. H. Dempster, Georg Pflug, and Gautam Mitra*

Robust Libor Modelling and Pricing of Derivative Products, *John Schoenmakers*

Structured Credit Portfolio Analysis, Baskets & CDOs, *Christian Bluhm and Ludger Overbeck*

Understanding Risk: The Theory and Practice of Financial Risk Management, *David Murphy*

Proposals for the series should be submitted to one of the series editors above or directly to:
CRC Press, Taylor & Francis Group
4th, Floor, Albert House
1-4 Singer Street
London EC2A 4BQ
UK

Chapman & Hall/CRC FINANCIAL MATHEMATICS SERIES

Quantitative Fund Management

Edited by

M. A. H. Dempster

Centre for Financial Research, University of Cambridge
& Cambridge Systems Associates Limited
Cambridge, U. K

Gautam Mitra

Centre for The Analysis of Risk and Optimisation Modelling Applications: Carisma
Brunel University
Uxbridge West London, U. K.

Georg Pflug

Department of Statistics and Decision Support Systems
University of Vienna
Vienna, Austria

CRC Press
Taylor & Francis Group
Boca Raton   London   New York

CRC Press is an imprint of the
Taylor & Francis Group, an **informa** business
A CHAPMAN & HALL BOOK

Chapman & Hall/CRC
Taylor & Francis Group
6000 Broken Sound Parkway NW, Suite 300
Boca Raton, FL 33487-2742

© 2009 by Taylor & Francis Group, LLC
Chapman & Hall/CRC is an imprint of Taylor & Francis Group, an Informa business

No claim to original U.S. Government works
Printed in the United States of America on acid-free paper
10 9 8 7 6 5 4 3 2 1

International Standard Book Number-13: 978-1-4200-8191-6 (Hardcover)

Library of Congress Cataloging-in-Publication Data

Dempster, M. A. H. (Michael Alan Howarth), 1938-
    Quantitative fund management / Michael Dempster, Georg Pflug, and Gautam Mitra.
        p. cm.
    Includes bibliographical references and index.
    ISBN 978-1-4200-8191-6 (alk. paper)
    1. Portfolio management--Mathematical models. 2. Investment analysis--Mathematical models. I. Pflug, Georg Ch., 1951- II. Mitra, Gautam, 1947- III. Title.

HG4529.5.D465 2008
332.63'2042--dc22   2008014075

Visit the Taylor & Francis Web site at
http://www.taylorandfrancis.com

and the CRC Press Web site at
http://www.crcpress.com

Contents

Editors

Dr. M.A.H. Dempster was educated at Toronto (BA), Oxford (MA), and Carnegie Mellon (MS, PhD) Universities and is currently professor emeritus at the Statistical Laboratory, Centre for Mathematical Sciences, in the University of Cambridge. Previously he was director of the Centre for Financial Research, director of research and director of the doctoral programme at the Judge Business School. He was also professor of mathematics and director of the Institute for Studies in Finance in the University of Essex, R. A. Jodrey Research Professor of Management and Information Sciences and professor of Mathematics, Statistics and Computing Science at Dalhousie University and fellow and University Lecturer in Industrial Mathematics at Balliol College, Oxford. Since 1996 he has been the managing director of Cambridge Systems Associates Limited, a financial services consultancy and software company. His teaching experience is in a broad range of mathematical, decision, information and managerial sciences, and he has supervised over 40 doctoral students in these fields. He has held visiting positions at Stanford, Berkeley, Princeton, Toronto, Rome and IIASA. He has also been a consultant to numerous leading international commercial and industrial organizations and several governments and is in demand as a speaker and executive educator around the world.

As well as mathematical and computational finance and economics, his present research interests include optimization and nonlinear analysis, stochastic systems, algorithm analysis, decision support and applications software. He is author of over 100 published research articles and reports and is author, editor or translator of 11 books including *Introduction to Optimization Methods* (with P. R. Adby), *Large Scale Linear Programming* (2 vols., with G. B. Danzig and M. Kallio), *Stochastic Programming, Deterministic and Stochastic Scheduling* (with J. K. Lenstra and A. H. G. Rinnooy Kan), *Mathematics of Derivative Securities* (with S. R. Pliska) and *Risk Management: Value at Risk and Beyond*. He is founding joint Editor-in-Chief of *Quantitative Finance* with J. Doyne Farmer and presently shares this position with J.-P. Bouchaud. He was formerly on the editorial boards of the *Review of Economic Studies, Journal of Economic Dynamics and Control, Mathematical Finance* and *Computational Economics* and is currently an associate editor of *Stochastics, Computational Finance* and the *Journal of Risk Management in Financial Institutions*. He received the D. E. Shaw Best Paper Award at Computational Intelligence in Financial Engineering 1999 and became an honorary fellow of the UK Institute of Actuaries in 2000. In 2004 The Mathematical Programming Society recognized him as a Pioneer of Stochastic Programming.

Dr. Gautam Mitra qualified as an electrical engineer at Jadavpur University, Kolkata. He then joined London University, where he completed first an MSc followed by a PhD in computer methods in operational research. Dr. Mitra joined the faculty of Mathematics, Statistics and OR at Brunel University in 1974; since 1988 he has held a chair of Computational Optimisation and Modelling, and between 1990–2001 he was head of the Department of Mathematical Sciences. In 2001, under the strategic research initiative of Brunel University, Professor Mitra established Centre for the Analysis of Risk and Optimisation Modelling (CARISMA) and has led it as its Director. Professor Mitra's research interests cover mathematical optimisation and, more recently, financial modelling and risk quantifications. In the year 2003 Brunel University, in recognition of his academic contributions, honoured him as a distinguished professor.

Professor Mitra, prior to his academic career at Brunel University, worked with SCICON, SIA and ICL; in all these organisations he was involved in the (software) development of mathematical programming optimisation and modelling systems. He was the director of advanced NATO Advanced Study Research Institute, which took place in Val-d-Isere (1987) on the topic of Mathematical Models for Decision Support. Professor Mitra has also had substantial involvement with industry-based projects in his role as a director of UNICOM Seminars and OptiRisk Systems. He is the author of two textbooks, four edited books and over ninety-five journal papers. His Web sites are http://www.carisma.brunel.ac.uk and http://www.optirisk-systems.com

Dr. Georg Pflug studied law, mathematics and statistics at the University of Vienna. He was assistant professor at the University of Vienna, professor at the University of Giessen, Germany and is currently full professor at the University of Vienna and Head of the Computational Risk Management Group. He will be Dean of the Faculty of Business, Economics and Statistics in the period 2008–2010. Georg Pflug held visting positions University of Bayreuth, Michigan State University, University of California at Davis, Université de Rennes, Technion Haifa and Princeton University. He is also part time research scholar at the International Institute of Applied Systems Analysis, Laxenburg, Austria.

He is currently editor-in-chief of *Statistics and Decisions* and associate editor of *Stochastic Programming Electronic Publication Series, Central European Journal of Operations Research, Austrian Journal of Statistics, Mathematical Methods of OR, Computational Optimization and Applications,* and *Computational Management Science.* Georg Pflug is author of 4 books, editor of 6 books, and has written more than 70 publications in refereed journals, such as: *Annals of Statistics, Annals of OR, Probability Theory, J. Statist. Planning and Inference, J. ACM, Parallel Computing, The Computer Journal, Math. Programming, Mathematics of Optimization, SIAM J. on Optimization, Computational Optimization and Applications, J. Applied Probability, Stoch. Processes and Applications, Graphs and Combinatorics, J. Theoretical Computer Science, Journal of Banking and Finance, Quantitative Finance* and others.

Contributors

Carlo Acerbi
Abaxbank
Milano, Italy

Massimo Bernaschi
Istituto per le Applicazioni del Calcolo
   'Mauro Picone'—CNR
Roma, Italy

Marida Bertocchi
Department of Mathematics, Statistics,
   Computer Science and Applications
Bergamo University
Bergamo, Italy

Maya Briani
Istituto per le Applicazioni del Calcolo
   'Mauro Picone'—CNR
Roma, Italy

Chris Charalambous
Department of Business Administration
University of Cyprus
Nicosia, Cyprus

Nicos Christofides
Centre for Quantitative Finance
Imperial College
London, U.K.

Stefano Ciliberti
CNRS
Université Paris Sud
Orsay, France

Eleni D. Constantinide
Department of Business Administration
University of Cyprus
Nicosia, Cyprus

Kenneth Darby-Dowman
The Centre for the Analysis of Risk and
   Optimisation Modelling (CARISMA)
Brunel University
London, U.K.

M. A. H. Dempster
Centre for Financial Research, Judge
   Business School
University of Cambridge & Cambridge
   Systems Associates Ltd.
Cambridge, U.K.

Gabriel Dondi
ETH Zurich
Zurich, Switzerland
and
swissQuant Group AG
Zurich, Switzerland

Jitka Dupačová
Faculty of Mathematics and Physics,
   Department of Probability and
   Mathematical Statistics
Charles University
Prague, Czech Republic

Igor V. Evstigneev
School of Economic Studies
University of Manchester
Manchester, U.K.

Frank J. Fabozzi
Yale School of Management
Yale University
New Haven, Connecticut

E. Fagiuoli
Dipartimento di Informatica
Sistemistica e Comunicazione, Università
  degli Studi di Milano-Bicocca
Milano, Italy

Sergio Focardi
Yale School of Management
Yale University
New Haven, Connecticut

Hans P. Geering
ETH Zurich
Zurich, Switzerland

M. Germano
Pioneer Investment Management Ltd.
Dublin, Ireland

Rosella Giacometti
Department of Mathematics, Statistics,
  Computer Science and Applications
Bergamo University
Bergamo, Italy

Florian Herzog
ETH Zurich
Zurich, Switzerland
and
swissQuant Group AG
Zurich, Switzerland

Yuan-Hung Hsuku
Department of Financial Operations
National Kaohsiung First University of
  Science and Technology
Taiwan, Republic of China

Gerd Infanger
Department of Management Science and
  Engineering
Stanford University
Stanford, California
and
Infanger Investment Technology, LLC
Mountain View, California

Caroline Jonas
The Intertek Group
Paris, France

Michal Kaut
Molde University College
Molde, Norway

Simon Keel
ETH Zurich
Zurich, Switzerland
and
swissQuant Group AG
Zurich, Switzerland

Imre Kondor
Collegium Budapest
Budapest, Hungary

Pavlo A. Krokhmal
Department of Mechanical and Industrial
  Engineering
The University of Iowa
Iowa City, Iowa

Spiros H. Martzoukos
Department of Business Administration
University of Cyprus
Nicosia, Cyprus

E. A. Medova
Centre for Financial Research, Judge
    Business School
University of Cambridge & Cambridge
    Systems Associates Ltd.
Cambridge, U.K.

Marc Mézard
CNRS
Université Paris Sud
Orsay, France

Gautam Mitra
The Centre for the Analysis of Risk and
    Optimisation Modelling (CARISMA)
Brunel University
London, U.K.

John M. Mulvey
Bendheim Center of Finance
Princeton University
Princeton, New Jersey

Marco Papi
Istituto per le Applicazioni del Calcolo
    'Mauro Picone'—CNR
Roma, Italy

Georg Pflug
Department of Statistics and Decision
    Support Systems
University of Vienna
Vienna, Austria

Traian A. Pirvu
Department of Mathematics
The University of British Columbia
Vancouver, British Columbia, Canada

Jan Polívka
Faculty of Mathematics and Physics,
    Department of Probability and
    Mathematical Statistics
Charles University
Prague, Czech Republic

Svetlozar T. Rachev
School of Economics and Business
    Engineering
University of Karlsruhe
Karlsruhe, Germany
and
Department of Statistics and Applied
    Probability
University of California
Santa Barbara, California

M. I. Rietbergen
Securitisation & Asset Monetisation Group
Morgan Stanley
London, U.K.

Diana Roman
The Centre for the Analysis of Risk and
    Optimisation Modelling (CARISMA)
Brunel University
London, U.K.

F. Sandrini
Pioneer Investment Management Ltd.
Dublin, Ireland

Klaus R. Schenk-Hoppé
Leeds University Business School and
  School of Mathematics
University of Leeds
Leeds, U.K.

Lorenz M. Schumann
swissQuant Group AG
Zurich, Switzerland

M. Scrowston
Pioneer Investment Management Ltd.
Dublin, Ireland

F. Stella
Dipartimento di Informatica
Sistemistica e Comunicazione, Università
  degli Studi di Milano-Bicocca
Milano, Italy

Cenk Ural
Lehman Brothers
New York, New York
and
Blackrock, Inc.
New York, New York

A. Ventura
Dipartimento di Informatica
Sistemistica e Comunicazione, Università
  degli Studi di Milano-Bicocca
Milano, Italy

Davide Vergni
Istituto per le Applicazioni del Calcolo
  'Mauro Picone'—CNR
Roma, Italy

Hercules Vladimirou
Centre for Banking and Financial
  Research, School of Economics and
  Management
University of Cyprus
Nicosia, Cyprus

Stein W. Wallace
Molde University College
Molde, Norway

David Wozabal
Department of Statistics and Decision
  Support Systems
University of Vienna
Vienna, Austria

Stavros A. Zenios
Centre for Banking and Financial
  Research, School of Economics and
  Management
University of Cyprus
Nicosia, Cyprus

N. Zhang
ABN Amro
London, U.K.

Zhuojuan Zhang
Blackrock, Inc.
New York, New York

Introduction to Quantitative Fund Management

A T THE TENTH TRIENNIAL INTERNATIONAL CONFERENCE on stochastic programming held at the University of Arizona in October 2004, it was observed that the fund management industry as a whole was far from the leading edge of research in financial planning for asset allocation, asset liability management, debt management and other financial management problems at the strategic (long term) level. This gap is documented in the timely survey of quantitative equity management by Fabozzi, Focardi and Jonas which forms the first chapter of this book. It was therefore agreed to bring out a special issue of *Quantitative Finance* to partially address the imbalance between research and practice by showcasing leading edge applicable theory and methods and their use for practical problems in the industry. A call for papers went out in August and October of 2005. As an outcome of this, we were able to compile a first special issue with the papers forming the ten chapters in Part 1 of this book. In fact, the response to the call was so good that a second special issue focusing on tactical financial planning and risk management is contained in the ten chapters of Part 2.

Taken together, the twenty chapters of this volume constitute the first collection to cover quantitative fund management at both the dynamic strategic and one period tactical levels. They consider optimal portfolio choice for wealth maximization together with integrated risk management using axiomatically defined risk measures. Solution techniques considered include novel applications to quantitative fund management of stochastic control, dynamic stochastic programming and related optimization techniques. A number of chapters discuss actual implemented solutions to fund management problems including equity trading, pension funds, mortgage funding and guaranteed investment products. All the contributors are well known academics or practitioners. The remainder of this introduction gives an overview of their contributions.

In Part I of the book on *dynamic financial planning* the survey by Fabozzi *et al.* (Chapter 1) finds that, at least in the equity world, the interest in quantitative techniques is shifting from basic Markowitz mean-variance portfolio optimization to risk management and trading applications. This trend is represented here with the chapter by Fagiuoli, Stella and Vetura (Chapter 5). The remaining chapters in Part 1 cover novel aspects of lifetime individual consumption investment problems, fixed mix portfolio rebalancing allocation strategies (including Cover-type universal portfolios), debt management for funding mortgages and national debt, and guaranteed return fund construction. Of the ten chapters in Part 1, one is the mentioned survey, three are theoretical, two concern proofs of concept for practical trading or fund management strategies and the remaining four concern real-world implementations for major financial institutions.

Chapter 2 by Pirvu expands on the classical consumption investment problem of Merton to include a value-at-risk constraint. The portfolio selection problem over a finite horizon is a stochastic control problem which is reduced to pathwise nonlinear optimization through the use of the stochastic Pontryagin maximum principal. Numerical results are given and closed form solutions obtained for special cases such as logarithmic utility. The third chapter by Hsuku extends the classical Merton problem in a different direction to study the positive effects of adding derivatives to investors' choices. The model utilizes a recursive utility function for consumption and allows predictable variation of equity return volatility. Both of these theoretical studies concern realistically incomplete markets in which not all uncertainties are priced.

The next three chapters mainly treat variants of the fixed-mix rebalance dynamic asset allocation strategy. The first of these (Chapter 4) by Dempster, Evstigneev and Schenk-Hoppé shows under very general stationary ergodic return assumptions that such a strategy, which periodically rebalances a portfolio to fixed proportions of the current portfolio value, grows exponentially on almost every path even in the presence of suitable transactions costs. Chapter 5 in this group by Fagiuoli, Stella and Ventura develops, and tests on stock data from four major North American indices, an online algorithm for equity trading based on Cover's non-parametric universal portfolios in the situation when some market state information is also available. Chapter 6 by Mulvey, Ural and Zhang discusses return enhancing additions to both fixed mix rebalance strategies and optimal dynamic allocation strategies obtained by dynamic stochastic programming in the context of work for the U.S. Department of Labor. In particular, positive return performance is demonstrated from diversification to non-traditional asset classes, leverage, and overlay strategies which require no investment capital outlay.

The next two chapters concern debt management problems which use dynamic stochastic programming to optimally fund mortgage lending and government spending requirements respectively. These are asset liability management problems in which assets are specified and decisions focus on liabilities, namely, when and how to issue bonds. The first, Chapter 7 by Infanger, is an exemplary study conducted for Freddie Mac which shows that significant extra profits can be made by employing dynamic models relative to static Markovitz mean-variance optimization or traditional duration and convexity matching of assets (mortgage loans) and liabilities (bonds). In addition, efficient out-of-sample simulation evaluation of the robustness of the recommended optimal funding strategies is described, but not historical backtesting. Chapter 8 by Bernaschi, Briani, Papi and Vergni concentrates on yield curve modelling for a dynamic model for funding Italian public debt by government bond issuance. The idea of this contribution, important in an EU context, is to model the basic ECB yield curve evolution together with an orthogonal national idiosyncratic component.

The last two chapters of Part 1 describe the use of dynamic stochastic programming techniques to design guaranteed return pension funds which employ dynamic asset allocations to balance fund return versus guarantee shortfall. Chapter 9 by Hertzog, Dondi, Keel, Schumann and Geering treats this asset liability management problem using a deterministic evolution of the guarantee liability, while Chapter 10 by Dempster, Germano, Medova, Rietbergen, Sandrini and Scrowston treats guarantee shortfall with respect to a stochastic liability which is evaluated from the forward ECB yield curve simulation used to price bonds in the dynamic portfolio. Both chapters employ historical backtesting of their models for respectively a hypothetical Swiss pension fund and (a simplified version of) actual funds backing guaranteed return products of Pioneer Investments.

Taken together, the ten chapters of Part 1 give a current snapshot of state-of-the-art applications of dynamic stochastic optimization techniques to long term financial planning. These techniques range from new pathwise (Pirvu) and standard dynamic programming (Hsuku) methods of stochastic control, through sub-optimal, but easily understood and implemented policies (Dempster *et al.*, Fagiouli *et al.*, Mulvey *et al.*) to dynamic stochastic programming techniques involving the forward simulation of many risk factors (Mulvey *et al.*, Infanger, Bernaschi *et al.*, Hertzog *et al.*, Dempster *et al.*). Although there is currently widespread interest in these approaches in the fund management industry, more than a decade after their commercial introduction they are still in the early stages of adoption by practitioners, as the survey of Fabozzi *et al.* shows. This volume will hopefully contribute to the recognition and wider acceptance of stochastic optimization techniques in financial practice.

Part 2 of this volume on *portfolio construction and risk management* concerns the *tactical* level of financial planning. Most funds, with or without associated liabilities—and explicitly or implicitly—employ a three level hierarchy for financial planning. The top *strategic* level considers asset classes and risk management over longer term horizons and necessarily involves *dynamics* (the topic of Part 1). The middle *tactical* level of the financial planning hierarchy concerns portfolio construction and risk management at the individual security or fund manager level over the period up to the next portfolio rebalance. This is the focus of the ten contributions of the second part of the book. The third and bottom *operational* level of the financial planning hierarchy is actual *trading* which, with the rise of hedge funds, and as the survey of quantitative equity management by Fabozzi *et al.* in Chapter 1 demonstrates, is becoming increasingly informed by tactical models and considerations beyond standard Markowitz *mean-variance optimization* (MVO). This interaction is the evident motivation for many of the chapters in Part 2 with their emphasis on non-Gaussian returns, new risk-return tradeoffs and robustness of benchmarks and portfolio decisions. The first two chapters are based on insights gained from actual commercial applications, while of the remaining eight chapters all but one, which is theoretically addressing an important practical issue, test new theoretical contributions on market data. Another theme of *all* the contributions in this part is that their concern is with techniques which are *scenario*—rather than *analytically*—*based* (although the purely theoretical chapter uses a limiting analytical approximation). This theme reflects the necessity for nontrivial computational approaches when the classical independent Gaussian return paradigm is set aside in favour of non-equity instruments and shorter term (e.g. daily or weekly) returns.

The first chapter of Part 2, Chapter 11 by Dempster, Germano, Medova, Rietbergen, Sandrini, Scrowston and Zhang treats the problem of benchmarking fund performance using *optimal* fixed mix rebalancing strategies (a theme of Part 1) and tests it relative to earlier work on optimal portfolios for guaranteed return funds described in Chapter 10.

Chapter 12 by Acerbi provides a timely and masterful survey of the recent literature on coherent risk measures, including practical linear programming models for portfolios constructed by their minimization. This theme is elaborated further by Krokhmal in Chapter 13 which treats higher moment coherent risk measures. It examines their theoretical properties and performance when used in portfolio construction relative to standard mean variance and expected shortfall *conditional value at risk* (CVaR) optimization.

The next three chapters treat the robustness properties of the numerical minimization of CVaR using linear programming as employed in practice, for example, for bond portfolios. The first, Chapter 14 by Ciliberti, Kondor and Mezard, uses limiting continuous approximations suggested by statistical physics to define a critical threshold for the ratio of the number of assets to the number of historical observations beyond which the expected shortfall (CVaR) risk measure is not well-defined—a phase-change phenomenon first noted by Kondor and co-authors. Next Kaut, Vladimirou, Wallace and Zenios examine in Chapter 15 the stability of portfolio solutions to this problem with respect to estimation (from historical data) errors. They conclude that sensitivity to estimation errors in the mean, volatility, skew and correlation all have about the same non-negligible impact, while error in kurtosis has about half that of the other statistics. Finally, Chapter 16 by Dupačova and Polívka discusses stress-testing the CVaR optimization problem using the contamination scenario technique of perturbation analysis. They also show that similar techniques may be applied to the minimal analytical *value at risk* (VaR) problem for the Gaussian case, but are not applicable to the corresponding historical scenario based problem.

The next group of three chapters extend the treatment of portfolio construction and risk management beyond the usual simple tradeoff of volatility risk and return embodied in MVO. Chapter 17 by Giacometti, Bertocchi, Rachev and Fabozzi shows that the Black-Litterman Bayesian approach to portfolio construction, incorporating both market and practitioner views, can be extended to Student-t and stable return distributions and VaR and CVaR risk measures. Pflug and Wozabal consider in Chapter 18 the *robust optimization* problem of finding optimal portfolios in the Knightian situation when the distributions underlying returns are not perfectly known. They develop and test an algorithm for this situation based on two level convex optimization. In the last chapter in this group, Chapter 19, Roman, Darby-Dowman and Mitra consider the multi-objective problem of simultaneously trading off expected return with *two* risk measures based on variance and expected shortfall (CVaR). In tests with FTSE 100 index securities they find that an optimal balance with the two risk measures dominates those using either alone.

The final chapter in Part 2, Chapter 20 by Charalambous, Christofides, Constantinide and Martzoukos, treats the basic requirement for pricing exotic and over-the-counter options—fitting vanilla option market price data—using *non-recombining* (binary) trees, a special case of the multi-period scenario trees used in Part 1 for strategic portfolio management. The authors' approach dominates the usual recombining tree (lattice) in that it can easily handle transactions costs, liquidity constraints, taxation, non-Markovian dynamics, etc. The authors demonstrate its practicality using a penalty method and quasi-Newton unconstrained optimization and its excellent fit to the volatility surface—crucial for hedging and risk control.

The ten chapters of Part 2 provide an up-to-date overview of current research in tactical portfolio construction and risk management. Their emphasis on general return distributions and tail risk measures is appropriate to the increasing penetration of hedge fund trading techniques into traditional fund and asset liability management. We hope that this treatment of tactical problems (and its companion strategic predecessor) will make a valuable contribution to the future practical use of systematic techniques in fund management.

M.A.H. DEMPSTER, GAUTAM MITRA and GEORG C. PFLUG
Cambridge, London & Vienna

Dynamic Financial Planning

CHAPTER 1

Trends in Quantitative Equity Management: Survey Results

FRANK J. FABOZZI, SERGIO FOCARDI and CAROLINE JONAS

CONTENTS

1.1   INTRODUCTION

I N THE SECOND HALF OF THE 1990s, there was so much skepticism about quantitative fund management that Leinsweber (1999), a pioneer in applying advanced techniques borrowed from the world of physics to fund management, wrote an article entitled: 'Is quantitative investment dead?' In the article, Leinweber defended quantitative fund management and maintained that in an era of ever faster computers and ever larger databases, quantitative investment was here to stay. The skepticism towards quantitative fund management, provoked by the failure of some high-profile quantitative funds, was related to the fact that investment professionals felt that capturing market inefficiencies could best be done by exercising human judgement.

Despite mainstream academic theory that had held that markets are efficient and unpredictable, the asset managers' job has always been to capture market inefficiencies for their clients. At the academic level, the notion of efficient markets has been progressively relaxed. Empirical evidence that began to be accumulated in the 1970s led to the gradual acceptance of the notion that financial markets are somewhat predictable and that systematic market inefficiencies can be detected (see Granger 1992 for a review to various models that accept departures from efficiency). Using the variance ratio test, Lo and MacKinlay (1988) disproved the random walk hypothesis. Additional insights return predictability was provided by Jegadeesh and Titman (1993), who established the existence of momentum phenomena. Since then, a growing number of studies have accumulated evidence that there are market anomalies that can be systematically exploited to earn excess profits after considering risk and transaction costs (see Pesaran 2005 for an up-to-date presentation of the status of market efficiency). Lo (2004) proposed replacing the Efficient Market Hypothesis with the Adaptive Market Hypothesis arguing that market inefficiencies appear as the market adapts to changes in a competitive environment.

The survey study described in this paper had as its objective to reveal to what extent the growing academic evidence that asset returns are predictable and that predictability can be exploited to earn a profit have impacted the way equity assets are being managed. Based on an Intertek 2003 survey on a somewhat different sample of firms, Fabozzi et al. (2004) revealed that models were used primarily for risk management, with many firms eschewing forecasting models. The 2006 survey reported in this chapter sought to reveal to what extent modelling has left the risk management domain to become full-fledged asset management methodology. Anticipating the results discussed below, the survey confirms that quantitative fund management is now an industrial reality, successfully competing with traditional asset managers for funds. Milevsky (2004) observes that the methods of quantitative finance have now been applied in the field of personal wealth management.

We begin with a brief description of the field research methodology and the profile of responding firms. Section 1.3 discusses the central finding, that is, that models are being used to manage an increasing amount of equity asset value. Section 1.4 discusses the changing role of modelling in equity portfolio management, from decision-support systems to a fully automated portfolio construction and trading system, and from passive management to active management. Section 1.5 looks at the forecasting models most commonly used in the industry and discusses the industry's evaluation of the techniques. Section 1.6 looks at the use (or lack of use) of high-frequency data and the motivating factors. Section 1.7 discusses risk measures being used and Section 1.8 optimization methodologies. The survey reveals a widespread use of optimization, which is behind the growing level of automation in fund management. The wide use of models has created a number of challenges: survey respondents say that differentiating quantitative products and improving on performance are a challenge. Lastly, in looking ahead, we discuss the issue of the role of models in market efficiency.

1.2 METHODOLOGY

The study is based on survey responses and conversations with industry representatives in 2006. In all, managers at 38 asset management firms managing a total of €3.3 trillion ($4.3 trillion) in equities participated in the survey. Participants include persons responsible for quantitative equity management and quantitative equity research at large and medium-sized firms in North America and Europe.

The home market of participating firms is 15 from North America (14 from U.S. 1 from Canada) and 23 from Europe (U.K. 7, Germany 5, Switzerland 4, Benelux 3, France 2 and Italy 2). Equities under management by participating firms range from €5 bn to €800 bn.

While most firms whose use of quantitative methods is limited to performance analysis or risk measurement declined to participate in this study (only 5 of the 38 participating firms reported no equity funds under quantitative management), the study does reflect the use of quantitative methods in equity portfolio management at firms managing a total of €3.3 trillion ($4.3 trillion) in equities; 63% of the participating firms are among the largest asset managers in their respective countries. It is fair to say that these firms represent the way a large part of the industry is going with respect to the use of quantitative methods in equity portfolio management. (Note that of the 38 participants in this survey, 2 responded only partially to the questionnaire. For some questions, there are therefore 36 (not 38) responses.)

1.3 GROWTH IN EQUITY ASSETS UNDER QUANTITATIVE MANAGEMENT

The skepticism relative to the future of quantitative management at the end of the 1990s has given way and quantitative methods are now playing a large role in equity portfolio management. Twenty-nine percent (11/38) of the survey participants report that more than 75% of their equity assets are being managed quantitatively. This includes a wide spectrum of firms, with from €5 billion to over €500 billion in equity assets under management. Another 58% (22/38) report that they have some equities under quantitative management, though for most of these (15/22) the percentage of equities under quantitative management is less than 25%—often under 5%—of total equities under management. Thirteen percent (5/38) report no equities under quantitative management. Figure 1.1 represents the distribution of percentage of equities under quantitative management at different intervals for responding firms.

Relative to the period 2004–2005, the amount of equities under quantitative management has grown at most firms participating in the survey. Eighty-four percent of the respondents (32/38) report that the percentage of equity assets under quantitative management has either increased with respect to 2004–2005 (25/38) or has remained stable at about 100% of equity assets (7/38). The percentage of equities under quantitative management was down at only one firm and stable at five.

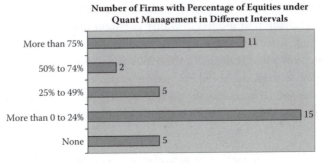

Number of Firms with Percentage of Equities under Quant Management in Different Intervals

Interval	Number of Firms
More than 75%	11
50% to 74%	2
25% to 49%	5
More than 0 to 24%	15
None	5

FIGURE 1.1 Distribution of the percentage of equities under quant management.

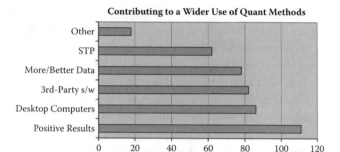

FIGURE 1.2 Score attributed to each factor contributing to a wider use of quant methods.

One reason given by respondents to explain the growth in equity assets under quantitative management is the flows into existing quantitative funds. A source at a large U.S. asset management firm with more than 50% of its equities now under quantitative management said, 'The firm has three distinct equity products: value, growth and quant. Quant is the biggest and is growing the fastest.' The trend towards quantitative management is expected to continue.

According to survey respondents, the most important factor contributing to a wider use of quantitative methods in equity portfolio management is the positive result obtained with these methods. Half of the participants rated positive results as the single most important factor contributing to the widespread use of quantitative methods. Other factors contributing to a wider use of quantitative methods in equity portfolio management are, in order of importance attributed to them by participants, the computational power now available on the desktop, more and better data, and the availability of third-party analytical software and visualization tools. Figure 1.2 represents the distribution of the score attributed to each factor. Participants were asked to rate from 1 to 5 in order of importance, 5 being the most important. Given the sample of 36 firms that responded, the maximum possible score is 180.

Sources identified the prevailing in-house culture as the most important factor holding back a wider use of quantitative methods (this evaluation obviously does not hold for firms that can be described as quantitative): more than one third (10/27) of the respondents at other than quant-oriented firms considered this the major blocking factor. Figure 1.3 represents the distribution of the total score attributed to each factor.

The positive evaluation of models in equity portfolio management is in contrast with the skepticism of some 10 years ago. A number of changes have occurred. First, expectations are now more realistic. In the 1980s and 1990s, traders were experimenting with methodologies from advanced science in hopes of making huge excess returns. Experience of the last 10 years has shown that models can indeed deliver but that their performance must be compatible with a well-functioning market.[1]

---

[1] There was a performance decay in quantitatively managed equity funds in 2006–2007. Many attribute this decaying performance to the fact that there are now more portfolio managers using the same factors and the same data.

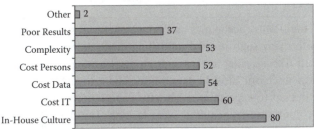

FIGURE 1.3 Score attributed to each factor holding back a wider use of quant methods.

Other technical reasons include a manifold increase in computing power and more and better data. Modellers have now available on their desktop computing power that, at the end of the 1980s, could be got only from multimillion dollar supercomputers. Data, including intraday data, can now be had (though the cost remains high) and are in general 'cleaner' and more complete. Current data include corporate actions, dividends, and fewer errors—at least in developed-country markets.

In addition, investment firms (and institutional clients) have learned how to use models throughout the investment management process. Models are now part of an articulated process that, especially in the case of institutional investors, involves satisfying a number of different objectives, such as superior information ratios.

1.4 CHANGING ROLE FOR MODELS IN EQUITY PORTFOLIO MANAGEMENT

The survey reveals that quantitative models are now used in active management to find alphas (i.e. sources of excess returns), either relative to a benchmark or absolute. This is a considerable change with respect to the past when quantitative models were used primarily to manage risk and to select parsimonious portfolios for passive management.

Another finding of this study is the growing amount of funds managed automatically by computer programs. The once futuristic vision of machines running funds automatically without the intervention of a portfolio manager is becoming a reality on a large scale: 55% of the respondents (21/38) report that at least part of their equity assets are now being managed automatically with quantitative methods; another three plan to automate at least a portion of their equity portfolios within the next 12 months. The growing automation of the equity investment process indicates that that there is no missing link in the technology chain that leads to automatic quantitative management. From return forecasting to portfolio formation and optimization, all the needed elements are in place.

Until recently, optimization represented the missing technology link in the automation of portfolio engineering. Considered too brittle to be safely deployed, many firms eschewed optimization, limiting the use of modelling to stock ranking or risk control functions. Advances in robust estimation methodologies and in optimization now allow a manager to construct portfolios of hundreds of stocks chosen in universes of thousands of stocks with little or no human intervention outside of supervising the models.

1.5 MODELLING METHODOLOGIES AND THE INDUSTRY'S EVALUATION

At the end of the 1980s, academics and researchers at specialized quant boutiques experimented with many sophisticated modelling methodologies including chaos theory, fractals and multi-fractals, adaptive programming, learning theory, complexity theory, complex nonlinear stochastic models, data mining and artificial intelligence. Most of these efforts failed to live up to expectations. Perhaps expectations were too high. Or perhaps the resources or commitment required were lacking. Derman (2001) provides a lucid analysis of the difficulties that a quantitative analyst has to overcome. As observed by Derman, though modern quantitative finance uses some of the techniques of physics, a wide gap remains between the two disciplines.

The modelling landscape revealed by the survey is simpler and more uniform. Regression analysis and momentum modelling are the most widely used techniques: respectively, 100% and 78% of the survey respondents say that these techniques are being used at their firms. Other modelling methods being widely used include cash flow analysis and behavioural modelling. Forty-seven percent (17/36) of the participating firms model cash flows; 44% (16/36) use behavioural modelling. Figure 1.4 represents the distribution of modelling methodologies among participants.

Let us observe that regression models used today have undergone a substantial change since the first multifactor models such as Arbitrage Pricing Theory (APT) were introduced. Classical multifactor models such as APT are static models embodied in linear regression between returns and factors at the same time:

$$r_i = \alpha_i + \sum_{j=1}^{p} \beta_{ij} f_j + \varepsilon_i.$$

Models of this type allow managers to measure risk but not to forecast returns, unless the factors are forecastable. Sources at traditional asset management firms typically use factor models to control risk or build stock screening systems. A source doing regression on factors to capture the risk-return trade-off of assets said, 'Factor models are the most intuitive and most comprehensive models for explaining the sources of risk.'

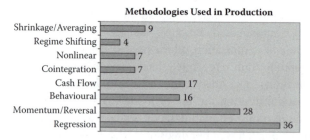

FIGURE 1.4 Distribution of modelling methodologies among participants.

However, modern regression models are dynamic models where returns at time $t + 1$ are regressed on factors at time t:

$$r_{i,t+1} = \alpha_i + \sum_{j=1}^{p} \beta_{ij} f_{j,t} + \varepsilon_{i,t}.$$

Models of this type are forecasting models insofar as the factors at time t are predictors of returns at time behaviour $t + 1$. In these models, individual return processes might exhibit zero autocorrelation but still be forecastable from other variables.

Predictors might include financial and macroeconomic factors as well as company specific parameters such as financial ratios. Predictors might also include human judgment, for example analyst estimates, or technical factors that capture phenomena such as momentum. A source at a quant shop using regression to forecast returns said, 'Regression on factors is the foundation of our model building. Ratios derived from financial statements serve as one of the most important components for predicting future stock returns. We use these ratios extensively in our bottom-up equity model and categorize them into five general categories: operating efficiency, financial strength, earnings quality (accruals), capital expenditures and external financing activities.'

Momentum and reversals are the second most widely used modelling technique among survey participants. In general, momentum and reversals are used as a strategy not as a model of asset returns. Momentum strategies are based on forming portfolios choosing the highest/lowest returns, where returns are estimated on specific time windows. Survey participants gave these strategies overall good marks but noted that (1) they do not always perform so well, (2) they can result in high turnover (though some use constraints/penalties to deal with this problem) and (3) identifying the timing of reversals is tricky.

Momentum was first reported in Jegadeesh and Titman (1993) in the U.S. market. Jegadeesh and Titman (2002) confirm that momentum continued to exist in the 1990s in the US market throughout the 1990s. Karolyi and Kho (2004) examined different models for explaining momentum and introduced a new bootstrap test. Karolyi and Kho conclude that no random walk or autoregressive model is able to explain the magnitude of momentum empirically found; they suggest that models with time varying expected returns come closer to explaining the empirical magnitude of momentum.

Momentum and reversals are presently explained in the context of local models updated in real time. For example, momentum as described in Jegadeesh and Titman (1993) is based on the fact that stock prices can be represented as independent random walks when considering periods of the length of one year. However, it is fair to say that there is no complete agreement on the econometrics of asset returns that would justify momentum and reversals and stylized facts on a global scale, and not as local models. It would be beneficial to know more about the econometrics of asset returns that sustain momentum and reversals.

Behavioural phenomena are considered to play an important role in asset predictability; as mentioned, 44% of the survey respondents say they use behavioural modelling. Behavioural modellers attempt to capture phenomena such as departures from rationality on the part of investors (e.g. belief persistence), patterns in analyst estimates, and corporate executive investment/disinvestment behaviour. Behavioural finance is related to momentum in that the latter is often attributed to various phenomena of persistence in analyst estimates and investor perceptions.

A source at a large investment firm that has incorporated behavioural modelling into its active equity strategies commented, 'The attraction of behavioural finance is now much stronger than it was just five years ago. Everyone now acknowledges that markets are not efficient, that there are behavioural anomalies. In the past, there was the theory that was saying that markets are efficient while market participants such as the proprietary trading desks ignored the theory and tried to profit from the anomalies. We are now seeing a fusion of theory and practice.'

We remark that the term behavioural modelling is often used rather loosely. Full-fledged behavioural modelling exploits a knowledge of human psychology to identify situations where investors are prone to show behaviour that leads to market inefficiencies. The tendency now is to call 'behavioural' any model that exploits market inefficiency. However, implementing true behavioural modelling is a serious challenge. Even firms with very large, powerful quant teams say that 'considerable work is required to translate [departures from rationality] into a set of rules for identifying stocks as well as entry and exit points for a quantitative stock selection process.'

As for other methodologies used in return forecasting, sources cited nonlinear methods and co-integration. Nonlinear methods are being used to model return processes at 19% (7/36) of the responding firms. The nonlinear method most widely used among survey participants is classification and regression trees (CART). The advantage of CART is its simplicity and the ability of CART methods to be cast in an intuitive framework.

A source using CART as a central part of the portfolio construction process in enhanced index and longer-term value-based portfolios said, 'CART compresses a large volume of data into a form which identifies its essential characteristics, so the output is easy to understand. CART is non-parametric—which means that it can handle an infinitely wide range of statistical distributions—and nonlinear so as a variable selection technique it is particularly good at handling higher-order interactions between variables.'

Only 11% (4/36) of the respondents use nonlinear regime-shifting models; at most firms, judgment is used to assess regime change. Obstacles to modelling regime shifts include the difficulty in detecting the precise timing of a regime switch and the very long time series required for true estimation.

A source at a firm where regime-shifting models have been experimented with commented, 'Everyone knows that returns are conditioned by market regimes, but the potential for overfitting when implementing regime-switching models is great. If you could go back with fifty years of data—but we have only some ten years of data and this is not enough to build a decent model.'

Co-integration is being used by 19% (7/36) of the respondents. Co-integration models the short-term dynamics (direction) and long-run equilibrium (fair value). A perceived plus of co-integration is the transparency that it provides: the models are based on economic and finance theory and calculated from economic data.

1.6   USING HIGH-FREQUENCY DATA

High frequency data (HFD) are being used at only 14% of the responding firms (5/36), to identify profit opportunities and improve forecasts. Another three plan to use HFD within the next 12 months. A source at a large investment firm that is using HFD said, 'We use high-frequency data in event studies. The objective is to gain an understanding of the mechanisms of the market.' A source which is planning to use high-frequency data in the coming 12 months remarked, 'We believe that high-frequency data will allow us to evaluate exactly when it is optimal to trade, for example at close, VWAP, or midday, and to monitor potential market impact of our trades and potential front-running of our brokers.' (VWAP stands for volume-weighted average price.)

Though it is believed that HFD could be useful, cost of data is the blocking factor. Survey participants voiced concerns that the cost of data will hamper the development of models in the future. One source observes, 'The quasi monopolistic positioning of data vendors allows them to charge prices that are incompatible with the revenues structure of all but the biggest firms.' Other reasons cited by the sources not using HFD are a (perceived) unattractive noise-to-signal ratio and resistance to HFD-based strategies on the part of institutional investors.

1.7   MEASURING RISK

Risk is being measured at all the responding firms. Risk measures most widely used among participants include variance (97% or 35/36), Value at Risk (VaR) (67% or 24/36) and downside risk measures (39% or 14/36), Conditional VaR (CVaR), and extreme value theory (EVT) are used at 4 (11%) and 2 (6%) firms, respectively. The considerable use of asymmetric risk measures such as downside risk can be ascribed to the growing popularity of financial products with guaranteed returns. Many firms compute several risk measures: the challenge here is to merge the different risk views into a coherent risk assessment. Figure 1.5 represents the distribution of risk measures used by participants.

It is also interesting to note that among survey participants, there is a heightened attention to model risk. Model averaging and shrinkage techniques are being used by one-fourth (9/36) of the survey participants. The recent take-up of these techniques is related to the fact that most firms are now using multiple models to forecast returns, a trend that is up compared to two or three years ago. Other techniques to mitigate model risk, such as

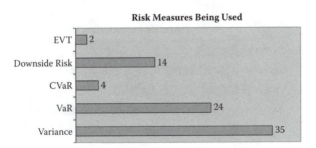

FIGURE 1.5 Distribution of risk measures adopted by participants.

random coefficient models, are not used much in the industry. In dealing with model risk we must distinguish between averaging model results and averaging models themselves. The latter technique, embodied in random coefficient models, is more difficult and requires more data.

1.8 OPTIMIZATION

Another area where much has changed recently is optimization. According to sources, optimization is now performed at 92% (33/36) of the participating firms, albeit in some cases only rarely. Mean-variance is the most widely used technique among survey participants: it is being used by 83% (30/36) of the respondents. It is followed by utility optimization (42% or 15/36) and, more recently, robust optimization (25% or 9/36). Only one firm mentioned that it is using stochastic optimization. Figure 1.6 represents the distribution of optimization methods.

The wider use of optimization is a significant development compared to just a few years ago when many sources reported that they eschewed optimization: the difficulty of identifying the forecasting error was behind the then widely held opinion that optimization techniques were too brittle and prone to 'error maximization.' The greater use of optimization is due to advances in large-scale optimization coupled with the ability to include constraints and robust methods for estimation and optimization itself. It is significant: portfolio formation strategies rely on optimization. With optimization now feasible, the door is open to a fully automated investment process. In this context, it is noteworthy that 55% of the survey respondents report that at least a portion of their equity assets is being managed by a fully automated process.

Optimization is the engineering part of portfolio construction. Most portfolio construction problems can be cast in an optimization framework, where optimization is applied to obtain the desired optimal risk-return profile. Optimization is the technology behind the current offering of products with specially engineered returns, such as guaranteed returns. However, the offering of products with particular risk-return profiles requires optimization methodologies that go well beyond the classical mean-variance optimization. In particular one must be able to (1) work with real-world utility functions and (2) apply constraints to the optimization process.

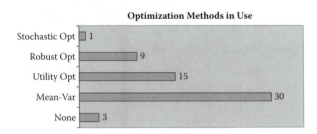

FIGURE 1.6 Distribution of optimization methods adopted by participants.

1.9 CHALLENGES

The growing diffusion of models is not without challenges. Survey participants noted three:

- increasing difficulty in differentiating products;
- difficulty in marketing quant funds, especially to non-institutional investors; and
- performance decay.

Quantitative equity management has now become so widespread that a source at a long-established quantitative investment firm remarked, 'There is now a lot of competition from new firms entering the space [of quantitative investment management]. The challenge is to continue to distinguish ourselves from competition in the minds of clients.'

With many quantitative funds based on the same methodologies and using the same data, the risk is to construct products with the same risk-return profile. The head of active equities at a large quantitative firm with more than a decade of experience in quantitative management remarked, 'Everyone is using the same data and reading the same articles: it's tough to differentiate.'

While sources report that client demand is behind the growth of (new) pure quantitative funds, some mentioned that quantitative funds might be something of a hard sell. A source at a medium-sized asset management firm servicing both institutional clients and high net worth individuals said, 'Though clearly the trend towards quantitative funds is up, quant approaches remain difficult to sell to private clients: they remain too complex to explain, there are too few stories to tell, and they often have low alpha. Private clients do not care about high information ratios.'

Markets are also affecting the performance of quantitative strategies. A recently released report from the Bank for International Settlements (2006) noted that this is a period of historically low volatility. What is exceptional about this period, observes the report, is the simultaneous drop in volatility in all variables: stock returns, bond spread, rates and so on. While the role of models in reducing volatility is unclear, what is clear is that models produce a rather uniform behaviour. Quantitative funds try to differentiate themselves either finding new unexploited sources of return forecastability, for example novel ways of looking at financial statements, or using optimization creatively to engineer special risk-return profiles.

A potentially more serious problem is performance decay. Survey participants remarked that model performance was not so stable. Firms are tackling these problems in two ways. First, they are protecting themselves from model breakdown with model risk mitigation techniques, namely by averaging results obtained with different models. It is unlikely that all models breakdown in the same way in the same moment, so that averaging with different models allows managers to diversify model risk. Second, there is an on-going quest for new factors, new predictors, and new aggregations of factors and predictors. In the long run, however, something more substantial might be required: this is the subject of the next and final section.

1.10   LOOKING AHEAD

Looking ahead, we can see a number of additional challenges. Robust optimization, robust estimation and the integration of the two are probably on the research agenda of many firms. As asset management firms strive to propose innovative products, robust and flexible optimization methods will be high on the R & D agenda. In addition, as asset management firms try to offer investment strategies to meet a stream of liabilities (i.e., measured against liability benchmarking), multistage stochastic optimization methods will become a priority for firms wanting to compete in this arena. Pan *et al.* (2006) call 'Intelligent Finance' the new field of theoretical finance at the confluence of different scientific disciplines. According to the authors, the theoretical framework of intelligent finance consists of four major components: financial information fusion, multilevel stochastic dynamic process models, active portfolio and total risk management, and *financial strategic analysis*.

The future role of high-frequency data is not yet clear. HFD are being used (1) to improve on model quality thanks to the 2000-fold increase in sample size they offer with respect to daily data and (2) to find intraday profit opportunities. The ability to improve on model quality thanks to HFD is the subject of research. It is already known that quantities such as volatility can be measured with higher precision using HFD. Using HFD, volatility ceases to be a hidden variable and becomes the measurable realized volatility, introduced by Torbin *et al.* (2003). If, and how, this increased accuracy impacts models whose time horizon is in the order of weeks or months is a subject not entirely explored. It might be that in modelling HFD one captures short-term effects that disappear at longer time horizons.

Regardless of the frequency of data sampling, modellers have to face the problem of performance decay that is the consequence of a wider use of models. Classical financial theory assumes that agents are perfect forecasters in the sense that they know the stochastic processes of prices and returns. Agents do not make systematic predictable mistakes: their action keeps the market efficient. This is the basic idea underlying rational expectations and the intertemporal models of Merton.

Practitioners (and now also academics) have relaxed the hypothesis of the universal validity of market efficiency; indeed, practitioners have always being looking for asset mispricings that could produce alpha. As we have seen, it is widely believed that mispricings are due to behavioural phenomena, such as belief persistence. This behaviour creates biases in agent evaluations—biases that models attempt to exploit in applications such as momentum strategies.

However, the action of models tends to destroy the same sources of profit that they are trying to exploit. This fact receives attention in applications such as measuring the impact of trades. In almost all current implementations, measuring the impact of trades means measuring the speed at which models constrain markets to return to an unprofitable efficiency. To our knowledge, no market impact model attempts to measure the opposite effect, that is, the eventual momentum induced by a trade.

It is reasonable to assume that the diffusion of models will reduce the mispricings due to behavioural phenomena. However, one might reasonably ask whether the action of models will ultimately make markets more efficient, destroying any residual profitability in excess of market returns, or if the action of models will create new opportunities that can be exploited by other models, eventually by a new generation of models based on an accurate analysis of model biases. It is far from being obvious that markets populated by agents embodied in mathematical models will move toward efficiency. In fact, models might create biases of their own. For example, momentum strategies (buy winners, sell losers) are a catalyst for increased momentum, farther increasing the price of winners and depressing the price of losers.

This subject has received much attention in the past as researchers studied the behaviour of markets populated by boundedly rational agents. While it is basically impossible, or at least impractical, to code the behaviour of human agents, models belong to a number of well-defined categories that process past data to form forecasts. Studies, based either on theory or on simulation, have attempted to analyse the behaviour of markets populated by agents that have bounded rationality, that is, filter past data to form forecasts.[2] One challenge going forward will be to understand what type of inefficiencies are produced by markets populated by automatic decision-makers whose decisions are based on past data. It is foreseeable that simulation and artificial markets will play a greater role as discovery devices.

REFERENCES

Bank for International Settlements, The recent behaviour of financial market volatility. *BIS Release Paper*, 29 August, 2006.

Derman, E., A guide for the perplexed quant. *Quant. Finan.*, 2001, 1(5), 476–480.

Fabozzi, F.J., Focardi, S.M. and Jonas, C.L., Trends in quantitative asset management in Europe. *J. Portfolio Manag.*, 2004, 31(4), 125–132 (Special European Section).

Granger, C.W.J., Forecasting stock market prices: Lessons for forecasters. *Int. J. Forecast.*, 1992, **8**, 3–13.

Jegadeesh, N. and Titman, S., Returns to buying winners and selling losers: Implications for stock market efficiency. *J. Finan.*, 1993, **48**(1), 65–92.

Jegadeesh, N. and Titman, S., Cross-sectional and time-series determinants of momentum returns. *Rev. Financ. Stud.*, 2002, 15(1), 143–158.

Kahneman, D., Maps of bounded rationality: Psychology for behavioral economics. *Am. Econ. Rev.*, 2003, **93**(5), 1449–1475.

Karolyi, G.A. and Kho, B.-C., Momentum strategies: Some bootstrap tests. *J. Empiric. Finan.*, 2004, **11**, 509–536.

LeBaron, B., Agent-based computational finance. In *Handbook of Computational Economics*, edited by L. Tesfatsion and K. Judd, 2006 (North-Holland: Amsterdam).

Leinweber, D., Is quantitative investing dead? *Pensions & Investments*, 8 February, 1999.

Lo, A., The adaptive markets hypothesis. *J. Portfolio Manag.*, 2004, 30 (Anniversary Issue), 15–29.

Lo, A. and MacKinley, C., Stock market prices do not follow random walks: Evidence from a simple specification test. *Rev. Financ. Stud.*, 1988, **1**, 41–66.

Milevsky, M., A diffusive wander through human life. *Quant. Finan.*, 2004, 4(2), 21–23.

---

[2] See Sargent (1994) and Kahneman (2003) for the the theoretical underpinning of bounded rationality from the statistical and behavioural point of view, respectively. See LeBaron (2006) for a survey of research on computational finance with boundedly rational agents.

Pan, H., Sornette, D. and Kortanek, K., Intelligent finance—an emerging direction. *Quant. Finan.*, 2006, **6**(4), 273–277.

Pesaran, M.H., Market efficiency today. *Working Paper 05.41*, 2005 (Institute of Economic Policy Research).

Sargent, J.T., *Bounded Rationality in Macroeconomics*, 1994 (Oxford University Press: New York).

Torbin, A., Bollerslev, T., Diebold, F.X. and Labys, P., Modeling and forecasting realized volatility. *Econometrica*, 2003, **71**(2), 579–625.

CHAPTER 2

Portfolio Optimization under the Value-at-Risk Constraint

TRAIAN A. PIRVU

CONTENTS

2.1 INTRODUCTION

MANAGERS LIMIT THE RISKINESS of their traders by imposing limits on the risk of their portfolios. Lately, the Value-at-Risk (VaR) risk measure has become a tool used to accomplish this purpose. The increased popularity of this risk measure is due to the fact that VaR is easily understood. It is the maximum loss of a portfolio over a given horizon, at a given confidence level. The Basle Committee on Banking Supervision requires U.S. banks to use VaR in determining the minimum capital required for their trading portfolios.

In the following we give a brief description of the existing literature. Basak and Shapiro (2001) analyse the optimal dynamic portfolio and wealth-consumption policies of utility maximizing investors who must manage risk exposure using VaR. They find that VaR risk managers pick a larger exposure to risky assets than non-risk managers, and consequently incur larger losses when losses occur. In order to fix this deficiency they choose another risk measure based on the risk-neutral expectation of a loss. They call this risk measure *Limited Expected Loss* (LEL). One drawback of their model is that the portfolios VaR is never re-evaluated after the initial date, making the problem a static one. In a similar setup, Berkelaar *et al.* (2002) show that, in equilibrium, VaR reduces market volatility, but in some cases raises the probability of extreme losses. Emmer *et al.* (2001) consider a dynamic model with *Capital-at-Risk* (a version of VaR) limits. However, they assume that portfolio proportions are held fixed during the whole investment horizon and thus the problem becomes a static one as well.

Cuoco *et al.* (2001) develop a more realistic dynamically consistent model of the optimal behaviour of a trader subject to risk constraints. They assume that the risk of the trading portfolio is re-evaluated dynamically by using the conditioning information, and hence the trader must satisfy the risk limit continuously. Another assumption they make is that when assessing the risk of a portfolio, the proportions of different assets held in the portfolio are kept unchanged. Besides the VaR risk measure, they consider a coherent risk measure *Tail Value at Risk* (TVaR), and establish that it is possible to identify a dynamic VaR risk limit equivalent to a dynamic TVaR limit. Another of their findings is that the risk exposure of a trader subject to VaR and TVaR limits is always lower than that of an unconstrained trader.

Pirvu (2005) started with the model of Cuoco *et al.* (2001). We find the optimal growth portfolio subject to these risk measures. The main finding is that the optimal policy is a projection of the optimal portfolio of an unconstrained log agent (the Merton proportion) onto the constraint set with respect to the inner product induced by the volatility matrix of the risky assets. Closed-form solutions are derived even when the constraint set depends on the current wealth level.

Cuoco and Liu (2003) study the dynamic investment and reporting problem of a financial institution subject to capital requirements based on self-reported VaR estimates. They show that optimal portfolios display a local three-fund property. Leippold *et al.* (2002) analyse VaR-based regulation rules and their possible distortion effects on financial markets. In partial equilibrium the effectiveness of VaR regulation is closely linked to the *leverage effect*, the tendency of volatility to increase when the prices decline.

Vidovic *et al.* (2003) considered a model with time-dependent parameters, but the risk constraints were imposed in a static fashion.

Yiu (2004) looks at the optimal portfolio problem, when an economic agent is maximizing the utility of her intertemporal consumption over a period of time under a dynamic VaR constraint. A numerical method is proposed to solve the corresponding HJB equation. They find that investment in risky assets is optimally reduced by the VaR constraint. Atkinson and Papakokinou (2005) derive the solution to the optimal portfolio and consumption subject to CaR and VaR constraints using stochastic dynamic programming.

This paper extends Pirvu (2005) by allowing for intertemporal consumption. We address an issue raised by Yiu (2004) and Atkinson and Papakokinou (2005) by considering a market with random coefficients. It was also suggested as a new research direction by Cuoco *et al.* (2001). To the best of our knowledge this is the first work in portfolio choice theory with CRRA-type preferences, time-dependent market coefficients, incomplete financial markets, and dynamically consistent risk constraints in a Brownian motion framework.

2.1.1 Our Contribution

We propose a new approach with the potential for numerical applications. The main idea is to consider, on every probabilistic path, an auxiliary deterministic control problem, which we analyse. The existence of a solution of the deterministic control problem does not follow from classical results. We establish it and we also show that first-order necessary conditions are also sufficient for optimality. We prove that a solution of this deterministic control problem is the optimal portfolio policy for a given path. The advantage of our method over classical methods is that it allows for a better numerical treatment.

The remainder of this chapter is organized as follows. Section 2.2 describes the model, including the definition of the Value-at-Risk constraint. Section 2.3 formulates the objective and shows the limitations of the stochastic dynamic programming approach in this context. Section 2.4 treats the special case of logarithmic utility. The problem of maximizing expected logarithmic utility of intertemporal consumption is solved in closed form. This is done by reducing it to a nonlinear program, which is solved pathwise. One finding is that, at the final time, the agent invests the least proportion of her wealth in stocks. The optimal policy is a projection of the optimal portfolio and consumption of an unconstrained agent onto the constraint set. Section 2.5 treats the case of power utility, in the totally unhedgable market coefficients paradigm (see Example 2.7.4, p. 305 of Karatzas and Shreve 1998). The stochastic control portfolio problem is transformed into a deterministic control problem. The solution is characterized by the Pontryagin maximum principle (first-order necessary conditions). Section 2.6 contains an appropriate discretiza-tion of the deterministic control problem. It leads to a nonlinear program that can be solved by standard methods. It turns out that the necessary conditions of the discretized problem converge to the necessary conditions of the continuous problem. For numerical experiments, one can use NPSOL, a software package for solving constrained optimization problems that employs a sequential quadratic programming (SQP) algorithm. We end this section with some numerical experiments. The conclusions are summarized in Section 2.7. We conclude the paper with an appendix containing the proofs of the lemmas.

2.2 MODEL DESCRIPTION

2.2.1 The Financial Market

Our model of a financial market, based on a filtered probability space $(\Omega, \{\mathcal{F}_t\}_{0 \le t \le \infty}, \mathcal{F}, \mathbb{P})$ satisfying the usual conditions, consists of $m + 1$ assets. The first, $\{S_0(t)\}_{t \in [0,\infty]}$, is a *riskless bond* with a positive interest rate r, i.e. $dS_0(t) = S_0(t)r \, dt$. The remaining m are *stocks* and evolve according to the following stochastic differential equation:

$$dS_i(t) = S_i(t)\left[\alpha_i(t)dt + \sum_{j=1}^{n}\sigma_{ij}(t)dW_j(t)\right],$$

$$0 \le t \le \infty, \; i = 1, \ldots, m,$$

(2.1)

where the process $\{W(t)\}_{t\in[0,\infty)} = \{(W_j(t))_{j=1,\ldots,n}\}_{t\in[0,\infty)}$ is an n-dimensional standard Brownian motion. Here, $\{\alpha(t)\}_{t\in[0,\infty)} = \{(\alpha_i(t))_{i=1,\ldots,m}\}_{t\in[0,\infty)}$ is an $\mathbb{R}^m$-valued *mean rate of return* process, and $\{\sigma(t)\}_{t\in[0,\infty)} = \{(\sigma_{ij}(t))_{i=1,\ldots,m}^{j=1,\ldots,n}\}_{t\in[0,\infty)}$ is an $m \times n$ matrix-valued *volatility* process. We impose the following regularity assumptions on the coefficient processes $\alpha(t)$ and $\sigma(t)$.

- All the components of the process $\{\alpha(t)\}_{t\in[0,\infty)}$ are assumed positive, continuous and $\{\mathcal{F}_t\}$-adapted.
- The matrix-valued volatility process $\{\sigma(t)\}_{t\in[0,\infty)}$ is assumed continuous, $\{\mathcal{F}_t\}$-adapted and with linearly independent rows for all $t \in [0,\infty)$, a.s.

The last assumption precludes the existence of a redundant asset and arbitrage opportunities. The *rate of excess* return is the $\mathbb{R}^m$-valued process $\{\mu(t)\}_{t\in[0,\infty)} = \{(\mu_i(t))_{i=1,\ldots,m}\}_{t\in[0,\infty)}$, with $\mu_i(t) = \alpha_i(t) - r$, which is assumed positive. This also covers the case of an incomplete market if $n > m$ (more sources of randomness than stocks).

2.2.2 Consumption, Trading Strategies and Wealth

In this model the agent is allowed to consume. The intermediate consumption process, denoted $\{C(t)\}_{t\in[0,\infty)}$, is assumed positive, and $\{\mathcal{F}_t\}$-progressively measurable. Let $\{(\zeta(t), c(t))\}_{t\in[0,\infty)} = \{(\zeta_i(t))_{i=1,\ldots,m}, c(t)\}_{t\in[0,\infty)}$ be an $\mathbb{R}^{m+1}$-valued *portfolio-proportion* process. At time t its components are the proportions of the agent's wealth invested in stocks, $\zeta(t)$, and her consumption rate, $c(t)$. An $\mathbb{R}^{m+1}$-valued *portfolio-proportion* process is called *admissible* if it is $\{\mathcal{F}_t\}$-progressively measurable and satisfies

$$\int_0^t |\zeta^T(u)\mu(u)|du + \int_0^t \|\zeta^T(u)\sigma(u)\|^2\,du$$

$$+ \int_0^t c(u)du < \infty, \quad \text{a.s.,} \quad \forall t \in [0,\infty),$$

(2.2)

where, as usual, $\|\cdot\|$ is the standard Euclidean norm in $\mathbb{R}^m$. Given $\{(\zeta(t), c(t))\}_{t\in[0,\infty)}$ is a portfolio-proportion process, the leftover wealth $X^{\zeta,c}(t)(1 - \sum_{i=1}^m \zeta_i(t))$ is invested in the riskless bond $S_0(t)$. It may be that this quantity is negative, in which case we are borrowing at rate $r > 0$. The dynamics of the wealth process $\{X^{\zeta,c}(t)\}_{t\in[0,\infty)}$ of an agent using the portfolio-proportion process $\{(\zeta(t), c(t))\}_{t\in[0,\infty)}$ is given by the following stochastic differential equation:

$$dX^{\zeta,c}(t) = X^{\zeta,c}(t)\big((\zeta^T(t)\alpha(t) - c(t))dt + \zeta^T(t)\sigma(t)dW(t)\big)$$

$$+ \left(1 - \sum_{i=1}^{m}\zeta_i(t)\right)X^{\zeta,c}(t)r\,dt$$

$$= X^{\zeta,c}(t)\big(r - c(t) + \zeta^T(t)\mu(t)\big)dt$$

$$+ \zeta^T(t)\sigma(t)dW(t)).$$

Let us define the p-quadratic correction to the saving rate r:

$$Q_p(t,\zeta,c) \stackrel{\triangle}{=} r - c + \zeta^T\mu(t) + \frac{p-1}{2}\left\|\zeta^T\sigma(t)\right\|^2. \qquad (2.3)$$

The above stochastic differential equation has a unique strong solution if (2.2) is satisfied and is given by the explicit expression

$$X^{\zeta,c}(t) = X(0)\exp\left\{\int_0^t Q_0(u,\zeta(u),c(u))du + \int_0^t \zeta^T(u)\sigma(u)dW(u)\right\}. \qquad (2.4)$$

The initial wealth $X^{\zeta,c}(0) = X(0)$ takes values in $(0,\infty)$ and is exogenously given.

2.2.3 Value-at-Risk Limits

For the purposes of risk measurement, one can use an approximation of the distribution of the investor's wealth at a future date. A detailed explanation of why this practice should be employed can be found on p. 8 of Cuoco *et al.* (2001) (see also p. 18 of Leippold *et al.* (2002)). Given a fixed time instance $t \geq 0$, and a length $\tau > 0$ of the measurement horizon $[t, t + \tau]$, the projected distribution of wealth from trading and consumption is usually calculated under the assumptions that

1. the portfolio proportion process $\{(\zeta(u), c(u))\}_{u\in[t,t+\tau]}$, as well as
2. the market coefficients $\{(\alpha(u)\}_{u\in[t,t+\tau]})$ and $\{(\sigma(u)\}_{u\in[t,t+\tau]}$,

will stay constant and equal their present value throughout $[t, t+\tau]$. If τ is small, for example $\tau = 1$ trading day, the market coefficients will not change much and this supports assumption 2. The wealth's dynamics equation yields *the projected wealth* at $t+\tau$:

$$X^{\zeta,c}(t+\tau) = X^{\zeta,c}(t)\exp\{Q_0(t,\zeta(t),c(t))\tau$$
$$+ \zeta^T(t)\sigma(t)(W(t+\tau) - W(t))\},$$

whence *the projected wealth* loss on the time interval $[t, t+\tau]$ is

$$X^{\zeta,c}(t) - X^{\zeta,c}(t+\tau) = X^{\zeta,c}(t)[1 - \exp\{Q_0(t,\zeta(t),c(t))\tau$$
$$+ \zeta^T(t)\sigma(t)(W(t+\tau) - W(t))\}].$$

The random variable $\zeta^T(t)\sigma(t)(W(t+\tau) - W(t))$ is, conditionally on $\mathcal{F}_t$, normally distributed with mean zero and standard deviation $\|\zeta^T(t)\sigma(t)\|\sqrt{\tau}$. Let the confidence parameter $\alpha \in (0, 1/2]$ be exogenously specified. The α-percentile of the projected loss $X^{\zeta,c}(t) - X^{\zeta,c}(t + \tau)$ conditionally on $\mathcal{F}_t$ is

$$X^{\zeta,c}(t)\left[1 - \exp\left\{Q_0(t, \zeta(t), c(t))\tau + N^{-1}(\alpha)\|\zeta^T(t)\sigma(t)\|\sqrt{\tau}\right\}\right],$$

where $N(\cdot)$ denotes the standard cumulative normal distribution function. This prompts the *Value-at-Risk* (VaR) of projected loss

$$\mathrm{VaR}(t, \zeta, c, x) \overset{\triangle}{=} x\left[1 - \exp\left\{Q_0(t, \zeta, c)\tau + N^{-1}(\alpha)\|\zeta^T\sigma(t)\|\sqrt{\tau}\right\}\right]^+.$$

Let $a_V \in (0,1)$ be an exogenous risk limit. The *Value-at-Risk constraint* is that the agent at every time instant $t \geq 0$ must choose a portfolio proportion $(\zeta(t), c(t))$ that would result in a *relative* VaR of the projected loss on $[t, t+\tau]$ less than a_V. This, strictly speaking, is the set of all admissible portfolios which, for all $t \geq 0$, belong to $F_V(t)$, defined by

$$F_V(t) \overset{\triangle}{=} \left\{(\zeta, c) \in \mathbb{R}^m \times [0, \infty); \frac{\mathrm{VaR}(t, \zeta, c, x)}{x} \leq a_V\right\}. \tag{2.5}$$

The fraction VaR/x rather than VaR is employed, whence the name *relative* VaR. If one imposes VaR in absolute terms, the constraint set depends on the current wealth level and this makes the analysis more involved (see Cuoco *et al.* 2001; Pirvu 2005). For a given path ω let us denote $\omega^{(t)} = (\omega_s)_{s \leq t}$ as the projection up to time t of its trajectory. One can see that, for a fixed ω^t, the set $F_V(t)$ is compact and convex, being the level set of a convex, unbounded function $f_V(t, \zeta, c)$,

$$F_V(t) = \left\{(\zeta, c) \in \mathbb{R}^m \times [0, \infty); f_V(t, \zeta, c) \leq \log\frac{1}{1 - a_V}\right\},$$

where

$$f_V(t, \zeta, c) \overset{\triangle}{=} -Q_0(t, \zeta, c)\tau - N^{-1}(\alpha)\|\zeta^T\sigma(t)\|\sqrt{\tau}. \tag{2.6}$$

The function f_V, although quadratic in ζ and linear in c, may still fail to be convex in (ζ, c) if $\alpha \geq 1/2$, thus $F_V(t)$ may not be a convex set (see Figure 2.1, p. 10 of Cuoco *et al.* 2001). However, the choice of $\alpha \in (0, (1/2)]$ makes $N^{-1}(\alpha) \leq 0$ and this yields convexity.

2.3 OBJECTIVE

Let finite time horizon T and the discount factor δ (the agent's impatient factor) be primitives of the model. Given $X(0)$, the agent seeks to choose an admissible portfolio-proportion process such that $(\zeta(t), c(t)) \in F_V(t)$ for all $0 \leq t \leq T$, and the expected value of her CRRA utility of intertemporal consumption and final wealth,

$$\int_0^T e^{-\delta t} U_p(C(t)) dt + e^{-\delta T} U_p(X^{\zeta,c}(T)), \tag{2.7}$$

is maximized over all admissible portfolios processes satisfying the same constraint. Here, $U_p(x) \triangleq x^p/p$, with $p < 1$ the coefficient of relative risk aversion (CRRA). Let us assume for the moment that the market coefficients are constants. In this case the constraint set $F_V(t)$ does not change over time and we denote it by F_V. Then one can use dynamic programming techniques to characterize the optimal portfolio and consumption policy. The problem is to find a solution to the HJB equation. Define the optimal value function as

$$V(x, t) = \max_{(\zeta, c) \in F_V} \mathbb{E}_t \left[\int_t^T e^{-\delta t} U_p(C(u)) du + e^{-\delta T} U_p(X^{\zeta,c}(T)) \right],$$

where $\mathbb{E}_t$ is the conditional operator given the information known up to time t and $X^{\zeta,c}(t) = x$. The HJB equation is

$$\max_{(\zeta, c) \in F_V} J(t, x, \zeta, c) = 0,$$

where

$$J(t, x, c, \zeta) \triangleq e^{-\delta t} U_p(cx) + \frac{\partial V}{\partial t} + x\left(r - c + \zeta^T \mu\right) \frac{\partial V}{\partial x}$$
$$+ \frac{\|\zeta^T \sigma\|^2 x^2}{2} \frac{\partial^2 V}{\partial x^2},$$

with the boundary condition $V(x, T) = e^{-\delta T} U_p(x)$. The value function V inherits the concavity of the utility functions U_p. Being jointly concave in (ζ, c), the function J is maximized over the set F_V at a unique point $(\bar{\zeta}, \bar{c})$. Moreover, this point should lie on the boundary of F_V and one can derive first-order optimality conditions by means of Lagrange multipliers. Together with the HJB equation this yields a highly nonlinear PDE that is hard to solve numerically (a numerical scheme was proposed by Yiu (2004), but no convergence result was reported). In the following we approach the portfolio optimization problem by reducing it to a deterministic control problem. We are then able to obtain explicit solutions for logarithmic utility.

2.4   LOGARITHMIC UTILITY

Let us consider the case where the agent is deriving utility from intermediate consumption only. It is straightforward to extend it to also encompass the utility of the final wealth. In light of (2.4),

$$\log X^{\zeta,c}(t) = \log X(0) + \int_0^t Q_0(s, \zeta(s), c(s)) ds$$

$$+ \int_0^t \zeta^T(s)\sigma(s) dW(s). \tag{2.8}$$

What facilitates the analysis is the decomposition of the utility from intertemporal consumption into a signal, a Lebesque integral and noise, which comes at an Itô integral rate. The decomposition is additive and the expectation operator cancels the noise. Indeed,

$$\int_0^T e^{-\delta t} \log C(t) dt = \int_0^T e^{-\delta t} \log (c(t) X^{\zeta,c}(t)) dt$$

$$= \frac{1 - e^{-\delta T}}{\delta} \log X(0) + \int_0^T e^{-\delta t} \log c(t) dt$$

$$+ \int_0^T \int_0^t e^{-\delta t} Q_0(s, \zeta(s), c(s)) ds \, dt$$

$$+ \int_0^T e^{-\delta t} \int_0^t \zeta^T(s)\sigma(s) dW(s) dt.$$

By Fubini's theorem

$$\int_0^T \int_0^t e^{-\delta t} Q_0(s, \zeta(s), c(s)) ds \, dt = \int_0^T \left(\int_s^T e^{-\delta t} Q_0(s, \zeta(s), c(s)) dt \right) ds$$

$$= \int_0^T \frac{e^{-\delta t} - e^{-\delta T}}{\delta} Q_0(t, \zeta(t), c(t)) dt,$$

hence

$$\int_0^T e^{-\delta t} \log C(t) dt = \frac{1 - e^{-\delta T}}{\delta} \log X(0)$$

$$+ \int_0^T e^{-\delta t} \left(\log c(t) + \frac{1}{\delta}(1 - e^{-\delta(T-t)}) Q_0(t, \zeta(t), c(t)) \right) dt \quad (2.9)$$

$$+ \int_0^T e^{-\delta t} \int_0^t \zeta^T(s)\sigma(s) dW(s) dt.$$

The linearly independent rows assumption on the matrix-valued volatility process yields the existence of the inverse $(\sigma(t)\sigma^T(t))^{-1}$ and so equation

$$\sigma(t)\sigma^T(t)\zeta_M(t) = \mu(t) \tag{2.10}$$

uniquely defines the stochastic process $\{\zeta_M(t)\}_{t\in[0,\infty)} = \{(\sigma(t)(\sigma^T(t))^{-1}\mu(t)\}_{t\in[0,\infty)}$, called the *Merton-ratio process*. It has the property that it maximizes (in the absence of portfolio constraints), the rate of growth, and the log-optimizing investor would invest exactly using the components of $\zeta_M(t)$ as portfolios proportions (see Section 3.10 of Karatzas and Shreve 1991). By (2.10)

$$\left\| \zeta_M^T(t)\sigma(t) \right\|^2 = \zeta_M^T(t)\mu(t) = \mu^T(t)(\sigma(t)\sigma^T(t))^{-1}\mu(t). \tag{2.11}$$

The following integrability assumption is rather technical, but it guarantees that a local martingale (Itô integral) is a (true) martingale (see p. 130 of Karatzas and Shreve 1991). Let us assume that

$$\mathbb{E} \int_0^T \left\| \zeta_M^T(u)\sigma(u) \right\|^2 du < \infty. \tag{2.12}$$

This requirement, although imposed on the market coefficients (see Equation (2.11)), is also inherited for all portfolios satisfying the Value-at-Risk constraint.

Lemma 2.1: *For every* $(\zeta(t), c(t)) \in F_V(t)$ *the process* $\int_0^t \zeta^{Ts}(s)\sigma(s)\, dW(s)$, $t \in [0,T]$, *is a martingale, hence* $\mathbb{E} \int_0^t \zeta^T(s)\sigma(s)\, dW(s) = 0$.

Proof: See Appendix 2.A. □

In light of this lemma, the expectation of the noise vanishes, i.e.

$$\mathbb{E} \int_0^T e^{-\delta t} \int_0^t \zeta^T(s)\sigma(s)\mathrm{d}W(s)\mathrm{d}t = 0,$$

after interchanging the order of integration. Thus, taking expectation in the additive utility decomposition (2.9),

$$\begin{aligned}
\mathbb{E} \int_0^T e^{-\delta t} \log C(t)\mathrm{d}t &= \frac{1 - e^{-\delta T}}{\delta} \log X(0) \\
&\quad + \mathbb{E} \int_0^T e^{-\delta t} \left(\log c(t) + \frac{1}{\delta}\left(1 - e^{-\delta(T-t)}\right) \right. \\
&\quad \left. \times Q_0(t, \zeta(t), c(t)) \right) \mathrm{d}t.
\end{aligned} \tag{2.13}$$

Therefore, to maximize

$$\mathbb{E} \int_0^T e^{-\delta t} \log C(t)\mathrm{d}t$$

over the constraint set, it suffices to maximize

$$g(t, \zeta, c) \triangleq \log c(t) + \frac{1}{\delta}\left(1 - e^{-\delta(T-t)}\right) Q_0(t, \zeta(t), c(t)) \tag{2.14}$$

pathwise over the constraint set. For a fixed path ω and a time instance t, we need to solve

(P1)   maximize $g(t, \zeta, c)$,

$$\text{subject to } f_V(t, \zeta, c) \stackrel{\triangle}{=} - Q_0(t, \zeta, c)\tau - N^{-1}(\alpha)\|\zeta^T \sigma(t)\|\sqrt{\tau} \le \log\frac{1}{1 - a_V}.$$

The optimal policy for an agent maximizing her logarithmic utility of intertemporal consumption without the risk constraint is to hold the proportion $\{(\zeta_M(t), c_M(t))\}_{t\in[0,T]}$, where $c_M(t) \stackrel{\triangle}{=} \delta/[1 - e^{-\delta(T-t)}]$ (the optimum of (P1) without the constraint is $(\zeta_M(t), c_M(t))$).

Lemma 2.2: *The solution of (P1) is given by*

$$\bar{\zeta}(t) = (1 \wedge (\beta(t) \vee 0))\zeta_M(t), \tag{2.15}$$

$$\begin{aligned}\bar{c}(t) = u(t, (1 \wedge \beta(t)))c_M(t)1_{\{\beta(t) > 0\}} \\ + \left(r + \frac{1}{\tau}\log\frac{1}{1 - a_V}\right)1_{\{\beta(t)\le 0\}},\end{aligned} \tag{2.16}$$

where $\beta(t)$ is the root of the equation

$$f_V(t, z\zeta_M(t), u(t, z)c_M(t)) = \log\frac{1}{1 - a_V} \tag{2.17}$$

in the variable z, with

$$u(t, z) \stackrel{\triangle}{=} 1 + \frac{\sqrt{\tau}\|\zeta_M^T(t)\sigma(t)\|}{N^{-1}(\alpha)}(1 - z). \tag{2.18}$$

Proof: See Appendix 2.A.   □

Theorem 2.3: *To maximize the logarithmic utility of intertemporal consumption,*

$$\mathbb{E}\int_0^T e^{-\delta t}\log C(t)\,dt,$$

over processes $(\zeta(t), c(t)) \in F_V(t), 0 \le t \le T$, *the optimal portfolio is* $\{(\bar{\zeta}(t), \bar{c}(t))\}_{t\in[0,T]}$.

Proof: This is a direct consequence of (2.13) and Lemma 2.2.   □

Remark 1: Since at the final time $c_M(T) = \infty$ and $\bar{c}(t)$ is bounded we must have $\beta(T) \leq 0$, so $\bar{\zeta}(T) = 0$, and this means that, at the final time, the agent invests the least proportion (in absolute terms) of her wealth in stocks. By (2.15) and (2.16) it follows that $\bar{\zeta}(t) \leq \zeta_M(t)$, and $\bar{c}(t) \leq c_M(t)$, for any $0 \leq t \leq T$, which means that the constrained agent is consuming and investing less in the risky assets than the unconstrained agent. Let T_1 and T_2 be two final time horizons, $T_1 > T_2$. Because $c_M(t, T_1) < c_M(t, T_2)$, from Equations (2.15) and (2.16) we conclude that $\beta(t, T_1) > \beta(t, T_2)$, hence $\bar{\zeta}(t, T_1) > \bar{\zeta}(t, T_2)$, and $\bar{c}(t, T_1) > \bar{c}(t, T_2)$. Therefore, long-term agents can afford to invest more in the stock market and consume more than short-term agents (in terms of proportions).

2.5 NONLOGARITHMIC UTILITY

Let us recall that we want to maximize the expected CRRA utility $(U_p(x) = x^p/p, p \neq 0)$ from intertemporal consumption and terminal wealth,

$$\mathbb{E} \int_0^T e^{-\delta t} U_p(C(t)) dt + \mathbb{E} e^{-\delta T} U_p(X^{\zeta,c}(T)), \qquad (2.19)$$

over portfolio-proportion processes satisfying the Value-at-Risk constraint, i.e. $(\zeta(t), c(t)) \in F_V(t)$, $0 \leq t \leq T$. One cannot obtain an additive decomposition into signal and noise as in the case of logarithmic utility. However, a multiplicative decomposition can be performed. By (2.7),

$$
\begin{aligned}
U_p(X^{\zeta,c}(t)) &= \frac{X^p(0)}{p} \exp\left(\int_0^t pQ_0(s, \zeta(s), c(s)) ds + \int_0^t p\zeta^T(s)\sigma(s) dW(s) \right) \\
&= \frac{X^p(0)}{p} \exp\left(\int_0^t \left(pQ_p(s, \zeta(s), c(s)) - \frac{1}{2}p^2 \|\zeta^T(s)\sigma(s)\|^2 \right) ds \right. \\
&\quad \left. + \int_0^t p\zeta^T(s)\sigma(s) dW(s) \right) = \frac{X^p(0)}{p} N^{\zeta,c}(t) Z^\zeta(t),
\end{aligned}
$$

where

$$N^{\zeta,c}(t) \triangleq \exp\left(\int_0^t pQ_p(s, \zeta(s), c(s)) ds \right), \qquad (2.20)$$

$$
\begin{aligned}
Z^\zeta(t) \triangleq \exp\left(-\frac{1}{2} \int_0^t p^2 \|\zeta^T(s)\sigma(s)\|^2 ds \right. \\
\left. + \int_0^t p\zeta^T(s)\sigma(s) dW(s) \right),
\end{aligned}
\qquad (2.21)
$$

with Q_p defined in (2.3). By taking expectation,

$$\mathbb{E}U_p(X^{\zeta,c}(t)) = \frac{X^p(0)}{p}\mathbb{E}(N^{\zeta,c}(t)Z^{\zeta}(t)). \tag{2.22}$$

The process $N^{\zeta,c}(t)$ is the signal and $Z^{\zeta}(t)$, a stochastic exponential, is the noise. Stochastic exponentials are local martingales, but if we impose the assumption

$$\mathbb{E}\left[\exp\left(\frac{p^2}{2}\int_0^T \left\|\zeta_M^T(u)\sigma(u)\right\|^2 du\right)\right] < \infty \tag{2.23}$$

on market coefficients (see Equation (2.11)), the process $Z^{\zeta}(t)$ is a (true) martingale for all portfolio processes satisfying the constraint, as the following lemma shows.

Lemma 2.4: *For every $(\zeta(t), c(t)) \in F_V(t)$ the process $Z^{\zeta}(t)$, $t\in[0, T]$, is a martingale, hence $\mathbb{E}Z^{\zeta}(t) = 1$.*

Proof: See Appendix 2.A. □

As for utility from intertemporal consumption,

$$\begin{aligned}
\int_0^T e^{-\delta t} U_p(C(t))dt &= \int_0^T e^{-\delta t} U_p(X^{\zeta,c}(t)c(t))dt \\
&= \frac{X^p(0)}{p}\int_0^T e^{-\delta t} c^p(t)N^{\zeta,c}(t)Z^{\zeta}(t)dt.
\end{aligned} \tag{2.24}$$

We claim that

$$\mathbb{E}\left(\int_0^T e^{-\delta t} c^p(t)N^{\zeta,c}(t)(Z^{\zeta}(t) - Z^{\zeta}(T))dt\right) = 0.$$

Indeed, by conditioning and Lemma 2.4 we obtain

$$\begin{aligned}
\mathbb{E}(c^p(t)N^{\zeta,c}(t)(Z^{\zeta}(t) - Z^{\zeta}(T))) &= \mathbb{E}(\mathbb{E}[c^p(t)N^{\zeta,c}(t)(Z^{\zeta}(t) - Z^{\zeta}(T))|\mathcal{F}_t]) \\
&= \mathbb{E}(c^p(t)N^{\zeta,c}(t)\mathbb{E}[(Z^{\zeta}(t) - Z^{\zeta}(T))|\mathcal{F}_t]) \\
&= 0,
\end{aligned}$$

and Fubini's theorem proves the claim. Hence, combined with (2.24), we obtain

$$\mathbb{E}\int_0^T e^{-\delta t} U_p(C(t))dt = \frac{X^p(0)}{p}\mathbb{E}\left(Z^{\zeta}(T)\int_0^T e^{-\delta t} c^p(t)N^{\zeta,c}(t)dt\right). \tag{2.25}$$

The decomposition for the total expected utility (Equations (2.22) and (2.25)) is

$$\mathbb{E}\int_0^T e^{-\delta t}U_p(C(t))dt + \mathbb{E}e^{-\delta T}U_p\left(X^{\zeta,c}(T)\right)$$
$$= \frac{X^p(0)}{p}\mathbb{E}(Z^\zeta(T)Y^{\zeta,c}(T)), \tag{2.26}$$

where the signal $Y^{\zeta,c}(T)$ is given by

$$Y^{\zeta,c}(T) = \int_0^T e^{-\delta t}c^p(t)N^{\zeta,c}(t)dt + e^{-\delta T}N^{\zeta,c}(T), \tag{2.27}$$

with $N^{\zeta,c}(t)$ defined in (2.20). It appears natural at this point to maximize $Y^{\zeta,c}(T)$ pathwise over the constraint set. For a given path ω, the existence of an optimizer $\{(\bar{\zeta}(t,\omega),\bar{c}(t,\omega))\}_{t\in[0,T]}$ is given by Lemma 2.5. Note that $N^{\zeta,c}(t,\omega)$ depends on the trajectory of $(\zeta(\cdot,\omega),c(\cdot,\omega))$ on $[0,t]$ so one is faced with a deterministic control problem. From now on, to keep the notation simple we drop ω. In the language of deterministic control we can write (2.27) as a cost functional $I\,[x,u]$ given in the form

$$I\,[x,u] = g(x(T)) + \int_0^T f_0(t,x(t),u(t))dt, \quad g(x)\overset{\triangle}{=}e^{-\delta T}x, \tag{2.28}$$

where $u = (\zeta,c)$ is the control, x is the state variable, and the function

$$f_0(t,x,u)\overset{\triangle}{=}e^{-\delta t}c^p x \tag{2.29}$$

is defined on the set

$$A = \{(t,x,u)|(t,x)\in[0,T]\times(0,K],u(t)\in F_V(t)\}\subset\mathbb{R}^{m+3}. \tag{2.30}$$

The dynamics of the state variable is given by the differential equation

$$\frac{dx}{dt} = f(t,x(t),u(t)), \quad 0\le t\le T, \tag{2.31}$$

with the boundary condition $x(0) = 1$, where

$$f(t,x,u)\overset{\triangle}{=}x\left(pr - pc + p\zeta^T\mu(t) + \frac{p(p-1)}{2}\|\zeta^T\sigma(t)\|^2\right). \tag{2.32}$$

The constraints are $(t,x(t))\in[0,T]\times(0,K]$ and $u(t)\in F_V(t)$. Due to the compactness of the set $F_V(t)$, $0\le t\le T$, it follows that $K < \infty$. A pair (x,u) satisfying the above conditions is called admissible. The problem of finding the maxima of $I\,[x,u]$ within all admissible pairs (x,u) is called the Bolza control problem. Classical existence theory for deterministic control does not apply to the present situation and we proceed with a direct proof of existence.

Lemma 2.5: *There exists a solution* $\{\bar{u}(t)\}_{0\leq t\leq T} \stackrel{\triangle}{=} \{(\bar{\zeta}(t), \bar{c}(t))\}_{0\leq t\leq T}$ *for the Bolza control problem defined above.*

Proof: See Appendix 2.A. □

An optimal solution $\{\bar{u}(t)\}_{0\leq t\leq T} \stackrel{\triangle}{=} \{(\bar{\zeta}(t), \bar{c}(t))\}_{0\leq t\leq T}$ is characterized by a system of forward backward equations (also known as the Pontryagin maximum principle). Let $\tilde{\lambda} = (\lambda_0, \lambda_1)$ be the adjoint variable and

$$H(t, x, u, \tilde{\lambda}) = \lambda_0 f_0(t, x, u) + \lambda_1 f(t, x, u)$$

the Hamiltonian function. The necessary conditions for the Bolza control problem (the Pontryagin maximum principle) can be found in Cesari (1983) (Theorem 2.5.1.i). In general, they are not sufficient for optimality. We prove that, in our context, the necessary conditions are also sufficient, as the following lemma shows.

Lemma 2.6: *A pair* $\bar{x}(t), \bar{u}(t) = (\bar{\zeta}(t), \bar{c}(t)) \in F_V(t), 0 \leq t \leq T$, *is optimal, i.e. it gives the maximum for the functional* $I[x, u]$, *if and only if there is an absolutely continuous non-zero vector function of Lagrange multipliers* $\bar{\lambda} = (\lambda_0, \lambda_1), 0 \leq t \leq T$, *with* λ_0 *a constant,* $\lambda_0 \geq 0$, *such that the function* $M(t) \stackrel{\triangle}{=} H(t, \bar{x}(t), \bar{u}(t), \bar{\lambda}(t))$ *is absolutely continuous and one has*

1. *adjoint equations:*

$$\frac{dM}{dt} = H_t(t, \bar{x}(t), \bar{u}(t), \bar{\lambda}(t)) \text{ a.e.;} \tag{2.33}$$

$$\frac{d\lambda_1}{dt} = -H_x(t, \bar{x}(t), \bar{u}(t), \bar{\lambda}(t)) \text{ a.e.;} \tag{2.34}$$

2. *maximum condition:*

$$\bar{u}(t) \in \arg\max_{v \in F_V(t)} H(t, \bar{x}(t), v, \bar{\lambda}(t)) \text{ a.e.;} \tag{2.35}$$

3. *transversality:*

$$\lambda_1(T) = \lambda_0 g'(\bar{x}(T)). \tag{2.36}$$

Proof: See Appendix 2.A. □

The following technical requirement on the market coefficients is sufficient to make $\{(\bar{\zeta}(t), \bar{c}(t))\}_{t \in [0,T]}$ an optimal portfolio process for maximizing the CRRA utility under the Value-at-Risk constraint, as Theorem 2.7 shows. We assume that market coefficients are totally unhedgeble, i.e. the mean rate of return process $\{\alpha(t)\}_{t\in[0,T]}$ and the matrix-valued volatility process $\{\sigma(t)\}_{t\in[0,T]}$ are adapted to filtration $\{\check{\mathcal{F}}_t\}_{0\leq t\leq\infty}$ generated by a Brownian motion independent of the Brownian motion driving the stocks (see Equation (2.1)).

Theorem 2.7: *A solution for maximizing*

$$\mathbb{E}\int_0^T e^{-\delta t}U_p(C(t))\,dt + \mathbb{E}\,e^{-\delta T}U_p(X(T))$$

over processes $(\zeta(t), c(t)) \in F_V(t)$, $0\leq t\leq T$, *is a process* $\{(\bar{\zeta}(t), \bar{c}(t))\}_{t\in[0,T]}$, *which, on every* ω, *solves* (2.33), (2.34), (2.35) *and* (2.36).

Proof: Lemma 2.5 gives the existence, on every ω, of $\{(\bar{\zeta}(t), \bar{c}(t))\}_{t\in[0,T]}$, optimal for the Bolza control problem, i.e. it maximizes $Y^{\zeta,c}(T)$ defined in (2.27) over $F_V(t)$, $0\leq t\leq T$. According to Lemma 2.5 it should solve (2.33), (2.34), (2.35) and (2.36) pathwise and these equations are sufficient for optimality. Let $(\zeta(t), c(t)) \in F_V(t)$ be another control. Let $\check{Z}^\zeta(t)$, $Z^{\bar{\zeta}}(t)$ and $Y^{\zeta,c}(T)$, $Y^{\bar{\zeta},\bar{c}}(T)$ be as in (2.21) and (2.27). The processes $\{Z^{\bar{\zeta}}(t)\}_{0\leq t\leq T}$, $\{Z^{\check{\zeta}}(t)\}_{0\leq t\leq T}$ are martingales by Lemma 2.4. Moreover, the independence of $\{\check{\mathcal{F}}_t\}_{0\leq t\leq\infty}$ and $\{\mathcal{F}_t\}_{0\leq t\leq\infty}$ implies

$$\mathbb{E}\left[Z^\zeta(T)|\check{\mathcal{F}}(T)\right] = \mathbb{E}\left[Z^{\bar{\zeta}}(T)|\check{\mathcal{F}}(T)\right] = 1.$$

Lemma 2.6 shows that $Y^{\bar{\zeta},\bar{c}}(T)$ is measurable with respect to $\check{\mathcal{F}}(T)$. Therefore, by (2.26) and iterated conditioning

$$\mathbb{E}\int_0^T e^{-\delta t}U_p(\bar{C}(t))dt + \mathbb{E}e^{-\delta T}U_p(X^{\bar{\zeta},\bar{c}}(T))$$

$$= \frac{X^p(0)}{p}\mathbb{E}\big(Z^{\bar{\zeta}}(T)Y^{\bar{\zeta},\bar{c}}(T)\big)$$

$$= \frac{X^p(0)}{p}\mathbb{E}\big(\mathbb{E}\big[Z^{\bar{\zeta}}(T)Y^{\bar{\zeta},\bar{c}}(T) \mid \check{\mathcal{F}}(T)\big]\big)$$

$$= \frac{X^p(0)}{p}\mathbb{E}\big(Y^{\bar{\zeta},\bar{c}}(T)\mathbb{E}\big[Z^{\bar{\zeta}}(T) \mid \check{\mathcal{F}}(T)\big]\big)$$

$$= \frac{X^p(0)}{p}\mathbb{E}Y^{\bar{\zeta},\bar{c}}(T).$$

Since $(\bar{\zeta}(t), \bar{c}(t))$ maximizes $Y^{\zeta,c}(T)$ over the constraint set,

$$\mathbb{E}\int_0^T e^{-\delta t} U_p(C(t))dt + \mathbb{E}e^{-\delta T}U_p(X^{\zeta,c}(T))$$

$$= \frac{X^p(0)}{p}\mathbb{E}(Z^\zeta(T)Y^{\zeta,c}(T))$$

$$\leq \frac{X^p(0)}{p}\mathbb{E}(Z^\zeta(T)Y^{\bar\zeta,\bar c}(T))$$

$$= \frac{X^p(0)}{p}\mathbb{E}(\mathbb{E}[Z^\zeta(T)Y^{\bar\zeta,\bar c}(T) \mid \check{\mathcal{F}}(T)])$$

$$= \frac{X^p(0)}{p}\mathbb{E}(Y^{\bar\zeta,\bar c}(T)\mathbb{E}[Z^\zeta(T) \mid \check{\mathcal{F}}(T)])$$

$$= \frac{X^p(0)}{p}\mathbb{E}Y^{\bar\zeta,\bar c}(T).$$

Remark 1: Let the interest rate and the discount factor be stochastic processes. In the formulae of $f_V(t, \zeta, c)$ and $Q_p(t, \zeta, c)$ the constant r is replaced by $r(t)$. All the results remain true if we assume that $\{r(t)\}_{0\leq t\leq\infty}$ and $\{\delta(t)\}_{0\leq t\leq\infty}$ are non-negative uniformly bounded continuous processes adapted to $\check{F}(t)$. We have considered the case of Value-at-Risk (VaR) in defining the risk constraint. The same methodology applies if one considers other measures of risk, as long as the corresponding constraint set is convex and compact.

2.6   NUMERICAL SOLUTION

Theorem 2.7 shows that the solution for every path of a Bolza control problem yields the optimal portfolio proportion for a VaR-constrained agent. The solution exists and is characterized by a system of forward backward equations that are also sufficient for optimality. In this section, by an appropriate discretization of control and state variables, the Bolza control problem is transformed into a finite-dimensional nonlinear program that can be solved by standard sequential quadratic programming (SQP) methods. The first step is to transform the Bolza problem into a Mayer control problem by introducing a new state variable x^0, with the boundary condition $x^0(0) = 0$ and an additional differential equation,

$$\frac{dx^0}{dt} = f_0(t, x(t), u(t)).$$

The cost functional is then $I[x, u] = x^0(T) + g(x(T))$ (see Equation (2.28)). Let us denote as $y = (x^0, x)$ the vector of state variables that satisfy the differential equation

$$\frac{dy}{dt} = \tilde{f}(t, x(t), u(t)),$$

with $\tilde{f} = (f_0, f)$ (see Equations (2.29) and (2.31)). The following discretization scheme is taken from Stryk (1993). The novel feature here is that the necessary first-order conditions of the discretized problem converge to the necessary first-order conditions of the continuous problem.

A partition of the time interval

$$0 = t_1 < t_2 < \cdots < t_N = T$$

is chosen. The parameters Y of the nonlinear program are the values of the control and state variables at the grod points t_j, $j = 1, \ldots, N$, and the final time $t_N = T$,

$$Y = (u(t_1), \ldots, u(t_N), y(t_1), \ldots, y(t_N), t_N) \in R^{4N+1}.$$

The controls are chosen as piecewise linear interpolating functions between $u(t_j)$ and $u(t_{j+1})$, for $t_j \leq t < t_{j+1}$,

$$u_{\text{app}}(t) = u(t_j) + \frac{t - t_j}{t_{j+1} - t_j}(u(t_{j+1}) - u(t_j)).$$

The states are chosen as continuously differentiable functions and piecewise cubic Hermite polynomials between $y(t_j)$ and $y(t_{j+1})$, with $\dot{y}_{\text{app}}(s) = \tilde{f}(x(s), u(s), s)$ at $s = t_j, t_{j+1}$,

$$y_{\text{app}}(t) = \sum_{k=0}^{3} d_k^j \left(\frac{t - t_j}{h_j}\right)^k, \tag{2.37}$$

$$t_j \leq t < t_{j+1}, \ j = 1, \ldots, N - 1,$$

$$d_0^j = y(t_j), \quad d_1^j = h_j \tilde{f}_j, \tag{2.38}$$

$$d_2^j = -3y(t_j) - 2h_j \tilde{f}_j + 3y(t_{j+1}) - h_j \tilde{f}_{j+1}, \tag{2.39}$$

$$d_3^j = 2y(t_{j+1}) + h_j \tilde{f}_j - 2y(t_{j+1}) + h_j \tilde{f}_{j+1}, \tag{2.40}$$

where

$$\tilde{f}_j \triangleq \tilde{f}(x(t_j), u(t_j), t_j), \quad h_j \triangleq t_{j+1} - t_j.$$

The reader can learn more about this in common textbooks of Numerical Analysis such as, Stoer and Bulrisch (1983). This way of discretizing has two advantages. The number of parameters of the nonlinear program is reduced because $\dot{y}(t_j), j = 1, \ldots, N$, are not parameters and the number of constraints is reduced because the constraints $\dot{y}_{app}(t_j) = \tilde{f}(x(t_j), u(t_j), t_j)$, $j = 1, \ldots, N$, are already fulfilled.

We impose the Value-at-Risk constraint (see Equation (2.6)) at the grid points

$$f_V(t_j, u(t_j)) \leq \log \frac{1}{1 - a_V}, \quad u = (\zeta, c), j = 1, \ldots, N. \tag{2.41}$$

Another constraint imposed is the so-called collocation constraint

$$\tilde{f}(x(\tilde{t}_j), u(\tilde{t}_j), \tilde{t}_j) = \dot{y}_{app}(\tilde{t}_j), \quad j = 1, \ldots, N,$$

or componentwise

$$f_0(x(\tilde{t}_j), u(\tilde{t}_j), \tilde{t}_j) = \dot{x}^0_{app}(\tilde{t}_j), \quad j = 1, \ldots, N, \tag{2.42}$$

and

$$f(x(\tilde{t}_j), u(\tilde{t}_j), \tilde{t}_j) = \dot{x}_{app}(\tilde{t}_j), \quad j = 1, \ldots, N, \tag{2.43}$$

where $\tilde{t}_j \stackrel{\triangle}{=} (t_j + t_{j+1})/2$, and the boundary condition $y(0) = (0,1)$. The nonlinear program is to maximize $I[y_N, t_N]$ subject to constraints (2.41), (2.42) and (2.43). It can be solved using NPSOL, a set of Fortran subroutines developed by Gill $et\ al.$ (1986). NPSOL uses a sequential quadratic programming (SQP) algorithm, in which the search direction is the solution of a quadratic programming (QP) subproblem. The Lagrangian of the nonlinear program is

$$L(Y, \phi, v) = I[y_N, t_N] + \sum_{j=1}^{N} v_j \left(f_V(t_j, u(t_j)) - \log \frac{1}{1 - a_V} \right)$$

$$+ \sum_{j=1}^{N-1} \phi_j^0 (f_0(x(\tilde{t}_j), u(\tilde{t}_j), \tilde{t}_j) - \dot{x}^0_{app}(\tilde{t}_j))$$

$$+ \sum_{j=1}^{N-1} \phi_j^1 (f(x(\tilde{t}_j), u(\tilde{t}_j), \tilde{t}_j) - \dot{x}_{app}(\tilde{t}_j)),$$

where $v = (v_1, \ldots, v_N) \in R^N$, $\phi^0 = (\phi_1^0, \ldots, \phi_{N-1}^0) \in R^{N-1}$ and $\phi^1 = (\phi_1^1, \ldots, \phi_{N-1}^1) \in R^{N-1}$ are the shadow prices. Let us denote $\zeta(t_i) \stackrel{\triangle}{=} \zeta_i$, $c(t_i) \stackrel{\triangle}{=} c_i$ and $x(t_i) \stackrel{\triangle}{=} x_i$. A solution of the nonlinear program satisfies the necessary first-order conditions of Karush, Kuhn and Tucker

$$\frac{\partial L}{\partial \zeta_i} = 0, \quad \frac{\partial L}{\partial c_i} = 0, \quad \frac{\partial L}{\partial x_i} = 0, \tag{2.44}$$

$$i = 1, \ldots, N.$$

The necessary first-order optimality conditions of the continuous problem are obtained in the limit from (2.44) as follows. Let $h \stackrel{\triangle}{=} \max\{h_j = t_{j+1} - t_j : j = 1, \ldots, N - 1\}$ be the norm of the partition. Letting $h \to 0$, after some calculations (see p. 5 of Stryk 1993) it is shown that, at $t = t_i$,

$$
\frac{\partial L}{\partial \zeta_i} \longrightarrow \frac{3}{2} \phi_i^1 \frac{\partial f(x(t_i), u(t_i), t_i)}{\partial \zeta} + \frac{3}{2} \phi_i^0 \frac{\partial f_0(x(t_i), u(t_i), t_i)}{\partial \zeta}
$$
$$
+ v_i \frac{\partial f_V(u(t_i), t_i)}{\partial \zeta},
$$

$$
\frac{\partial L}{\partial c_i} \longrightarrow \frac{3}{2} \phi_i^1 \frac{\partial f(x(t_i), u(t_i), t_i)}{\partial c} + \frac{3}{2} \phi_i^0 \frac{\partial f_0(x(t_i), u(t_i), t_i)}{\partial c}
$$
$$
+ v_i \frac{\partial f_V(u(t_i), t_i)}{\partial c},
$$

$$
\frac{\partial L}{\partial x_i} \longrightarrow \frac{3}{2} \dot{\phi}_i^1 + \frac{3}{2} \phi_i^1 \frac{\partial f(x(t_i), u(t_i), t_i)}{\partial x} + \frac{3}{2} \phi_i^0 \frac{\partial f_0(x(t_i), u(t_i), t_i)}{\partial x}.
$$

Therefore, $\partial L / \partial \zeta_i = 0$ and $\partial L / \partial c_i = 0$ converge to an equation equivalent to the maximum condition (2.35), and $\partial L / \partial x_i = 0$ converges to the adjoint Equation (2.34). This discretization scheme gives good estimates for the adjoint variables.

In the following we perform some numerical experiments. We consider one stock following a geometric Brownian motion with drift $\alpha_1 = 0.12$ and volatility $\sigma = 0.2$. The choices for the horizon τ and the confidence level α are largely arbitrary, although the Basle Committee proposals of April 1995 prescribed that VaR computations for the purpose of assessing bank capital requirements should be based on a uniform horizon of 10 trading days (two calendar weeks) and a 99% confidence level (Jorion 1997). We take $\tau = 1/25$, $\alpha = 0.01$, the interest rate $r = 0.05$ and the discount factor $\delta = 0.1$ (Figure 2.1).

2.7   CONCLUDING REMARKS

Let us summarize the results. This chapter examines in a stochastic paradigm the portfolio choice problem under a risk constraint, which is applied dynamically consistent at every time instant. The classical stochastic control methods, Dynamic Programming and the Martingale Method, are not very effective in this context. The latter works if the risk constraint is imposed in a static way. The Dynamic Programming approach (as shown in Section 2.3) leads to a highly nonlinear PDE. If the agent has CRRA preferences we propose a new method that relies on a decomposition of the utility into signal and noise. We neglect the noise (the expectation operator takes care of this) and this leads to a deterministic control problem on every path. We have reported explicit analytical solutions for the case of logarithmic utility even if the market coefficients are random processes. In this case, on every path the deterministic control problem is just a time-dependent constrained nonlinear program. The explicit solution shows that constrained agents consume and invest less in stocks than unconstrained agents, and long-term agents invest and consume more than short-term agents. These effects support the use of dynamically consistent risk constraints. If the utility is non-logarithmic CRRA we have to analyse a Bolza control problem on every path. We still allow the market coefficients to be

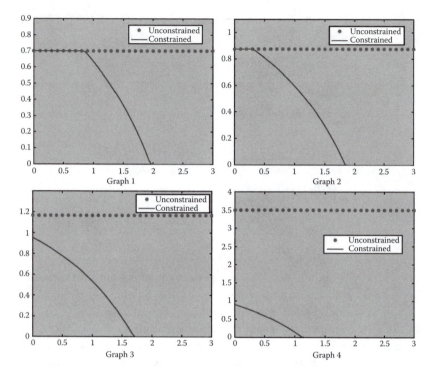

FIGURE 2.1 Asset allocation with and without VaR constraints, for the utility maximization of intertemporal consumption and final wealth. The graphs corresponds to different values of CRRA, $p : p = -1.5$ (Graph 1), $p = -1$ (Graph 2), $p = -0.5$ (Graph 3) and $p = 0.5$ (Graph 4). The x axis represents the time and the y axis the proportion of wealth invested in stocks. Note that the Merton line and, as time goes by, the portfolio value, increase, hence the VaR constraint becomes binding and reduces the investment in the risky asset. At the final time the agent is investing the least in stocks (in terms of proportions). When p increases, i.e. when the agent becomes less risk-averse, the effect of the VaR constraint becomes more significant.

random, but independent of the Brownian motion driving the stocks. Theorem 2.7 shows that a solution of the deterministic control problem is an optimal policy. Although the existence of an optimal policy is known if the constraint set is convex (Cvitanic and Karatzas 1992), it does not necessarily yield the existence for the Bolza problem. Standard existence theorems do not apply, but we manage to give a direct proof of existence in Lemma 2.5. The solution of the Bolza problem must solve a system of forward backward equations (the first-order necessary conditions) and this is also sufficient for optimality. In Section 2.6 we suggest a numerical treatment of the Bolza problem. The reduction of the stochastic control problem to a deterministic problem relies on the structure of the CRRA preferences. It would be interesting to extend this to other classes of preferences, because it turns out to be very effective for the case of dynamically consistent risk constraints.

ACKNOWLEDGEMENTS

The author would like to thank Ulrich Haussmann, Philip Loewen, Karsten Theissen, and two anonymous referees for helpful discussions and comments. This work was supported by NSERC under research grant 88051 and NCE grant 30354 (MITACS).

REFERENCES

Atkinson, C. and Papakokinou, M., Theory of optimal consumption and portfolio selection under a Capital-at-Risk (CaR) and a Value-at-Risk (VaR) constraint. *IMA J. Mgmt. Math.*, 2005, **16**, 37–70.

Basak, S. and Shapiro, A., Value-at-risk-based risk management: Optimal policies and asset prices. *Rev. Finan. Stud.*, 2001, **14**, 371–405.

Berkelaar, A., Cumperayot, P. and Kouwenberg, R., The effect of VaR-based risk management on asset prices and volatility smile. *Eur. Finan. Mgmt.*, 2002, **8**, 139–164; **1**, 65–78.

Cesari, L., *Optimization—Theory and Applications Problems with Ordinary Differential Equations*, 1983 (Springer: New York).

Cuoco, D. and Liu, H., An analysis of VaR-based capital requirements. Preprint, Finance Department, The Wharton School, University of Pennsylvania, 2003.

Cuoco, D., Hua, H. and Issaenko, S., Optimal dynamic trading strategies with risk limits. Preprint, Yale International Center for Finance, 2001.

Cvitanic, J. and Karatzas, I., Convex duality in convex portfolio optimization. *Ann. Appl. Probabil.*, 1992, **3**, 652–681.

Delbaen, F. and Schachermayer, W., A general version of the fundamental theorem of asset pricing. *Math. Ann.*, 1994, **300**, 463–520.

Emmer, S., Klüppelberg, C. and Korn, R., Optimal portfolios with bounded capital at risk. *Math. Finan.*, 2001, **11**, 365–384.

Gill, P.E., Murray, W., Saunders, M.A. and Wright, M.H., User's guide for NPSOL (version 4.0). Report SOL 86-2, Department of Operations Research, Stanford University, 1986.

Jorion, P., *Value at Risk*, 1997 (Irwin: Chicago).

Karatzas, I. and Shreve, S.E., *Brownian Motion and Stochastic Calculus*, 2nd ed., 1991 (Springer: New York).

Karatzas, I. and Shreve, S.E., *Methods of Mathematical Finance*, 1998 (Springer: New York).

Leippold, M., Trojani, F. and Vanini, P., Equilibrium impact of value-at-risk. Preprint, Swiss Banking Institute, University of Zurich, Switzerland, 2002.

Pirvu, T.A., Maximizing portfolio growth rate under risk constraints. Doctoral Dissertation, Department of Mathematics, Carnegie Mellon University, 2005.

Stoer, J. and Bulrisch, R., *Introduction to Numerical Analysis*, 2nd ed., 1983 (Springer: New York).

Stryk, O., Numerical solution of optimal control problems by direct collocation. In *Optimal Control—Calculus of Variations, Optimal Control Theory and Numerical Methods*. International Series of Numerical Mathematics, Vol. 111, 1993, pp. 129–143 (Birkhauser: Basle and Boston).

Vidovic, D.G., Lavassani, L.A., Li, X. and Ware, A., Dynamic portfolio selection under capital-at-risk. University of Calgary Yellow Series, Report 833, 2003.

Yiu, K.F.C., Optimal portfolios under a Value-at-Risk constraint. *J. Econ. Dynam. Contr.*, 2004, **28**, 1317–1334.

APPENDIX 2.A

Proof (proof of Lemma 2.1): In order to prove the martingale property of $\int_0^t \zeta^T(s)\sigma(s)\mathrm{d}W(s)$ it suffices to show that

$$\mathbb{E}\int_0^T \left\|\zeta^T(u)\sigma(u)\right\|^2 \mathrm{d}u < \infty. \tag{A2.1}$$

Note that

$$\left\|\zeta^T(t)\mu(t)\right\| = \left\|\zeta^T(t)\sigma(t)\sigma^T(t)\zeta_M(t)\right\|$$
$$\leq \left\|\zeta^T(t)\sigma(t)\right\| \cdot \left\|\zeta_M^T(t)\sigma(t)\right\|. \tag{A2.2}$$

For $(\zeta(t), c(t)) \in F_V(t)$, we have

$$\left(r - c(t) + \zeta^T(t)\mu(t) - \frac{1}{2}\left\|\zeta^T(t)\sigma(t)\right\|^2\right)\tau$$
$$+ N^{-1}(\alpha)\left\|\zeta^T(t)\sigma(t)\right\|\sqrt{\tau}$$
$$\geq \log(1 - a_V).$$

This combined with (A2.2) yields

$$\left\|\zeta^T(t)\sigma(t)\right\| \leq k_1 \vee \left\|\zeta_M^T(t)\sigma(t)\right\|, \tag{A2.3}$$

for some positive constant k_1, where, as usual, $a \vee b = \max(a, b)$. In light of assumption (2.12) the inequality (A2.1) follows. □

Proof (proof of Lemma 2.2): The proof relies on the method of Lagrange multipliers. The concave function

$$g(t, \zeta, c) \stackrel{\triangle}{=} \log c + \frac{1}{\delta}(1 - e^{-\delta(T-t)})Q_0(t, \zeta, c) \tag{A2.4}$$

is maximized over $(\zeta, c) \in \mathbb{R}^m \times [0, \infty)$ by $(\zeta_M(t), c_M(t))$, where $\zeta_M(t) \stackrel{\triangle}{=} (\sigma(t)\sigma^T(t))^{-1}\mu(t)$ and $c_M(t) \stackrel{\triangle}{=} \delta/[1 - e^{-\delta(T-t)}]$. If this point is in the Value-at-Risk constraint set, then is the optimal solution of (P 1). Otherwise, the concave function g is maximized over the compact, convex set $F_V(t)$ at a unique point $(\bar{\zeta}(t), \bar{c}(t))$; moreover, this point must be on the boundary of $F_V(t)$. Hence, it solves the optimization problem

(P2)   maximize $g(t, \zeta, c)$,

$$\text{subject to } f_V(t, \zeta, c) \stackrel{\triangle}{=} - Q_0(t, \zeta, c)\tau - N^{-1}(\alpha)\left\|\zeta^T\sigma(t)\right\|\sqrt{\tau} = \log\frac{1}{1 - a_V}.$$

The function f_V is not differentiable when ζ is the zero vector. Let us assume that the optimal $\bar{\zeta}(t)$ is not the zero vector. According to the *Lagrange multiplier* theorem, either $\nabla f_V(t, \bar{\zeta}(t), \bar{c}(t))$ is the zero vector or there is a positive λ such that

$$\nabla g(t, \bar{\zeta}(t), \bar{c}(t)) = \lambda \nabla f_V(t, \bar{\zeta}(t), \bar{c}(t)). \tag{A2.5}$$

The first case cannot occur and computations show that $\bar{\zeta}(t)$ is parallel to $\zeta_M(t)$. This implies that the optimal solution $(\bar{\zeta}(t), \bar{c}(t)) = (\bar{\lambda}_1\zeta_M(t), \bar{\lambda}_2 c_M(t))$, with $\bar{\lambda}_1, \bar{\lambda}_2$ the solution of

(P3) maximize $l(\lambda_1, \lambda_2)$,

$$\text{subject to } f_V(t, \lambda_1 \zeta_M(t), \lambda_2 c_M(t)) = \log \frac{1}{1 - a_V},$$

where

$$l(\lambda_1, \lambda_2) \overset{\triangle}{=} g(t, \lambda_1 \zeta_M(t), \lambda_2 c_M(t)). \tag{A2.6}$$

The concave function l is maximized over $\mathbb{R}^2$ at $(1,1)$. We know that this point is not in the constraint set (this is because we have assumed $(\zeta_M(t), c_M(t)) \notin F_V(t)$), hence $\bar{\lambda}_1 < 1, \bar{\lambda}_2 < 1$ and either $\nabla f_V(t, \bar{\lambda}_1 \zeta_M(t), \bar{\lambda}_2 c_M(t)) = 0$ or $\nabla l(\bar{\lambda}_1, \bar{\lambda}_2) = \gamma \nabla f_V(t, \bar{\lambda}_1 \zeta_M(t), \bar{\lambda}_2 c_M(t))$, for some positive *Lagrange multiplier* γ. The first case cannot occur and by eliminating γ we obtain $\bar{\lambda}_2 = u(t, \bar{\lambda}_1)$, where u was defined in (2.18). Henceforth, $\bar{\lambda}_1$ is the unique root of the equation

$$f_V(t, z\zeta_M(t), u(t, z)c_M(t)) = \log \frac{1}{1 - a_V} \tag{A2.7}$$

in the variable z. It may be the case that the root of this equation is negative, in which case

$$(\bar{\zeta}(t), \bar{c}(t)) = \left(\mathbf{0}_m, r + \frac{1}{\tau} \log \frac{1}{1 - a_V} \right),$$

where $\mathbf{0}_m$ is the m-dimensional vector of zeros.

Proof (proof of Lemma 2.4): Assumption (2.23) combined with (A2.3) and the Novikov condition (see Karatzas and Shreve 1991, p. 199, Corollary 5.13) make the process $Z^{\zeta}(t)$ a martingale. □

Proof (proof of Lemma 2.5): According to (2.31) and (2.32),

$$x(t) = \exp \left(\int_0^t f_1(t, c(u), \zeta(u)) du \right), \tag{A2.8}$$

with

$$f_1(t, c, \zeta) \overset{\triangle}{=} pr - pc + p\zeta^T \mu(t) + \frac{p(p-1)}{2} \|\zeta^T \sigma(t)\|^2.$$

Let us recall that, for $u(t) = (\zeta(t), c(t)) \in F_V(t)$,

$$\|\zeta^T(t)\sigma(t)\| \le k_1 \vee \|\zeta_M^T(t)\sigma(t)\|, \tag{A2.9}$$

hence $\|\zeta^T(t)\sigma(t)\|$ is uniformly bounded on $[0, T]$ due to the continuity of market coefficients (see Equation (2.11)). Moreover, one can conclude that $\|\zeta(t)\| \leq K_1$ and $c(t) \leq K_1$ for some constant K_1. Gronwall's lemma gives $x(t) \leq K_2$ and $|\dot{x}(t)| \leq K_2$ on $[0, T]$ (here, $\dot{x}(t) = dx/dt$). Let (x_n, u_n) be a maximizing sequence, i.e. $I[x_n, u_n] \longrightarrow \sup I[x, u]$. The above arguments show that the sequence x_n is uniformly bounded and equicontinuous, thus by the Arzela–Ascoli theorem it converges uniformly to some function $\tilde{x}$. According to Komlos' Lemma (see Lemma A1.1 of Delbaen and Schachermayer 1994) we can find some sequences of convex combinations $\bar{\zeta}_n \in \text{conv}(\zeta_n, \zeta_{n+1},)$ and $\bar{c}_n \in \text{conv}(c_n, c_{n+1},)$ that converge a.e. to some measurable functions $\bar{\zeta}$ and $\bar{c}$. Moreover, $\bar{u}(t) \overset{\triangle}{=} (\bar{\zeta}(t), \bar{c}(t)) \in F_V(t)$, $0 \leq t \leq T$, due to the convexity and compactness of the set $F_V(t)$. Let us denote $\bar{x}_n$ as the sequence of state variables corresponding to these controls, i.e.

$$\bar{x}_n(t) = \exp\left(\int_0^t f_1(t, \bar{c}_n(s), \bar{\zeta}(s)) ds\right), \quad 0 \leq t \leq T$$

(see Equation (A2.8)). Let us assume $p > 0$; the case $p < 0$ can be treated similarly. Due to the concavity of the function f_1, $\ln \bar{x}_n(t) \geq \text{conv}(\ln x_n(t), \ln x_{n+1}(t), \ldots)$, where the convex combination is that defining $\bar{\zeta}_n, \bar{c}_n$. If $\bar{y}_n \overset{\triangle}{=} \exp(\text{conv}(\ln x_n(t), \ln x_{n+1}(t), \ldots))$, then $\bar{x}_n(t) \geq \bar{y}_n(t)$ and $\bar{y}_n(t) \longrightarrow \tilde{x}(t)$, i.e. $\bar{y}_n(t) - x_n(t) \longrightarrow 0$ for $t \in [0, T]$. By the dominated convergence theorem $\bar{x}_n(t) \longrightarrow \bar{x}(t)$, $0 \leq t \leq T$, a.e., where

$$\bar{x}(t) = \exp\left(\int_0^t f_1(t, \bar{c}(s), \bar{\zeta}(s)) ds\right), \quad 0 \leq t \leq T.$$

From Fatou's lemma, the dominated convergence theorem and the concavity of the function f_0 in u (see Equation (2.29)), it follows that

$$I[\bar{x}, \bar{u}] \geq \limsup I[\bar{x}_n, \bar{u}_n] \geq \limsup I[\bar{y}_n, \bar{u}_n]$$
$$= \limsup I[x_n, \bar{u}_n] = \sup I[x, u].$$

Proof (proof of Lemma 2.6): By Theorem 5.1.i of Cesari (1983), $\{(\bar{\zeta}(t), \bar{c}(t))\}_{t \in [0, T]}$ should solve (2.33), (2.34), (2.35) and (2.36). For sufficiency, let us consider $\lambda_0 = 1$, and define the maximized Hamiltonian

$$H^*(t, x, \bar{\lambda}) \overset{\triangle}{=} \max_{v \in F_V(t)} H(t, x, v, \bar{\lambda}).$$

Let $(x(t), u(t))$ be another admissible pair. Since the Hamiltonian is linear in x, by the adjoint equation for λ_1 and the maximum condition

$$
\begin{aligned}
H(t, \bar{x}(t), \bar{u}(t), \bar{\lambda}(t)) &- H(t, x(t), u(t), \bar{\lambda}(t)) \\
&\geq H^*(t, \bar{x}(t), \bar{\lambda}(t)) - H^*(t, x, \bar{\lambda}(t)) \\
&= \lambda_1'(t)(x(t) - \bar{x}(t)),
\end{aligned}
\tag{A2.10}
$$

one has

$$
\begin{aligned}
I[\bar{x}, \bar{u}] - I[x, u] = &\int_0^T (H(t, \bar{x}(t), \bar{u}(t), \bar{\lambda}(t)) - H(t, x(t), u(t), \bar{\lambda}(t)))\mathrm{d}t \\
&+ \int_0^T \lambda_1(t)(\dot{x}(t) - \dot{\bar{x}}(t))\mathrm{d}t - g(x(T)) + g(\bar{x}(T)).
\end{aligned}
$$

The inequality (A2.10) and the transversality condition (2.36) yield

$$
\begin{aligned}
I[\bar{x}, \bar{u}] - I[x, u] \geq &\int_0^T \lambda_1'(t)(x(t) - \bar{x}(t))\mathrm{d}t \\
&+ \int_0^T \lambda_1(t)(\dot{x}(t) - \dot{\bar{x}}(t))\mathrm{d}t - g(x(T)) + g(\bar{x}(T)) \\
&= \lambda_1(T)(x(T) - \bar{x}(T)) - g(x(T)) + g(\bar{x}(T)) \\
&= g'(\bar{x}(T))(x(T) - \bar{x}(T)) - g(x(T)) + g(\bar{x}(T)) \\
&= g(x(T)) - g(\bar{x}(T)) - g(x(T)) + g(\bar{x}(T)) = 0,
\end{aligned}
$$

proving the optimality of $(\bar{x}(t), \bar{u}(t))$. □

CHAPTER 3

Dynamic Consumption and Asset Allocation with Derivative Securities

YUAN-HUNG HSUKU

CONTENTS

3.1 INTRODUCTION

T HIS PAPER FINDS THE OPTIMAL consumption and dynamic asset allocation of stocks, bonds and derivatives for long-term investors in contrast to the standard optimal dynamic asset allocation strategies involving only stocks and bonds. The chapter explores and attempts to understand the effect of introducing a non-redundant derivative security on an already-existing stock—in particular, on the volatility of stock returns. Instead of a single-period (two-date) result, we also delve into the optimal intertemporal consumption as well as dynamic asset allocation strategies under a stochastic investment opportunity set.

We show that these long-term investors have access to an environment where investment opportunities vary over time with stochastic volatility. There is abundant empirical evidence that the conditional variance of stock returns is not constant over time. Merton (1969, 1971, 1973a) shows that if investment opportunities vary over time, then multi-period investors will have a very different optimal consumption rule and portfolio strategies than those of single-period investors. If multi-period investors hope to maintain a stable long-run consumption stream, then they may seek to hedge their exposures to future shifts in the time-varying investment opportunity set, and this creates extra intertemporal hedging demands for financial assets.

Following Merton's (1969, 1971) introduction of the standard intertemporal consumption and investment model, it has been studied extensively in the finance literature and has become a classical problem in financial economics. The literature on the broad set of issues covering intertemporal consumption and investment or optimal dynamic asset allocation strategies restricts investors' access to bond and stock markets only, while excluding the derivatives market. Haugh and Lo (2001), Liu and Pan (2003) and Ilhan *et al.* (2005) remove this restriction and introduce derivative securities in the financial market. Cont *et al.* (2005) consider the problem of hedging a contingent claim in a market where prices of traded assets can undergo jumps, by trading in the underlying asset and a set of traded options, and they work in a continuous time setting. However, they give an expression for the hedging strategy which minimizes the variance of the hedging error, while our model provides an investor who aims to maximize expected utility and gives an expression for the dynamic asset allocation strategies. They take a look at optimal dynamic or buy-and-hold portfolio strategies for an investor who can control not just the holding of stocks and bonds, but also derivatives. Our work is related to their research and makes several extensions.

The classical option pricing framework of Black and Scholes (1973) and Merton (1973) is based on a replication argument in a continuous trading model with no transaction costs. The presence of transaction costs, however, invalidates the Black–Scholes arbitrage argument for option pricing, since continuous revision requires that the portfolio be rebalanced infinitely, implying infinite transaction costs (Chalasani and Jha 2001). As a result, some of the recent literature has begun to work on transactions costs. However, this is beyond the scope of this chapter which follows Black and Scholes (1973) and Merton (1973) in assuming zero transaction costs for stocks and options trading.

First, Cont *et al.* (2005) solve the optimal investment or asset allocation problem of a representative investor whose investment opportunity includes not only the usual riskless bond and risky stock, but also derivatives on the stock. In their model, investors are assumed to have a specified utility defined over wealth at a single terminal date. This chapter extends this setting to consider a model in which a long-term investor chooses consumption as well as an optimal portfolio including the riskless bond, risky stock and derivatives on the stock. We then maximize a utility function defined over intermediate consumption rather than terminal wealth.

Abstraction from the choice of consumption over time implies that investors value only wealth at a single terminal date, no consumption takes place before the terminal date, and all portfolio returns are reinvested until that date. The assumption that investors derive utility only from terminal wealth and not from intermediate consumption simplifies the analysis through avoidance of an additional source of non-linearity in the differential equation. However, many long-term investors seek a stable consumption path over a long horizon. This simplification makes it hard to apply the above-mentioned papers' results to the realistic problem facing an investor saving for the future. Very often, intermediate consumption can be used as an indicator of marginal utility, especially in the asset pricing related literature (Campbell and Viceira 1999).

The second extension of this chapter is in contrast to Liu and Pan (2003), in which investors are assumed to have power utility, in contrast to Ilhan *et al.* (2005), in which investors maximize expected exponential utility of terminal wealth and also in contrast to Haugh and Lo (2001), who set the special cases of CRRA and CARA preferences. In this model we assume that investors have continuous-time recursive preferences introduced by Duffie and Epstein (1992b). This allows us not only to provide the effects of risk aversion, but also to separate cleanly an investor's elasticity of intertemporal substitution in consumption from the coefficient of relative risk aversion. This is because power utility functions restrict risk aversion to be the reciprocal of the elasticity of intertemporal substitution, but in fact these parameters have very different effects on optimal consumption and portfolio choice (Campbell and Viceira 1999; Bhamra and Uppal 2003; Chacko and Viceira 2005).

Under the model settings of Liu and Pan (2003), the mean–variance allocation to stocks, i.e. the ratio between expected stock excess returns and stock return volatility, is constant, while our work more realistically reflects the real-world situation with time-varying allocation components. Liu and Pan (2003) consider the Heston (1993) and Stein and Stein (1991) models of stochastic volatility, in which volatility follows a mean-reverting process and stock returns are a linear function of volatility. Our setting is the more general assumption that expected stock returns are an affine function of volatility. In this setting, the Liu and Pan (2003) result is the special case where we constrain the intercept of the affine function to be zero. Therefore, we provide a dynamic asset allocation in both stocks and derivatives, in contrast both to Ilhan *et al.* (2005), who restrict to a static position in derivative securities, and to Haugh and Lo (2001), who construct a buy-and-hold portfolio of stocks, bonds and options that involves no trading once established at the start of the investment horizon. A buy-and-hold portfolio with a derivative securities strategy may come closest to optimal dynamic asset-allocation policies involving only stocks and bonds, as concluded by Haugh and Lo (2001). We know that the problem of dynamic investment policies, i.e. asset-allocation rules, arise from standard dynamic optimization problems in which an investor maximizes the expected utility as shown by Haugh and Lo (2001). At the same time they pose the following problem: given an investor's optimal dynamic investment policy for two assets, stocks and bonds, construct a 'buy-and-hold' portfolio—a portfolio that involves no trading once it is established—of stocks, bonds and options at the start of the investment horizon. They state that this comes closest to the optimal dynamic policy (Haugh and Lo 2001), but their strategies differ from our dynamic asset allocation strategies which involve not only stocks, but also derivative securities.

Chacko and Viceira (2005) examine the optimal consumption and portfolio choice problem of long-horizon investors who only have access to a riskless asset with constant return and a risky asset (stocks), without introducing any derivative securities. Comparing with Chacko and Viceira (2005), our generalized model considers that holding derivative securities complicates the asset allocation strategies for long-horizon investors. If one does not hold any derivative securities as in Chacko and Viceira (2005), the assumption of imperfect instantaneous correlation between risky stock returns and its stochastic volatility means that the intertemporal hedging component of the risky stock can only provide partial hedging ability for multi-period investors when facing the time-varying investment opportunity set. In this chapter, where we introduce non-redundant derivative securities written on the risky stock in the incomplete financial market under this optimal dynamic asset allocation, the derivative securities in the asset allocation can provide differential exposures to stochastic volatility and make the market complete. The derivative securities can also supplement the deficient hedging ability of the intertemporal hedging component of the risky stock, because of the nonlinear nature of derivative securities.

The chapter will obtain a solution to this problem which is exact for investors with unit elasticity of intertemporal substitution of consumption, and approximate otherwise. The chapter is organized as follows. Section 3.2 describes the model and environment assumed in this chapter. Section 3.3 develops the model of optimal consumption policy and dynamic asset allocation strategies. We also extend the model with constant expected excess returns and constant volatility risk premiums to time-varying instantaneous expected excess returns in relation to the risky stock and time-varying stochastic volatility risk premium. Section 3.4 provides the analyses of optimal consumption policy and dynamic asset allocation strategies. Finally, conclusions are given in Section 3.5.

3.2   THE MODEL

3.2.1   Investment Opportunity Set

In this chapter we assume that wealth comprises investments in traded assets only. There are two prime assets available for trading in the economy. One of the assets is an instantaneously riskless bond (B_t) with a constant interest rate of r. Its instantaneous return is

$$\frac{dB_t}{B_t} = r\,dt. \tag{3.1}$$

The second prime asset is a risky stock that represents the aggregate equity market. Its instantaneous total return dynamics are given by

$$\frac{dS_t}{S_t} = \mu\,dt + \sqrt{V_t}\,dZ_S, \tag{3.2}$$

where S_t denotes the price of the risky asset at time t, $\sqrt{V_t}$ is the time-varying instantaneous standard deviation of the return on the risky asset, and dZ_S is the increment of a standard Brownian motion. We assume that the short rate is constant in order to focus on the stochastic volatility of the risky asset. From these asset return dynamics, we have the assumption of constant expected excess return on the risky asset over the riskless asset, i.e. $(\mu - r)$; this assumption will be relaxed in Section 3.3.2. We denote time variation with the subscript 't' and let the conditional variance of the risky asset vary stochastically over time.

From the following the investment opportunity is time-varying. We assume that the instantaneous variance process is

$$dV_t = \kappa(\theta - V_t)dt + \sigma\sqrt{V_t}\left(\rho\, dZ_S + \sqrt{1 - \rho^2}\, dZ_V\right), \tag{3.3}$$

where the parameter $\theta > 0$ describes the long-term mean of the variance, $\kappa \in (0, 1)$ is the reversion parameter of the instantaneous variance process, i.e. this parameter describes the degree of mean reversion, and ρ is the correlation between the two Brownian motions, which is assumed to be negative to capture the asymmetric effect (Black 1976, Glosten *et al.* 1993). This negative correlation assumption, together with the mean-reversion of the stock return volatility, can capture major important features of the empirical literature of the equity market.

In the traditional theory of derivative pricing (Black and Scholes 1973; Merton 1973b), derivative assets like options are viewed as redundant securities, the payoffs of which can be replicated by portfolios of primary assets. Thus, the market is generally assumed to be complete without the options. In this chapter we introduce derivative securities that allow the investor to include them in dynamic asset allocation strategies. If only a risky stock and a riskless bond are available for trading, then the financial market here is incomplete. The nonlinear nature of derivative securities serves to complete the market. This follows from our setting in which stock returns are not instantaneously perfectly correlated with their time-varying volatility. In this chapter the derivative securities written on the stock are non-redundant assets. In our setting derivative securities can provide differential exposure to the imperfect instantaneous correlation and make the market complete.

Following Sircar and Papanicolaou (1999) and Liu and Pan (2003), in the above setting the non-redundant derivative $O_t = p(S_t, V_t)$, that is the function (p) of the prices of the stock (S_t) and on the volatility of stock returns (V_t) at time t, will have the following price dynamics:

$$\begin{aligned} dO_t = &\left[(\mu - r)(p_s S_t + \rho\sigma p_v) + \lambda\sigma\sqrt{1 - \rho^2}p_v + rO_t\right]dt \\ &+ (p_s S_t + \rho\sigma p_v)\sqrt{V_t}\, dZ_S + (\sigma\sqrt{1 - \rho^2}p_v)\sqrt{V_t}\, dZ_V, \end{aligned} \tag{3.4}$$

where λ determines the stochastic volatility risk premium and p_s and p_v are measures of the derivative's price sensitivity to small changes in the underlying stock price and volatility, respectively.

3.2.2   Preferences

We assume that the investor's preferences are recursive and of the form described by Duffie and Epstein (1992a, b). Recursive utility is a generalization of the standard and time-separable power utility function that separates the elasticity of intertemporal substitution of consumption from the relative risk aversion. This means that the power utility is just a special case of the recursive utility function in which the elasticity of the intertemporal substitution parameter is the inverse of the relative risk aversion coefficient. The value function of the problem (J) is to maximize the investor's expected lifetime utility. We adopt the Duffie and Epstein (1992b) parameterization

$$J = E_t \left[\int_t^\infty f(C_\tau, J_\tau) d\tau \right], \tag{3.5}$$

where the utility $f(C_\tau, J_\tau)$ is a normalized aggregator of an investor's current consumption (C_τ) and has the following form:

$$f(C, J) = \beta \left(1 - \frac{1}{\varphi} \right)^{-1} (1 - \gamma) J \left[\left(\frac{C}{((1 - \gamma)J)^{1/(1-\gamma)}} \right)^{1-(1/\varphi)} - 1 \right], \tag{3.6}$$

with γ is the coefficient of relative risk aversion, β is the rate of time preference and φ is the elasticity of intertemporal substitution. They are all larger than zero. The normalized aggregator $f(C_\tau, J_\tau)$ takes the following form when φ approaches one:

$$f(C, J) = \beta(1 - \gamma)J \left[\log(C) - \frac{1}{1 - \gamma} \log((1 - \gamma)J) \right].$$

The investor's objective is to maximize expected lifetime utility by choosing consumption and the proportions of wealth to invest across the risky stock and the derivative securities subject to the following intertemporal budget constraint:

$$dW_t = \left\{ n_t(\mu - r)W_t + \pi_t \left[(\mu - r) \left(\frac{p_s S_t}{O_t} + \rho\sigma \frac{p_v}{O_t} \right) + \lambda\sigma\sqrt{1 - \rho^2} \frac{p_v}{O_t} \right] W_t + rW_t - C_t \right\} dt$$

$$+ n_t \sqrt{V_t} W_t \, dZ_S + \pi_t \left[\left(\frac{p_s S_t}{O_t} + \rho\sigma \frac{p_v}{O_t} \right) \sqrt{V_t} W_t \, dZ_S + \sigma\sqrt{1 - \rho^2} \frac{p_v}{O_t} \sqrt{V_t} W_t \, dZ_V \right],$$

$$\tag{3.7}$$

where W_t represents the investor's total wealth, n_t and π_t are the fractions of the investor's financial wealth allocated to the stock and the derivative securities at time t, respectively, and C_t represents the investor's instantaneous consumption.

3.3 OPTIMAL CONSUMPTION POLICY AND DYNAMIC ASSET ALLOCATION STRATEGIES

3.3.1 Optimal Consumption and Dynamic Asset Allocation Strategies

The principle of optimality leads to the following Bellman equation for the utility function. In the above setting, the Bellman equation becomes

$$
\begin{aligned}
0 = \sup_{n,\pi,C} & \left\{ f(C_\tau, J_\tau) + J_W \left(n_t(\mu - r)W_t + \pi_t \left[(\mu - r) \right. \right. \right. \\
& \times \left(\frac{p_s S_t}{O_t} + \rho\sigma\frac{p_v}{O_t} \right) + \lambda\sigma\sqrt{1-\rho^2}\frac{p_v}{O_t} \right] W_t + rW_t - C_t \right) \\
& + J_V[\kappa(\theta - V_t)] + \frac{1}{2}J_{WW}W_t^2 V_t \left((n_t)^2 + 2n_t\pi_t \left(\frac{p_s S_t}{O_t} + \rho\sigma\frac{p_v}{O_t} \right) \right. \\
& + (\pi_t)^2 \left[\left(\frac{p_s S_t}{O_t} + \rho\sigma\frac{p_v}{O_t} \right)^2 + \left(\sigma\sqrt{1-\rho^2}\frac{p_v}{O_t} \right)^2 \right] \right) + \frac{1}{2}J_{VV}\sigma^2 V_t + J_{WV}W_t\sigma V_t \\
& \times \left(n_t\rho + \pi_t \left[\left(\frac{p_s S_t}{O_t} + \rho\sigma\frac{p_v}{O_t} \right)\rho + \sigma(1-\rho^2)\frac{p_v}{O_t} \right] \right) \right\},
\end{aligned} \tag{3.8}
$$

where J_W, J_V denote the derivatives of the value function J with respect to wealth W and stochastic volatility V, respectively. We use a similar notation for higher derivatives as well.

The first-order conditions for the optimization in (3.8) are

$$
C_t = J_W^{-\varphi} J^{(1-\varphi\gamma)/(1-\gamma)} \beta^\varphi (1-\gamma)^{(1-\varphi\gamma)/(1-\gamma)}, \tag{3.9}
$$

$$
\begin{aligned}
n_t = & -\frac{J_W}{J_{WW}W_t}\frac{(\mu-r)}{V_t} - \frac{J_{WV}}{J_{WW}W_t}\rho\sigma \\
& + \frac{J_W}{J_{WW}W_t}\frac{\lambda}{V_t}\frac{(p_s S_t + \rho\sigma p_v)}{\sigma\sqrt{1-\rho^2}p_v} + \frac{J_{WV}}{J_{WW}W_t}\frac{(p_s S_t + \rho\sigma p_v)}{p_v},
\end{aligned} \tag{3.10}
$$

$$
\pi_t = -\frac{J_W}{J_{WW}W_t}\frac{\lambda}{\sigma\sqrt{1-\rho^2}}\frac{O_t}{p_v}\frac{1}{V_t} - \frac{J_{WV}}{J_{WW}W_t}\frac{O_t}{p_v}. \tag{3.11}
$$

Equation (3.9), which demonstrates the rule of optimal consumption, stems from the envelope condition, $f_c = J_w$, once the value function is obtained. Equations (3.10) and (3.11) show that the optimal portfolio allocation in the stock has four components, while the derivative allocation has two. In the optimal asset allocation in both the risky stock and the derivative, their first terms are the mean–variance portfolio weights. These are the myopie demands for an investor who only invests in a single period horizon or under a constant investment opportunity set. The second terms of both the stock and the derivative are the intertemporal hedging demands that characterize demand arising from the desire to hedge against changes in the investment opportunity set induced by the stochastic volatility. These terms are determined by the instantaneous rate of changes in relation to the value function. Without introducing the derivative security, the optimal asset allocation of the stock will contain only the first two terms. Aside from the mean–variance weights of the optimal derivative allocation, the derivative plays a role that allows the investor to insure against changes in the stochastic volatility and the time-varying investment opportunity set. The second term, i.e. the intertemporal hedging demand, of the optimal asset allocation on the stock also partially provides a similar rule, so that the introduction of the derivative makes the stock allocation interact with the derivative allocation and is expressed in the third and the fourth terms of the stock demand.

The first-order conditions for our problem are not in fact explicit solutions unless we know the complicated form of the value function. After substituting the first-order conditions back into the Bellman equation and rearranging them, we conjecture that there exists a solution of the functional form $J(W_t, V_t) = I(V_t)(W_t^{1-\gamma}/(1-\gamma))$. We first restrict to the special case of $\varphi = 1$. We then substitute this form into Equation (3.A1) of Appendix 3.A, and the resulting ordinary differential equation will have a solution of the form $I = \exp(Q_0 + Q_1 V_t + Q_2 \log V_t)$. Rearranging this equation, we have three equations for Q_2, Q_1, and Q_0 after collecting terms in $1/V_t$, V_t, and 1. We provide the full details in Appendix 3.A. Hence, we obtain the form of the value function and the optimal consumption rule and dynamic asset allocation strategies for investing in the stock and the derivative security when $\varphi = 1$. The value function is

$$J(W_t, V_t) = I(V_t) \frac{W_t^{1-\gamma}}{1-\gamma} = \exp(Q_0 + Q_1 V_t + Q_2 \log V_t) \frac{W_t^{1-\gamma}}{1-\gamma}. \qquad (3.12)$$

The investor's optimal consumption–wealth ratio and dynamic asset allocation strategies for investing in the stock and the derivative are

$$\frac{C_t}{W_t} = \beta, \qquad (3.13)$$

$$n_t = \frac{1}{\gamma} \frac{\mu - r}{V_t} + \frac{1}{\gamma} \left(Q_1 + Q_2 \frac{1}{V_t} \right) \rho \sigma - \frac{1}{\gamma} \frac{\lambda}{V_t} \frac{(p_s S_t + \rho \sigma p_v)}{\sigma \sqrt{1 - \rho^2} p_v}$$

$$- \frac{1}{\gamma} \left(Q_1 + Q_2 \frac{1}{V_t} \right) \frac{(p_s S_t + \rho \sigma p_v)}{p_v}, \qquad (3.14)$$

$$\pi_t = \frac{1}{\gamma}\frac{\lambda}{V_t}\frac{1}{\sigma\sqrt{1-\rho^2(p_v/O_t)}} + \frac{1}{\gamma}\left(Q_1 + Q_2\frac{1}{V_t}\right)\frac{O_t}{p_v}. \tag{3.15}$$

Equation (3.13) demonstrates the invariance of the optimal log consumption–wealth ratio to changes in volatility. It equals investors' rate of time preference with unit elasticity of intertemporal substitution. The substitution effect and the intertemporal income effect on consumption resulting from a change in investment opportunity set perfectly cancel each other out, and the consumption of a fixed proportion of investors' wealth each period is optimal. This is why such a consumption policy is normally referred to as being myopic (Chacko and Viceira 2005).

In the more general case $\varphi \neq 1$, there is no exact analytical solution. However, we can still find an approximate analytical solution following the methods described in Campbell and Viceira (2002) and Chacko and Viceira (2005). The basic idea behind the use of approximate analytical methods is that of formulating a general problem, on the condition that we can find a particular case that has a known solution, and then using that particular case and its solution as a starting point for computing approximate solutions to nearby problems. In the context of our problem, the insight we obtain is the solution for the recursive utility function when $\varphi = 1$ which provides a convenient starting point for performing the expansion.

Without the restriction of $\varphi = 1$, the Bellman equation can be expressed as the following equation by substituting (3.9) into (3.A1) and again conjecturing that there exists a solution of the functional form $J(W_t, V_t) = I(V_t)(W_t^{1-\gamma}/(1-\gamma))$:

$$0 = -\frac{\beta^\varphi}{1-\varphi}I^{1+[(1-\varphi)/(1-\gamma)]} + \frac{\varphi}{1-\varphi}\beta I + \frac{1}{2}\frac{1}{\gamma}I\frac{1}{V_t}\left[(\mu-r)^2 + \lambda^2\right]$$

$$+ \frac{1}{\gamma}I_V\sigma\left[(\mu-r)\rho + \lambda\sqrt{1-\rho^2}\right] + Ir + I_V\frac{1}{1-\gamma}[\kappa(\theta-V_t)]$$

$$+ \frac{1}{2}I_{VV}\frac{1}{1-\gamma}\sigma^2 V_t + \frac{1}{2}\frac{1}{\gamma}\frac{(I_V)^2}{I}\sigma^2 V_t. \tag{3.16}$$

To simplify, we can make the transformation $I(V_t) = \Phi(V_t)^{-[(1-\gamma)/(1-\varphi)]}$ and have a non-homogeneous ordinary differential equation. Unfortunately, this non-homogeneous ordinary differential equation cannot be solved in closed form. Our approach is to obtain the asymptotic approximation to Equation (3.B1) shown in Appendix 3.B, by taking a log-linear expansion of the consumption–wealth ratio around its unconditional mean as introduced in the papers of Campbell and Viceira (2002) and Chacko and Viceira (2005). We provide the full details of our model's approximate results in Appendix 3.B.

We are now able to obtain the form of the value function and the optimal consumption and dynamic asset allocation strategies toward investing in the risky stock and the derivative in the stochastic environment without the constraint $\varphi = 1$. The value function is

$$J(W_t, V_t) = I(V_t)\frac{W_t^{1-\gamma}}{1-\gamma} = \Phi(V_t)^{-[(1-\gamma)/(1-\varphi)]}\frac{W_t^{1-\gamma}}{1-\gamma}$$

$$= \exp\left[-\left(\frac{1-\gamma}{1-\varphi}\right)(\hat{Q}_0 + \hat{Q}_1 V_t + \hat{Q}_2 \log V_t)\right]\frac{W_t^{1-\gamma}}{1-\gamma}. \tag{3.17}$$

The investor's optimal instantaneous consumption–wealth ratio is

$$\frac{C_t}{W_t} = \beta^\varphi \exp(-\hat{Q}_0 - \hat{Q}_1 V_t - \hat{Q}_2 \log V_t). \tag{3.18}$$

The investor's optimal dynamic asset allocation strategies toward investing in the stock and the derivative are

$$n_t = \frac{1}{\gamma}\left((\mu - r) - \frac{(p_s S_t + \rho\sigma p_v)}{\sigma\sqrt{1-\rho^2}p_v}\lambda\right)\frac{1}{V_t} + \left(1 - \frac{1}{\gamma}\right)\left(\frac{\hat{Q}_1}{1-\varphi} + \frac{\hat{Q}_2}{1-\varphi}\frac{1}{V_t}\right)\rho\sigma$$

$$- \left(1 - \frac{1}{\gamma}\right)\left(\frac{\hat{Q}_1}{1-\varphi} + \frac{\hat{Q}_2}{1-\varphi}\frac{1}{V_t}\right)\frac{(p_s S_t + \rho\sigma p_v)}{p_v}, \tag{3.19}$$

$$\pi_t = \frac{1}{\gamma}\frac{\lambda}{\sigma\sqrt{1-\rho^2}(p_v/O_t)}\frac{1}{V_t} + \left(1 - \frac{1}{\gamma}\right)\left(\frac{\hat{Q}_1}{1-\varphi} + \frac{\hat{Q}_2}{1-\varphi}\frac{1}{V_t}\right)\frac{O_t}{p_v}. \tag{3.20}$$

We have now solved the approximate closed-form solution of the optimal consumption and dynamic asset allocation strategy. In the next section we provide a general extension of this model.

3.3.2 Optimal Consumption and Dynamic Asset Allocation Strategies with Time-Varying Risk Premiums

We next extend the model with constant expected excess returns of the risky stock and the constant volatility risk premiums of the derivative to a more general case in order to explore the optimal consumption rule and dynamic asset allocation strategies with time-varying instantaneous expected excess returns and time-varying stochastic volatility risk premium of the risky stock.

Following Chacko and Viceira (2005), we replace the assumption of the constant excess returns on the stock with one that allows for the expected excess returns on the stock to vary with volatility:

$$E\left[\left(\frac{dS_t}{S_t - r\,dt}\right)\bigg| V_t\right] = (\mu_0 + \mu_1 V_t)dt. \tag{3.21}$$

When $\mu_1 > 0$, the functional form implies that an increase in volatility increases risk and the expected excess return on the stock. Similarly, we replace the assumption of the constant volatility risk premium (λ) with one that allows the volatility risk premium to vary with volatility in the form $\lambda_0 + \lambda_1 V_t$. Ever since the seminal work of Engle (1982), discrete-time ARCH models have become a proven approach to modeling security price volatility. For a review of the substantial literature, see Bollerslev *et al.* (1992). Following Nelson (1990), it is understood that GARCH-type models have well-defined continuous-time limits. Therefore, it seems reasonable to model the risk premium dependent on the conditional variance. As a basis, some empirical papers assume an ARCH/GARCH-M model with a risk premium for stochastic volatility, which is a linear function of the conditional standard deviation or variance as in this chapter. Using Heston's (1993) option pricing formula to price currency options, Lamoureux and Lastrapes (1993) suggest a time-varying volatility risk premium. Empirical applications of the ARCH-M model have met with much success. Some studies (see, e.g. French *et al.* (1987), Chou (1988) and Campbell and Hentschel (1992)) have reported consistently positive and significant estimates of the risk premium. When $\lambda_1 > 0$, the functional form implies that an increase in volatility increases risk and also the stochastic volatility risk premium. When $\mu_1 = 0$ and $\lambda_1 = 0$, the results of this section will reduce to those of Section 3.3.1.

In the above setting, the non-redundant derivative ($O_t = p(S_t, V_t)$), which is a function (p) of the prices of the stock (S_t) and the volatility of stock returns (V_t) at time t, will have the following price dynamics:

$$dO_t = \left[(\mu_0 + \mu_1 V_t)(p_s S_t + \rho\sigma p_v) + (\lambda_0 + \lambda_1 V_t)\sigma\sqrt{1 - \rho^2}p_v + rO_t\right]dt$$
$$+ (p_s S_t + \rho\sigma p_v)\sqrt{V_t}\,dZ_S + (\sigma\sqrt{1 - \rho^2}p_v)\sqrt{V_t}\,dZ_V. \tag{3.22}$$

From these extensions, the Bellman equation can be expressed as

$$0 = -\frac{\beta^\varphi}{1 - \varphi}I^{1+[(1-\varphi)/(1-\gamma)]} + \frac{\varphi}{1 - \varphi}\beta I + \frac{1}{2}\frac{1}{\gamma}I\frac{1}{V_t}\left[(\mu_0 + \mu_1 V_t)^2 + (\lambda_0 + \lambda_1 V_t)^2\right]$$
$$+ Ir + \frac{1}{\gamma}I_V\sigma\left[(\mu_0 + \mu_1 V_t)\rho + (\lambda_0 + \lambda_1 V_t)\sqrt{1 - \rho^2}\right]$$
$$+ I_V\frac{1}{1 - \gamma}[\kappa(\theta - V_t)] + \frac{1}{2}I_{VV}\frac{1}{1 - \gamma}\sigma^2 V_t + \frac{1}{2}\frac{1}{\gamma}\frac{(I_V)^2}{I}\sigma^2 V_t. \tag{3.23}$$

We follow the same method in Section 3.3.1 to derive the optimal consumption and dynamic asset allocation strategies under these assumptions. The main steps of the derivation are guessing the same functional form for $J(W_t, V_t)$ and $I(V_t)$ as in Appendix 3.B. We then can reduce the Bellman equation to an ODE in $\Phi(V_t)$ as follows:

$$0 = -\beta^{\varphi}\Phi^{-1} + \varphi\beta + \frac{1}{2\gamma}(1-\varphi) \times \left[(\mu_0^2 + \lambda_0^2)\frac{1}{V_t} + 2(\mu_0\mu_1 + \lambda_0\lambda_1) + (\mu_1^2 + \lambda_1^2)V_t\right]$$

$$-\frac{1-\gamma}{\gamma}\frac{\Phi_V}{\Phi}\sigma\left[\mu_0\rho + \lambda_0\sqrt{1-\rho^2} + \left(\mu_1\rho + \lambda_1\sqrt{1-\rho^2}\right)V_t\right]$$

$$+ (1-\varphi)r - \frac{\Phi_V}{\Phi}\kappa(\theta - V_t) + \frac{1}{2}\left[\left(\frac{1-\gamma}{1-\varphi}+1\right)\left(\frac{\Phi_V}{\Phi}\right)^2 - \frac{\Phi_{VV}}{\Phi}\right]\sigma^2 V_t$$

$$+ \frac{1}{2}\frac{(1-\gamma)^2}{\gamma}\frac{1}{1-\varphi}\left(\frac{\Phi_V}{\Phi}\right)^2\sigma^2 V_t. \tag{3.24}$$

Following the steps in Equation (3.A8) to Equation (3.A10), we obtain the solution of the ODE in $\Phi(V_t)$ with $\Phi(V_t) = \exp(\tilde{Q}_0 + \tilde{Q}_1 V_t + \tilde{Q}_2 \log V_t)$, and can express Equation (3.24) as

$$0 = -\left\{\phi_0 + \phi_1\left[\varphi\log\beta - \tilde{Q}_0 - \tilde{Q}_1 V_t - \tilde{Q}_2\left(\log\theta + \frac{1}{\theta}V_t - 1\right)\right]\right\}$$

$$+ \varphi\beta + \frac{1}{2\gamma}(1-\varphi) \times \left[(\mu_0^2 + \lambda_0^2)\frac{1}{V_t} + 2(\mu_0\mu_1 + \lambda_0\lambda_1) + (\mu_1^2 + \lambda_1^2)V_t\right]$$

$$- \frac{1-\gamma}{\gamma}\left(\tilde{Q}_1 + \tilde{Q}_2\frac{1}{V_t}\right)\sigma \times \left[\mu_0\rho + \lambda_0\sqrt{1-\rho^2} + \left(\mu_1\rho + \lambda_1\sqrt{1-\rho^2}\right)V_t\right]$$

$$+ (1-\varphi)r - \left(\tilde{Q}_1 + \tilde{Q}_2\frac{1}{V_t}\right)\kappa(\theta - V_t) + \frac{1}{2}\left[\left(\frac{1-\gamma}{1-\varphi}+1\right)\left(\tilde{Q}_1 + \tilde{Q}_2\frac{1}{V_t}\right)^2\right.$$

$$\left. - \left(\tilde{Q}_1 + \tilde{Q}_2\frac{1}{V_t}\right)^2 + \tilde{Q}_2\frac{1}{V_t^2}\right]\sigma^2 V_t + \frac{1}{2}\frac{(1-\gamma)^2}{\gamma}\frac{1}{1-\varphi}\left(\tilde{Q}_1 + \tilde{Q}_2\frac{1}{V_t}\right)^2\sigma^2 V_t. \tag{3.25}$$

Rearranging this equation after collecting terms in $1/V_t$, V_t and 1 we have three equations for $\tilde{Q}_2$, $\tilde{Q}_1$, and $\tilde{Q}_0$:

$$\left[\frac{1}{2}\frac{1-\gamma}{1-\varphi}\sigma^2 + \frac{1}{2}\frac{(1-\gamma)^2}{\gamma}\frac{1}{1-\varphi}\sigma^2\right]\tilde{Q}_2^2$$

$$- \left(\frac{1-\gamma}{\gamma}\sigma\mu_0\rho + \frac{1-\gamma}{\gamma}\lambda_0\sqrt{1-\rho^2}\sigma + \kappa\theta - \frac{1}{2}\sigma^2\right)\tilde{Q}_2$$

$$+ \frac{1}{2\gamma}(1-\varphi)(\mu_0^2 + \lambda_0^2) = 0, \tag{3.26}$$

$$\left[\frac{1}{2}\frac{1-\gamma}{1-\varphi}\sigma^2+\frac{1}{2}\frac{(1-\gamma)^2}{\gamma}\frac{1}{1-\varphi}\sigma^2\right]\tilde{Q}_1^2+\left[\phi_1-\frac{1-\gamma}{\gamma}\sigma\left(\mu_1\rho+\lambda_1\sqrt{1-\rho^2}\right)+\kappa\right]\tilde{Q}_1$$

$$+\phi_1\frac{1}{\theta}\tilde{Q}_2+\frac{1}{2\gamma}(1-\varphi)\left(\mu_1^2+\lambda_1^2\right)=0, \tag{3.27}$$

$$\left[\phi_1\log\theta-\phi_1-\frac{1-\gamma}{\gamma}\sigma\left(\mu_1\rho+\lambda_1\sqrt{1-\rho^2}\right)+\kappa\right]\tilde{Q}_2$$

$$-\left(\frac{1-\gamma}{\gamma}\sigma\mu_0\rho+\frac{1-\gamma}{\gamma}\lambda_0\sqrt{1-\rho^2}\sigma+\kappa\theta\right)\tilde{Q}_1+\left[\frac{1-\gamma}{1-\varphi}\sigma^2+\frac{(1-\gamma)^2}{\gamma}\frac{1}{1-\varphi}\sigma^2\right]\tilde{Q}_1\tilde{Q}_2$$

$$+\phi_1\tilde{Q}_0-\phi_0-\phi_1\varphi\log\beta+\varphi\beta+\frac{1}{\gamma}(1-\varphi)(\mu_0\mu_1+\lambda_0\lambda_1)+(1-\varphi)r=0. \tag{3.28}$$

We are now able to obtain the value function and the investor's optimal consumption and dynamic asset allocation strategy for investing in the stock and derivative security with time-varying risk premiums. The value function is

$$J(W_t,V_t)=I(V_t)\frac{W_t^{1-\gamma}}{1-\gamma}=\Phi(V_t)^{-[(1-\gamma)/(1-\varphi)]}\frac{W_t^{1-\gamma}}{1-\gamma}$$

$$=\exp\left[-\left(\frac{1-\gamma}{1-\varphi}\right)\left(\tilde{Q}_0+\tilde{Q}_1V_t+\tilde{Q}_2\log V_t\right)\right]\frac{W_t^{1-\gamma}}{1-\gamma}. \tag{3.29}$$

The investor's optimal instantaneous consumption–wealth ratio is

$$\frac{C_t}{W_t}=\beta^\varphi\exp\left(-\tilde{Q}_0-\tilde{Q}_1V_t-\tilde{Q}_2\log V_t\right). \tag{3.30}$$

The investor's optimal dynamic asset allocation strategy toward investing in the stock and the derivative security is

$$n_t=\frac{1}{\gamma}\left(\mu_1+\mu_0\frac{1}{V_t}\right)+\left(1-\frac{1}{\gamma}\right)\left(\frac{\tilde{Q}_1}{1-\varphi}+\frac{\tilde{Q}_2}{1-\varphi}\frac{1}{V_t}\right)\rho\sigma$$

$$-\frac{1}{\gamma}\left(\lambda_1+\lambda_0\frac{1}{V_t}\right)\frac{(p_sS_t+\rho\sigma p_v)}{\sigma\sqrt{1-\rho^2}p_v}$$

$$-\left(1-\frac{1}{\gamma}\right)\left(\frac{\tilde{Q}_1}{1-\varphi}+\frac{\tilde{Q}_2}{1-\varphi}\frac{1}{V_t}\right)\frac{(p_sS_t+\rho\sigma p_v)}{p_v}$$

$$=\frac{1}{\gamma}\left(\mu_1+\mu_0\frac{1}{V_t}\right)-\frac{1}{\gamma}\left(\lambda_1+\lambda_0\frac{1}{V_t}\right)\frac{\rho}{\sqrt{1-\rho^2}}-\pi_t\frac{p_sS_t}{O_t}, \tag{3.31}$$

$$\pi_t = \left[\frac{1}{\gamma}\frac{1}{\sigma\sqrt{1-\rho^2}}\left(\lambda_1 + \lambda_0 \frac{1}{V_t}\right) + \left(1 - \frac{1}{\gamma}\right)\left(\frac{\tilde{Q}_1}{1-\varphi} + \frac{\tilde{Q}_2}{1-\varphi}\frac{1}{V_t}\right)\right]\frac{O_t}{P_v}. \quad (3.32)$$

We have solved for the approximate solution of the optimal consumption and dynamic asset allocation strategies with time-varying risk premiums. In the next section we discuss the optimal consumption rule and dynamic asset allocation of our results and the effects of the introduction of derivative securities. We also provide analyses of our results, in this section and of a numerical example shown in the figures.

3.4 ANALYSES OF THE OPTIMAL CONSUMPTION RULE AND DYNAMIC ASSET ALLOCATION STRATEGY

In Section 3.3.2 we have the most general results. Equation (3.30) shows that the optimal log consumption–wealth ratio is a function of stochastic volatility, with coefficients $\tilde{Q}_1/(1-\varphi)$ and $\tilde{Q}_2/(1-\varphi)$. While $\tilde{Q}_2$ is the solution to the quadratic Equation (3.26), $\tilde{Q}_1$ is the solution to Equation (3.27) given $\tilde{Q}_2$, and $\tilde{Q}_0$ is the solution to Equation (3.28) given $\tilde{Q}_1$ and $\tilde{Q}_2$. When $\gamma > 1$ for coefficient $\tilde{Q}_2$, Equation (3.26) has two real roots of opposite signs. In each quadratic equation, we would like to know which solution is good for our problem from the following criteria. First, we must ensure that the roots of the discriminant are real. We must then determine the sign of the roots that we will choose. Campbell and Viceira (1999, 2002), Campbell et al. (2004) and Chacko and Viceira (2005) show that only one of them maximizes the value function. This is also the only root that ensures $\tilde{Q}_2$ is equal to zero when $\gamma = \varphi = 1$, that is, in the log utility case. Under these criteria, the value function J is maximized only with the solution associated with the negative root of the discriminant of the quadratic Equation (3.26), i.e. the positive root of Equation (3.26). It can immediately be shown that $\tilde{Q}_2/(1-\varphi) > 0$.

By the same criteria, it can immediately be shown that $\tilde{Q}_1/(1-\varphi) < 0$. Thus, the ratio of consumption to wealth is shown to be an increasing function of volatility for those investors whose elasticity of intertemporal substitution of consumption (φ) is less than one. Conversely, this ratio is a decreasing function of volatility when the elasticity of intertemporal substitution of consumption for an investor is greater than one. (It is assumed that $-\tilde{Q}_1 - \tilde{Q}_2/V_t > 0$.) The relative importance of intertemporal income and substitution effects of volatility on consumption is thus demonstrated. For illustration, suppose the greater effect of volatility on consumption. The increase in volatility provides greater opportunities for investment, because derivative securities can provide volatility exposures. Moreover, assuming positively priced volatility risk, returns on the portfolio do not increase in volatility, yet there is an improvement in the expected return. A negative intertemporal substitution effect on consumption will result from these investment opportunities, since the latter are more favorable than at other times. A positive income effect also results, as an increase in expected returns decreases the marginal utility of consumption. The income effect dominates the substitution effect for investors whose

$\varphi < 1$ and their current consumption rises relative to wealth. Investors whose $\varphi > 1$ will see the substitution effect dominate and will therefore cut their current consumption relative to wealth.

The optimal dynamic asset allocation strategy for the risky stock has four components, as with the first equality of Equation (3.31) and in Figure 3.1. If we do not introduce any derivative security and the investor holds only the risky stock, then the optimal dynamic asset allocation for the risky stock will only have the first two components in the first equality of Equation (3.31), i.e. the myopic component and the intertemporal hedging component. First, the dependence of the myopic component is simple. It is an affine function of the reciprocal of the time-varying volatility and decreases with the coefficient of relative risk aversion. Since volatility is time-varying, the myopic component is time-varying too. The position of the myopic component can be either positive or negative, depending on μ_0, μ_1, and the level of volatility. We know that if $\mu_1 = 0$ and $\mu_0 = \mu - r$, then the result also nests the model results in Section 3.3.1. An extension of the replacement of the constant expected excess return with one that allows an expected excess return on the risky stock to vary with volatility implies that increased risk is rewarded with an increase in expected excess return when $\mu_1 > 0$. Hence, the investor will take a long position on the myopic component of the risky stock.

The intertemporal hedging component of the optimal dynamic asset allocation for the risky stock is an affine function of the reciprocal of the time-varying volatility, with

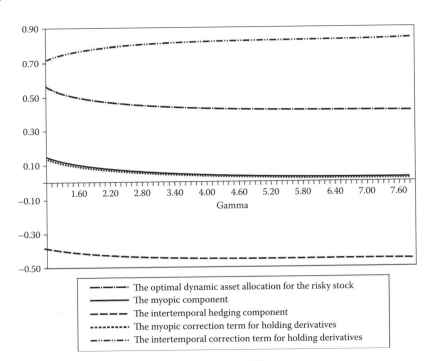

FIGURE 3.1 The optimal dynamic asset allocation strategy toward investing in the stock and its components in relation to γ.

coefficients $\tilde{Q}_1/(1-\varphi)$ and $\tilde{Q}_2/(1-\varphi)$. Since $\tilde{Q}_2/(1-\varphi)>0$, the sign of the coefficient of the intertemporal hedging demand coming from pure changes in time-varying volatility is positive when $\gamma > 1$ [this is true only for $\rho > 0$]. If we do not introduce any derivative security and instead hold only risky stock, then the intertemporal hedging component for the risky stock will consist of the correlation or asymmetric effect. The intertemporal hedging component of the optimal asset allocation for risky stock is affected by the instantaneous correlation between the two Brownian motions. If the instantaneous correlation is perfect, then markets are complete, without the need for holding any derivative securities. However, we allow for imperfect instantaneous correlation in the model. In particular, if $\rho < 0$, this means that the unexpected return of the risky asset is low (the market situation is bad), and then the state of the market uncertainty will be high. Since $\hat{Q}_2/(1-\varphi)>0$ when $\gamma > 1$, the negative instantaneous correlation implies the investor will have negative intertemporal hedging demand due to changes solely in the volatility of the risky asset, which lacks the hedging ability against an increase in volatility. Similar discussions are found in Liu (2005) and Chacko and Viceira (2005). However, in our generalized model the consideration of holding derivative securities complicates the asset allocation strategies for long-horizon investors.

The other two components of the optimal dynamic asset allocation for the risky stock are correction terms for holding the derivative. The first and the second component of these two correction terms (i.e. the third and the fourth terms of the optimal dynamic asset allocation for the risky stock) are the myopic correction term and the intertemporal correction term for the derivatives held, respectively. These terms are from the interaction between the derivative and its underlying stock. We can see that from the first equality to the second equality of Equation (3.31), the intertemporal hedging demand on the risky stock will be canceled by the correction term of holding the derivative. In the second equality of Equation (3.31), in its first term we show that the net demand for the risky stock will finally link to the risk-and-return tradeoff associated with price risk, because volatility exposures have been captured by holding the derivatives. The second term is the correction term of the correlation effect, and the third correction term is to correct for the delta effect of the derivative held. The relationships between these components with the degree of risk aversion (γ) are also seen in Figure 3.2.

The optimal dynamic asset allocation for the derivative depends on the proportion of derivative's price to its volatility exposure. This proportion measures the dollars expended on the derivative for each unit of volatility exposure. The less expenditure that a derivative provides for the same unit of volatility exposure, the more effective it is as a vehicle to hedge volatility risk. Therefore, a smaller portion of the investor's wealth need be allocated to the derivative (Liu and Pan 2003). Furthermore, the optimal dynamic asset allocation strategy for the derivative security can also be separated into two components: the myopic demand and the intertemporal hedging demand, which are also seen in Figure 3.3. The myopic demand is an affine function of the reciprocal of the time-varying volatility with coefficients of λ_0 and λ_1, and it depends on $(1/\gamma)$. The time-varying volatility also makes the myopic demand time-varying. The long or short position of the myopic derivative's

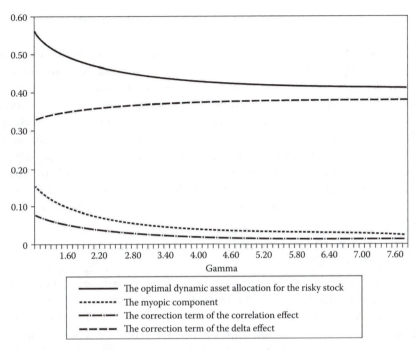

FIGURE 3.2 The optimal dynamic asset allocation strategy toward investing in the risky stock and its components in relation to γ.

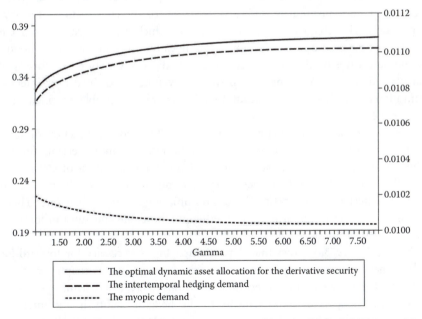

FIGURE 3.3 The optimal dynamic asset allocation strategy toward investing in the derivative security and its components in relation to γ.

demand depends on the volatility–risk premium parameters λ_0 and λ_1, and the level of volatility.

The second term of the optimal dynamic asset allocation of the derivative is the intertemporal hedging component, which depends on all the parameters that characterize investor preferences and the investment opportunity set. Without loss of generality we may assume that the option is volatility exposure is positive. In our setting, the instantaneous correlation between unexpected returns of the derivative and changes in volatility is perfect positive. In addition, since $\tilde{Q}_2/(1 - \varphi) > 0$ when $\gamma > 1$, for more risk-averse investors the intertemporal hedging demand is positive due to changes solely in the stochastic volatility. Investors who are more risk-averse than logarithmic investors have a positive intertemporal hedging demand for the derivative, because it tends to do better when there is an increase in volatility risk. The derivative provides hedging ability against an increase in volatility.

3.5 CONCLUSIONS

Ever since Merton (1969, 1971) introduced the standard intertemporal consumption and investment model, it has been studied extensively in the finance literature and has become a classical problem in financial economics. The literature on the broad set of issues of intertemporal consumption and investment or optimal dynamic asset allocation strategies deals with investors' access only to bond and stock markets and excludes the derivatives market. While a few recent studies include derivative securities in the investment portfolio, investors are assumed to have a specified utility defined over wealth at a single terminal date, abstraction from the choice of consumption over time, and the studies are restricted to a static position in derivative securities or the construction of a buy-and-hold portfolio. This chapter considers a model in which a long-term investor chooses consumption as well as optimal dynamic asset allocation including a riskless bond, risky stock and derivatives on the stock when there is predictable variation in return volatility. We then maximize a more general recursive utility function defined over intermediate consumption rather than terminal wealth to reflect the realistic problem facing an investor saving for the future.

We show that the optimal log consumption–wealth ratio is a function of stochastic volatility. Furthermore, the consumption–wealth ratio is an increasing function of volatility for investors whose elasticity of intertemporal substitution of consumption is smaller than one, while it is a decreasing function of volatility for investors whose elasticity of intertemporal substitution of consumption is larger than one. This result reflects the comparative importance of intertemporal income and substitution effects of volatility on consumption.

Merton (1971, 1973a) shows that dynamic hedging is necessary for forward-looking investors when investment opportunities are time-varying. In this chapter we show that considering derivative securities in portfolio decisions to create a dynamic asset allocation strategy brings benefits of improvements to the hedging ability in the intertemporal hedging component. When we introduce a non-redundant derivative written on the risky stock in the incomplete financial market in this optimal dynamic asset allocation, the derivative in the asset allocation can provide differential exposures to stochastic volatility and make the market complete. Derivative securities are a significant tool for expanding investors' dimensions for risk-and-return tradeoffs, as a vehicle to hedge the additional risk factor of stochastic volatility in the stock market. Non-myopic investors utilize derivative securities, which provide access to volatility risk, to capitalize upon the time-varying nature of their opportunity set.

REFERENCES

Bhamra, H. and Uppal, R., The role of risk aversion and intertemporal substitution in dynamic consumption-portfolio choice with recursive utility. EFA Annual Conference Paper No. 267, 2003.

Black, F., Studies of stock market volatility changes, in *Proceedings of the American Statistical Association, Business and Economic Statistics Section*, 1976, pp. 177–181.

Black, F. and Scholes, M.S., The pricing of options and corporate liabilities. *J. Polit. Econ.*, 1973, **81**, 637–654.

Bollerslev, T., Chou, R.Y. and Kroner, K.F., ARCH modeling in finance: A review of the theory and empirical evidence. *J. Econometrics*, 1992, **52**, 5–59.

Campbell, J.Y., Intertemporal asset pricing without consumption data. *Am. Econ. Rev.*, 1993, **83**, 487–512.

Campbell, J.Y. and Hentschel, L., No news is good news: An asymmetric model of changing volatility in stock returns. *J. Finan. Econ.*, 1992, **31**, 281–318.

Campbell, J.Y. and Viceira, L.M., Consumption and portfolio decisions when expected returns are time varying. *Q. J. Econ.*, 1999, **114**, 433–495.

Campbell, J.Y. and Viceira, L.M., Who should buy long-term bonds? *Am. Econ. Rev.*, 2001, **91**, 99–127.

Campbell, J.Y. and Viceira, L.M., *Strategic Asset Allocation: Portfolio Choice for Long-term Investors*, 2002 (Oxford University Press: Oxford).

Campbell, J.Y., Chacko, G., Rodriguez, J. and Viceira, L.M., Strategic asset allocation in a continuous-time VAR model. *J. Econ. Dynam. Contr.*, 2004, **28**, 2195–2214.

Chacko, G. and Viceira, L., Dynamic consumption and portfolio choice with stochastic volatility in incomplete markets. *Rev. Finan. Stud.*, 2005, **18**, 1369–1402.

Chalasani, P. and Jha, S., Randomized stopping times and American option pricing with proportional transaction cost. *Math. Finan.*, 2001, **11**, 33–77.

Chou, R.Y., Volatility persistence and stock valuations: Some empirical evidence using GARCH. *J. Appl. Econometrics*, 1988, **3**, 279–294.

Cont, R., Tankov, P. and Voltchkova, E., Hedging with options in models with jumps, in *Proceedings of the II Abel Symposium 2005 on Stochastic Analysis and Applications*, 2005.

Duffie, D. and Epstein, L.G., Stochastic differential utility. *Econometrica*, 1992a, **60**, 353–394.

Duffie, D. and Epstein, L.G., Asset pricing with stochastic differential utility. *Rev. Finan. Stud.*, 1992b, **5**, 411–436.

Engle, R.F., Autoregressive conditional heteroscedasticity with estimates of the variance of UK inflation. *Econometrica*, 1982, **50**, 987–1008.

French, K.R., Schwert, G.W. and Stambaugh, R.F., Expected stock returns and volatility. *J. Finan. Econ.*, 1987, **19**, 3–30.

Glosten, L.R., Jagannathan, R. and Runkle, D., On the relation between the expected value and the volatility of the nominal excess return on stocks. *J. Finan.*, 1993, **48**, 1779–1801.

Haugh, M. and Lo, A., Asset allocation and derivatives. *Quant. Finan.*, 2001, **1**, 45–72.

Heston, S.L., A closed-form solution for options with stochastic volatility with applications to bond and currency options. *Rev. Finan. Stud.*, 1993, **6**, 327–343.

Ilhan, A., Jonsson, M. and Sircar, R., Optimal investment with derivative securities. *Finan. Stochast.*, 2005, **9**, 585–595.

Lamoureux, C.G. and Lastrapes, W.D., Forecasting stock-return variances: Toward an understanding of stochastic implied volatilities. *Rev. Finan. Stud.*, 1993, **6**, 293–326.

Liu, J., Portfolio selection in stochastic environments. Stanford GSB Working Paper, 2005.

Liu, J. and Pan, J., Dynamic derivative strategies. *J. Finan. Econ.*, 2003, **69**, 401–430.

Merton, R.C., Lifetime portfolio selection under uncertainty: The continuous time case. *Rev. Econ. Statist.*, 1969, **51**, 247–257.

Merton, R.C., Optimum consumption and portfolio rules in a continuous-time model. *J. Econ. Theor.*, 1971, **3**, 373–413.

Merton, R.C., An intertemporal capital asset pricing model. *Econometrica*, 1973a, **41**, 867–887.

Merton, R.C., The theory of rational option pricing. *Bell J. Econ. Mgmt. Sci.*, 1973b, **4**, 141–183.

Nelson, D., ARCH models as diffusion approximations. *J. Econometrics*, 1990, **45**, 7–38.

Sircar, K.R. and Papanicolaou, G.C., Stochastic volatility, smile & asymptotics. *Appl. Math. Finan.*, 1999, **6**, 107–145.

Stein, E.M. and Stein, J., Stock price distributions with stochastic volatility: An analytic approach *Rev. Finan. Stud.*, 1991, **4**, 727–752.

APPENDIX 3.A: DERIVATION OF THE EXACT SOLUTION WHEN $\varphi = 1$

Substituting the first-order conditions (i.e. Equations (3.9)–(3.11)) back into the Bellman equation (i.e. Equation (3.8)) and rearranging we get

$$
0 = f(C(J), J) - J_W C(J) - \frac{1}{2} \frac{(J_W)^2}{J_{WW}} \frac{1}{V_t} \left[(\mu - r)^2 + \lambda^2 \right]
$$

$$
- \frac{J_W J_{WV}}{J_{WW}} \sigma \left[(\mu - r)\rho + \lambda\sqrt{1 - \rho^2} \right] + J_W r W_t + J_V \left[\kappa(\theta - V_t) \right]
$$

$$
+ \frac{1}{2} J_{VV} \sigma^2 V_t - \frac{1}{2} \frac{(J_{WV})^2}{J_{WW}} \sigma^2 V_t. \tag{3.A1}
$$

We conjecture that there exists a solution of the functional form $J(W_t, V_t) = I(V_t) \times [W_t^{1-\gamma}/(1 - \gamma)]$ when $\varphi = 1$, and substituting it into Equation (3.A1) we obtain

$$
0 = \left(\log \beta - \frac{1}{1 - \gamma} \log I - 1 \right) \beta I + \frac{1}{2} \frac{1}{\gamma} I \frac{1}{V_t} \left[(\mu - r)^2 + \lambda^2 \right]
$$

$$
+ \frac{1}{\gamma} I_V \sigma \left[(\mu - r)\rho + \lambda\sqrt{1 - \rho^2} \right] + Ir + I_V \frac{1}{1 - \gamma} \left[\kappa(\theta - V_t) \right]
$$

$$
+ \frac{1}{2} I_{VV} \frac{1}{1 - \gamma} \sigma^2 V_t + \frac{1}{2} \frac{1}{\gamma} \frac{(I_V)^2}{I} \sigma^2 V_t. \tag{3.A2}
$$

The above ordinary differential equation has a solution of the form $I = \exp(Q_0 + Q_1 V_t + Q_2 \log V_t$, and so Equation (3.A2) can be expressed as

$$0 = \left\{ \log \beta - \frac{1}{1-\gamma} \left[Q_0 + Q_1 V_t + Q_2 \left(\log \theta + \frac{1}{\theta} V_t - 1 \right) \right] - 1 \right\} \beta$$

$$+ \frac{1}{2} \frac{1}{\gamma} \frac{1}{V_t} [(\mu - r)^2 + \lambda^2] + \frac{1}{\gamma} \left(Q_1 + Q_2 \frac{1}{V_t} \right) \sigma \left[(\mu - r)\rho + \lambda \sqrt{1 - \rho^2} \right]$$

$$+ r + \left(Q_1 + Q_2 \frac{1}{V_t} \right) \frac{1}{1-\gamma} [\kappa(\theta - V_t)]$$

$$+ \frac{1}{2} \frac{1}{1-\gamma} \left[\left(Q_1 + Q_2 \frac{1}{V_t} \right)^2 - Q_2 \frac{1}{V_t^2} \right] \sigma^2 V_t + \frac{1}{2} \frac{1}{\gamma} \left(Q_1 + Q_2 \frac{1}{V_t} \right)^2 \sigma^2 V_t. \quad (3.A3)$$

Rearranging that equation, we have three equations for Q_2, Q_1, and Q_0 after collecting terms in $1/V_t$, V_t, and 1:

$$\left(\frac{1}{2} \frac{1}{1-\gamma} \sigma^2 + \frac{1}{2} \frac{1}{\gamma} \sigma^2 \right) Q_2^2 + \left[\frac{1}{\gamma} \sigma(\mu - r)\rho + \frac{1}{\gamma} \sigma \lambda \sqrt{1 - \rho^2} + \frac{1}{1-\gamma} \kappa \theta - \frac{1}{2} \frac{1}{1-\gamma} \sigma^2 \right] Q_2$$

$$+ \frac{1}{2} \frac{1}{\gamma} [(\mu - r)^2 + \lambda^2] = 0, \quad (3.A4)$$

$$\left(\frac{1}{2} \frac{1}{1-\gamma} \sigma^2 + \frac{1}{2} \frac{1}{\gamma} \sigma^2 \right) Q_1^2 - \left(\frac{1}{1-\gamma} \beta + \frac{1}{1-\gamma} \kappa \right) Q_1 - \frac{1}{1-\gamma} \beta \frac{1}{\theta} Q_2 = 0, \quad (3.A5)$$

$$\left(-\frac{1}{1-\gamma} \beta \log \theta + \frac{1}{1-\gamma} \beta - \frac{1}{1-\gamma} \kappa \right) Q_2 + \left[\frac{1}{\gamma} \sigma(\mu - r)\rho + \frac{1}{\gamma} \sigma \lambda \sqrt{1 - \rho^2} + \frac{1}{1-\gamma} \kappa \theta \right] Q_1$$

$$+ \left[\frac{1}{1-\gamma} \sigma^2 + \frac{1}{\gamma} \sigma^2 \right] Q_1 Q_2 - \frac{1}{1-\gamma} \beta Q_0 + \beta \log \beta - \beta + r = 0, \quad (3.A6)$$

and in Equation (3.A4), we have

$$Q_2 = \frac{-b \pm \sqrt{b^2 - 4ac}}{2a},$$

where

$$a = \frac{1}{2}\frac{1}{1-\gamma}\sigma^2 + \frac{1}{2}\frac{1}{\gamma}\sigma^2, \quad b = \frac{1}{\gamma}\sigma(\mu - r)\rho$$

$$+ \frac{1}{\gamma}\sigma\lambda\sqrt{1-\rho^2} + \frac{1}{1-\gamma}\kappa\theta - \frac{1}{2}\frac{1}{1-\gamma}\sigma^2,$$

and

$$c = \frac{1}{2}\frac{1}{\gamma}\left[(\mu - r)^2 + \lambda^2\right].$$

APPENDIX 3.B: DERIVATION OF THE APPROXIMATE RESULTS

For simplicity, we can make the transformation $I(V_t) = \Phi(V_t)^{-[(1-\gamma)/(1-\varphi)]}$ and obtain the following non-homogeneous ordinary differential equation:

$$0 = -\beta^\varphi \Phi^{-1} + \varphi\beta + \frac{1}{2\gamma}(1-\varphi)\left[(\mu - r)^2 + \lambda^2\right] \times \frac{1}{V_t}$$

$$- \frac{1-\gamma}{\gamma}\frac{\Phi_V}{\Phi}\sigma\left[(\mu - r)\rho + \lambda\sqrt{1-\rho^2}\right] + (1-\varphi)r - \frac{\Phi_V}{\Phi}\kappa(\theta - V_t)$$

$$+ \frac{1}{2}\left[\left(\frac{1-\gamma}{1-\varphi} + 1\right)\left(\frac{\Phi_V}{\Phi}\right)^2 - \frac{\Phi_{VV}}{\Phi}\right]\sigma^2 V_t + \frac{1}{2}\frac{(\gamma-1)^2}{\gamma}\frac{1}{1-\varphi}\left(\frac{\Phi_V}{\Phi}\right)^2 \sigma^2 V_t. \quad (3.B1)$$

From the transformation, we obtain the envelope condition of Equation (3.9):[1]

$$\frac{C_t}{W_t} = \beta^\varphi \Phi^{-1} = \exp\left\{\log\left(\frac{C_t}{W_t}\right)\right\} \equiv \exp\{c_t - w_t\}. \quad (3.B2)$$

---

[1] From Equation (3.9) we have $C_t = J_W^{-\varphi} J^{(1-\varphi\gamma)/(1-\gamma)}\beta^\varphi(1-\gamma)^{(1-\varphi\gamma)/(1-\gamma)}$, and we also conjecture that there exists a solution of the functional form $J(W_t, V_t) = I(V_t)[W_t^{1-\gamma}/(1-\gamma)]$, and we make the transformation $I(V_t) = \Phi(V_t)^{-(1-\gamma)/(1-\varphi)}$. We thus have $J = [W_t^{1-\gamma}/(1-\gamma)]\Phi^{-(1-\gamma)/(1-\varphi)}$ and $J_W = \Phi^{-(1-\gamma)/(1-\varphi)}W_t^{-\gamma}$. Therefore, we have

$$C_t = J_W^{-\varphi} J^{(1-\varphi\gamma)/(1-\gamma)}\beta^\varphi(1-\gamma)^{(1-\varphi\gamma)/(1-\gamma)}$$

$$= (\Phi^{-(1-\gamma)/(1-\varphi)} W_t^{-\gamma})^{-\varphi}\left(\frac{W_t^{1-\gamma}}{1-\gamma}\Phi^{-(1-\gamma)/(1-\varphi)}\right)^{(1-\varphi\gamma)/(1-\gamma)}\beta^\varphi(1-\gamma)^{(1-\varphi\gamma)/(1-\gamma)}$$

$$= \Phi^{-1}\beta^\varphi W_t.$$

Using a first-order Taylor expansion of $\exp\{c_t - w_t\}$ around the expectation of $(c_t - w_t)$, we can write

$$\beta^\varphi \Phi^{-1} \approx \exp\{E(c_t - w_t)\} + \exp\{E(c_t - w_t)\} \cdot [(c_t - w_t) - E(c_t - w_t)]$$

$$= \exp\{E(c_t - w_t)\} \cdot \{1 - E(c_t - w_t)\} + \exp\{E(c_t - w_t)\} \cdot (c_t - w_t)$$

$$\equiv \phi_0 + \phi_1(c_t - w_t). \tag{3.B3}$$

Substituting (3.B3) into Equation (3.B1) and assuming this equation has a solution of the form $\Phi(V_t) = \exp(\hat{Q}_0 + \hat{Q}_1 V_t + \hat{Q}_2 \log V_t)$, from this guessed solution an Equation (3.B2) we can show that

$$(c_t - w_t) = \log\left\{\beta^\varphi \left[\exp(\hat{Q}_0 + \hat{Q}_1 V_t + \hat{Q}_2 \log V_t)\right]^{-1}\right\}$$

$$= \varphi \log \beta - \hat{Q}_0 - \hat{Q}_1 V_t - \hat{Q}_2 \log V_t. \tag{3.B4}$$

As such, we can express Equation (3.B1) as

$$0 = -\left\{\phi_0 + \phi_1\left[\varphi \log \beta - \hat{Q}_0 - \hat{Q}_1 V_t - \hat{Q}_2\left(\log \theta + \frac{1}{\theta} V_t - 1\right)\right]\right\}$$

$$+ \varphi\beta + \frac{1}{2\gamma}(1-\varphi)\left[(\mu-r)^2 + \lambda^2\right]\frac{1}{V_t} - \frac{1-\gamma}{\gamma}\left(\hat{Q}_1 + \hat{Q}_2\frac{1}{V^t}\right)\sigma\left[(\mu-r)\rho + \lambda\sqrt{1-\rho^2}\right]$$

$$+ (1-\varphi)r - \left(\hat{Q}_1 + \hat{Q}_2\frac{1}{V_t}\right)\kappa(\theta - V_t)$$

$$+ \frac{1}{2}\left[\left(\frac{1-\gamma}{1-\varphi}+1\right)\left(\hat{Q}_1 + \hat{Q}_2\frac{1}{V_t}\right)^2 - \left(\hat{Q}_1 + \hat{Q}_2\frac{1}{V_t}\right)^2 + \hat{Q}_2\frac{1}{V_t^2}\right]\sigma^2 V_t$$

$$+ \frac{1}{2}\frac{(1-\gamma)^2}{\gamma}\frac{1}{1-\varphi}\left(\hat{Q}^1 + \hat{Q}_2\frac{1}{V^t}\right)^2\sigma^2 V_t. \tag{3.B5}$$

Rearranging the above equation we have the following three equations for $\hat{Q}_2$, $\hat{Q}_1$, and $\hat{Q}_0$:

$$\left[\frac{1}{2}\frac{1-\gamma}{1-\varphi}\sigma^2 + \frac{1}{2}\frac{(1-\gamma)^2}{\gamma}\frac{1}{1-\varphi}\sigma^2\right]\hat{Q}_2^2$$

$$- \left[\frac{1-\gamma}{\gamma}\sigma(\mu-r)\rho + \frac{1-\gamma}{\gamma}\lambda\sqrt{1-\rho^2}\sigma + \kappa\theta - \frac{1}{2}\sigma^2\right]\hat{Q}_2$$

$$+ \frac{1}{2\gamma}(1-\varphi)[(\mu-r)^2 + \lambda^2] = 0, \tag{3.B6}$$

$$\left[\frac{1}{2}\frac{1-\gamma}{1-\varphi}\sigma^2+\frac{1}{2}\frac{(1-\gamma)^2}{\gamma}\frac{1}{1-\varphi}\sigma^2\right]\hat{Q}_1^2+(\phi_1+\kappa)\hat{Q}_1+\phi_1\frac{1}{\theta}\hat{Q}_2=0, \tag{3.B7}$$

$$(\phi_1\log\theta-\phi_1+\kappa)\hat{Q}_2$$

$$-\left[\frac{1-\gamma}{\gamma}\sigma(\mu-r)\rho+\frac{1-\gamma}{\gamma}\lambda\sqrt{1-\rho^2}\sigma+\kappa\theta\right]\hat{Q}_1$$

$$+\left[\frac{1-\gamma}{1-\varphi}\sigma^2+\frac{(1-\gamma)^2}{\gamma}\frac{1}{1-\varphi}\sigma^2\right]\hat{Q}_1\hat{Q}_2+\phi_1\hat{Q}_0-\phi_0$$

$$-\phi_1\varphi\log\beta+\varphi\beta+(1-\varphi)r=0, \tag{3.B8}$$

where $\hat{Q}_2$ can be found from the quadratic Equation (3.B6), $\hat{Q}_1$ can be found from Equation (3.B7) given $\hat{Q}_2$, and $\hat{Q}_0$ can be found from Equation (3.B8), given $\hat{Q}_1$ and $\hat{Q}_2$.

CHAPTER 4

Volatility-Induced Financial Growth

M. A. H. DEMPSTER, IGOR V. EVSTIGNEEV and KLAUS R. SCHENK-HOPPÉ

CONTENTS

4.1 INTRODUCTION

C AN VOLATILITY, WHICH IS PRESENT in virtually every financial market and usually thought of in terms of a risky investment's downside, serve as an 'engine' for financial growth? Paradoxically, the answer to this question turns out to be positive.

To demonstrate this paradox, we examine the long-run performance of constant proportions investment strategies in a securities market. Such strategies prescribe rebalancing the investor's portfolio, depending on price fluctuations, so as to keep fixed proportions of wealth in all the portfolio positions. Assume that asset returns form a stationary ergodic process and asset prices grow (or decrease) at a common asymptotic rate ρ. It is shown in this chapter that if an investor employs *any* constant proportions strategy, then the value of his/her portfolio grows almost surely at a rate *strictly greater* than ρ, provided that the investment proportions are strictly positive and the stochastic price process is in a sense non-degenerate. The very mild assumption of non-degeneracy we impose requires at least some randomness, or volatility, of the price process. If this assumption is violated, then the market is essentially deterministic and the result ceases to hold. Thus, in the present context, the price volatility may be viewed as an endogenous source of acceleration of financial growth. This phenomenon might seem counterintuitive, especially in stationary markets (Evstigneev and Schenk-Hoppé 2002; Dempster *et al.* 2003), where the asset prices themselves, and not only their returns, are stationary. In this case, $\rho = 0$, i.e. each asset grows at zero rate, while any constant proportions strategy exhibits *growth at a strictly positive exponential rate with probability one*!

To begin with, we focus on the case where all the assets have the same growth rate ρ. The results are then extended to a model with different growth rates $\rho^1, \ldots, \rho^K$. In this setting, a constant proportions strategy with proportions $\lambda^1 > 0, \ldots, \lambda^K > 0$ grows almost surely at a rate strictly greater than $\sum_k \lambda^k \rho^k$ (see Theorem 4.1 in Section 4.2).

The phenomenon highlighted in this paper has been mentioned in the literature, but its analysis has been restricted to examples involving specialized models. The terms 'excess growth' (Fernholz and Shay 1982) and the more descriptive 'volatility pumping' (Luenberger 1998) have been used to name similar effects to those discussed here. Cover (1991) used the mechanism of volatility pumping in the construction of universal portfolios. These ideas have been discussed in connection with financial market data in Mulvey and Ziemba (1998), Mulvey (2001) and Dries *et al.* (2002). Such questions have typically been studied in the context of maximization of expected logarithmic utilities— 'log-optimal investments' (Kelly 1956; Breiman 1961; Algoet and Cover 1988; MacLean *et al.* 1992; Hakansson and Ziemba 1995; Li 1998; Aurell *et al.* 2000). In this chapter we ignore questions of optimality of trading strategies and do not use the related notion of expected utility involved in optimality criteria.[1]

Constant proportions strategies play an important role in various practical financial computations, see e.g. Perold and Sharpe (1995). The assumption of stationarity of asset returns is widely accepted in financial theory and practice allowing, as it does, expected exponential price growth and mean reversion, volatility clustering and very general intertemporal dependence, such as long memory effects, of returns. However, no general results justifying and explaining the fact of volatility-induced growth have been established up to now. In spite of the fundamental importance and generality of this fact, no results pertaining to an *arbitrary* constant proportions strategy (regardless of its optimality) and *any* securities market with stationary non-degenerate asset returns have been available in the literature. The purpose of this chapter is to fill this gap.

Most of our results are rather easy consequences of some general mathematical facts, and the mathematical aspects do not play a crucial role. The main contribution of the present work is that we pose and analyse a number of questions that have not been systematically analysed before. These questions are especially interesting because the common intuition

---

[1] In connection with the discussion of relevant literature, we can mention a strand of publications dealing with Parrondo games (Harmer and Abbott 1999). Models considered in those publications are based on the analysis of lotteries whose odds depend on the investor's wealth. It is pointed out that losing lotteries, being played in a randomized alternating order, can become winning. In spite of some similarity, there are no obvious direct links between this phenomenon and that studied in the present chapter.

currently prevailing in the mathematical finance community suggests wrong answers to them (see the discussion in Section 4.4). Therefore it is important to clarify the picture in order to reveal and correct misconceptions. This is a central goal in this study.

The chapter is organized as follows. In Section 4.2 we describe the model, formulate the assumptions and state the main results. Section 4.3 contains proofs of the results and a discussion of their intuitive meaning. In Sections 4.4 and 4.5 we analyse the phenomenon of volatility-induced growth from various angles, focusing primarily on the case of stationary prices. We answer a number of questions arising naturally in connection with the theory developed. In Section 4.6, we show how this theory can be extended to markets with small transaction costs. Section 4.7 analyses an example in which estimates for the size of transaction cost rates allowing volatility-induced growth can be established.

4.2   THE MODEL AND THE MAIN RESULTS

Consider a financial market with $K \geq 2$ *securities* (*assets*). Let $S_t := (S_t^1, \ldots, S_t^K)$ denote the vector of security prices at time $t = 0, 1, 2, \ldots$. Assume that $S_t^k > 0$ for each t and k, and define by

$$R_t^k := \frac{S_t^k}{S_{t-1}^k} \quad (k = 1, 2, \ldots, K, \ t = 1, 2, \ldots) \tag{4.1}$$

the (gross) *return on asset* k over the time period $(t-1, t]$. Let $R_t := (R_t^1, \ldots, R_t^K)$. At each time period t, an investor chooses a *portfolio* $h_t = (h_t^1, \ldots, h_t^K)$, where h_t^k is the number of units of asset k in the *portfolio* h_t. Generally, h_t might depend on the observed values of the price vectors $S_0, S_1, \ldots, S_t$. A sequence $H = (h_0, h_1, \ldots)$ specifying a portfolio $h_t = h_t(S_0, \ldots, S_t)$ at each time t as a measurable function of $S_0, S_1, \ldots, S_t$ is called a *trading strategy*. If not otherwise stated, we will consider only those trading strategies for which $h_t^k \geq 0$, thus excluding short sales of assets (h_t^k can take on all non-negative real values).

One can specify trading strategies in terms of *investment proportions* (or *portfolio weights*). Suppose that for each $t = 1, 2, \ldots$, we are given a vector $\lambda_t = (\lambda_t^1, \ldots, \lambda_t^K)$ in the unit simplex

$$\Delta := \left\{ \lambda = (\lambda^1, \ldots, \lambda^K) : \lambda^k \geq 0, \ \sum_{k=1}^K \lambda^k = 1 \right\}.$$

The vector λ_t is assumed to be a measurable function of $S_0, \ldots, S_t$. Given an *initial portfolio* h_0 (specified by a non-negative non-zero vector), we can construct a trading strategy H recursively by the formula

$$h_t^k = \lambda_t^k S_t h_{t-1} / S_t^k \quad (k = 1, 2, \ldots, K, \ t = 1, 2, \ldots). \tag{4.2}$$

Here the scalar product $S_t h_{t-1} = \sum_{k=1}^K S_t^k h_{t-1}^k$ expresses the value of the portfolio h_{t-1} in terms of the prices S_t^k at time t. An investor following the strategy (4.2) rebalances (without transaction costs) the portfolio h_{t-1} at time t so that the available wealth $S_t h_{t-1}$ is distributed across the assets $k = 1, 2, \ldots, K$ according to the proportions $\lambda_t^1, \ldots, \lambda_t^K$. It is immediate from (4.2) that

$$S_t h_t = S_t h_{t-1}, \quad t = 1, 2, \ldots, \tag{4.3}$$

i.e. the strategy H is *self-financing*. If a strategy is self-financing, then the relations (4.2) and

$$S_t^k h_t^k = \lambda_t^k S_t h_t, \quad t = 1, 2, \ldots, \tag{4.4}$$

are equivalent. If the vectors of proportions λ_t are *fixed* (do not depend on time and on the price process), i.e. $\lambda_t = \lambda = (\lambda^1, \ldots, \lambda^K) \in \Delta$, then the strategy H defined recursively by

$$h_t^k = \lambda^k S_t h_{t-1} / S_t^k \quad (k = 1, 2, \ldots, K, \ t = 1, 2, \ldots) \tag{4.5}$$

is called a *constant proportions strategy* (or a *fixed-mix strategy*) with vector of proportions $\lambda = (\lambda^1, \ldots, \lambda^K)$. If $\lambda^k > 0$ for each k, then H is said to be *completely mixed*.

We will assume that the price vectors S_t, and hence the return vectors R_t, are random, i.e. they change in time as stochastic processes. Then the trading strategy h_t, $t = 0, 1, 2, \ldots$, generated by the investment rule (4.2) and the value $V_t = S_t h_t$, $t = 0, 1, 2, \ldots$, of the portfolio h_t are stochastic processes as well. We are interested in the asymptotic behaviour of V_t as $t \to \infty$ for constant proportions strategies. We will assume:

(R) The vector stochastic process R_t, $t = 1, 2, \ldots$, is stationary and ergodic. The expected values $E|\ln R_t^k|$, $k = 1, 2, \ldots, K$, are finite.

Recall that a stochastic process $R_1, R_2, \ldots$ is called *stationary* if, for any $m = 0, 1, 2, \ldots$ and any measurable function $\phi(x_0, x_1, \ldots, x_m)$, the distribution of the random variable $\phi_t := \phi(R_t, R_{t+1}, \ldots, R_{t+m})$ $(t = 1, 2, \ldots)$ does not depend on t. According to this definition, all probabilistic characteristics of the process R_t are time-invariant. If R_t is stationary, then for any measurable function ϕ for which $E|\phi(R_t, R_{t+1}, \ldots, R_{t+m})| < \infty$, the averages

$$\frac{\phi_1 + \cdots + \phi_t}{t} \tag{4.6}$$

converge almost surely (a.s.) as $t \to \infty$ (Birkhoff's ergodic theorem—see, e.g. Billingsley 1965). If the limit of all averages of the form (4.6) is non-random (equal to a constant a.s.), then the process R_t is called *ergodic*. In this case, the above limit is equal a.s. to the expectation $E\phi_t$, which does not depend on t by virtue of stationarity of R_t.

An example of a stationary ergodic process is a sequence of independent identically distributed (i.i.d.) random variables. To avoid misunderstandings, we emphasize that

Brownian motion and a random walk are *not* stationary. According to the conventional probabilistic terminology, these Markov processes are (time) *homogeneous*.

We have $S_t^k = S_0^k R_1^k \ldots R_t^k$, where (according to (**R**)) the random sequence R_t^k is stationary. This assumption on the structure of the price process is a fundamental hypothesis commonly accepted in finance. Moreover, it is quite often assumed that the random variables $R_t^k, t = 1, 2, \ldots$ are independent, i.e. the price process S_t^k forms a *geometric random walk*. This postulate, which is much stronger than the hypothesis of stationarity of R_t^k, lies at the heart of the classical theory of asset pricing (Black, Scholes, Merton), see e.g. Luenberger (1998).

By virtue of Birkhoff's ergodic theorem, we have

$$\lim_{t \to \infty} \frac{1}{t} \ln S_t^k = \lim_{t \to \infty} \frac{1}{t} \sum_{j=1}^{t} \ln R_j^k = E \ln R_t^k \; \text{(a.s.)} \tag{4.7}$$

for each $k = 1, 2, \ldots, K$. This means that the price of each asset k has almost surely a well-defined and finite (asymptotic, exponential) *growth rate*, which turns out to be equal a.s. to the expectation $\rho^k := E \ln R_t^k$, the *drift* of this asset's price. The drift can be positive, zero or negative. It does not depend on t in view of the stationarity of R_t. Let $H = (h_0, h_1, \ldots)$ be a trading strategy. If the limit

$$\lim_{t \to \infty} \frac{1}{t} \ln(S_t h_t)$$

exists, it is called the (asymptotic, exponential) *growth rate of the strategy H.*

We now formulate central results of this paper—Theorems 4.1 and 4.2. In these theorems, $H = (h_0, h_1, \ldots)$ is a constant proportions strategy with some vector of proportions $\lambda = (\lambda^1, \ldots, \lambda^K) \in \Delta$ and a non-zero initial portfolio $h_0 \geq 0$. In Theorems 4.1 and 4.2, we assume that the following condition holds:

(**V**) With strictly positive probability,

$$\frac{S_t^k}{S_t^m} \neq \frac{S_{t-1}^k}{S_{t-1}^m} \quad \text{for some } 1 \leq k, m \leq K \text{ and } t \geq 1.$$

Theorem 4.1: *If all the coordinates λ^k of the vector λ are strictly positive, i.e. the strategy H is completely mixed, then the growth rate of the strategy H is almost surely equal to a constant which is strictly greater than $\sum_k \lambda^k \rho^k$, where ρ^k is the drift of asset k.*

Condition (**V**) is a very mild assumption of volatility of the price process. This condition does not hold if and only if, with probability one, the ratio S_t^k / S_t^m of the prices of any two assets k and m does not depend on t. Thus condition (**V**) fails to hold if and only if the *relative prices of the assets are constant in time (a.s.).*

We are primarily interested in the situation when all the assets under consideration have *the same* drift and hence a.s. the same asymptotic growth rate:

(**R1**) There exists a number ρ such that, for each $k = 1, \ldots, K$, we have $E \ln R_t^k = \rho$.

From Theorem 4.1, we immediately obtain the following result.

Theorem 4.2: *Under assumption* (**R1**), *the growth rate of the strategy H is almost surely strictly greater than the growth rate of each individual asset.*

In the context of Theorem 4.2, the volatility of the price process appears to be the only cause for any completely mixed constant proportions strategy to grow at a rate strictly greater than ρ, the growth rate of each particular asset. This result contradicts conventional finance theory, where the volatility of asset prices is usually regarded as an impediment to financial growth. The result shows that in the present context volatility serves as an endogenous source of its acceleration.

4.3   PROOFS OF THE MAIN RESULTS AND THEIR EXPLANATION

We first observe that if a strategy H is generated according to formula (4.2) by a sequence $\lambda_1, \lambda_2, \ldots$ of vectors of investment proportions, then

$$
\begin{aligned}
V_t = S_t h_t &= \sum_{m=1}^{K} S_t^m h_{t-1}^m = \sum_{m=1}^{K} \frac{S_t^m}{S_{t-1}^m} S_{t-1}^m h_{t-1}^m \\
&= \sum_{m=1}^{K} \frac{S_t^m}{S_{t-1}^m} \lambda_{t-1}^m S_{t-1} h_{t-1} = V_{t-1} \sum_{m=1}^{K} R_t^m \lambda_{t-1}^m \\
&= (R_t \lambda_{t-1}) V_{t-1}
\end{aligned}
\tag{4.8}
$$

for each $t \geq 2$, and so

$$
V_t = (R_t \lambda_{t-1})(R_{t-1} \lambda_{t-2}) \ldots (R_2 \lambda_1) V_1, \quad t \geq 2.
\tag{4.9}
$$

Proof of Theorem 4.1: By virtue of (4.9), we have

$$
V_t = \left[V_1 (R_1 \lambda)^{-1} \right] (R_1 \lambda)(R_2 \lambda) \ldots (R_t \lambda),
\tag{4.10}
$$

and so

$$
\lim_{t \to \infty} \frac{1}{t} \ln V_t = \lim_{t \to \infty} \frac{1}{t} \sum_{j=1}^{t} \ln(R_j \lambda) = E \ln(R_t \lambda) \text{ (a.s.)}
\tag{4.11}
$$

by virtue of Birkhoff's ergodic theorem. It remains to show that if assumption (**V**) holds, then $E \ln(R_t \lambda) > \sum_{k=1}^{K} \lambda^k \rho^k$. To this end observe that condition (**V**) is equivalent to the following one.

(**V1**) For some $t \geq 1$ (and hence, by virtue of stationarity, for each $t \geq 1$), the probability

$$P\{R_t^k \neq R_t^m \quad \text{for some } 1 \leq k, m \leq K\}$$

is strictly positive.

Indeed, we have $S_t^k/S_t^m \neq S_{t-1}^k/S_{t-1}^m$ if and only if $S_t^k/S_{t-1}^k \neq S_t^m/S_{t-1}^m$, which can be written as $R_t^k \neq R_t^m$. Denote by δ_t the random variable that is equal to 1 if the event $\{R_t^k \neq R_t^m$ for some $1 \leq k, m \leq K\}$ occurs and 0 otherwise. Condition (**V**) means that $P\{\max_{t \geq 1} \delta_t = 1\} > 0$, while (**V1**) states that, for some t (and hence for each t), $P\{\delta_t = 1\} > 0$. The latter property is equivalent to the former because

$$\left\{ \max_{t \geq 1} \delta_t = 1 \right\} = \bigcup_{t=1}^{\infty} \{\delta_t = 1\}.$$

By using Jensen's inequality and (**V1**), we find that

$$\ln \sum_{k=1}^{K} R_t^k \lambda^k > \sum_{k=1}^{K} \lambda^k (\ln R_t^k)$$

with strictly positive probability, while the non-strict inequality holds always. Consequently,

$$E \ln(R_t \lambda) > \sum_{k=1}^{K} \lambda^k E(\ln R_t^k) = \sum_{k=1}^{K} \lambda^k \rho^k, \tag{4.12}$$

which completes the proof. □

The above considerations yield a rigorous proof of the fact of volatility induced growth. But what is the intuition, the underlying fundamental reason for it? We have only one explanation, which is nothing but a repetition in one phrase of the idea of the above proof. If $R_t^1, \ldots, R_t^K$ are random returns of assets $k = 1, 2, \ldots, K$, then the asymptotic growth rates of these assets are $E \ln R_t^k$, while the asymptotic growth rate of a constant proportions strategy is $E \ln(\sum \lambda^k R_t^k)$, which is strictly greater than $\sum \lambda^k E \ln(R_t^k)$ by Jensen's inequality—because the function $\ln x$ is strictly concave.

It would be nice, however, to give a general common-sense explanation of volatility-induced growth, without using such terms as a 'strictly convex function,' 'Jensen's inequality,' etc. One can, indeed, find in the literature explanations of examples of volatility pumping based on the following reasoning (see e.g. Fernholz and Shay 1982;

Luenberger 1998). The reason for growth lies allegedly in the fact that constant proportions always force one to 'buy low and sell high'—the common sense dictum of all trading. Those assets whose prices have risen from the last rebalance date will be overweighted in the portfolio, and their holdings must be reduced to meet the required proportions and to be replaced in part by assets whose prices have fallen and whose holdings must therefore be increased. Obviously, for this mechanism to work the prices must change in time; if they are constant, one cannot get any profit from trading.

We have, alas, repeated this reasoning ourselves (e.g. in Evstigneev and Schenk-Hoppé 2002 and in an earlier version of the present chapter), but somewhat deeper reflection on this issue inevitably leads to the conclusion that the above argument does not explain everything and raises more questions than it gives answers. For example, what is the meaning of 'high' and 'low?' If the price follows a geometric random walk, the set of its values is generally unbounded, and for every value there is a larger value. One can say that 'high' and 'low' should be understood in relative terms, based on the comparison of the prices today and yesterday. Fine, but what if the prices of *all* the assets increase or decrease simultaneously? Thus, the above argument, to be made valid, should be at least relativized, both with respect to time and the assets.

However, a more substantial lacuna in such reasoning is that it does not reflect the assumption of *constancy* of investment proportions. This leads to the question: what will happen if the 'common sense dictum of all trading' is pushed to the extreme and the portfolio is rebalanced so as to sell all those assets that gain value and buy only those ones which lose it? Assume, for example, that there are two assets, the price S_t^1 of the first (riskless) is always 1, and the price S_t^2 of the second (risky) follows a geometric random walk, so that the gross return on it can be either 2 or 1/2 with equal probabilities. Suppose the investor sells the second asset and invests all wealth in the first if the price S_t^2 goes up and performs the converse operation, betting all wealth on the risky asset, if S_t^2 goes down. Then the sequence $\lambda_t = (\lambda_t^1, \lambda_t^2)$ of the vectors of investment proportions will be i.i.d. with values $(0,1)$ and $(1,0)$ taken on with equal probabilities. Furthermore, λ_{t-1} will be independent of R_t. By virtue of (4.9), the growth rate of the portfolio value for this strategy is equal to $E\ln(R_t\lambda_{t-1}) = [\ln(0 \cdot 1 + 1 \cdot 2) + \ln(0 \cdot 1 + 1 \cdot \frac{1}{2}) + \ln(1 \cdot 1 + 0 \cdot 2) + \ln(1 \cdot 1 + 0 \cdot \frac{1}{2})]/4 = 0$, which is the same as the growth rate of each of the two assets $k = 1, 2$ and is strictly less than the growth rate of any completely mixed constant proportions strategy.

4.4 STATIONARY MARKETS: PUZZLES AND MISCONCEPTIONS

Consider a market where the price process S_t (and not only the process of asset returns R_t) is ergodic and stationary and where $E|\ln S_t^k| < \infty$. This situation is a special case of stationary returns, because if the vector process S_t is stationary, then the process R_t is stationary as well. In this case the growth rate of each asset is zero,

$$E\ln R_t^k = E\ln S_t^k - E\ln S_{t-1}^k = 0,$$

while, as we have seen, any completely mixed constant proportions strategy grows at a strictly positive exponential rate. The assumption of stationarity of asset prices, perhaps after some detrending, seems plausible when modelling currency markets (Kabanov 1999; Dempster *et al.* 2003). Then the 'prices' are determined by exchange rates of all the currencies with respect to some selected reference currency.

We performed a casual experiment, asking numerous colleagues (in private, at seminars and at conferences) to promptly guess the correct answer to the following question.

Question 4.1: *Suppose vectors of asset prices $S_t = (S_t^1, \ldots, S_t^K)$ fluctuate randomly, forming a stationary stochastic process (assume even that S_t are i.i.d.). Consider a fixed-mix self-financing investment strategy prescribing rebalancing one's portfolio at each of the dates $t = 1, 2, \ldots$, so as to keep equal investment proportions of wealth in all the assets. What will happen with the portfolio value in the long run, as $t \to \infty$? What will be its tendency: (a) to decrease; (b) to increase; or (c) to fluctuate randomly, converging in one sense or another to a stationary process?*

The audience of our respondents was quite broad and professional, but practically *nobody* succeeded in guessing the correct answer, (b). Among those who expressed a clear guess, *nearly all* selected (c). There were also a couple of respondents who decided to bet on (a).

Common intuition suggests that if the market is stationary, then the portfolio value V_t for a constant proportions strategy must converge in one sense or another to a stationary process. The usual intuitive argument in support of this conjecture appeals to the self-financing property (3). The self-financing constraint seems to exclude possibilities of unbounded growth. This argument is also substantiated by the fact that in the deterministic case both the prices and the portfolio value are constant. This way of reasoning makes the answer (c) to the above question more plausible *a priori* than the others.

It might seem surprising that the wrong guess (c) has been put forward even by those who have known about examples of volatility pumping for a long time. The reason for this might lie in the non-traditional character of the setting where not only the asset returns but the prices themselves are stationary. Moreover, the phenomenon of volatility-induced growth is more paradoxical in the case of stationary prices, where growth emerges 'from nothing.' In the conventional setting of stationary returns, volatility serves as the cause of an acceleration of growth, rather than its emergence from prices with zero growth rates.

A typical way of understanding the correct answer to Question 4.1 is to reduce it to something well known that is apparently relevant. A good candidate for this is the concept of *arbitrage*. Getting something from nothing as a result of an arbitrage opportunity seems to be similar to the emergence of growth in a stationary setting where there are no obvious sources for growth.

As long as we deal with an infinite time horizon, we would have to consider some kind of *asymptotic* arbitrage (e.g. Ross 1976; Huberman 1982; Kabanov and Kramkov 1994; Klein and Schachermayer 1996). However, all known concepts of this kind are much weaker than what we would need in the present context. According to our results, growth is exponentially fast, unbounded wealth is achieved with probability one, and the effect of growth is demonstrated for specific (constant proportions) strategies. None of these properties can be directly deduced from asymptotic arbitrage.

Thus there are no convincing arguments showing that volatility-induced growth in stationary markets can be derived from, or explained by, asymptotic arbitrage over an infinite time horizon. But what can be said about relations between stationarity and arbitrage over finite time intervals? As is known, there are no arbitrage opportunities (over a finite time horizon) if and only if there exists an equivalent martingale measure. A stationary process can be viewed as an 'antipodal concept' to the notion of a martingale. This might lead to the conjecture that in a stationary market arbitrage is a typical situation. Is this true or not? Formally, the question can be stated as follows.

Question 4.2: *Suppose the process* $S_t = (S_t^1, \ldots, S_t^K)$ *of the vectors of asset prices is stationary, and moreover, assume that the vectors* S_t *are i.i.d. Furthermore, suppose the first asset* $k = 1$ *is riskless and its price* S_t^1 *is equal to one. Does this market have arbitrage opportunities over a finite time horizon?*

When asking this question, we assume that the market is frictionless and that there are no portfolio constraints. In particular, all short sales are allowed. An arbitrage opportunity over a time horizon $t = 0, \ldots, T$ is understood in the conventional sense. It means the existence of a self-financing trading strategy $(h_0, \ldots, h_T)$ such that $S_0 h_0 = 0$, $S_T h_T \geq 0$ a.s. and $P\{S_T h_T > 0\} > 0$.

The answer to this question, as well as to the previous one, is practically never guessed immediately. Surprisingly, the answer depends, roughly speaking, on whether the distribution of the random vector $\tilde{S}_t := (S_t^2, \ldots, S_t^K)$ of prices of the risky assets is continuous or discrete. For example, if $\tilde{S}_t$ takes on a finite number of values, then an arbitrage opportunity exists. If the distribution of $\tilde{S}_t$ is continuous, there are no arbitrage opportunities. More precisely, the result is as follows. Let G be the support of the distribution of the random vector $\tilde{S}_t$ (which is assumed to be non-degenerate) and let $F := cl\ co\ G$ be the closure of the convex hull of G. Denote by $\partial_r F$ the relative boundary of F, i.e. the boundary of the convex set F in the smallest linear manifold containing F.

Theorem 4.3: *If* $P\{\tilde{S}_t \in \partial_r F\} = 0$, *then for any* T *there are no arbitrage opportunities over the time horizon of length* T. *If* $P\{\tilde{S}_t \in \partial_r F\} > 0$, *then for each* T *there is an arbitrage opportunity over the time horizon of length* T.

For a proof of this result see Evstigneev and Kapoor (2006).

4.5 GROWTH ACCELERATION, VOLATILITY REDUCTION AND NOISE-INDUCED STABILITY

The questions we analyse in this section stem from an example of volatility pumping considered originally by Fernholz and Shay (1982) and later others (e.g. Luenberger 1998). The framework for this example is the well-known continuous-time model developed by Merton and others, in which the price processes S_t^k $(t \geq 0)$ of two assets $k = 1, 2$ are supposed to be solutions to the stochastic differential equations $dS_t^k / S_t^k = \mu_k dt + \sigma_k dW_t^k$, where the W_t^k are independent (standard) Wiener processes and $S_0^k = 1$. As is well known, these equations admit explicit solutions $S_t^k = \exp[\mu_k t - (\sigma_k^2/2)t + \sigma_k W_t^k]$. Given some $\theta \in (0,1)$, the value V_t of the constant proportions portfolio prescribing investing the proportions θ and $1 - \theta$ of wealth into assets $k = 1, 2$ is the solution to the equation

$$\frac{dV_t}{V_t} = [\theta\mu_1 + (1 - \theta)\mu_2]dt + \theta\sigma_1 dW_t^1 + (1 - \theta)\sigma_2 dW_t^2.$$

Equivalently, V_t can be represented as the solution to the equation $dV_t/V_t = \bar{\mu}dt + \bar{\sigma}dW_t$, where $\bar{\mu} := \theta\mu_1 + (1 - \theta)\mu_2$, $\bar{\sigma}^2 := (\theta\sigma_1)^2 + [(1 - \theta)\sigma_2]^2$ and W_t is a standard Wiener process. Thus, $V_t = \exp[\bar{\mu}t - (\bar{\sigma}^2/2)t + \bar{\sigma}W_t]$, and so the growth rate and the volatility of the portfolio value process V_t are given by $\bar{\mu} - (\bar{\sigma}^2/2)$ and $\bar{\sigma}$. In particular, if $\mu_1 = \mu_2 = \mu$ and $\sigma_1 = \sigma_2 = \sigma$, then the growth rate and the volatility of V_t are equal to

$$\mu - (\bar{\sigma}^2/2) \quad \text{and} \quad \bar{\sigma} = \sigma\sqrt{\theta^2 + (1 - \theta)^2} < \sigma, \tag{4.13}$$

while for each individual asset the growth rate and the volatility are $\mu - (\sigma^2/2)$ and σ, respectively.

Thus, in this example, the use of a constant proportions strategy prescribing investing in a mixture of two assets leads (due to diversification) to an increase of the growth rate and to a simultaneous decrease of the volatility. When looking at the expressions in (4.13), the temptation arises even to say that the volatility reduction is the *cause* of volatility-induced growth. Indeed, the growth rate $\mu - (\bar{\sigma}^2/2)$ is greater than the growth rate $\mu - (\sigma^2/2)$ because $\bar{\sigma} < \sigma$. This suggests speculation along the following lines. Volatility is something like energy. When constructing a mixed portfolio, it converts into growth and therefore itself decreases. The greater the volatility reduction, the higher the growth acceleration.

Do such speculations have any grounds in the general case, or do they have a justification only in the above example? To formalize this question and answer it, let us return to the discrete time-framework we deal with in this paper. Suppose there are two assets with i.i.d. vectors of returns $R_t = (R_t^1, R_t^2)$. Let $(\xi, \eta) := (R_1^1, R_1^2)$ and assume, to avoid technicalities, that the random vector (ξ, η) takes on a finite number of values and is strictly positive. The value V_t of the portfolio generated by a fixed-mix strategy with proportions x and $1 - x$ $(0 < x < 1)$ is computed according to the formula

$$V_t = V_1 \prod_{j=2}^{t} [xR_j^1 + (1 - x)R_j^2], \quad t \geq 2,$$

see (4.9). The growth rate of this process and its volatility are given, respectively, by the expectation $E \ln \zeta_x$ and the standard deviation $\sqrt{\operatorname{Var} \ln \zeta_x}$ of the random variable $\ln \zeta_x$, where $\zeta_x := \ln[x\xi + (1 - x)\eta]$. We know from the above analysis that the growth rate increases when mixing assets with the same growth rate. What can be said about volatility? Specifically, let us consider the following question.

Question 4.3: *(a) Suppose* $\mathrm{Var}\ln\xi = \mathrm{Var}\ln\eta$. *Is it true that* $\mathrm{Var}\ln[x\xi + (1-x)\eta]$ $\leq \mathrm{Var}\ln\xi$ *when* $x \in (0,1)$? *(b) More generally, is it true that* $\mathrm{Var}\ln[x\xi + (1-x)\eta] \leq$ $\max(\mathrm{Var}\ln\xi, \mathrm{Var}\ln\eta)$ *for* $x \in (0,1)$?

Query (b) asks whether the logarithmic variance is a quasi-convex functional. Questions (a) and (b) can also be stated for volatility defined as the square root of logarithmic variance. They will have the same answers because the square root is a strictly monotone function. Positive answers to these questions would substantiate the above conjecture of volatility reduction—negative, refute it.

It turns out that in general (without additional assumptions on ξ and η) the above questions 4.3(a) and 4.3(b) have *negative* answers. To show this consider two i.i.d. random variables U and V with values 1 and $a > 0$ realized with equal probabilities. Consider the function

$$f(y) := \mathrm{Var}\ln[yU + (1-y)V], \quad y \in [0,1]. \tag{4.14}$$

By evaluating the first and the second derivatives of this function at $y = 1/2$, one can show the following. There exist some numbers $0 < a_- < 1$ and $a_+ > 1$ such that the function $f(y)$ attains its minimum at the point $y = 1/2$ when a belongs to the closed interval $[a_-, a_+]$ and it has a local maximum (!) at $y = 1/2$ when a does not belong to this interval. The numbers a_- and a_+ are given by

$$a_\pm = 2e^4 - 1 \pm \sqrt{(2e^4 - 1)^2 - 1},$$

where $a_- \approx 0.0046$ and $a_+ \approx 216.388$. If $a \in [a_-, a_+]$, the function $f(y)$ is convex, but if $a \notin [a_-, a_+]$, its graph has the shape illustrated in Figure 4.1.

Fix any a for which the graph of $f(y)$ looks like the one depicted in Figure 4.1. Consider any number $y_0 < 1/2$ which is greater than the smallest local minimum of $f(y)$ and define $\xi := y_0 U + (1 - y_0)V$ and $\eta := y_0 V + (1 - y_0)U$. ($U$ and V may be interpreted as 'factors' on which the returns ξ and η on the two assets depend.) Then $\mathrm{Var}\ln[(\xi + \eta)/2] > \mathrm{Var}\ln\xi = \mathrm{Var}\ln\eta$, which yields a negative answer both to (a) and (b). In this

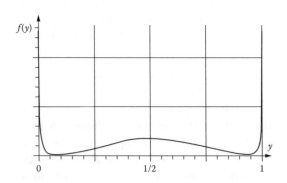

FIGURE 4.1 Graph of the function $f(y)$ in Equation (4.14) for $a = 10^4$.

example ξ and η are dependent. It would be of interest to investigate questions (a) and (b) for general independent random variables ξ and η. It can be shown that the answer to (b) is positive if one of the variables ξ and η is constant. But even in this case the function $Var \ln[x\xi + (1 - x)\eta]$ is not necessarily convex: it may have an inflection point in $(0, 1)$, which can be easily shown by examples involving two-valued random variables.

Thus it can happen that a mixed portfolio may have a greater volatility than each of the assets from which it has been constructed. Consequently, the above conjecture and the 'energy interpretation' of volatility are generally not valid. It is interesting, however, to find additional conditions under which assertions regarding volatility reduction hold true. In this connection, we can assert the following fact.

Theorem 4.4: *Let U and V be independent random variables bounded above and below by strictly positive constants. If U is not constant, then $Var \ln[yU + (1 - y)V] < Var \ln U$ for all $y \in (0,1)$ sufficiently close to 1.*

Volatility can be regarded as a quantitative measure of instability of the portfolio value. The above result shows that small independent noise can reduce volatility. This result is akin to a number of known facts about *noise-induced stability*, e.g. Abbott (2001) and Mielke (2000). An analysis of links between the topic of the present work and results about stability under random noise might constitute an interesting theme for further research.

Proof of Theorem 4.4: This can be obtained by evaluating the derivative of the function $f(y)$ defined by (4.14) at $y = 1$. Basic computations show that

$$f'(1) = 2(EV)(-Ee^{-Z}Z + Ee^{-Z} \cdot EZ), \tag{4.15}$$

where $Z = \ln U$. The assertion of the theorem is valid because $f'(1) > 0$. The verification of this inequality is based on the following fact. If $\phi(z)$ is a function on $(-\infty, +\infty)$ with $\phi'(z) > 0$, then

$$E[Z\phi(Z)] > (EZ)E\phi(Z) \tag{4.16}$$

for any non-constant bounded random variable Z. This fact follows from Jensen's inequality applied to the strictly convex function $\psi(x) := \int_0^x \phi(z)dz$ and from the relation $\psi(y) - \psi(x) \geq \phi(x)(y - x)$ (to obtain (16) put $x := Z$, $y := EZ$ and take the expectation). By applying (16) to $\phi(z) := -e^{-z}$, we find that the expression in (4.15) is positive. □

4.6 VOLATILITY-INDUCED GROWTH UNDER SMALL TRANSACTION COSTS

We now assume that, in the market under consideration (see Section 4.2), there are transaction costs for buying and selling assets. When selling x units of asset k at time t, one receives the amount $(1 - \varepsilon_-^k)S_t^k x$ (rather than $S_t^k x$ as in the frictionless case). To buy x units of asset k, one has to pay $(1 + \varepsilon_+^k)S_t^k x$. The numbers $\varepsilon_-^k, \varepsilon_+^k \geq 0$, $k = 1, 2, \ldots, K$

(the *transaction cost rates*) are assumed given. In this market, a trading strategy $H = (h_0, h_1, \ldots)$ is *self-financing* if

$$\sum_{k=1}^{K} (1 + \varepsilon_+^k) S_t^k (h_t^k - h_{t-1}^k)_+ \leq \sum_{k=1}^{K} (1 - \varepsilon_-^k) S_t^k (h_{t-1}^k - h_t^k)_+, \quad t \geq 1, \qquad (4.17)$$

where $x_+ = \max\{x, 0\}$. Inequality (4.17) means that asset purchases can be made only at the expense of asset sales.

Relation (4.17) is equivalent to

$$\sum_{k=1}^{K} S_t^k (h_t^k - h_{t-1}^k) \leq -\sum_{k=1}^{K} \varepsilon_+^k S_t^k (h_t^k - h_{t-1}^k)_+ - \sum_{k=1}^{K} \varepsilon_-^k S_t^k (h_{t-1}^k - h_t^k)_+$$

(which follows from the identity $x_+ - (-x)_+ = x$). Therefore, if the portfolio h_t differs from the portfolio h_{t-1} in at least one position k for which $\varepsilon_+^k \neq 0$ and $\varepsilon_-^k \neq 0$, then $S_t h_t < S_t h_{t-1}$. Thus, under transaction costs, portfolio rebalancing typically leads to a loss of wealth. The number $S_t h_t / S_t h_{t-1} \leq 1$ is called the *loss coefficient* (of the portfolio strategy H at time t).

We say that $H = (h_0, h_1, \ldots)$ is a *constant proportions strategy* with vector of proportions $\lambda = (\lambda^1, \ldots, \lambda^K) \in \Delta$ if $S_t^k h_t^k = \lambda^k S_t h_t$ for all $k = 1, 2, \ldots, K$ and $t = 1, 2, \ldots$ (cf. (4)). Let $\delta \in (0, 1)$ be a constant. Given a vector of proportions $\lambda = (\lambda^1, \ldots, \lambda^K) \in \Delta$ and a non-zero initial portfolio $h_0 \geq 0$, define recursively

$$h_t^k = \frac{(1 - \delta) \lambda^k S_t h_{t-1}}{S_t^k} \quad (k = 1, 2, \ldots, K, \; t = 1, 2, \ldots). \qquad (4.18)$$

This rule defines a trading strategy with constant investment proportions $\lambda^1, \ldots, \lambda^K$ and a constant loss coefficient $1 - \delta$. We will call it the (δ, λ)-*strategy*.

Theorem 4.5 below extends the results of Theorems 4.1 and 4.2 to the model with transaction costs. As before, we assume that hypotheses (**R**) and (**V**) hold.

Theorem 4.5: *Let $\lambda = (\lambda^1, \ldots, \lambda^K) \in \Delta$ be a strictly positive vector. If $\delta \in (0,1)$ is small enough, then the (δ, λ)-strategy H defined by (4.18) has a growth rate strictly greater than $\sum_{k=1}^{K} \lambda^k \rho^k$ (a.s.), and so if $\rho^1 = \cdots = \rho^K = \rho$, then the growth rate of H is strictly greater than ρ (a.s.). Further, if the transaction cost rates $\varepsilon_-^k, \varepsilon_+^k \geq 0$, $k = 1, 2, \ldots, K$, are small enough (in particular, if they do not exceed $\delta/2$), then the strategy H is self-financing.*

The purpose of this theorem is to demonstrate that the results on volatility-induced growth remain valid under small transaction costs. In contrast with a number of the questions we considered previously, the answer to the question we pose here is quite predictable and does not contradict intuition. We deal in Theorem 4.5 with constant proportions strategies of a special form—those for which the loss rate is constant (and small enough). We are again *not* concerned with the question of optimality of such strategies for one criterion or another.

Proof of Theorem 4.5: We first observe that the growth rate of the strategy H is equal to $E \ln[(1 - \delta)R_t\lambda]$. This fact is proved exactly like (4.11) (simply multiply the vectors of proportions in (4.8), (4.9), (4.10) and (4.11) by $(1 - \delta)$). According to (4.12), we have $E \ln(R_t\lambda) > \sum_{k=1}^{K} \lambda^k \rho^k$. This inequality will remain valid if λ is replaced by $(1 - \delta)\lambda$, provided $\delta \in (0, 1)$ is small enough. Fix any such $\delta \in (0, 1)$. Denote by ε the greatest of the numbers $\varepsilon_-^k, \varepsilon_+^k$. It remains to show that H is self-financing when $\varepsilon \le \delta/2$. To this end we note that inequality (4.17) is implied by

$$\sum_{k=1}^{K} (1 + \varepsilon) S_t^k \left(h_t^k - h_{t-1}^k \right)_+ \le \sum_{k=1}^{K} (1 - \varepsilon) S_t^k \left(h_{t-1}^k - h_t^k \right)_+,$$

which is equivalent to

$$\varepsilon \sum_{k=1}^{K} \left| S_t^k h_t^k - S_t^k h_{t-1}^k \right| \le S_t (h_{t-1} - h_t). \tag{4.19}$$

In view of (4.18), the right-hand side of the last inequality is equal to $\delta S_t h_{t-1}$, and the left-hand side can be estimated as follows:

$$\varepsilon \sum_{k=1}^{K} \left| (1 - \delta)\lambda^k S_t h_{t-1} - S_t^k h_{t-1}^k \right| \le \varepsilon \sum_{k=1}^{K} (1 - \delta)\lambda^k S_t h_{t-1} + \varepsilon \sum_{k=1}^{K} S_t^k h_{t-1}^k$$

$$= \varepsilon(1 - \delta) S_t h_{t-1} + \varepsilon S_t h_{t-1} \le 2\varepsilon S_t h_{t-1}.$$

Consequently, if $0 \le \varepsilon \le \delta/2$, then the strategy H is self-financing. $\square$

4.7 GROWTH UNDER TRANSACTION COSTS: AN EXAMPLE

In this section we consider an example (a binomial model) in which quantitative estimates for the size of the transaction costs needed for the validity of the result on volatility-induced growth can be provided. Suppose that there are two assets $k = 1, 2$: one riskless and one risky. The price of the former is constant and equal to 1. The price of the latter follows a geometric random walk. It can either jump up by $u > 1$ or down by u^{-1} with equal probabilities. Thus both securities have zero growth rates.

Suppose the investor pursues the constant proportions strategy prescribing to keep 50% of wealth in each of the securities. There are no transaction costs for buying and selling the riskless asset, but there is a transaction cost rate for buying and selling the risky asset of $\varepsilon \in (0, 1)$. Assume the investor's portfolio at time $t - 1$ contains v units of cash; then the value of the risky position of the portfolio must be also equal to v. At time t, the riskless position of the portfolio will remain the same, and the value of the risky position will become either uv or $u^{-1}v$ with equal probability. In the former case, the investor rebalances his/her portfolio by selling an amount of the risky asset worth A so that

$$v + (1 - \varepsilon)A = vu - A. \tag{4.20}$$

by selling an amount of the risky asset of value A in the current prices, the investor receives $(1 - \varepsilon)A$, and this sum of cash is added to the riskless position of the portfolio. After rebalancing, the values of both portfolio positions must be equal, which is expressed in (4.20). From (4.20) we obtain $A = v(u - 1)(2 - \varepsilon)^{-1}$. The positions of the new (rebalanced) portfolio, measured in terms of their current values, are equal to $v + (1 - \varepsilon)A = v[1 + (1 - \varepsilon)(2 - \varepsilon)^{-1}(u - 1)]$. In the latter case (when the value of the risky position becomes u^{-1}), the investor buys some amount of the risky asset worth B, for which the amount of cash $(1 + \varepsilon)B$ is needed, so that

$$v - (1 + \varepsilon)B = u^{-1}v + B.$$

From this, we find $-B = v(u^{-1} - 1)(2 + \varepsilon)^{-1}$, and so $v - (1 + \varepsilon)B = v[1 + (1 + \varepsilon)(2 + \varepsilon)^{-1}(u^{-1} - 1)]$.

Thus, the portfolio value at each time t is equal to its value at time $t - 1$ multiplied by the random variable ξ such that $P\{\xi = g'\} = P\{\xi = g''\} = 1/2$, where $g' := 1 + (1 + \varepsilon)(2 + \varepsilon)^{-1}(u^{-1} - 1)$ and $g'' := 1 + (1 - \varepsilon) \times (2 - \varepsilon)^{-1}(u - 1)$. Consequently, the asymptotic growth rate of the portfolio value, $E\ln \xi = (1/2)(\ln g' + \ln g'')$, is equal to $(1/2) \ln \phi(\varepsilon, u)$, where

$$\phi(\varepsilon, u) := \left[1 + (1 + \varepsilon)\frac{u^{-1} - 1}{2 + \varepsilon}\right]\left[1 + (1 - \varepsilon)\frac{u - 1}{2 - \varepsilon}\right].$$

We have $E\ln \xi > 0$, i.e. the phenomenon of volatility induced growth takes place, if $\phi(\varepsilon, u) > 1$. For $\varepsilon \in [0, 1)$, this inequality turns out to be equivalent to the following very simple relation: $0 \leq \varepsilon < (u - 1)(u + 1)^{-1}$. Thus, given $u > 1$, the asymptotic growth rate of the fixed-mix strategy under consideration is greater than zero if the transaction cost rate ε is less than $\varepsilon^*(u) := (u - 1)(u + 1)^{-1}$. We call $\varepsilon^*(u)$ the *threshold transaction cost rate*. Volatility-induced growth takes place—in the present example, where the portfolio is rebalanced in every one period[2]—when $0 \leq \varepsilon < \varepsilon^*(u)$.

The volatility σ of the risky asset under consideration (the standard deviation of its logarithmic return) is equal to $\ln u$. In the above considerations, we assumed that σ—or equivalently, u—is fixed, and we examined $\phi(\varepsilon, u)$ as a function of ε. Let us now examine $\phi(\varepsilon, u)$ as a function of u when the transaction cost rate ε is fixed and strictly positive. For the derivative of $\phi(\varepsilon, u)$ with respect to u, we have

$$\phi_u'(\varepsilon, u) = \frac{1 + \varepsilon}{4 - \varepsilon^2}\left[\frac{1 - \varepsilon}{1 + \varepsilon} - u^{-2}\right].$$

If $u = 1$, then $\phi_u'(\varepsilon, 1) < 0$. Thus if the volatility of the risky asset is small, the performance of the constant proportions strategy at hand is *worse* than the performance of each individual asset. This fact *refutes the possible conjecture that 'the higher the volatility,*

[2] For the optimal timing of rebalancing in markets with transaction costs see, e.g. Aurell and Muratore (2000).

the higher the induced growth rate.' Further, the derivative $\phi'_u(\varepsilon, u)$ is negative when $u \in [0, u*(\varepsilon))$, where $u*(\varepsilon) := (1 - \varepsilon)^{-1/2}(1 + \varepsilon)^{1/2}$. For $u = u*(\varepsilon)$ the asymptotic growth rate of the constant proportions strategy at hand attains its minimum value. For those values of u which are greater than $u*(\varepsilon)$, the growth rate increases, and it tends to infinity as $u \to \infty$. Thus, although the assertion 'the greater the volatility, the greater the induced growth rate' is not valid always, it is valid (in the present example) under the additional assumption that the volatility is large enough.

4.8 CONCLUSION

In this chapter we have established the surprising result that when asset returns are stationary ergodic, their volatility, together with any fixed-mix trading strategy, generates a portfolio growth rate in excess of the individual asset growth rates. As a consequence, even if the growth rates of the individual securities all have mean zero, the value of a fixed-mix portfolio tends to infinity with probability one. By contrast with the 25 years in which the effects of 'volatility pumping' have been investigated in the literature by example, our results are quite general. They are obtained under assumptions which accommodate virtually all the empirical market return properties discussed in the literature. We have in this chapter also dispelled the notion that the demonstrated acceleration of portfolio growth is simply a matter of 'buying lower and selling higher.' The example of Section 4.3 shows that our result depends critically on rebalancing to an arbitrary *fixed* mix of portfolio proportions. Any such mix defines the relative magnitudes of individual asset returns realized from volatility effects. This observation and our analysis of links between growth, arbitrage and noise-induced stability suggest that financial growth driven by volatility is a subtle and delicate phenomenon.

ACKNOWLEDGEMENTS

We are indebted to numerous colleagues who had enough patience to participate in our 'intuition tests,' the results of which are discussed in Section 4.4 of this chapter. Our comparative analysis of volatility-induced growth and arbitrage (also contained in Section 4.4) was motivated by Walter Schachermayer's comments at a conference on Mathematical Finance in Paris in 2003. Albert N. Shiryaev deserves our thanks for a helpful discussion of some mathematical questions related to this work. Financial support by the Swiss National Centre of Competence in Research 'Financial Valuation and Risk Management' (NCCR FINRISK) is gratefully acknowledged.

REFERENCES

Abbott, D., Overview: Unsolved problems of noise and fluctuations. *Chaos*, 2001, **11**, 526–538.
Algoet, P.H. and Cover, T.M., Asymptotic optimality and asymptotic equipartition properties of log-optimum investment. *Ann. Probab.*, 1988, **16**, 876–898.
Aurell, E., Baviera, R., Hammarlid, O., Serva, M. and Vulpiani, A., A general methodology to price and hedge derivatives in incomplete markets. *Int. J. Theoret. Appl. Finance*, 2000, **3**, 1–24.
Aurell, E. and Muratore, P., Financial friction and multiplicative Markov market games. *Int. J. Theoret. Appl. Finance*, 2000, **3**, 501–510.

Billingsley, P., *Ergodic Theory and Information*, 1965 (Wiley: New York).

Breiman, L., Optimal gambling systems for favorable games, in *Fourth Berkeley Symposium on Mathematical Statistics and Probability*, Vol. 1, 1961, pp. 65–78.

Cover, T.M., Universal portfolios. *Math. Finance*, 1991, **1**, 1–29.

Dempster, M.A.H., Evstigneev, I.V. and Schenk-Hoppé, K.R., Exponential growth of fixed-mix strategies in stationary asset markets. *Finance Stochast.*, 2003, **7**, 263–276.

Dries, D., Ilhan, A., Mulvey, J., Simsek, K.D. and Sircar, R., Trend-following hedge funds and multiperiod asset allocation. *Quant. Finance*, 2002, **2**, 354–361.

Evstigneev, I.V. and Kapoor, D., Arbitrage in stationary markets. *Discussion Paper 0608*, 2006, (School of Economic Studies, University of Manchester).

Evstigneev, I.V. and Schenk-Hoppé, K.R., From rags to riches: On constant proportions investment strategies. *Int. J. Theoret. Appl. Finance*, 2002, **5**, 563–573.

Fernholz, R. and Shay, B., Stochastic portfolio theory and stock market equilibrium. *J. Finance*, 1982, **37**, 615–624.

Föllmer, H. and Schied, A., *Stochastic Finance—An Introduction in Discrete Time*, 2002 (Walter de Gruyter: Berlin).

Hakansson, N.H. and Ziemba, W.T., Capital growth theory, in *Handbooks in Operations Research and Management Science, Volume 9, Finance*, edited by R.A. Jarrow, V. Maksimovic and W.T. Ziemba, pp. 65–86, 1995 (Elsevier: Amsterdam).

Harmer, G.P. and Abbott, D., Parrondo's paradox. *Stat. Sci.*, 1999, **14**, 206–213.

Huberman, G., A simple approach to arbitrage pricing theory. *J. Econ. Theory*, 1982, **28**, 183–191.

Kabanov, YuM., Hedging and liquidation under transaction costs in currency markets. *Finance Stochast.*, 1999, **3**, 237–248.

Kabanov, YuM. and Kramkov, D.A., Large financial markets: Asymptotic arbitrage and contiguity. *Theory Probab. Appl.*, 1994, **39**, 222–228.

Kelly, J.L., A new interpretation of information rate. *Bell Sys. Techn. J.*, 1956, **35**, 917–926.

Klein, I. and Schachermayer, W., Asymptotic arbitrage in non-complete large financial markets. *Theory Probab. Appl.*, 1996, **41**, 927–334.

Li, Y., Growth-security investment strategy for long and short runs. *Manag. Sci.*, 1998, **39**, 915–924.

Luenberger, D., *Investment Science*, 1998 (Oxford University Press: New York).

MacLean, L.C., Ziemba, W.T. and Blazenko, G., Growth versus security in dynamic investment analysis. *Manag. Sci.*, 1992, **38**, 1562–1585.

Mielke, A., Noise induced stability in fluctuating bistable potentials. *Phys. Rev. Lett.*, 2000, **84**, 818–821.

Mulvey, J.M., Multi-period stochastic optimization models for long-term investors, in *Quantitative Analysis in Financial Markets*, edited by M. Avellaneda, Vol. 3, pp. 66–85, 2001 (World Scientific: Singapore).

Mulvey, J.M. and Ziemba, W.T., (Eds), *Worldwide Asset and Liability Modeling*, 1998 (Cambridge University Press: Cambridge, UK).

Perold, A.F. and Sharpe, W.F., Dynamic strategies for asset allocation. *Financ. Analysts J.*, 1995, **51**, 149–160.

Ross, S.A., The arbitrage theory of asset pricing. *J. Econ. Theory*, 1976, **13**, 341–360.

CHAPTER 5

Constant Rebalanced Portfolios and Side-Information

E. FAGIUOLI, F. STELLA and A. VENTURA

CONTENTS

5.1   INTRODUCTION

T HE PORTFOLIO SELECTION PROBLEM is one of the basic problems within the research area of computational finance. It has been studied intensively throughout the last 50 years, producing several relevant contributions described in the specialized literature. Portfolio selection originates from the seminal paper of Markowitz (1952), who introduced and motivated the *mean-variance investment framework*. This conventional approach to portfolio selection, which has received increasing attention, consists of two separate steps. The first step concerns distributional assumptions about the behaviour of stock prices, while the second step is related to the selection of the optimal portfolio depending on some objective function and/or utility function defined according to the investor's goal. This conceptual model has proved in the past to be useful in spite of the many drawbacks that have been pointed out by finance practitioners, private investors and researchers. Indeed, the first step, related to distributional assumptions concerning the behaviour of stock prices, encounters many difficulties because the future evolution of stock prices is notoriously difficult to predict, while the selection of a distribution class inevitably brings a measure of arbitrariness. These problems become even more evident and dramatic in the case where there are reasons to believe that the process that governs stock price behaviour changes over time.

A different approach to portfolio selection, to overcome the main limitations and problems related to the mean variance approach, has been proposed by Cover (1991a, b). The main characteristic of Cover's approach to portfolio selection is that no distributional assumptions on the sequence of price relatives are required. Indeed, within Cover's investment framework, portfolio selection is based completely on the sequence of past prices, which is taken *as is*, with little, if any, statistical processing. No assumptions are made concerning the family of probability distributions that describes the stock prices, or even concerning the existence of such distributions. To emphasize this independence from statistical assumptions, such portfolios are called *universal portfolios*. It has been shown that such portfolios possess important theoretical properties concerning their asymptotic behaviour and exhibit reasonable finite time behaviour. Indeed, it is well known (Bell and Cover 1980; Algoet and Cover 1988; Cover 1991a) that if the price relatives are independent and identically distributed, the optimal growth rate of wealth is achieved by a *constant rebalanced portfolio*, i.e. an investment strategy that keeps fixed through time, trading period by trading period, the distribution of wealth among a set of assets. In recent years, constant rebalanced portfolios have received increasing attention (Auer and Warmuth 1995; Herbster and Warmuth 1995; Helmbold *et al.* 1996; Singer 1997; Browne 1998; Vovk and Watkins 1998; Borodin *et al.* 2000; Gaivoronski and Stella 2000, 2003) and have also been studied in the case where transaction costs are involved (Blum and Kalai 1998; Evstigneev and Schenk-Hoppè 2002).

It is worth noting that the best constant rebalanced portfolio, as well as the universal portfolio, are designed to deal with the portfolio selection problem in the case where no additional information is available concerning the stock market. However, it is common practice that investors, fund managers and private investors adjust their portfolios, i.e. rebalance, using various sources of information concerning the stock market which can be conveniently summarized by the concept of *side-information*. A typical example of side-information originates from sophisticated trading strategies that often develop signalling algorithms that individuate the nature of the investment opportunity about to be faced. In this context, the side-information is usually considered to be a causal function of past stock market performance. Cover and Ordentlich (1996) presented the *state constant rebalanced portfolio*, i.e. a sequential investment algorithm that achieves, to first order in the exponent, the same wealth as the best side-information-dependent investment strategy determined in hindsight from observed market and side-information outcomes. The authors, at each trading period $t = \{1, \ldots, n\}$, used a state constant rebalanced portfolio investment algorithm that invests in the market using one of k distinct portfolios $\mathbf{x}(1), \ldots, \mathbf{x}(k)$ depending on the current state of side-information y_t. They established a set of allowable investment actions (sequence of portfolio choices $\mathbf{x}_t$), and sought to achieve the same asymptotic growth rate of wealth as the best action in this set, not in any stochastic sense, but uniformly over all possible sequences of price relatives and side-information states.

In this chapter we study and analyse the topic proposed by Cover and Ordentlich (1996). Attention is focused on the interplay between constant rebalanced portfolios and side-information. A mathematical framework is proposed for dealing with constant rebalanced portfolios in the case where side-information is available concerning the stock market. The mathematical framework introduces a new investment strategy, namely the *mixture best constant rebalanced portfolio*, which directly exploits the available side-information to outperform, in terms of the achieved wealth, the *best constant rebalanced portfolio* determined in hindsight, i.e. by assuming perfect knowledge of future stock prices. We provide a mathematical comparison of the achieved wealth by means of the best constant rebalanced portfolio and its counterpart, namely the mixture best constant rebalanced portfolio. The mixture best constant rebalanced portfolio is shown to outperform the best constant rebalanced portfolio by an exponential factor in terms of the achieved wealth. In addition, we present an online investment algorithm, namely the *mixture successive constant rebalanced portfolio with side-information*, that relies on the mixture best constant rebalanced portfolio and side-information. The proposed online investment algorithm assumes the existence of an oracle, which, by exploiting the available side-information, is capable of predicting, with different levels of accuracy (Han and Kamber 2001), the state of the stock market for the next trading period. The empirical performance of the online investment algorithm is investigated using a set of numerical experiments concerning four major stock market data sets, namely the Dow Jones Industrial Average, the Standard and Poor's 500, the Toronto Stock Exchange (Borodin *et al.* 2000) and the New York Stock Exchange (Cover 1991b; Helmbold *et al.* 1996). The results obtained emphasize the relevance of the proposed sequential investment strategy and underline the central role of the quality of the side-information in outperforming the best constant rebalanced portfolio.

The remainder of the chapter is organized as follows. In Section 5.2 we introduce the notation and main definitions concerning the stock market, the price relative, the constant rebalanced portfolio and the successive constant rebalanced portfolio (Gaivoronski and Stella 2000). Side-information, the mixture best constant rebalanced portfolio and the mixture successive constant rebalanced portfolio are introduced and analysed in Section 5.3. Section 5.3 is concerned with the theoretical framework for online investment in the case where side-information is available and provides the theoretical analysis and comparison between the best constant rebalanced portfolio, the mixture best constant rebalanced portfolio and the mixture successive constant rebalanced portfolio. Finally, Section 5.4 presents and comments on the results of a set of numerical experiments concerning some of the main financial market data sets described in the specialized literature (Cover 1991b; Helmbold *et al.* 1996; Borodin *et al.* 2000).

5.2   CONSTANT REBALANCED PORTFOLIOS

Following Cover (1991b), a *stock market vector* is represented as a vector

$$\mathbf{z} = (z_1, \ldots, z_m),$$

such that $z_i \geq 0$, $\forall i = 1, \ldots, m$, where m is the number of stocks and z_i is the *price relative*, i.e. it represents the ratio of the price at the end of the trading period to the price at the beginning of the trading period.

A *portfolio* is described by the vector

$$\mathbf{x} = (x_1, \ldots, x_m),$$

such that

$$\mathbf{x} \in \mathbf{X} = \left\{ \mathbf{x} \mid x_i \geq 0, \ \forall i = 1, \ldots, m, \ \sum_{i=1}^{m} x_i = 1 \right\}.$$

The portfolio $\mathbf{x}$ is an allocation of the current wealth across the stocks in the sense that x_i represents the fraction of wealth invested in the ith stock.

By assuming that $\mathbf{x}$ and $\mathbf{z}$ represent, respectively, the portfolio and the stock market vector for one investment period, the *wealth relative* (i.e. the ratio of the wealth at the end of the trading period to the wealth at the beginning of the trading period), given by

$$S = \mathbf{x}^T \mathbf{z},$$

represents the factor by which the wealth increases/decreases in one investment period using portfolio $\mathbf{x}$.

The problem of portfolio selection consists of selecting a portfolio $\mathbf{x}$ that would maximize S in some sense. Financial theory has developed various notions of optimality for the portfolio selection problem. One possibility is to maximize the expected value of S subject to a constraint on the variance as proposed by the Sharpe–Markowitz theory of investment (Markowitz 1952), which deals with the long-term behaviour of fixed portfolios. However, the *mean-variance investment framework* does not take into proper account the possibilities of frequent portfolio rebalances, which are one of the most important features characterizing a stock market.

To overcome this limitation, another possibility for the portfolio selection problem was proposed and described by Cover (1991a, b) that exploits the concept of the *constant rebalanced portfolio* (CRP), i.e. a portfolio such that, after each trading period, it is arranged in order to keep constant the fraction of wealth invested in every stock. By considering an arbitrary non-random sequence of n stock market vectors $\mathbf{z}^{(1)}, \ldots, \mathbf{z}^{(n)}$, a CRP $\mathbf{x}$ achieves wealth

$$S(\mathbf{x}, n) = \prod_{t=1}^{n} \mathbf{x}^T \mathbf{z}^{(t)},$$

where we assume that the initial wealth ($t=0$) is normalized to 1 ($S(x,0) = 1$).

Within the class of *constant rebalanced portfolios* (CRPs), the best of such portfolios determined in hindsight, namely the *best constant rebalanced portfolio* (BCRP), i.e. the CRP computed by assuming perfect knowledge of future stock prices, possesses interesting properties. Indeed, Cover (1991a, b) showed that the wealth achieved by means of the

BCRP is not inferior to that achieved by the best stock, to that associated with the value line and to that associated with the arithmetic mean.

These properties have motivated increasing interest in the study and analysis of the main features of this investment strategy and the use of this portfolio as the reference benchmark to evaluate and compare sequential investment strategies.

Let us now formally introduce the BCRP $\overset{*}{\mathbf{x}}^{(n)}$ that solves the following optimization problem:

$$\max_{\mathbf{x} \in X} S(\mathbf{x}, n). \tag{5.1}$$

The vector $\overset{*}{\mathbf{x}}^{(n)}$ maximizes the wealth $S(\mathbf{x},n)$ across the stock market vector sequence $\mathbf{z}^{(1)}, \ldots, \mathbf{z}^{(n)}$ and therefore it is defined as the BCRP for the stock market vector sequence $\mathbf{z}^{(1)}, \ldots, \mathbf{z}^{(n)}$.

The portfolio $\overset{*}{\mathbf{x}}^{(n)}$ cannot be used, however, for actual stock selection, at trading period n, because it explicitly depends on the sequence $\mathbf{z}^{(1)}, \ldots, \mathbf{z}^{(n)}$ which becomes known only after the expiration of this time interval.

Therefore, a reasonable objective might be to construct a sequence of portfolios $\{\mathbf{x}^{(t)}\}$, i.e. a sequential investment strategy such that, at trading periods $t = 2, \ldots, n$, the portfolio $\mathbf{x}^{(t)}$, used for stock selection, depends on the sequence $\mathbf{z}^{(1)}, \ldots, \mathbf{z}^{(t-1)}$. Let us denote by $S(\{\mathbf{x}^{(t)}\}, n)$ the wealth generated after n trading periods by successive application of the sequence of portfolios $\{\mathbf{x}^{(t)}\}$, then

$$S(\{\mathbf{x}^{(t)}\}, n) = \prod_{t=1}^{n} \mathbf{x}^{(t)^T} \mathbf{z}^{(t)}.$$

It would be desirable if such a sequential investment strategy $\{\mathbf{x}^{(t)}\}$ would yield wealth in some sense *close* to the wealth obtained by means of the BCRP $\overset{*}{\mathbf{x}}^{(n)}$. One such strategy was proposed by Cover (1991b) under the name of the *universal portfolio* (UP) and consists of selecting the investment portfolio as follows:

$$\widehat{\mathbf{x}}^{(1)} = \left(\frac{1}{m}, \ldots, \frac{1}{m}\right), \qquad \widehat{\mathbf{x}}^{(t+1)} = \frac{\int_X \mathbf{x} S(\mathbf{x}, t) \, d\mathbf{x}}{\int_X S(\mathbf{x}, t) \, d\mathbf{x}}. \tag{5.2}$$

The UP (5.2) has been shown by Cover (1991a) to possess a very interesting property: it has the same exponent, to first order, as the BCRP. Formally, by letting

$$S(\{\widehat{\mathbf{x}}^{(t)}\}, n) = \prod_{t=1}^{n} \widehat{\mathbf{x}}^{(t)^T} \mathbf{z}^{(t)}$$

be the wealth achieved by means of the UP, then it has been shown that

$$\frac{1}{n} \log S(\{\widehat{\mathbf{x}}^{(t)}\}, n) - \frac{1}{n} \log S(\overset{*}{\mathbf{x}}^{(n)}, n) \to 0,$$

with the following inequality holding:

$$S(\{\widehat{\mathbf{x}}^{(t)}\}, n) \geq S(\overset{*}{\mathbf{x}}^{(n)}, n) C_n n^{-(m-1)/2},$$

where C_n tends to some limit along subsequences for which

$$W(\overset{*}{\mathbf{x}}^{(n)}, n) = \frac{1}{n} \log S(\overset{*}{\mathbf{x}}^{(n)}, n) \rightarrow W(\overset{*}{\mathbf{x}}^{(n)})$$

for some strictly concave function $W(\mathbf{x})$ (see theorem 6.1 of Cover 1991a).

Another example of an investment strategy that exploits the definition of BCRP has been proposed by Gaivoronski and Stella (2000). This strategy, called the *successive constant rebalanced portfolio* (SCRP), selects the investment portfolio as follows:

$$\widetilde{\mathbf{x}}^{(1)} = \left(\frac{1}{m}, \ldots, \frac{1}{m}\right), \qquad \widetilde{\mathbf{x}}^{(t+1)} = \arg\max_{\mathbf{x}\in\mathbf{X}} S(\mathbf{x}, t),$$

where $\mathbf{X} = \{\mathbf{x} : x_i \geq 0, \forall i = 1, \ldots, m, \sum_{i=1}^{m} x_i = 1\}$. The SCRP $\{\widetilde{\mathbf{x}}^{(t)}\}$ possesses interesting properties. Indeed, its asymptotic wealth $S(\{\widetilde{\mathbf{x}}^{(t)}\}, n)$ coincides with the wealth obtained by means of the BCRP to first order in the exponent, i.e.

$$\frac{1}{n} \log S(\{\widetilde{\mathbf{x}}^{(t)}\}, n) - \frac{1}{n} \log S(\overset{*}{\mathbf{x}}^{(n)}, n) \rightarrow 0,$$

with the following inequality holding:

$$S(\{\widetilde{\mathbf{x}}^{(t)}\}, n) \geq S(\overset{*}{\mathbf{x}}^{(n)}, n) C(n-1)^{-2K^2/\delta}, \tag{5.4}$$

where $K = \sup_{t,\mathbf{x}\in\mathbf{X}} \|\nabla_x[\ln(\mathbf{x}^T\mathbf{z}^{(t)})]\|$, while C and δ are constants.

5.3 ONLINE INVESTMENT WITH SIDE-INFORMATION

BCRP, UP and SCRP are investment strategies designed to deal with the portfolio selection problem in the case where no additional information is available concerning the stock market. However, it is common practice that investors, fund managers and private investors adjust their portfolios, i.e. rebalance, using various sources of information concerning the stock market, which can be conveniently summarized by the concept of side-information.

A typical example of side-information originates from sophisticated trading strategies that often develop signalling algorithms that individuate the nature of the investment opportunity about to be faced. In this context, side-information is usually considered to be a causal function of past stock market performance. Therefore, the availability of side-information concerning the stock market calls for the definition of a new investment benchmark, other than the BCRP, capable of appropriately exploiting side-information about the stock market. In this direction, Cover and Ordentlich (1996) proposed the state constant rebalanced portfolio, which is capable of appropriately exploiting the available side-information concerning the stock market. The state constant rebalanced portfolio achieves, to first order in the exponent, the same wealth as the best side-information-dependent investment strategy determined in hindsight from the observed stock market vector and side-information outcomes.

Following the main ideas of Cover and Ordentlich (1996) we propose and describe a theoretical framework for online investment in the case where side-information is available concerning the stock market. We propose a mathematical model for the stock market in the presence of side-information and define a new investment benchmark that appropriately exploits the available side-information.

The proposed mathematical model assumes that, at any trading period t, the stock market can be in one of H possible states belonging to $\mathbf{H} = \{ 1, \ldots, H \}$. The stock market state at trading period t influences the stock market vector $\mathbf{z}^{(t)}$. Formally, the model assumes that the infinite sequence of stock market vectors $\mathbf{z}^{(1)}, \ldots, \mathbf{z}^{(\infty)}$ is a realization from a *mixture* consisting of H components. Each mixture's component is associated with a given stock market state. Therefore, any infinite sequence of stock market vectors $\mathbf{z}^{(1)}, \ldots, \mathbf{z}^{(\infty)}$ partitions into H mutually exclusive subsets $\mathbf{Z}_1, \ldots, \mathbf{Z}_H$, each subset $\mathbf{Z}_h$ containing those stock market vectors associated with the corresponding stock market state h. It should be emphasized that the partitioning $\mathbf{Z}_1, \ldots, \mathbf{Z}_H$ is assumed to be *data independent*, i.e. it does not depend on the given infinite sequence of stock market vectors $\mathbf{z}^{(1)}, \ldots, \mathbf{z}^{(\infty)}$, but it only depends on the mixture's components. The same model applies to finite stock market vector sequences $\mathbf{z}^{(1)}, \ldots, \mathbf{z}^{(n)}$. However, in order to avoid confusion, for any given finite stock market vector sequence $\mathbf{z}^{(1)}, \ldots, \mathbf{z}^{(n)}$, we will let $\mathbf{Z}_h^{(n)}$ be the subset containing only those stock market vectors $\mathbf{z}^{(t)}$, $t = 1, \ldots, n$, associated with the stock market state h. Furthermore, we let $n_h^{(n)}$ be the cardinality of $\mathbf{Z}_h^{(n)}$, $\sum_{h=1}^{H} n_h^{(n)} = n$ and assume that $n_h^{(n)} \to \infty$ as $n \to \infty$, $\forall\, h \in \mathbf{H}$.

This stock market model requires the definition of a new investment benchmark, the *mixture best constant rebalanced portfolio* (MBCRP), in the case where side-information is available, and can be used to make an inference about the current stock market state. To introduce the mathematical framework for dealing with the proposed stock market model and side-information, let us take into account the generic stock market state h and let $\overset{*}{\mathbf{x}}_h^{(n)}$ solve the optimization problem (5.1) for those stock market vectors $\mathbf{z}^{(1)}, \ldots, \mathbf{z}^{(n)}$ belonging to subset $\mathbf{Z}_h^{(n)}$, i.e. $\overset{*}{\mathbf{x}}_h^{(n)}$ maximizes the *logarithmic wealth relative* associated with stock market state h

$$\log S_h(\mathbf{x}, n) = \sum_{t=1}^{n} \mathbb{I}_{\mathbf{z}^{(t)} \in \mathbf{Z}_h^{(n)}} \log \left(\mathbf{x}^T \mathbf{z}^{(t)} \right). \tag{5.5}$$

Let us now expand (5.5) in a Taylor series, up to second order, centred at $\overset{*}{\mathbf{x}}_h^{(n)}$ to obtain

$$\log S_h(\mathbf{x}, n) = \sum_{t=1}^{n} \mathbb{1}_{\mathbf{z}^{(t)} \in Z_h^{(n)}} \left[\log \left(\overset{*(n)}{\mathbf{x}_h}^T \mathbf{z}^{(t)} \right) + \frac{\left(\mathbf{x} - \overset{*(n)}{\mathbf{x}_h} \right)^T \mathbf{z}^{(t)}}{\overset{*(n)}{\mathbf{x}_h}^T \mathbf{z}^{(t)}} \right.$$

$$- \frac{\left(\mathbf{x} - \overset{*(n)}{\mathbf{x}_h} \right)^T \left(\mathbf{z}^{(t)} (\mathbf{z}^{(t)})^T \right) \left(\mathbf{x} - \overset{*(n)}{\mathbf{x}_h} \right)}{2 \left(\overset{*(n)}{\mathbf{x}_h}^T \mathbf{z}^{(t)} \right)^2}$$

$$\left. + \frac{\left[\left(\mathbf{x} - \overset{*(n)}{\mathbf{x}_h} \right)^T \mathbf{z}^{(t)} \right]^3}{3 \left(\epsilon^T \mathbf{z}^{(t)} \right)^3} \right]$$

for a given vector ϵ between $\mathbf{x}$ and $\overset{*(n)}{\mathbf{x}_h}$. Equation (5.6), by exploiting (5.5) and by isolating the contribution for each of its four terms, can be rewritten as

$$\log S_h(\mathbf{x}, n) - \log S_h(\overset{*(n)}{\mathbf{x}_h}, n)$$

$$= \sum_{t=1}^{n} \mathbb{1}_{\mathbf{z}^{(t)} \in Z_h^{(n)}} \left[\frac{\left(\mathbf{x} - \overset{*(n)}{\mathbf{x}_h} \right)^T \mathbf{z}^{(t)}}{\overset{*(n)}{\mathbf{x}_h}^T \mathbf{z}^{(t)}} \right.$$

$$\left. - \frac{\left(\mathbf{x} - \overset{*(n)}{\mathbf{x}_h} \right)^T \left(\mathbf{z}^{(t)} (\mathbf{z}^{(t)})^T \right) \left(\mathbf{x} - \overset{*(n)}{\mathbf{x}_h} \right)}{2 \left(\overset{*(n)}{\mathbf{x}_h}^T \mathbf{z}^{(t)} \right)^2} + \frac{\left[\left(\mathbf{x} - \overset{*(n)}{\mathbf{x}_h} \right)^T \mathbf{z}^{(t)} \right]^3}{3 \left(\epsilon^T \mathbf{z}^{(t)} \right)^3} \right]. \qquad (5.7)$$

Now, according to Gaivoronski and Stella (2000), the following conditions hold:

- *Asymptotic independence*

$$\lim_{n} \inf \gamma_{\min} \left(\sum_{t=1}^{n} \mathbb{1}_{\mathbf{z}^{(t)} \in Z_h^{(n)}} \frac{\mathbf{z}^{(t)} (\mathbf{z}^{(t)})^T}{\|\mathbf{z}^{(t)}\|^2} \right) \geq \delta_h > 0,$$

where by $\gamma_{\min}(A)$ we denote the smallest eigenvalue of matrix A;
- *Uniform boundedness*

$$0 < z^- \leq z_i^{(t)} \leq z^+, \quad \forall t, i.$$

Theorem 5.2 in Gaivoronski and Stella (2000) ensures that (5.5) is strictly concave on $\mathbf{X}$ uniformly over n and therefore the following inequality holds:

$$\log S_h(\mathbf{x}, n) - \log S_h \left(\overset{*(n)}{\mathbf{x}_h}, n \right) \leq \sum_{t=1}^{n} \mathbb{1}_{\mathbf{z}^{(t)} \in Z_h^{(n)}} \frac{\left(\mathbf{x} - \overset{*(n)}{\mathbf{x}_h} \right)^T \mathbf{z}^{(t)}}{\overset{*(n)}{\mathbf{x}_h}^T \mathbf{z}^{(t)}} - \frac{\delta_h}{2} \left\| \mathbf{x} - \overset{*(n)}{\mathbf{x}_h} \right\|^2. \qquad (5.8)$$

Now, from optimality conditions, applied to the BCRP $\overset{*(n)}{\mathbf{x}_h}$, we know that, for each stock market state h and for any CRP $\mathbf{x}$, the following condition holds:

$$\sum_{t=1}^{n} \mathbb{1}_{\mathbf{z}^{(t)} \in Z_h^{(n)}} \frac{\left(\mathbf{x} - \overset{*(n)}{\mathbf{x}_h}\right)^T \mathbf{z}^{(t)}}{\overset{*(n)}{\mathbf{x}_h}{}^T \mathbf{z}^{(t)}} \leq 0, \tag{5.9}$$

and, therefore, by combining (5.8) with (5.9) we can write

$$\log S_h\left(\overset{*(n)}{\mathbf{x}_h}, n\right) - \log S_h(\mathbf{x}, n) \geq \frac{\delta_h}{2} \left\| \mathbf{x} - \overset{*(n)}{\mathbf{x}_h} \right\|^2. \tag{5.10}$$

Finally, by recalling (5.8) and using the relationship between δ_h and $n_h^{(n)}$, we obtain the inequality

$$\log S_h\left(\overset{*(n)}{\mathbf{x}_h}, n\right) - \log S_h(\mathbf{x}, n) \geq \frac{1}{2} n_h^{(n)} \gamma_h \left\| \mathbf{x} - \overset{*(n)}{\mathbf{x}_h} \right\|^2, \tag{5.11}$$

where γ_h is the lim inf of the minimum eigenvalue of the matrix

$$\frac{1}{n_h^{(n)}} \sum_{t=1}^{n} \mathbb{1}_{\mathbf{z}^{(t)} \in Z_h^{(n)}} \frac{\mathbf{z}^{(t)} \mathbf{z}^{(t)T}}{\left\| \mathbf{z}^{(t)} \right\|^2}.$$

Let us now return to the proposed mathematical framework for dealing with stock markets and side-information, i.e. let us take into account the case where the stock market can be in one of H possible states belonging to $\mathbf{H} = \{1, \ldots, H\}$. In such a framework, given a sequence of stock market vectors $\mathbf{z}^{(1)}, \ldots, \mathbf{z}^{(n)}$, where each stock market vector is associated with a given stock market state, the achieved logarithmic wealth, by means of the investment strategy $\{\overset{*(n)}{\mathbf{x}_H}\}$ that, at each trading period t, exploits knowledge of the current stock market state h and applies the corresponding BCRP $\overset{*(n)}{\mathbf{x}_h}$ according to (5.5), is given by

$$\log S\left(\{\overset{*(n)}{\mathbf{x}_H}\}, n\right) = \sum_{h=1}^{H} \log S_h\left(\overset{*(n)}{\mathbf{x}_h}, n\right), \tag{5.12}$$

whereas the logarithmic wealth achieved by means of any CRP $\mathbf{x}$ is given by $\log S(\mathbf{x}, n) = \sum_{t=1}^{n} \log(\mathbf{x}^T \mathbf{z}^{(t)})$.

The difference between the logarithmic wealth achieved by means of the investment strategy $\{\overset{*(n)}{\mathbf{x}_H}\}$ and the logarithmic wealth achieved by means of any other CRP $\mathbf{x}$ is given by

$$\log S\left(\{\overset{*(n)}{\mathbf{x}_H}\}, n\right) - \log S(\mathbf{x}, n) = \sum_{h=1}^{H} \log S_h\left(\overset{*(n)}{\mathbf{x}_h}, n\right) - \sum_{t=1}^{n} \log\left(\mathbf{x}^T \mathbf{z}^{(t)}\right),$$

and from (5.11) we can write

$$\log S\left(\left\{\mathbf{x_H}^{*(n)}\right\}, n\right) - \log S(\mathbf{x}, n) \geq \frac{1}{2} \sum_{h=1}^{H} n_h^{(n)} \gamma_h \left\| \mathbf{x} - \mathbf{x}_h^{*(n)} \right\|^2.$$

Therefore, given the investment strategy $\{\mathbf{x_H}^{*(n)}\}$ and the BCRP $\mathbf{\dot{x}}^{(n)}$, the following inequality holds:

$$\log S\left(\left\{\mathbf{x_H}^{*(n)}\right\}, n\right) - \log S(\mathbf{\dot{x}}^{(n)}, n) \geq \frac{1}{2} \sum_{h=1}^{H} n_h^{(n)} \gamma_h \left\| \mathbf{\dot{x}}^{*(n)} - \mathbf{x}_h^{*(n)} \right\|^2. \tag{5.13}$$

The bound (5.13) strongly motivates interest in investment strategies depending on knowledge of the current stock market state. Indeed, coming back to wealth, the following condition holds:

$$S\left(\left\{\mathbf{x_H}^{*(n)}\right\}, n\right) \geq S(\mathbf{\dot{x}}^{*(n)}, n) \exp\left(\frac{1}{2} \sum_{h=1}^{H} n_h^{(n)} \gamma_h \left\| \mathbf{\dot{x}}^{*(n)} - \mathbf{x}_h^{*(n)} \right\|^2 \right), \tag{5.14}$$

and, therefore, the investment strategy $\{\mathbf{x_H}^{*(n)}\}$ achieves a wealth that is exponential with respect to the wealth achieved by means of the BCRP $\mathbf{\dot{x}}^{*(n)}$.

Let us now give the formal definition of the MBCRP investment strategy $\{\mathbf{x_H}^{*(n)}\}$, that, at each trading period t, exploits knowledge of the current stock market state h to select and to apply the corresponding BCRP $\mathbf{x}_h^{*(n)}$.

Definition 5.3.1 (*Mixture best constant rebalanced portfolio*): Given a stock market characterized by the states in $\mathbf{H} = \{1, \ldots, H\}$, the MBCRP is defined as the following investment strategy, i.e. the set of BCRPs:

$$\left\{\mathbf{x_H}^{*(n)}\right\} = \left\{\mathbf{x}_1^{*(n)}, \ldots, \mathbf{x}_H^{*(n)}\right\}, \tag{5.15}$$

where each portfolio $\mathbf{x}_h^{*(n)}$ is the BCRP, according to (5.5), associated with stock market state h. The MBCRP is the investment strategy that, at each trading period t, knows the stock market state h and therefore invests using the corresponding BCRP $\mathbf{x}_h^{*(n)}$.

The MBCRP possesses an interesting property; it outperforms by an exponential factor the corresponding BCRP in terms of the wealth achieved as stated by the following theorem.

Theorem 5.3.2: *The wealth accumulated after n trading periods from the MBCRP $\{\mathbf{x_H}^{*(n)}\}$ outperforms the wealth of the corresponding BCRP $\mathbf{\dot{x}}^{*(n)}$ by an exponential factor. Formally,*

$$S\left(\left\{\mathbf{x_H}^{*(n)}\right\}, n\right) \geq S(\mathbf{\dot{x}}^{*(n)}, n) \exp\left(\frac{1}{2} \sum_{h=1}^{H} n_h^{(n)} \gamma_h \left\| \mathbf{\dot{x}}^{*(n)} - \mathbf{x}_h^{*(n)} \right\|^2 \right). \tag{5.16}$$

Proof: The proof follows directly from Equation (5.14).   □

Let us now introduce an online investment strategy, namely the *mixture successive constant rebalanced portfolio* (MSCRP), which approximates the MBCRP and relies on the *successive constant rebalanced portfolio* (SCRP) online investment strategy introduced and analysed by Gaivoronski and Stella (2000).

Definition 5.3.3 (*Mixture successive constant rebalanced portfolio*): Given a stock market characterized by the states in $\mathbf{H} = \{1, \ldots, H\}$, the MSCRP is defined as the following investment strategy, i.e. the set of SCRPs:

$$\{\widetilde{\mathbf{x}}_{\mathbf{H}}^{(n)}\} = \{\{\widetilde{\mathbf{x}}_1^{(t)}\}, \ldots, \{\widetilde{\mathbf{x}}_H^{(t)}\}\}, \tag{5.17}$$

where each portfolio $\{\widetilde{\mathbf{x}}_h^{(t)}\}$ is the SCRP associated with the stock market state h at trading period t. The MSCRP is the investment strategy that, at each trading period t, knows the stock market state h and therefore invests using the corresponding SCRP $\widetilde{\mathbf{x}}_h^{(t)}$.

The relationship between the accumulated wealth, after n trading periods, by means of the MBCRP benchmark and by means of the MSCRP online investment strategy, is clarified by the following theorem.

Theorem 5.3.4: *For each stock market characterized by states in* $\mathbf{H} = \{1, \ldots, H\}$, *the MSCRP (5.17) is universal with respect to the MBCRP (5.15). Formally,*

$$\frac{1}{n} \log S(\{\widetilde{\mathbf{x}}_{\mathbf{H}}^{(n)}\}, n) - \frac{1}{n} \log S(\{\mathbf{x}_{\mathbf{H}}^{*(n)}\}, n) \to 0.$$

Proof: From property (5.4) of the SCRP (5.3) online investment strategy, for any given stock market state h, the following inequality holds:

$$S(\widetilde{\mathbf{x}}_h^{*(n)}, n) \leq S(\{\widetilde{\mathbf{x}}_h^{(t)}\}, n) C^{(h)} \left(n_h^{(n)} - 1\right)^{Q^{(h)}}, \tag{5.18}$$

where $C^{(h)}$ and $Q^{(h)}$ are constants depending on the stock market state h.

Then, by combining Equation (5.12) with Equation (5.18), it is possible to write the inequality

$$S(\{\mathbf{x}_{\mathbf{H}}^{*(n)}\}, n) = \prod_{h=1}^{H} S(\mathbf{x}_h^{*(n)}, n) \leq \prod_{h=1}^{H} S(\{\widetilde{\mathbf{x}}_h^{(t)}\}, n) C^{(h)} \left(n_h^{(n)} - 1\right)^{Q^{(h)}}, \tag{5.19}$$

and, thus, by taking the logarithm of both sides we obtain

$$\log S(\{\mathbf{x}_{\mathbf{H}}^{*(n)}\}, n) \leq \sum_{h=1}^{H} \left(\log S(\{\widetilde{\mathbf{x}}_h^{(t)}\}, n) + Q^{(h)} \log \left(n_h^{(n)} - 1\right) + \log C^{(h)}\right),$$

which can be rewritten as

$$\log S(\{\tilde{\mathbf{x}}_{\mathbf{H}}^{*(n)}\}, n) - \log S(\{\tilde{\mathbf{x}}_{\mathbf{H}}^{(n)}\}, n) \le \sum_{h=1}^{H} \left(Q^{(h)} \log \left(n_h^{(n)} - 1 \right) + \log C^{(h)} \right).$$

Therefore, dividing both sides by n, it is possible to conclude that

$$\frac{1}{n} \log S(\{\tilde{\mathbf{x}}_{\mathbf{H}}^{(n)}\}, n) - \frac{1}{n} \log S(\{\tilde{\mathbf{x}}_{\mathbf{H}}^{*(n)}\}, n) \rightarrow 0,$$

which completes the proof. □

Let us now compare the MSCRP with the corresponding BCRP in terms of the accumulated wealth after n trading periods.

Theorem 5.3.5: *For each stock market characterized by the states in $\mathbf{H} = \{1, \ldots, H\}$, the accumulated wealth after n trading periods by means of the MSCRP $\{\tilde{\mathbf{x}}_{\mathbf{H}}^{(n)}\}$ outperforms the accumulated wealth by means of the corresponding BCRP $\tilde{x}^{(n)}$ by an exponential factor. Formally, we have that, for $n \rightarrow \infty$, the following inequality holds:*

$$S(\{\tilde{\mathbf{x}}_{\mathbf{H}}^{(n)}\}, n) \ge S(\tilde{x}^{*(n)}, n) \exp \left(\frac{1}{2} \sum_{h=1}^{H} \left(n_h^{(n)} \gamma_h \| \tilde{x}^{*(n)} - \tilde{x}_h^{*(n)} \|^2 \right. \right.$$

$$\left. \left. - Q^{(h)} \log \left(n_h^{(n)} - 1 \right) - \log C^{(h)} \right) \right).$$

Proof: Inequality (5.16) from theorem 5.3.2 states that

$$S(\{\tilde{\mathbf{x}}_{\mathbf{H}}^{*(n)}\}, n) \ge S(\tilde{x}^{*(n)}, n) \exp \left(\frac{1}{2} \sum_{h=1}^{H} n_h^{(n)} \gamma_h \| \tilde{x}^{*(n)} - \tilde{x}_h^{*(n)} \|^2 \right),$$

therefore using (5.19) it is possible to write

$$\prod_{h=1}^{H} S(\{\tilde{\mathbf{x}}_h^{(t)}\}, n) C^{(h)} \left(n_h^{(n)} - 1 \right)^{Q^{(h)}}$$

$$\ge S(\{\tilde{\mathbf{x}}_{\mathbf{H}}^{*(n)}\}, n) \ge S(\tilde{x}^{*(n)}, n) \exp \left(\frac{1}{2} \sum_{h=1}^{H} n_h^{(n)} \gamma_h \| \tilde{x}^{*(n)} - \tilde{x}_h^{*(n)} \|^2 \right),$$

and, by taking the logarithm and reordering, we obtain the inequality

$$\log S\big(\{\tilde{\mathbf{x}}_{\mathbf{H}}^{(n)}\}, n\big) - \log S\big(\overset{*}{\mathbf{x}}^{(n)}, n\big)$$

$$\geq \frac{1}{2}\sum_{h=1}^{H} n_h^{(n)} \gamma_h \big\|\overset{*}{\mathbf{x}}^{(n)} - \overset{*}{\mathbf{x}}_h^{(n)}\big\|^2$$

$$-\sum_{h=1}^{H} Q^{(h)} \log\big(n_h^{(n)} - 1\big) - \log C^{(h)}.$$

Now, by taking the limit $n \to \infty$, where for each stock market state h we assume that $n_h^{(n)} \to \infty$, we have

$$S\big(\{\tilde{\mathbf{x}}_{\mathbf{H}}^{(n)}\}, n\big) \geq S\big(\overset{*}{\mathbf{x}}^{(n)}, n\big) \exp\Bigg(\frac{1}{2}\sum_{h=1}^{H}\Big(n_h^{(n)} \gamma_h \big\|\overset{*}{\mathbf{x}}^{(n)} - \overset{*}{\mathbf{x}}_h^{(n)}\big\|^2$$

$$- Q^{(h)} \log\big(n_h^{(n)} - 1\big) - \log C^{(h)}\Big)\Bigg).$$

Furthermore, we can write

$$S\big(\overset{*}{\mathbf{x}}^{(n)}, n\big) \exp\Bigg(\frac{1}{2}\sum_{h=1}^{H}\Big(n_h^{(n)} \gamma_h \big\|\overset{*}{\mathbf{x}}^{(n)} - \overset{*}{\mathbf{x}}_h^{(n)}\big\|^2$$

$$- Q^{(h)} \log\big(n_h^{(n)} - 1\big) - \log C^{(h)}\Big)\Bigg)$$

$$\approx S\big(\overset{*}{\mathbf{x}}^{(n)}, n\big) \exp\Bigg(\frac{1}{2}\sum_{h=1}^{H} n_h^{(n)} \gamma_h \big\|\overset{*}{\mathbf{x}}^{(n)} - \overset{*}{\mathbf{x}}_h^{(n)}\big\|^2\Bigg),$$

and therefore it is possible to write the following inequality:

$$S\big(\{\tilde{\mathbf{x}}_{\mathbf{H}}^{(n)}\}, n\big) \geq S\big(\overset{*}{\mathbf{x}}^{(n)}, n\big) \exp\Bigg(\frac{1}{2}\sum_{h=1}^{H} n_h^{(n)} \gamma_h \big\|\overset{*}{\mathbf{x}}^{(n)} - \overset{*}{\mathbf{x}}_h^{(n)}\big\|^2\Bigg),$$

which completes the proof. □

According to theorems 5.3.2, 5.3.4 and 5.3.5 the MSCRP investment strategy (definition 5.3.3) possesses interesting theoretical properties in the case where side-information is available concerning the stock market. However, the MSCRP cannot be used directly to invest in the stock market because, for each trading period t, it assumes perfect knowledge about the current stock market state h, which is indeed known just after the current trading period t has expired. Therefore, it would be desirable to develop an algorithm that, by using past information about the stock market, makes predictions about the stock market state h for the next trading period t. More precisely, the algorithm to be developed should exploit the available side-information to provide predictions about the stock market state associated with the next trading period t.

In order to clarify how the prediction task concerning the stock market state h for the next trading period t can be formulated and therefore to complete the proposed mathematical framework for online investment with side-information, we adopt the Bayesian paradigm.

In particular, we let $\mu \in \Re$ be the side-information vector and $p(Y|h)$ the state conditional probability density function for Y, i.e. the probability density function for the side-information vector Y conditioned on h being the stock market state. Furthermore, we let $P(h)$ be the *a priori* probability that the stock market is in state h. Then, the posterior probability $P(h|Y)$ that the stock market is in state h given the side-information vector Y can be computed using the celebrated Bayes' rule (Duda and Hart 1973) as follows:

$$P(h|Y) = \frac{p(Y|h)P(h)}{p(Y)}. \qquad (5.20)$$

The Bayes' rule (5.20) offers a precious theoretical model for exploiting the available side-information and therefore to make an inference concerning the stock market state for the next trading period. However, the quantities on the right-hand side of (5.20), i.e. the state conditional probability density $p(Y|h)$ as well as the *a priori* probability $P(h)$, are unknown and must be estimated by combining *a priori* knowledge with the available data. Several computational approaches and algorithms have been proposed and developed in recent years and published in the specialized literature (Duda and Hart 1973; Bernardo and Smith 2000; Duda *et al.* 2001; Zaffalon 2002; Congdon 2003; Zaffalon and Fagiuoli 2003) to deal with the problem of the unknown likelihood $p(Y|h)$ and prior $P(h)$, and treatment of such problem is out of the scope of this chapter.

In this chapter we assume the existence of an *oracle* that, at each trading period t and using the available side-information vector Y, is capable of making an inference concerning the true stock market state for trading period $t + 1$ with different accuracy levels q (Han and Kamber 2001). In each trading period t the oracle first accesses the available side-information vector Y and then makes inference $\hat{h}$ concerning the stock market state h for the next trading period $t + 1$. The oracle is assumed to provide predictions with different levels of accuracy, i.e. it is assumed that the following condition holds:

$$P(\hat{h} = h) = q, \quad \forall h \in \mathbf{H}. \qquad (5.21)$$

It is worth noting that condition (5.21) states that the oracle is capable of providing predictions that, in the long run, are associated with a misclassification error equal to $1 - q$ with q defined as the *oracle's accuracy level*.

Let us now describe an online investment algorithm that relies on the MSCRP investment strategy and which, to make an inference concerning the stock market state for each trading period, uses the predictions provided by means of the considered oracle.

Algorithm 5.1 *MSCRP with side-information*

1. At the beginning of the first trading period take

$$\widetilde{\mathbf{x}}^{(1)} = \left(\frac{1}{m}, \ldots, \frac{1}{m} \right).$$

2. At the end of trading period $t = 1, \ldots$ use side-information Y to make an inference concerning the stock market state at trading period $t + 1$. Assume the oracle makes an inference in favour of stock market state h.

3. Let $\mathbf{z}^{(1)}, \ldots, \mathbf{z}^{(t)}$ be the stock market vectors available up to trading period t. Compute $\widetilde{\mathbf{x}}^{(t+1)}$ as the solution of the following optimization problem:

$$\max_{\mathbf{x} \in \mathbf{X}} \log S_h(\mathbf{x}, t),$$

where $\log S_h(\mathbf{x}, t) = \sum_{s=1}^{t} \mathbb{1}_{\mathbf{z}^{(s)} \in \mathbf{Z}_h^{(t)}} \log(\mathbf{x}^T \mathbf{z}^{(s)})$.

The *MSCRP with side-information* algorithm exploits the side-information Y available at trading period t and makes inference $\widehat{h} = h$ for the next stock market state, i.e. at trading period $t + 1$. Then, the *MSCRP with side-information* algorithm exploits the forecast $\widehat{h} = h$ and invests in the stock market by means of the corresponding SCRP $\widetilde{\mathbf{x}}_h^{(t)}$. Therefore, the *MSCRP with side-information* algorithm approximates the MSCRP sequential investment strategy. The theoretical properties of the *MSCRP with side-information* algorithm clearly depend on the capability of correctly assessing the stock market state for the next trading period and therefore depend on the oracle's accuracy level q (5.21).

5.4 NUMERICAL EXPERIMENTS

This section illustrates and comments on the results obtained from a set of numerical experiments concerning the MSCRP with side-information algorithm in the case where four major stock market data sets described in the specialized literature are considered. The four stock market data sets are the Dow Jones Industrial Average, the Standard and Poor's 500, the Toronto Stock Exchange (Borodin *et al.* 2000) and the New York Stock Exchange (Cover 1991b; Helmbold *et al.* 1996).

The first data set consists of the 30 stocks belonging to the Dow Jones Industrial Average (DJIA) for the 2-year period (507 trading periods, days) starting from Jan 14, 2001 to Jan 14, 2003. The second data set consists of the 25 stocks from the Standard and Poors 500 (S&P500) having the largest market capitalization during the period starting from Jan 2, 1998 to Jan 31, 2003 (1276 trading periods, days). The Toronto Stock Exchange (TSE) data set consists of 88 stocks for the 5-year period (1259 trading periods, days) starting from Jan 4, 1994 to Dec 31, 1998. Finally, the New York Stock Exchange (NYSE) data set consists of 5651 daily prices (trading periods) for 35 stocks for the 22-year period starting from July 3, 1962 to Dec 31, 1984.

We assume that each stock market is characterized by two states ($\mathbf{H} = \{1, 2\}$) with state $h = 1$ associated with trading periods for which the average value of price relatives over all stocks is greater than one, whereas state $h = 2$ is associated with the remaining trading periods. This partitioning can be conveniently interpreted as generating bull ($h = 1$) and bear ($h = 2$) trading periods. It represents a first proposal and much effort will be devoted to finding alternative partitioning criteria to improve the effectiveness of the MSCRP with side-information online investment algorithm.

Numerical experiments were performed without the help of any inference device in the sense that no computational devices were used to predict the stock market state h for the next trading period. The inference concerning the stock market state h for the next trading period is obtained by exploiting predictions provided by means of an oracle associated with different accuracy levels q (5.21).

It should be noted that an accuracy level equal to one ($q = 1$) corresponds to perfect knowledge of the stock market state h before the investment step takes place, thus leading to invest according to the corresponding optimal SCRP $\{\tilde{\mathbf{x}}_h^{(t)}\}$. The results associated with the numerical experiments in the case where the oracle is assumed to have perfect knowledge of the stock market state ($q = 1$) are reported in Table 5.1.

The data reported in Table 5.1 demonstrate the effectiveness of the MSCRP with side-information: indeed, the achieved wealth significantly outperforms that achieved by means of the corresponding BCRP algorithm. However, it is unrealistic to assume perfect knowledge concerning the stock market state and a further investigation is required. It is of central relevance to analyse how the MSCRP with side-information behaves in the case where different values of the accuracy level q are considered. To this extent, a set of numerical experiments for the DJIA, S&P500, TSE and the NYSE stock market data sets was planned and performed by considering different values for the oracle's accuracy level q. The numerical experiments were organized in such a way that, at each trading period, given the oracle's accuracy level q, we extract a random number r from a uniform distribution in the interval $[0,1]$. If the extracted random number r is less than or equal to the selected oracle's accuracy level q, i.e. if $r \leq q$, then the stock market state h for the next trading period is correctly predicted, otherwise the stock market state is wrongly assessed. The results are summarized for each stock market data set using the mean value of achieved wealth by means of the MSCRP with side-information. The mean value of the wealth is computed using the described random sampling procedure in the case where 1000 samples are extracted for each trading period and different values of the oracle's accuracy

TABLE 5.1 Wealth for the MSCRP with Side-Information Assuming Perfect Knowledge ($q = 1$)

	BCRP	MBCRP	MSCRP
DJIA	1.24	54.70	21.27
S&P500	4.07	10 527.00	3750.00
TSE	6.78	591.61	45.08
NYSE	250.59	3.62E+10	1.02E+10

level q belonging to the interval $[0,1]$. The *stock market state frequency*, i.e. the percentage of trading periods associated with each stock market state h, is reported in Table 5.2.

The results of the numerical experiments are summarized using a graphical representation of the *extra wealth* obtained using the MSCRP with side-information over the corresponding BCRP, defined as

$$\omega = \frac{S(\{\tilde{\mathbf{x}}_{\mathbf{H}}^{(n)}\}, n)}{S(\overset{*}{\mathbf{x}}^{(n)}, n)}, \tag{5.22}$$

with respect to the oracle's accuracy level q and the trading period t for each data set (Figures 5.1–5.4).

Figures 5.1 to 5.4, where the minimum value of ω (5.22) is zero, clearly show the exponential nature of the extra wealth obtained for the MSCRP with side-information over the corresponding BCRP. It is possible to observe the exponential nature of the extra wealth ω for an increasing number t of trading periods and depending on the oracle's accuracy level q.

The oracle's accuracy level required to achieve extra wealth ω (5.22) equal to one, i.e. the value of q such that the MSCRP with side-information achieves the same wealth as the

TABLE 5.2 Stock Market State Frequency

	$h = 1$	$h = 2$
DJIA	0.48	0.52
S&P500	0.52	0.48
TSE	0.58	0.42
NYSE	0.53	0.47

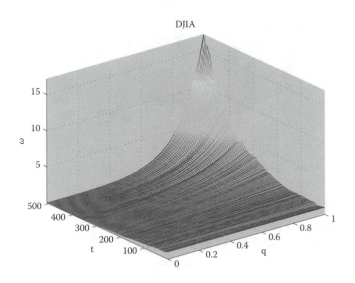

FIGURE 5.1 MSCRP extra wealth for the DJIA data set.

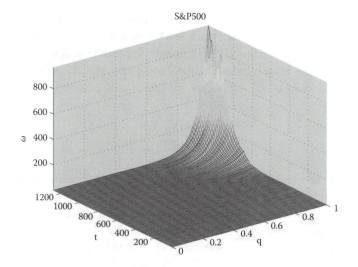

FIGURE 5.2 MSCRP extra wealth for the S&P500 data set.

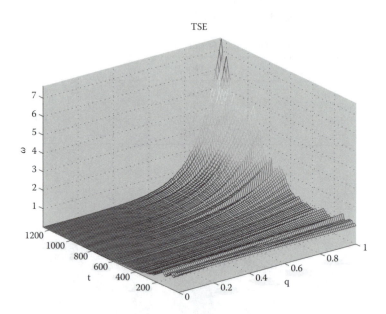

FIGURE 5.3 MSCRP extra wealth for the TSE data set.

corresponding BCRP, is defined as the *oracle's parity accuracy level* and is reported, for each stock market data set, in Table 5.3.

The data reported in Table 5.3 allow us to assess the effectiveness of the MSCRP with side-information depending on the oracle accuracy level q required. It is interesting to investigate the relationship between the extra wealth ω and the variables: number of

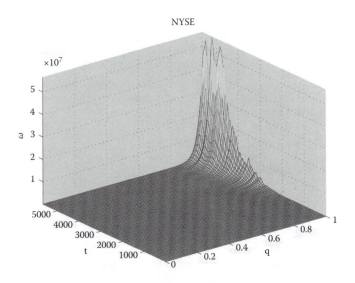

FIGURE 5.4 MSCRP extra wealth for the NYSE data set.

TABLE 5.3 Oracle's Parity Accuracy Levels

	q
DJIA	0.60
S&P500	0.59
TSE	0.77
NYSE	0.55

TABLE 5.4 Parameter Estimates for Model (5.23) and R^2 Values

	a_0	a_1	a_2	a_3	R^2	N
DJIA	0.4272	−0.0086	−0.9871	0.0148	0.9791	51 207
S&P500	1.0097	−0.0084	−1.8281	0.0144	0.9850	128 876
TSE	0.3058	−0.0056	−0.8444	0.0075	0.9814	127 159
NYSE	−3.8365	−0.0041	7.7298	0.0073	0.9880	570 751

trading periods t and oracle accuracy level q. To this end the linear regression model for the logarithm of the extra wealth ω with four d.o.f.,

$$\log(\omega) = a_0 + a_1 t + a_2 q + a_3 tq, \qquad (5.23)$$

was fitted using a different number of data points N for each stock market data set. Parameter estimates for model (5.23) together with the corresponding R^2 values are reported in Table 5.4.

Finally, Figures 5.5–5.8 plot the estimated value, using regression model (5.23), of the logarithm of the extra wealth $\log(\omega)$ with respect to trading period t and oracle accuracy level q.

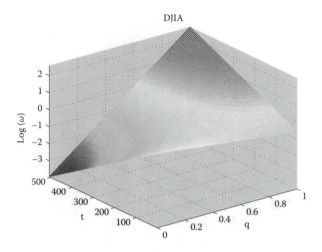

FIGURE 5.5 Estimated logarithm of the extra wealth for the DJIA data set.

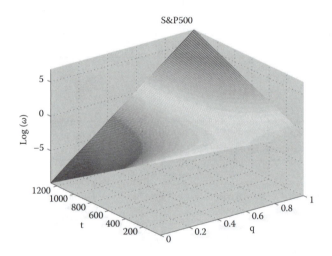

FIGURE 5.6 Estimated logarithm of the extra wealth for the S&P500 data set.

5.5   CONCLUSIONS AND FURTHER RESEARCH DIRECTIONS

This chapter deals with online portfolio selection in the case where side-information is available concerning the stock market. Its theoretical achievements strongly support the further study and investigation of the class of MBCRP investment strategies. The experimental evidence that the MSCRP with side-information investment algorithm outperforms, in terms of the achieved wealth, the corresponding BCRP investment algorithm by an exponential factor is a basic result of the present work. However, this empirical achievement strongly depends on the oracle's accuracy level. Therefore, the

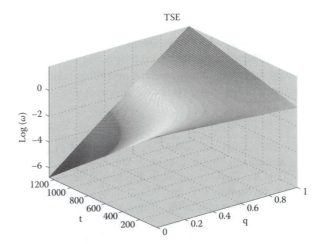

FIGURE 5.7 Estimated logarithm of the extra wealth for the TSE data set.

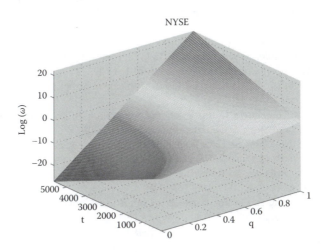

FIGURE 5.8 Estimated logarithm of the extra wealth for the NYSE data set.

problem is shifted to the study and development of efficient and reliable computational devices for predicting the stock market state, at each trading period, by exploiting the available side-information. A second issue of interest, which probably has significant interplay with the development of reliable prediction models, is the choice of the stock market states. These motivations orient the next step of this research to the study and analysis of stock market state partitioning criteria as well as to the study and development of efficient models for the prediction of the stock market state by exploiting the available side-information.

ACKNOWLEDGEMENTS

The authors are grateful to the anonymous referees whose insightful comments enabled us to make significant improvements to the chapter. Special thanks go to Professor Georg Pflug for his comments and suggestions concerning the mathematical notation, which allowed us to significantly improve the clarity of the chapter.

REFERENCES

Algoet, P.H. and Cover, T.M., Asymptotic optimality and asymptotic equipartition properties of log-optimum investment. *Ann. Probabil.*, 1988, **2**, 876–898.

Auer, P. and Warmuth, M.K., Tracking the best disjunction, in *36th Annual Symposium on Foundations of Computer Science.*, 1995, pp. 312–321.

Bell, R. and Cover, T.M., Competitive optimality of logarithmic investment. *Math. Oper. Res.*, 1980, **5**, 161–166.

Bernardo, J.M. and Smith, A.F.M., *Bayesian Theory*, 2000 (Wiley: New York).

Blum, A. and Kalai, A., Universal portfolios with and without transaction costs. *Mach. Learn.*, 1998, **30**, 23–30.

Borodin, A., El-Yaniv, R.. and Gogan, V., On the competitive theory and practice of portfolio selection, in *Proceedings of the Latin American Theoretical Informatics* (Latin), 2000.

Browne, S., The return on investment from proportional portfolio strategies. *Adv. Appl. Probabil.*, 1998, **30**, 216–238.

Congdon, P., *Applied Bayesian Modelling*, 2003 (Wiley: New York).

Cover, T.M., *Elements of Information Theory*, 1991a (Wiley: New York); Chapter 15.

Cover, T.M., Universal portfolios. *Math. Finan.*, 1991b, **1**, 1–29.

Cover, T.M. and Ordentlich, E., Universal portfolios with side information. *IEEE Trans. Inf. Theory*, 1996, **42**, 348–363.

Duda, R.O. and Hart, P.E., *Pattern Classification and Scene Analysis*, 1973 (Wiley: New York).

Duda, R.O., Hart, P.E. and Stork, D.G., *Pattern Classification*, 2001 (Wiley-Interscience: New York).

Evstigneev, I.V. and Schenk-Hoppè, K.R., From rags to riches: On constant proportions investment strategies *J. Theor. Appl. Finan.*, 2002, **5**, 563–573.

Gaivoronski, A. and Stella, F., A stochastic nonstationary optimization for finding universal portfolios. *Ann. Oper. Res.*, 2000, **100**, 165–188.

Gaivoronski, A. and Stella, F., On-line portfolio selection using stochastic programming. *J. Econ. Dynam. Contr.*, 2003, **27**, 1013–1014.

Han, J. and Kamber, M.T., *Data Mining: Concepts and Techniques*, 2001 (Morgan-Kaufmann: San Francisco, CA).

Helmbold, D.P., Schapire, R.E., Singer, Y. and Warmuth, M.K., On-line portfolio selection using multiplicative updates, in *International Conference on Machine Learning.*, 1996, pp. 243–251.

Herbster, M. and Warmuth, M., Tracking the best expert. *12th International Conference on Machine Learning.*, 1995, pp. 286–294.

Markowitz, H., Portfolio selection. *J. Finan.*, 1952, **7**, 77–91.

Singer, Y., Switching portfolios. *J. Neural Syst.*, 1997, **8**, 445–455.

Vovk, V. and Watkins, C., Universal portfolio selection, in *11th Annual Conference on Computational Learning Theory*, 1998, pp. 12–23.

Zaffalon, M., The naive credal classifier. *J. Statist. Plann. Infer.*, 2002, **105**, 5–21.

Zaffalon, M. and Fagiuoli, E., Tree-based credal networks for classification. *Reliable Comput.*, 2003, **9**, 487–509.

CHAPTER 6

Improving Performance for Long-Term Investors: Wide Diversification, Leverage and Overlay Strategies

JOHN M. MULVEY, CENK URAL and ZHUOJUAN ZHANG

CONTENTS

6.1 INTRODUCTION

ALTERNATIVE ASSETS, INCLUDING HEDGE FUNDS, private equity and venture capital, have gained in popularity due to the low return of equities since early 2000 and the commensurate search by institutional investors to improve performance in order to regain lost funding surpluses. In theory, alternative assets possess small dependencies with traditional assets such as stocks, bonds and the general level of economic activity—GDP, earnings and interest rates. However, a number of alternative asset categories have shown greater dependencies than originally perceived. We focus on a special class of dynamic (multi-stage) strategies for improving performance in the face of these issues.

To set the stage for our analysis, we describe the historical returns and risks of two prototypical benchmarks for long-term investors in the United States: (1) 70% S&P 500 and 30% long government bonds (70/30); and (2) 60% S&P 500 and 40% government bonds (60/40). These two strategies have proven resilient over the last five decades with reasonable performance during most economic conditions. We conduct our historical evaluation over the recent 12-year period—1 January 1994 to 31 December 2005.[1] This interval is divided into two distinct periods—the first six years (high equity = 23.5% annual returns), and the last six years (low equity = -1.1% annual returns). For the tests, we employ the so-called fixed-mix rule; the portfolio composition is rebalanced to the target mix (either 70/30 or 60/40) at the beginning of each time period (monthly). We will show the benefits of this rule in subsequent sections.

Unfortunately, both the 70/30 and 60/40 strategies greatly under performed their long-term averages in the second six-year period 2000–2005, leading to a massive drop in the surpluses of pension plans and other institutional investors.

To evaluate performance, we employ two standard measures of risks—volatility, and maximum drawdown. Other measures—value at risk, conditional/tail value at risk, and certainty equivalent returns—are closely related to these. Risk-adjusted returns are indicated by Sharpe and return to drawdown ratios. As a significant issue, the under performance of the fixed-mix 70/30 and 60/40 benchmarks as well as similar approaches over the past six years has caused severe difficulties for institutional investors, especially pension trusts. The large drawdown values—28.9% and 22.8%, respectively, during 2000–2005 pinpoint the problems better than volatility—10.4% and 8.9%, respectively, for the two benchmarks (Table 6.1).

How can an investor improve upon these results? First, she might discover assets that provide higher returns than either the S&P 500 or long government bonds. Categories such as real estate investment trusts (REIT) have done just that over the past decade.[2] Investors continue to search for high performing assets.

Once a set of asset categories is chosen, there is a decision regarding the best asset allocation. Much has been written about financial optimization models. Rather than performing an optimization, we can search for novel diversifying assets. In this case, we might accept equal or lower expected returns in exchange for an improved risk profile. To this end, investors interested in achieving wide diversification might turn to assets such as foreign equity, emerging market equity and debt, and so on. Given wide diversification, we can apply leverage to achieve higher returns and lower (or equal) risks than the 70/30 or the 60/40 mix. Mulvey (2005) discusses increasing diversification and associated leverage for improving risk adjusted returns.

Alternatively, a savvy investor might dynamically modify her asset mix as conditions warrant—moving into equity when certain indictors are met or reducing equity exposure when other conditions occur. Such an investor applies more complex decision rules than fixed-mix. Numerous fundamental and technical approaches are employed in this quest. In Section 6.4, we show that many dynamic strategies can be incorporated within the context of a multi-stage stochastic program.

---

[1] Performance results for several alternative asset categories are unavailable or suspect before 1994.
[2] As always, these assets are not guaranteed to provide superior results in the future.

TABLE 6.1 Historical Returns for 70/30 and 60/40 Fixed-Mix Benchmarks and Variants (Large Drawdowns Are Common over the Period)

	S&P 500	100%		70%	60%		
	LB Agg.		100%	30%	40%		
	S&P equal weighted index					70%	60%
	LB 20-year STRIPS index					30%	40%
1994–2005	Geometric return	10.5%	6.8%	9.7%	9.3%	12.5%	12.4%
	Standard deviation	14.8%	4.5%	10.4%	9.1%	11.5%	10.8%
	Sharpe ratio	0.45	0.66	0.56	0.61	0.76	0.80
	Maximum drawdown	44.7%	5.3%	28.9%	22.8%	15.5%	11.6%
	Return/drawdown	0.24	1.29	0.33	0.41	0.81	1.07
1994–1999	Geometric return	23.5%	5.9%	18.2%	16.4%	14.4%	13.5%
	Standard deviation	13.6%	4.0%	10.0%	8.9%	11.2%	10.8%
	Sharpe ratio	1.37	0.24	1.32	1.30	0.85	0.79
	Maximum drawdown	15.4%	5.2%	10.2%	8.5%	11.8%	11.6%
	Return/drawdown	1.53	1.14	1.77	1.92	1.22	1.16
2000–2005	Geometric return	−1.1%	7.7%	1.8%	2.7%	10.6%	11.3%
	Standard deviation	15.2%	5.1%	10.4%	8.9%	11.8%	10.8%
	Sharpe ratio	−0.25	0.99	−0.09	−0.01	0.67	0.79
	Maximum drawdown	44.7%	5.3%	28.9%	22.8%	15.5%	10.4%
	Return/drawdown	−0.03	1.47	0.06	0.12	0.69	1.09

The developed 'overlay' securities/strategies prove beneficial for both fixed-mix and dynamic mix investors. To define an overlay in a simplified setting, we start with a single-period static model. (A multi-stage version appears in the Appendix.) An un-levered, long-only portfolio model allocates the investor's initial capital, C, to a set of assets $\{I\}$ via decision variables $x_i \geq 0$ so as to optimize the investor's random wealth at the horizon:

$$[SP] \qquad \text{Maximize} \quad U(Z_1, Z_2),$$
$$\text{Subject to} \qquad Z_1 \equiv E(\widetilde{w}) \quad \text{and} \quad Z_2 \equiv Risk(\widetilde{w}).$$

$$\sum_{i \in I} x_i = C, \tag{6.1}$$

$$\widetilde{w} = \sum_{i \in I} \widetilde{r}_i x_i. \tag{6.2}$$

The generic utility function $U(\bullet)$ consists of two terms—expected return and a risk function. The latter encompasses most implemented approaches, including volatility, downside, value-at-risk, conditional-value-at-risk and expected von Neumann Morgenstern utility (Bell 1995). Random asset returns are identified as $\widetilde{r}_i$. It is a simple matter to address traditional leverage: we add a borrowing variable $y \geq 0$ and replace Equations (6.1) and (6.2) with

$$\sum_{i \in I} x_i \leq C + y, \tag{6.3}$$

$$\tilde{w} = \sum \tilde{r}_i x_i - \tilde{r}_b y, \qquad (6.4)$$

where the borrowing rate is $\tilde{r}_b$, with perhaps an upper limit on the amount of borrowing $y \leq u_y$. To improve risk-adjusted performance, the borrowing rate needs to be low enough so the optimal dual variable on the $y \geq 0$ constraint equals zero (i.e. the constraint is non-binding).

In contrast to traditional leverage, the overlays (called securities, positions, assets) do not require a capital outlay. For example, two creditworthy investors might establish a forward contract on currencies between themselves. Herein, the net returns—positive or negative—are simply added to the investor's horizon wealth. Under selective conditions, futures markets approximate this possibly favourable environment. In the static model, we expand the decision variables to include the overlays, $x_j \geq 0$ for $j \in J$. The relevant constraints, replacing (6.3) and (6.4), are (6.1) and

$$\tilde{w} = \sum_{i \in I} \tilde{r}_i x_i + \sum_{j \in J} \widetilde{(r_j - 1)} x_j, \qquad (6.5)$$

Importantly, due to the nature of futures markets, the overlay variables can refer to a wide variety of underlying strategies—long-only, short-only, or long–short.[3] Thus, the overlay variables $x_j \geq 0$ for $j \in J$ indicate the presence of a particular futures market contract (long, short or dynamic strategy), and its size.

In this chapter, we evaluate the overlays within a classical trend-following rule (Mulvey *et al.* 2004); alternative rules are worthy of future tests. For risk management purposes, we limit at each time period the designated notional value of the overlays to a small multiple—say $1 \leq m \leq 4$—of investor's capital: $\sum_{j \in J} x_j \leq m * C$. Since capital is not directly allocated for the overlays, the resulting portfolio problem falls into the domain of risk allocation/budgeting. The static portfolio model may be generalized in a manner to multi-stage planning models (Appendix). However, as we will see, some of the standard features of asset performance statistics must be re-evaluated in a multi-stage environment.

6.2   FIXED-MIX PORTFOLIO MODELS AND REBALANCING GAINS

The next three sections take up multi-stage investment models via fixed-mix rules. First, we discuss general issues relating to the fixed-mix rule; then we measure the advantages of the overlays for improving performance within a fixed-mix context. To start, we describe the advantages of fixed-mix over a static, buy-and-hold approach.

The topic of re-balancing gains (also called excess growth or volatility pumping) as derived from the fixed-mix decision rule is well understood from a theoretical perspective. The fundamental solutions were developed by Merton (1969) and Samuelson (1969) for long-term investors. Further work was done by Fernholz and Shay (1982) and Fernholz

---

[3] An overlay asset must include a form of investment strategy, since the investment must be re-evaluated before or at the expiration date of the futures or forward contract.

(2002). Luenberger (1997) presents a clear discussion. We illustrate how rebalancing the portfolio to a fixed-mix creates excess growth. Suppose that the stock price process P_t is lognormal so that it can be represented by the equation

$$dP_t = \alpha P_t dt + \sigma P_t dz_t, \qquad (6.6)$$

where α is the rate of return of P_t and σ^2 is its variance, z_t is Brownian motion with mean 0 and variance t.

The risk-free asset follows the same price process with rate of return equal to r and standard deviation equal to 0. If we represent the price process of risk-free asset by B_t,

$$dB_t = rB_t dt. \qquad (6.7)$$

When we integrate (6.6), the resulting stock price process is

$$P_t = P_0 e^{(\alpha - \sigma^2/2)t + \sigma z_t}. \qquad (6.8)$$

Clearly, the growth rate $\gamma := \alpha - \sigma^2/2$ is the most relevant measure for long-run performance. For simplicity, we assume equality of growth rates across all assets. This assumption is *not* required for generating excess growth, but it makes the illustration easier to understand.

Let us assume that the market consists of n stocks with stock price processes $P_{1,t}, \ldots, P_{n,t}$ each following the lognormal price process. A fixed-mix portfolio has a wealth process W_t that can be represented by the equation

$$\frac{dW_t}{W_t} = \frac{\eta_1 dP_{1,t}}{P_{1,t}} + \cdots + \frac{\eta_n dP_{n,t}}{P_{n,t}}, \qquad (6.9)$$

where $\eta_1, \ldots, \eta_n$ are the fixed weights given to each stock (proportion of capital allocated to each stock) which sum up to one:

$$\sum_{i=1}^{n} \eta_i = 1. \qquad (6.10)$$

The *fixed-mix* strategy in continuous time always applies the same weights to stocks over time. The instantaneous *rate of return* of the fixed-mix portfolio at anytime is the weighted average of the instantaneous rates of returns of the stocks in the portfolio.

In contrast, a buy-and-hold portfolio is one where there is no rebalancing and therefore the number of shares for each stock does not change over time. This portfolio can be represented by the wealth process W_t following

$$dW_t = m_1 dP_{1,t} + \cdots + m_n dP_{n,t}, \qquad (6.11)$$

where $m_1, \ldots, m_n$ depicts the number of shares for each stock.

Again for simplicity, let us assume that there is one stock and a risk-free instrument in the market. This case is sufficient to demonstrate the concept of excess growth in a fixed-mix portfolio as originally presented in Fernholz and Shay (1982). Assume that we invest η portion of our wealth in the stock and the rest $(1-\eta)$ in the risk-free asset. Then the wealth process W_t with these constant weights over time can be expressed as

$$\frac{dW_t}{W_t} = \frac{\eta dP_t}{P_t} + \frac{(1-\eta)dB_t}{B_t}, \tag{6.12}$$

where P_t is the stock price process and B_t is the risk-free asset value.

When we substitute the dynamic equations for P_t and B_t, we get

$$\frac{dW_t}{W_t} = (r + \eta(\alpha - r))dt + \eta\sigma dz_t. \tag{6.13}$$

As before, we assume the growth rate of the stock and the risk-free asset are equal. Hence

$$\alpha - \sigma^2/2 = r. \tag{6.14}$$

From Equation (6.13), we can see that the rate of return of the portfolio, α_ω, is

$$\alpha_w = r + \eta(\alpha - r). \tag{6.15}$$

From (6.14) this rate of return is equal to

$$\alpha_w = r + \eta\sigma^2/2. \tag{6.16}$$

The variance of the resulting portfolio return is

$$\sigma_w^2 = \eta^2\sigma^2. \tag{6.17}$$

Hence the growth rate of the fixed-mix portfolio becomes

$$\gamma_w = \frac{\alpha_w - \sigma_w^2}{2} = r + \frac{(\eta - \eta^2)\sigma^2}{2}. \tag{6.18}$$

This quantity is greater than r for $0 < \eta < 1$. As it is greater than r, which is the growth rate of individual assets, the portfolio growth rate has an excess component, which is $(\eta - \eta^2)\sigma^2/2$. Excess growth is due to rebalancing the portfolio constantly to a fixed-mix. The strategy moves capital out of stock when it performs well and moves capital into stock when it performs poorly. By moving capital between the two assets in the portfolio, a higher growth rate than each individual asset is achievable. See Dempster et al. (2007) for a more general discussion of this phenomenon.

The buy-and-hold investor with equal returning assets lacks the excess growth component. Therefore, buy-and-hold portfolios will under-perform fixed-mix portfolios in various cases. We can easily see from (6.16) that the excess growth component is larger when σ takes a higher value. In this sense, the volatility of an asset is considered not as a risk but rather can be an opportunity to create excess growth in a re-balanced portfolio although of course from (6.17) portfolio return volatility also scales with σ versus σ^2 for rate of return. Surprisingly perhaps, there is no need for *mean reversion* in the stock price processes. Higher performance is obtained through greater volatility in individual assets. Accordingly as we will see in the next section, Sharpe ratios may not provide adequate information about the marginal impact of including an asset category within a fixed-mix, re-balanced portfolio.

6.3 EMPIRICAL RESULTS WITH HISTORICAL DATA AND FIXED-MIX STRATEGIES

In this section, we describe the results of applying several fixed-mix decision rules to data over the 12-year historical period summarized in Table 6.1. The purpose of these empirical tests is to set benchmarks, to find suitable mixes of assets, and to illustrate the advantages of the overlay variables, as compared with solely traditional assets. Of course dynamic decision rules such as the multi-stage stochastic programs discussed in Section 6.5, may be implemented in practice. Here again, the overlays prove to be beneficial for improving risk adjusted returns. In the historical results the portfolio is re-balanced monthly via the fixed-mix rule.

First, we show that re-balancing gains were readily attainable over the turbulent period 1994 to 2005 by deploying assets so as to attain wide diversification and leverage. As described above, we follow the fixed-mix strategy—re-balancing the portfolio at the beginning of each month. The strategies work best when the investor incurs small transaction costs such as for tax-exempt and tax-deferred accounts. Index funds and exchange traded funds present ideal securities since they are highly liquid and can be moved with minimal transaction costs.

Table 6.2 depicts the returns and volatilities for a set of 12 representative asset categories—both traditional and alternative—over the designated 12-year period—1994 to 2005. Annual geometric returns and the two risk measure values are shown. We focus on general asset categories rather than sub-categories, such as, small/medium/large equities, in order to evaluate general benefits. Clearly, further diversification is possible via other investment categories.

Over the period, annual returns range from low = 2.6% (for currencies) to high = 13.1% (for real estate investment trusts, REITs). Many assets display disparate behaviour over the two six-year sub-periods: The Goldman Sachs commodity index (GSCI) and NAREIT had their worst showing during 1994–1999—the lowest returns and highest drawdown values, whereas EAFE and S&P 500 had the opposite results. As a general observation, investors should be ready to encounter sharp drops in individual asset categories. Drawdown for half of the categories lies in the range 26% to 48% (Table 6.2).

Two of the highest historical Sharpe ratios occur in the hedge fund categories: (1) the CSFB hedge fund index (0.87); and (2) the Tremont long/short index (0.78). In both cases, returns are greater than the S&P 500 index with much lower volatility. This performance

TABLE 6.2 Summary Statistics for Asset Categories

	S&P 500	LB Agg.	EAFE	T-bills	NAREIT	GSCI	Hedge Fund index	Man. Fut. index	Currency index	Tremont L/S	S&P EWI	20-year Strips
1994–2005												
Return	10.5%	6.8%	6.8%	3.8%	13.1%	10.0%	10.7%	6.4%	2.6%	11.9%	12.5%	10.6%
Std Dev	14.8%	4.5%	14.5%	0.5%	13.1%	20.1%	7.9%	12.1%	6.8%	10.3%	15.4%	15.5%
Sharpe R.	0.45	0.66	0.20	0.00	0.71	0.31	0.87	0.21	−0.18	0.78	0.56	0.43
M. drawdown	44.7%	5.3%	47.5%	0.0%	26.3%	48.3%	13.8%	17.7%	28.7%	15.0%	30.3%	22.8%
Ret/drawdown	0.24	1.29	0.14	N/A	0.50	0.21	0.77	0.36	0.09	0.79	0.41	0.46
1994–1999												
Return	23.5%	5.9%	12.3%	4.9%	6.5%	4.7%	14.1%	5.5%	0.1%	18.5%	17.1%	7.2%
Std Dev	13.6%	4.0%	13.8%	0.2%	12.0%	17.4%	9.9%	11.5%	6.7%	11.6%	13.7%	14.7%
Sharpe R.	1.37	0.24	0.54	0.00	0.13	−0.01	0.93	0.05	−0.73	1.18	0.89	0.16
M. drawdown	15.4%	5.2%	15.0%	0.0%	26.3%	48.3%	13.8%	17.7%	20.4%	11.4%	19.9%	22.8%
Ret/drawdown	1.53	1.14	0.82	N/A	0.25	0.10	1.02	0.31	0.00	1.62	0.86	0.32
2000–2005												
Return	−1.1%	7.7%	1.5%	2.7%	20.0%	15.7%	7.4%	7.2%	5.1%	5.6%	8.1%	14.0%
Std Dev	15.2%	5.1%	15.1%	0.5%	13.9%	22.5%	5.1%	12.7%	7.0%	8.6%	17.0%	16.3%
Sharpe R.	−0.25	0.99	−0.08	0.00	1.24	0.57	0.92	0.35	0.34	0.34	0.31	0.69
M. drawdown	44.7%	5.3%	47.5%	0.0%	15.3%	35.4%	7.7%	13.9%	15.3%	15.0%	30.3%	19.1%
Ret/drawdown	−0.03	1.47	0.03	N/A	1.31	0.44	0.96	0.52	0.33	0.37	0.27	0.73

S&P 500 = Standard and Poor 500 index, LB Bond = Lehman long aggregate bond index, EAFE = Morgan Stanley equity index for Europe, Australia, and the Far East, NAREIT = National Association of Real Estate Investment Trusts, GSCI = Goldman Sachs commodity index, Hedge Fund Index = Tremont hedge fund aggregate index, Man. Fut. Index = Tremont managed futures index, Tremont L/S = Tremont long/short equity index, S&P 500 equal weighted = Rydex S&P 500 equal weighted index, 20-year Strips = 20-year US government zero coupon bonds.

has led to increasing interest in hedge funds. Many experts believe that the median future returns for hedge funds are likely to be lower than historical values—due in part to the large number of managers entering the arena. And as we will see, in fact low volatility may be a detriment for increasing overall portfolio performance by means of excess growth. Similarly, there are advantages to combining assets with inferior Sharpe ratios and reasonable returns, when these inferior values are caused by higher volatility.

Interestingly, by comparing the capital-weighted S&P 500 index with an equal-weighted S&P 500,[4] we see that the returns for the latter are higher than the former, and with higher volatility. The capital weighed index approximates a buy-and-hold portfolio. Extra volatility improves overall portfolio growth for the equal weighted index (as expected from the theoretical results of the previous section). To a degree, the equal-weighted index achieves rebalancing gains, but also displays a tilt to mid-size over the largest companies in the S&P 500 index.

A similar issue pertains to the long-government bond index versus the 20-year strip index. The strip category has a lower Sharpe ratio (0.43 versus 0.66) due to the extra volatility embedded in the index. Strips are penalized by higher volatility. In contrast, Table 6.1 depicts the superior performance of the equal-weighted equity/strip portfolios over the traditional equity/bond portfolios. For the modified portfolio, not only is the Sharpe ratio higher for the fixed-mix 70/30 portfolio, but the excess returns are higher due to the higher volatility of the portfolio components—12.5% geometric return versus 11.9% for the static portfolio. The modified 70/30 mix, although much better than the traditional 70/30 mix, moderately under performs during 2000–2005 as compared with the earlier period—10.6% versus 14.4%, respectively. The modified 60/40 mix performs better over the second time period, for a slightly more robust result due, in part, to the rebalancing gains obtained from the fixed-mix rule.

What else can be done to increase performance vis-à-vis the 70/30 and 60/40 benchmarks? As a first idea, we might try adding leverage to the benchmarks.[5] While the returns increase with leverage (Table 6.3), the two risk measures also increase so that risk adjusted returns remains modest—Sharpe ratios around 0.55 and return/drawdown around 0.30. Increasing leverage does not improve the situation. The large drawdown values persist during the 2000–2005 period.

As the next idea, we strive to achieve much wider diversification among the asset categories in our portfolio. To this end, we assemble an equally weighted mix (10% each) across 10 asset categories. The resulting fixed-mix portfolio takes a neutral view of any particular asset category, except that we disfavour assets with ultra low volatility (t-bills). The resulting portfolio displays much better performance over the full 12-year period and the two sub-periods (Table 6.4). In particular, the widely diversified portfolio can be levered to achieve 10% to 15% returns with reasonable drawdowns (under 15%). The risk adjusted returns are much better than the previous benchmarks (with or without leverage). Clearly, there are advantages to wide diversification and leverage in a fixed-mix portfolio.

---

[4] Rydex Investments sponsors an exchange traded fund with equal weights on the S&P 500 index. See Mulvey (2005).
[5] We charge t-bill rates here for leverage. Most investors will be required to pay additional fees.

TABLE 6.3 Historical Results of Leverage Applied to the Fixed-Mix 70/30 Asset Mix (Higher Returns Are Possible, but with Higher Volatility)

	Leverage	0%	20%	50%	100%
	S&P 500 equal weighted index	70%	70%	70%	70%
	LB Agg.	30%	30%	30%	30%
1994–2005	Geometric return	9.7%	10.7%	12.2%	14.6%
	Standard deviation	10.4%	12.5%	15.6%	20.8%
	Sharpe ration	0.56	0.55	0.54	0.52
	Maximum drawdown	28.9%	34.8%	42.9%	54.6%
	Return/drawdown	0.33	0.31	0.29	0.27
1994–1999	Geometric return	18.2%	20.9%	24.9%	31.7%
	Standard deviation	10.0%	12.0%	15.0%	20.0%
	Sharpe ration	1.32	1.33	1.33	1.34
	Maximum drawdown	10.2%	12.4%	15.7%	21.1%
	Return/drawdown	1.77	1.68	1.59	1.50
2000–2005	Geometric return	1.8%	1.4%	0.9%	-0.3%
	Standard deviation	10.4%	12.5%	15.7%	21.0%
	Sharpe ration	-0.09	-0.10	-0.12	-0.15
	Maximum drawdown	28.9%	34.8%	42.9%	54.6%
	Return/drawdown	0.06	0.04	0.02	-0.01

TABLE 6.4 Historical Results of Leverage Applied to a Widely Diversified Fixed-Mix Asset Mix (Each Asset Takes 10% Allocation—Excellent Risk-Adjusted Performance)

	Leverage	0%	20%	50%	100%
	LB Agg.	10%	10%	10%	10%
	EAFE	10%	10%	10%	10%
	NAREIT	10%	10%	10%	10%
	GSCI	10%	10%	10%	10%
	Hedge fund index	10%	10%	10%	10%
	CSFB managed futures index	10%	10%	10%	10%
	Currency index	10%	10%	10%	10%
	Tremont long/short	10%	10%	10%	10%
	S&P 500 equal weighted index	10%	10%	10%	10%
	LB 20-year STRIPS index	10%	10%	10%	10%
1994–2005	Geometric return	9.8%	10.9%	12.7%	15.6%
	Standard deviation	6.2%	7.4%	9.3%	12.4%
	Sharpe ration	0.96	0.96	0.95	0.95
	Maximum drawdown	6.4%	8.0%	10.4%	14.4%
	Return/drawdown	1.54	1.37	1.22	1.08
1994–1999	Geometric return	9.6%	10.5%	11.9%	14.1%
	Standard deviation	6.3%	7.5%	9.4%	12.5%
	Sharpe ration	0.75	0.75	0.74	0.73
	Maximum drawdown	6.4%	8.0%	10.4%	14.4%
	Return/drawdown	1.51	1.32	1.14	0.98
2000–2005	Geometric return	9.9%	11.3%	13.5%	17.2%
	Standard deviation	6.2%	7.4%	9.3%	12.4%
	Sharpe ration	1.16	1.16	1.16	1.16
	Maximum drawdown	4.7%	6.3%	8.7%	12.6%
	Return/drawdown	2.10	1.81	1.56	1.35

As the final improvement, we apply three overlay variables (commodities, currencies and fixed income) as defined in Section 6.1. These variables employ trend following rules based on the longstanding Mt. Lucas Management (MLM) index. Mulvey *et al.* (2004) evaluate the index with regard to re-balancing gains and related measures. The MLM index has produced equity-like returns with differential patterns over the past 30 years. However, an important change is made for our analysis. Rather than designating t-bills for the margin capital requirements, we assign the core assets (x_i variables) for the margin capital, consistent with the model defined in Section 6.1 (as would be the case for a multi-strategy hedge fund). Table 6.5 displays the results. Here, we lever the overlay variables at three values—20%, 50% and 100%—within a fixed-mix rule. In all cases, the overlays greatly improve the risk-adjusted returns over the historical period (Sharpe and return to drawdown ratios greater than 1 and 1.5, respectively). (See also Brennan and Schwartz 1998 where continuous dynamic programming is used to optimize a traditional three asset model with a Treasury bond futures overlay.) The performance is positively affected by the relatively high volatility of individual asset categories, increasing portfolio returns via re-balancing gains. The overlay variables with the fixed-mix rule markedly improved performance over the historical period.

TABLE 6.5 Historical Results of Overlay Variables—Fixed-Mix (Overlays Are More Efficient than Simple Leverage)

	LB Agg.	10%	10%	10%	10%
	EAFE	10%	10%	10%	10%
	NAREIT	10%	10%	10%	10%
	GSCI	10%	10%	10%	10%
	Hedge fund index	10%	10%	10%	10%
	CSFB managed futures index	10%	10%	10%	10%
	Currency index	10%	10%	10%	10%
	Tremont long/short	10%	10%	10%	10%
	S&P 500 equal weighted index	10%	10%	10%	10%
	LB 20-year STRIPS index	10%	10%	10%	10%
	Mt. Lucas commodity index	0%	20%	50%	100%
	Mt. Lucas currency index	0%	20%	50%	100%
	Mt. Lucas fixed income index	0%	20%	50%	100%
1994–2005	Geometric return	9.8%	12.4%	16.5%	23.0%
	Standard deviation	6.2%	7.1%	9.7%	15.3%
	Sharpe ration	0.96	1.21	1.30	1.26
	Maximum drawdown	6.4%	5.8%	9.2%	15.0%
	Return/drawdown	1.54	2.13	1.80	1.53
1994–1999	Geometric return	9.6%	12.6%	17.0%	24.3%
	Standard deviation	6.3%	6.7%	8.6%	13.5%
	Sharpe ration	0.75	1.14	1.40	1.43
	Maximum drawdown	6.4%	5.8%	9.2%	14.5%
	Return/drawdown	1.51	2.15	1.85	1.68
2000–2005	Geometric return	9.9%	12.3%	15.9%	21.7%
	Standard deviation	6.2%	7.6%	10.8%	16.9%
	Sharpe ration	1.16	1.26	1.23	1.12
	Maximum drawdown	4.7%	5.4%	8.0%	15.0%
	Return/drawdown	2.10	2.27	2.00	1.45

To summarize, the historical tests illustrate that (6.1) re-balancing gains are possible with assets displaying relatively high volatility within fixed-mix portfolios and that (6.2) including overlay variables and leverage via fixed-mix can result in excellent risk-adjusted performance (almost hedge fund acceptable—23% annual geometric returns). Herein to reduce data mining concerns, we did not change the asset proportions during the period, except to re-balance back to the target mix each month. Also, we did not optimize asset proportions on the historical data, again to minimize data mining. The next section takes up the advantages of applying overlays in multi-period (dynamic mix) optimization models.

6.4   A STOCHASTIC PROGRAMMING PLANNING MODEL

A stochastic program (SP) gives the investor greater opportunities to improve performance as a function of changing economic conditions. These models can be constructed in two basic ways: (1) asset only, or (2) asset and liability management. We focus on asset-only problems in this report.[6] It is generally agreed that the equity risk premium changes over longer time periods. In response, a number of researchers have developed equity valuation models. Bakshi and Chen (2005) designed a 'fair' equity valuation model based on three correlated stochastic processes: interest rates, projected earnings growth and actual earnings. The parameters of these processes are calibrated with market data (mostly historical prices of assets). They showed that future prices of equity assets revert on average to the calculated fair values. This type of analysis can be applied directly to a financial planning model based on a stochastic program.

We highlight here only the major features of a stochastic program. The appendix provides further details. Also, see Mulvey and Thorlacius (1998) and Mulvey et al. (2000). In multi-stage stochastic programs, the evolution of future uncertainties is depicted in terms of a scenario tree. Constructing such a tree requires attention to three critical issues: (1) the realism of the model equations, (2) calibration of the parameters and (3) procedures to extract the sample set of scenarios. The projection system should be evaluated with historical data (back-testing), as well as on an ongoing basis.

Our model employs a scenario generator that has been implemented widely for pension plans and insurance companies—the CAP:Link system (Mulvey et al. 2000). The system develops a close connection between the government spot rate and other economic and monetary factors such as GDP, inflation and currency movements. These connections are described in a series of references including Mulvey and Thorlacius (1998) and Mulvey et al. (1999, 2000). To illustrate, we describe a pair of linked stochastic processes for modelling the long and short interest rates. We assume that the rates link together through a correlated white noise term and by means of a stabilizing term that keeps the long–short spread under control. The resulting spot rates follow

---

[6] See Mulvey et al. (2000) for details of related issues in ALM.

$$dr_t = \kappa_r(\bar{r} - r_t)dt + \gamma_r(s_t - \bar{s})dt + \phi_r r_t^{1/2}dz_r,$$
$$dl_t = \kappa_l(\bar{l} - l_t)dt + \gamma_l(s_t - \bar{s})dt + \phi_l l_t^{1/2}dz_l,$$
$$s_t = r_t - l_t,$$

where r_t and l_t are the short and long interest rate, respectively, s_t is the spread between long and short interest rates, and white noise terms dz_r and dz_l are correlated. The model parameters include:

κ_r, κ_l	drift on short and long interest rates,
γ_r, γ_l	drift on the spread between long and short interest rates,
ϕ_r, ϕ_l	instantaneous volatility,
$\bar{r}$	mean reversion level of short rate,
$\bar{l}$	mean reversion level of long rate,
$\bar{s}$	mean reversion level of the spread between long and short interest rates.

The second step involves parameter estimation. We calibrate parameters as a function of a set of common factors over the multi-period horizon. For example, equity returns and bond returns link to interest rate changes and underlying economic factors. Mulvey et al. (1999) described an integrated parameter estimation approach. Also, Bakshi and Chen (2005) and Chen and Dong (2001) discuss a related approach based on market prices of assets.

Given the scenarios of traditional asset returns derived by means of the CAP:Link system, we obtain return scenarios of the overlay variables by assuming that they are conditionally normally distributed with the traditional asset categories according to historical relationships. This assumption is often employed in scenario generators as a form of a mixture model;[7] see Chatfield and Collins (1980) for a discussion of the general properties.

For our tests, we developed a condensed stochastic program in order to illustrate the issues for long-term investors. To this end, the resulting scenario tree is defined over a nine-year planning period, with three three-year time steps. The resulting problem consists of a modest (by current standards) nonlinear program. In this chapter, we employ a tree with 500 scenarios.[8] The corresponding stochastic program contains 22,507 decision variables and 22,003 linear constraints. On average, it takes 10–20 s to solve for each point on the efficient frontier using a PC. Much larger stochastic programs are readily solvable with modern computers. See Dempster and Consigli (1998), Dempster et al. (2003) and Zenios and Ziemba (2006) for examples.

Assume that an investor can invest her capital in the following core asset categories: treasury bills, S&P 500 index and 20-year zero coupon government bonds (STRIPS).

---

[7] The characteristics of the return series for the overlay strategies will be similar to those of the independent variables, such as mean reversion of interest rates and bond returns or fat tailed distributions.
[8] There are 10 arcs (states) pointing out of the first node, followed by 50 arcs pointing out of the second stage nodes.

TABLE 6.6 Summary Statistics of Scenario Data for Six Asset Classes across a Nine-Year Planning Period

	T-bills	S&P 500	20-year STRIPS	Commodity index	Currency index	Fixed income index
Expected return	3.17%	9.58%	8.13%	6.49%	3.29%	1.18%
Standard deviation	0.016	0.172	0.284	0.106	0.058	0.057
Sharpe ratio	0.000	0.374	0.175	0.313	0.021	−0.347
Correlation matrix						
T-bills	1.000	−0.017	−0.030	0.004	0.009	0.009
S&P 500	−0.017	1.000	−0.008	−0.165	0.110	0.284
20-year STRIPS	−0.030	−0.008	1.000	0.003	−0.024	−0.016
Commodity index	0.004	−0.165	0.003	1.000	0.002	−0.028
Currency index	0.009	0.110	−0.024	0.002	1.000	0.164
Fixed income index	0.009	0.284	−0.016	−0.028	0.164	1.000

In addition, she has the option to add the three previous overlay variables—commodity, currency and fixed income futures indices. As mentioned, in the scenario generator, we applied a conditional mean/covariance approach, based on historical relations between the core assets and the overlay variables. Remember that the overlays do not require any capital outlay. Table 6.6 lists the summary statistics of the generated scenario data for the six asset classes.

Next, under a multi-period framework, we can calculate sample efficient frontiers, with any two objectives, for example, portfolio expected geometric returns and portfolio volatility[9]—Z_1 and Z_2 (Appendix). The first stochastic program forces the investor to invest solely in traditional assets. In contrast, under the latter two stochastic programs, the investor is allowed to add the overlays up to a given bound: 200% of the wealth at any time period in the second case and 300% in the last case. Note that unlike the historical back tests (previous sections), the allocation among overlays is no longer a fixed-mix, but is determined by the recommendation of the optimization and therefore will vary dynamically across time periods and scenarios.

Figure 6.1 displays the illustrative efficient frontier under the described investment constraints.[10] As expected, the solutions possessing the higher overlay bound dominate those under the lower overlay bounds and, as before, the overlay results dominate the traditional strategy. The larger the overlay bound, the greater potential to obtain higher returns at the cost of higher volatilities.

For each efficient frontier, Table 6.7 lists descriptive statistics and asset allocations for the first period of three selected points on each frontier: the maximum return point, the minimum risk point and a compromise solution. Several observations are noteworthy. First, for all points under both strategies, the optimization chooses to invest either exclusively or dominantly in the commodity index for the first period. Second, although

---

[9] We advocate that the investor evaluate a wide range of risk measures. These two are employed for illustration purposes. Mulvey et al. (2007) discuss real-world, multi-objective issues.
[10] This model is a highly simplified version of an ALM system that has been implemented for the U.S. Department of Labor. The goals of the model are to assist pension plans in recovering their lost surpluses by optimizing assets in conjunction with managing liabilities (Mulvey et al. 2007). The unabridged system takes on the multi-objective environment discussed in the Appendix.

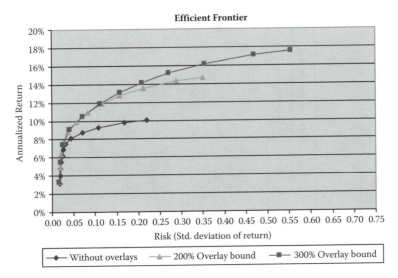

FIGURE 6.1 Efficient frontiers under three investment constraints.

the minimum risk points display the smallest volatility, they tend to have large expected maximum drawdown compared with other points on the efficient frontier. For example, for the minimum point under 300% overlay bound, the volatility is 2% while the expected maximum drawdown is 23%, and for the maximum point, although the volatility is 56%, the expected maximum drawdown is only 6%. Investors should choose a compromise point among the objectives that fits their risk appetite. In many cases, a compromise tends to possess reasonable trade-offs among the risk measures.

As with the historical back tests, the three overlay assets improve investment performance. There are strong advantages to a stochastic program for assisting in financial planning. However, in most cases, a stochastic program is more complex to implement than a simple decision rule such as fixed-mix. Mulvey *et al.* (2007) demonstrated that running a stylized stochastic program can be helpful in discovering novel, improved policy rules.

The results of the stochastic program show that the overlay (trend following approach) for commodities provides the best marginal risk/reward characteristics as compared with the other overlays (trend following for currencies and fixed income). In practice, alternative investment strategies should be considered. For example, Crownover (2006) shows that a combination of strategies (combined with a *z*-score approach) improves performance for currencies. These and related concepts can be readily applied via the discussed models.

6.5   SUMMARY AND FUTURE DIRECTIONS

This chapter shows that risk-adjusted performance can be enhanced by adding specialized overlays to multi-stage portfolios. We improved the returns in both the historical back-tests with the fixed-mix rule and the stochastic programs. In the former case, with wide

TABLE 6.7 Statistics for Selected Points on the Efficient Frontiers

Point on efficient frontier	Without overlays			200% overlay bound			300% overlay bound		
	Max.	Compromise	Min.	Max.	Compromise	Min.	Max.	Compromise	Min.
Statistics									
Exp. annualized geo. ret.	9.16%	7.78%	3.11%	13.19%	11.10%	3.27%	15.25%	13.04%	3.27%
Exp. annualized ret.	10.05%	8.12%	3.17%	14.70%	11.84%	3.33%	17.51%	14.17%	3.33%
Standard deviation	0.221	0.054	0.017	0.351	0.117	0.016	0.555	0.208	0.016
Mean (max drawdown)	0.028	0.000	0.000	0.035	0.016	0.137	0.055	0.038	0.231
Stdev (max drawdown)	0.060	0.000	0.000	0.092	0.050	0.067	0.133	0.087	0.059
Period 1 Alloc.									
T-bills	0.00%	0.00%	100.00%	0.00%	0.00%	100.00%	0.00%	0.00%	100.00%
S&P 500	100.00%	76.98%	0.00%	100.00%	100.00%	0.00%	100.00%	100.00%	0.00%
20-year STRIPS	0.00%	23.02%	0.00%	0.00%	0.00%	0.00%	0.00%	0.00%	0.00%
Commodity index	0.00%	0.00%	0.00%	200.00%	200.00%	199.63%	300.00%	300.00%	268.69%
Currency index	0.00%	0.00%	0.00%	0.00%	0.00%	0.20%	0.00%	0.00%	16.99%
Fixed income index	0.00%	0.00%	0.00%	0.00%	0.00%	0.09%	0.00%	0.00%	9.54%

diversification, the investor benefits by the overlays within a fixed-mix rule. In this context, we demonstrated that long-term investors can take advantage of the overlay's relatively high volatility. As mentioned, improvements are most appropriate for tax-advantaged investors such as pension trusts. Further, the re-balancing gains are available at the fund level (e.g. equal weighted S&P 500), and at the asset allocation level (for fixed-mix). The use of overlays provides significant opportunities when linked with core asset categories requiring capital outlays: (1) there are potential roll returns[11] via the futures markets—which reduce leverage costs, and (2) transaction costs can be often minimized through the liquid futures market. Importantly, the Sharpe ratios, while helpful at the portfolio level, can be misleading regarding the marginal benefit of an asset category within a fixed-mix portfolio context since volatility is penalized.

What are potential implementation barriers? First, the fixed-mix rule requires re-balancing the asset mix at the end of each time period. The investor must take a disciplined approach, even in the face of large swings in asset returns, and be able to invest in a wide range of assets. Also, transaction costs must be considered; Mulvey and Simsek (2002) discuss approaches for addressing transaction costs through no-trade zones. Another possible barrier involves institutional legal constraints. Additionally individual investors may be unable to deploy equity and related assets as margin capital for their futures positions.[12]

In the domain of stochastic programs, we saw that the overlays can be beneficial as well. However, the resulting model grows exponentially as a function of the number of time periods and scenarios. For our simple example, the nine-year model with three-year time periods allows minimal re-balancing. A more realistic stochastic program with a greater number of time periods, while larger, would improve the trade-off between re-balancing gains and the extra returns derived from dynamic asset allocation in the face of changing economic conditions. Current computational power and greater information regarding the economic environment and patterns of asset prices helps overcome this barrier.

A continuing research topic involves the search for assets with novel patterns of return (driven by factors outside the usual triple—interest rates, earnings and the general level of risk premium). An example might involve selling a limited amount of catastrophe insurance for hurricanes and earthquakes, or perhaps, taking on other weather related risks. Another example would be the numerous long–short equity strategies that have become available. While these securities/strategies are not currently treated as asset categories, numerous novel futures-market instruments are under development; several have been recently implemented by exchanges such as the CME and CBOT. Undoubtedly, some of these instruments will help investors improve their risk-adjusted performance by achieving wider diversification in conjunction with selective leverage.

---

[11] Positive roll returns are possible when the futures market is in contango and the investor has a short position, or when backwardation occurs and the investor has a long position in the futures market.

[12] The decision depends upon the arrangement with the investor's prime broker. Swaps are ideal in this regard.

REFERENCES

Bakshi, G.S. and Chen, Z., Stock valuation in dynamic economics. *J. Financ. Mkts*, 2005, **8**, 115–151.

Bell, D., Risk, return, and utility. *Mgmt. Sci*, 1995, **41**, 23–30.

Brennan, M.J. and Schwartz, E.S., The use of treasury bill futures in strategic asset allocation programs. In *Worldwide Asset and Liability Modeling*, edited by W.T. Ziemba and J.M. Mulvey, pp. 205–228, 1998 (Cambridge University Press: Cambridge).

Chatfield, C. and Collins, A.J., *Introduction to Multivariate Analysis*, 1980 (Chapman & Hall: New York).

Chen, Z. and Dong, M., Stock valuation and investment strategies. *Yale ICF Working Paper No. 00-46*, 2001.

Crownover, C., Dynamic capital allocation: Exploiting persistent patterns in currency performance. *Quant. Finance*, 2006, **6**(3), 185–191.

Dempster, M.A.H. and Consigli, G., The CALM stochastic programming model for dynamic asset-liability management. In *Worldwide Asset and Liability Modeling*, edited by W.T. Ziemba and J.M. Mulvey, pp. 464–500, 1998 (Cambridge University Press: Cambridge).

Dempster, M.A.H., Evstigneev, I.V. and Schenk-Hoppé, K.R., Volatility-induced financial growth. *Quant. Finance*, April 2007.

Dempster, M.A.H., Germano, M., Medova, E.A. and Villaverde, M., Global asset liability management. *Br. Actuarial J.*, 2003, **9**, 137–216.

Fernholz, R., *Stochastic Portfolio Theory*, 2002 (Springer-Verlag: New York).

Fernholz, R. and Shay, B., Stochastic portfolio theory and stock market equilibrium. *J. Finance*, 1982, **37**, 615–624.

Luenberger, D., *Investment Science*, 1997 (Oxford University Press: New York).

Merton, R.C., Lifetime portfolio selection under uncertainty: The continuous-time case. *Rev. Econ. Stat.*, 1969, **51**, 247–257.

Mulvey, J.M., Essential portfolio theory. *A Rydex Investment White Paper*, 2005, (also Princeton University report).

Mulvey, J.M., Gould, G. and Morgan, C., An asset and liability management system for Towers Perrin-Tillinghast. *Interfaces*, 2000, **30**, 96–114.

Mulvey, J.M., Kaul, S.S.N. and Simsek, K.D., Evaluating a trend-following commodity index for multi-period asset allocation. *J. Alternative Invest.*, 2004, **Summer**, 54–69.

Mulvey, J.M., Shetty, B. and Rosenbaum, D., Parameter estimation in stochastic scenario generation systems. *Eur. J. Operat. Res.*, 1999, **118**, 563–577.

Mulvey, J.M. and Simsek, K.D., Rebalancing strategies for long-term investors. In *Computational Methods in Decision-Making, Economics and Finance: Optimization Models*, edited by E.J. Kontoghiorghes, B. Rustem and S. Siokos, pp. 15–33, 2002 (Kluwer Academic Publishers: Netherlands).

Mulvey, J.M., Simsek, K.D., Zhang, Z., Fabozzi, F. and Pauling, W., Assisting defined-benefit pension plans. *Operat. Res.*, 2008.

Mulvey, J.M. and Thorlacius, A.E., The Towers Perrin global capital market scenario generation system: CAP Link. In *Worldwide Asset and Liability Modeling*, edited by W.T. Ziemba and J.M. Mulvey, 1998 (Cambridge University Press: Cambridge).

Samuelson, P.A., Lifetime portfolio selection by dynamic stochastic programming. *Rev. Econ. Stat.*, 1969, **51**, 239–246.

Zenios, S.A. and Ziemba, W.T., (Eds), *Handbook of Asset and Liability Modeling*, 2006 (North-Holland: Amsterdam).

APPENDIX 6.A: MATHEMATICAL MODEL FOR THE MULTI-STAGE STOCHASTIC PROGRAM

This appendix defines the asset-only investment problem as a multi-stage stochastic program. We define the set of planning periods as $T = \{0, 1, \ldots, \tau, \tau + 1\}$. We focus on the investor's position at the beginning of period $\tau + 1$. Decisions occur at the beginning of each time period. Under a multi-period framework, we assume that the portfolio is rebalanced at the beginning of each period.

Asset investment categories—assets requiring capital outlays—are defined by set $I = \{1, 2, \ldots, N\}$, with category 1 representing cash. The remaining categories can include broad investment groupings such as growth and value domestic and international stocks, long-term government and corporate bonds, and real estate. The overlay variables are defined by set $J = \{1, 2, \ldots, M\}$. Uncertainty is represented by a set of scenarios $s \in S$. The scenarios may reveal identical values for the uncertain quantities up to a certain period—i.e. they share common information history up to that period. We address the representation of the information structure through non-anticipativity constraints, which require that variables sharing a common history, up to time period t, must be set equal to each other.

For each $i \in I$, $j \in J$, $t \in T$ and $s \in S$, we define the following parameters and decision variables:

Parameters

$r_{i,t,s} = 1 + \rho_{i,t,s}$, where $\rho_{i,t,s}$ is return of traditional asset i in period t, under scenario s (e.g. Mulvey et al. (2000)).

$r_{j,t,s} = 1 + \rho_{j,t,s}$, where $\rho_{j,t,s}$ is return of overlay asset j in period t, under scenario s.

π_s Probability that scenario s occurs - $\sum_{s \in S} \pi_s = 1$.

$x^{\rightarrow}_{i,0,s}$ Amount allocated to traditional asset class i, at the end of period 0, under scenario s, before first rebalancing.

$\sigma_{i,t}$ Transaction costs for rebalancing asset i in period t (symmetric transaction costs are assumed).

B^{OL} Total overlay bound.

G^{TA} Target assets at the horizon.

Decision variables

$x_{i,t,s}$ Amount allocated to traditional asset class i, at the beginning of period t, under scenario s, after rebalancing.

$x^{\rightarrow}_{i,t,s}$ Amount allocated to traditional asset class i at the end of period t, under scenario s, before rebalancing.

$x^{BUY}_{i,t,s}$ Amount of traditional asset class i purchased for rebalancing in period t, under scenario s.

$x^{SELL}_{i,t,s}$ Amount of traditional asset class i sold for rebalancing in period t, under scenario s.

$x_{j,t,s}$ Amount allocated to overlay asset j, at the beginning of period t, under scenario s.

$x^{\mapsto}_{j,t,s}$ Amount allocated to overlay asset j at the end of period t, under scenario s.

$x^{\text{TA}\mapsto}_{t,s}$ Asset wealth at the end of time period t, under scenario s.

Given these definitions, we present the deterministic equivalent of the stochastic asset-only allocation problem:

Model [MSP] (general structure)

$$\text{Maximize} \qquad U\{Z_1, Z_2, \dots, Z_k\}, \qquad (6.A1)$$

where the goals are defined as functions of the decision variables (examples of various goals are shown below):

$$Z_k = f_k(x),$$

subject to:

$$\sum_{i \in I} x^{\mapsto}_{i,0,s} = x^{\text{TA}\mapsto}_{0,s} \quad \forall\, s \in S, \qquad (6.A2)$$

$$\sum_{i \in I} x^{\mapsto}_{i,t,s} + \sum_{i \in J} x^{\mapsto}_{j,t,s} = x^{\text{TA}\mapsto}_{t,s} \quad \forall\, s \in S,\ t = 1, \dots, \tau+1, \qquad (6.A3)$$

$$x^{\mapsto}_{i,t,s} = r_{i,t,s} x_{i,t,s} \quad \forall\, s \in S,\ t = 1, \dots, \tau,\ i \in I, \qquad (6.A4)$$

$$x^{\mapsto}_{j,t,s} = (r_{j,t,s} - 1) x_{j,t,s} \quad \forall\, s \in S,\ t = 1, \dots, \tau,\ j \in J, \qquad (6.A5)$$

$$\sum_{j \in J} x_{j,t,s} \leq B^{\text{OL}} x^{\text{TA}\mapsto}_{t-1,s} \quad \forall\, s \in S,\ t = 1, \dots, \tau \qquad (6.A6)$$

$$x_{i,t,s} = x^{\mapsto}_{i,t-1,s} + x^{\text{BUY}}_{i,t-1,s}(1 - \sigma_{i,t-1}) - x^{\text{SELL}}_{i,t-1,s} \quad \forall\, s \in S,\ i \neq 1,\ t = 1, \dots, \tau+1, \qquad (6.A7)$$

$$x_{1,t,s} = x^{\mapsto}_{1,t-1,s} + \sum_{i \neq 1} x^{\text{SELL}}_{i,t-1,s}(1 - \sigma_{i,t-1}) - \sum_{i \neq 1} x^{\text{BUY}}_{i,t-1,s}$$
$$- b_{t-1,s} + y^{\text{CONT}}_{t-1,s} \quad \forall\, s \in S,\ t = 1, \dots, \tau+1, \qquad (6.A8)$$

$$x_{i,t,s_1} = x_{i,t,s_2}, x_{j,t,s_1}$$

$$= x_{j,t,s_2} \quad \forall \; s_1 \text{ and } s_2 \text{ with identical past up to time } t, \qquad (6.A9)$$

$$Risk\{Z_1, Z_2, \ldots, Z_k\} \leq Risk_{max}. \qquad (6.A10)$$

The objective function (6.A1) depicts a generic multi-objective optimization problem. It can take several forms. For instance, we could employ the von Neumann–Morgenstern utility function of the final wealth. Alternatively, we could use the classical return-risk function $Z = \eta \cdot Mean(x_\tau^{TA\mapsto}) + (1 - \eta) \cdot Risk(x_\tau^{TA\mapsto})$, where $Mean(x_\tau^{TA\mapsto})$ is the expected total assets at the end of τ and $Risk(x_\tau^{TA\mapsto})$ is a risk measure of the total final wealth across all scenarios. The weight parameter η indicates the relative importance of risk as compared with expected wealth.

Constraint (6.A2) represents the initial total value of assets at the end of period 0. Constraint (6.A3) depicts wealth at the end of period t, aggregating assets in traditional asset classes and investment gains/losses from overlay strategies. The wealth accumulated at the end of period t before rebalancing in traditional asset class i is given by (6.A4). The wealth accumulation due to overlay variable j at the end of period t is depicted in (6.A5). Constraint (6.A6) sets the bound for overlays for each time period and across all scenarios. The flow balance constraints for all traditional asset classes except cash, for all periods, are given by (6.A7). (6.A8) represents the flow balance constraint for cash. Non-anticipativity constraints are represented by (6.A9), ensuring that the scenarios with the same past will have identical decisions up to that period.

Risk-based constraints appear in (6.A10). Here we list a few popular goals among numerous others. Especially, we set G^{TA} to be the target wealth for the investor at $\tau + 1$. The first goal is to maximize the expected final investor wealth at the horizon:

$$Z_1 = \sum_s \pi_s x_{\tau,s}^{TA\mapsto}.$$

Both the second and the third goals quantify the risk of missing the target wealth at the planning horizon. Goal 2 is the downside risk of the expected final wealth:

$$Z_2 = \sum_s \pi_s \left[\left(Z_1 - x_{\tau,s}^{TA\mapsto} \right)^+ \right]^2.$$

A similar goal is the downside risk of the expected final investor wealth with respect to target wealth G^{TA} at the horizon:

$$Z_3 = \sum_s \pi_s \left[\left(G^{TA} - x_{\tau,s}^{TA\mapsto} \right)^+ \right]^2.$$

Goal 3 is zero if and only if final wealth reaches the target under all scenarios.

The fourth and fifth goals focus on the timing of achieving the target wealth. Goal 4 measures the expected earliest time for the investor's assets to reach the target:

$$Z_4 = \sum_s \pi_s \inf\left\{t : x_{t,s}^{\text{TA}\mapsto} \geq G^{\text{TA}}\right\}.$$

The goal Z_4 could be greater than $\tau+1$ if the wealth could not reach the target before the planning horizon.

Aiming at measuring the risk of missing the target date of reaching the goal, we propose the fifth goal, the downside risk of the time to achieve the goal:

$$Z_5 = \sum_s \pi_s \left[\left(\inf\left\{t : x_{t,s}^{\text{TA}\mapsto} \geq G^{\text{TA}}\right\} - \tau\right)^+\right]^2.$$

The model could be readily modified to incorporate liability-related decisions and other investment strategies. For instance, the fixed-mix rule enforces the following constraint at each juncture:

$$\lambda_i = \frac{x_{i,t,s}}{x_{t,s}^{\text{TA}}}, \quad \text{for any time period t and under any scenario } s,$$

where $x_{t,s}$ is the total wealth at the beginning of period t and we define the proportion of wealth to be λ_i for each asset $i \in I$. Ideally, we would maintain the target proportion λ at all time periods and under every scenario. Rebalancing under a fixed-mix rule automatically 'buys low and sells high.' However, the fixed-mix constraints induce non-convexity into the stochastic program. Specially designed algorithms are needed to solve such a problem.

CHAPTER 7

Stochastic Programming for Funding Mortgage Pools

GERD INFANGER

CONTENTS

7.1   INTRODUCTION

H ISTORICALLY, THE BUSINESS OF CONDUITS, like Freddie Mac, Fannie Mae or Ginnie Mae, has been to purchase mortgages from primary lenders, pool these mortgages into mortgage pools, and securitize some if not all of the pools by selling the resulting Participation Certificates (PCs) to Wall Street. Conduits keep a fixed markup on the interest for their profit and roll over most of the (interest rate and prepayment) risk to the PC buyers. Recently, a more active approach, with the potential for significantly higher profits, has become increasingly attractive: instead of securitizing, funding the purchase of mortgage pools by issuing debt. The conduit firm raises the money for the mortgage purchases through a suitable combination of long- and short-term debt. Thereby, the conduit assumes a higher level of risk due to interest rate changes and prepayment risk but gains higher expected revenues due to the larger spread between the interest on debt and mortgage rates compared with the fixed markup by securitizing the pool.

The problem faced by the conduits is an asset-liability management problem, where the assets are the mortgages bought from primary lenders and the liabilities are the bonds issued. Asset liability problems usually are faced by pension funds and insurance companies. Besides assets, pension plans need to consider retirement obligations, which may depend on uncertain economic and institutional variables, and insurance companies need to consider uncertain pay-out obligations due to unforseen and often catastrophic events. Asset liability models are most useful when both asset returns and liability pay-outs are driven by common, e.g. economic, factors. Often, the underlying stochastic processes and decision models are multi-dimensional and require multiple state variables for their representation. Using stochastic dynamic programming, based on Bellman's (1957) dynamic programming principle, for solving such problems is therefore computationally difficult, well known as the 'curse of dimensionality.' If the number of state variables of the problem is small, stochastic dynamic programming can be applied efficiently. Infanger (2006) discusses a stochastic dynamic programming approach for determining optimal dynamic asset allocation strategies over an investment horizon with many re-balancing periods, where the value-to-go function is approximated via Monte Carlo sampling. The chapter uses an in-sample/out-of-sample approach to avoid optimization bias.

Stochastic programming can take into account directly the joint stochastic processes of asset and liability cash flows. Traditional stochastic programming uses scenario trees to represent possible future events. The trees may be constructed by a variety of scenario-generation techniques. The emphasis is on keeping the resulting tree thin but representative of the event distribution and on arriving at a computationally tractable problem, where obtaining a good first-stage solution rather than obtaining an entire accurate policy is the goal. Early practical applications of stochastic programming for asset liability management are reported in Kusy and Ziemba (1986) for a bank and in Carino *et al.* (1994) for an insurance company. Ziemba (2003) gives a summary of the stochastic programming approach for asset liability and wealth management. Early applications of stochastic programming for asset allocation are discussed in Mulvey and Vladimirou (1992), formulating financial networks, and Golub *et al.* (1995). Examples of early applications of stochastic programming for dynamic fixed-income strategies are Zenios (1993), discussing the management of mortgage-backed securities, Hiller and Eckstein (1993), and Nielsen and Zenios (1996). Wallace and Ziemba (2005) present recent applications of stochastic programming, including financial applications. Frauendorfer and Schüerle (2005) discuss the re-financing of mortgages in Switzerland.

Monte Carlo sampling is an efficient approach for representing multi-dimensional distributions. An approach, referred to as decomposition and Monte Carlo sampling, uses Monte Carlo (importance) sampling within a decomposition for estimating Benders cut coefficients and right-hand sides. This approach has been developed by Dantzig and Glynn (1990) and Infanger (1992). The success of the sampling within the decomposition approach depends on the type of serial dependency of the stochastic parameter processes, determining whether or not cuts can be shared or adjusted between different scenarios of a stage. Infanger (1994) and Infanger and Morton (1996) show that, for serial correlation of stochastic parameters (in the form of autoregressive processes), unless the correlation is limited to the right-hand side of the (linear) program, cut sharing is at best difficult for more than three-stage problems.

Monte Carlo pre-sampling uses Monte Carlo sampling to generate a tree, much like the scenario-generation methods referred to above, and then employs a suitable method for solving the sampled (and thus approximate) problem. We use Monte Carlo pre-sampling for representing the mortgage funding problem, and combine optimization and simulation techniques to obtain an accurate and tractable model. We also provide an efficient way to independently evaluate the solution strategy from solving the multi-stage stochastic program to obtain a valid upper bound on the objective. The pre-sampling approach provides a general framework of modeling and solving stochastic processes with serial dependency and many state variables; however, it is limited in the number of decision stages. Assuming a reasonable sample size for representing a decision tree, problems with up to four decision stages are meaningfully tractable. Dempster and Thorlacius (1998) discuss the stochastic simulation of economic variables and related asset returns. A recent review of scenario-generation methods for stochastic programming is given by Di Domenica *et al.* (2006), discussing also simulation for stochastic programming scenario generation.

In this chapter we present how multi-stage stochastic programming can be used for determining the best funding of a pool of similar fixed-rate mortgages through issuing bonds, callable and non-callable, of various maturities. We show that significant profits can be obtained using multi-stage stochastic programming compared with using a single-stage model formulation and compared with using duration and convexity hedging, strategies often used in traditional finance. For the comparison we use an implementation of Freddie Mac's interest rate model and prepayment function. We describe in Section 7.2 the basic formulation of funding mortgage pools and discuss the estimation of expected net present value and risk for different funding instruments using Monte Carlo sampling techniques. In Section 7.3 we discuss the single-stage model. In Section 7.4 we present the multi-stage model. Section 7.5 discusses duration and convexity and delta and gamma hedging. In Section 7.6 we discuss numerical results using practical data obtained from Freddie Mac. We compare the efficient frontiers from the single-stage and multi-stage models, discuss the different funding strategies and compare them with delta and gamma hedged strategies, and evaluate the different strategies using out-of-sample simulations. In particular, Section 7.6.5 presents the details of the out-of-sample evaluation of the solution strategy obtained from solving a multi-stage stochastic program. Section 7.7 reports on the solution of very large models and gives model sizes and solution times. Finally, Section 7.8 summarizes the results of the chapter.

While not explicitly discussed in this chapter, the problem of what fraction of the mortgage pool should be securitized, and what portion should be retained and funded through issuing debt can be addressed through a minor extension of the models presented. Funding decisions for a particular pool are not independent of all other pools already in the portfolio and those to be acquired in the future. The approach can of course be extended to address also the funding of a number of pools with different characteristics. While the chapter focuses on funding a pool of fixed-rate mortgages, the framework applies analogously to funding pools of adjustable-rate mortgages.

7.2   FUNDING MORTGAGE POOLS

7.2.1   Interest Rate Term Structure

Well-known interest rate term structure models in the literature are Vasicek (1977), Cox *et al.* (1985), Ho and Lee (1986), and Hull and White (1990), based on one factor, and Longstaff and Schwarz (1992) based on two factors.

Observations of the distributions of future interest rates are obtained using an implementation of the interest rate model of Luytjes (1993) and its update according to the Freddie Mac document. The model reflects a stochastic process based on equilibrium theory using random shocks for short rate, spread (between the short rate and the ten-year rate) and inflation.

We do not use the inflation part of the model and treat it as a two-factor model, where the short rate and the spread are used to define the yield curve. To generate a possible interest rate path we feed the model at each period with realizations of two standard normal random variables and obtain as output for each period a possible outcome of a yield curve of interest rates based on the particular realizations of the random shocks. Given a realization of the short rate and the spread, the new yield curve is constructed free of arbitrage for all calculated yield points.

We denote as $i_t(m)$, $t = 1, \ldots, T$, the random interest rate of a zero coupon bond of term m in period t.

7.2.2 The Cash Flows of a Mortgage Pool

We consider all payments of a pool of fixed-rate mortgages during its lifetime. Time periods t range from $t = 0, \ldots, T$, where T denotes the end of the horizon; e.g. $T = 360$ reflects a horizon of 30 years considering monthly payments. We let B_t be the balance of the principal of the pool at the end of period t. The principal capital B_0 is given to the homeowners at time period $t = 0$ and is regained through payments β_t and through prepayments α_t at periods $t = 1, \ldots, T$. The balance of the principal is updated periodically by

$$B_t = B_{t-1}(1 + \kappa_0) - \beta_t - \alpha_t, \quad t = 1, \ldots, T.$$

The rate κ_0 is the contracted interest rate of the fixed-rate mortgage at time $t = 0$. We define λ_t to be the payment factor at period $t = 1, \ldots, T$. The payment factor when multiplied by the mortgage balance yields the constant monthly payments necessary to pay off the loan over its remaining life, e.g.

$$\lambda_t = \kappa_0 / (1 - (1 + \kappa_0)^{t-T-1});$$

thus,

$$\beta_t = \lambda_t B_{t-1}.$$

The payment factor λ_t depends on the interest rate κ_0. For fixed-rate mortgages the quantity κ_0, and thus the quantities λ_t, are known with certainty. However, prepayments α_t, at periods $t = 1, \ldots, T$, depend on future interest rates and are therefore random parameters.

Prepayment models or functions represent the relationship between interest rates and prepayments. See, for example, Kang and Zenios (1992) for a detailed discussion of prepayment models and factors driving prepayments.

In order to determine α_t we use an implementation of Freddie Mac's prepayment function according to Lekkas and Luytjes (1993). Denoting the prepayment rates obtained from the prepayment function as γ_t, $t = 1, \ldots, T$, we compute the prepayments α_t in period t as

$$\alpha_t = \gamma_t B_{t-1}.$$

7.2.3 Funding through Issuing Debt

We consider funding through issuing bonds, callable and non-callable, with various maturities. Let ℓ be a bond with maturity m_ℓ, $\ell \in L$, where L denotes the set of bonds under consideration. Let $f_{\ell t}^\tau$ be the payment factor for period t, corresponding to a bond issued at period τ, $\tau \leq t \leq \tau + m_\ell$:

$$f_{\ell t}^{\tau} \equiv \begin{cases} +1, & \text{if } t - \tau = 0, \\ -(i_{\tau}(m_{\ell}) + s_{\tau\ell}), & \text{if } 0 < t - \tau < m_{\ell}, \\ -(1 + i_{\tau}(m_{\ell}) + s_{\tau\ell}), & \text{if } t - \tau = m_{\ell}, \end{cases}$$

where i_{τ} (m_{ℓ}) reflects the interest rate of a zero coupon bond with maturity m_{ℓ}, issued at period τ, and $s_{\tau\ell}$ denotes the spread between the zero coupon rate and the actual rate of bond l issued at τ. The spread $s_{\tau\ell}$ includes the spread of bullet bonds over zero coupon bonds (referred to as agency spread) and the spread of callable bonds over bullet bonds (referred to as agency call spread), and is computed according to the model specification given in the Freddie Mac document (Luytjes 1996).

Let M_{τ}^{ℓ} denote the balance of a bond ℓ at the time τ it is issued. The finance payments resulting from bond ℓ are

$$d_t^{\ell} = f_{\ell t}^{\tau} M_{\tau}^{\ell}, \quad t = \tau, \ldots, \tau + m_{\ell},$$

from the time of issue (τ) until the time it matures $(\tau + m_l)$ or, if callable, it is called. We consider the balance of the bullet from the time of issue until the time of maturity as

$$M_t^{\ell} = M_{\tau}^{\ell}, \quad t = \tau, \ldots, \tau + m_{\ell}.$$

7.2.4   Leverage Ratio

Regulations require that, at any time t, $t = 0, \ldots, T$, equity is set aside against debt in an amount such that the ratio of the difference of all assets minus all liabilities to all assets is greater than or equal to a given value μ. Let E_t be the balance of an equity (cash) account associated with the funding. The equity constraint requires that

$$\frac{B_t + E_t - M_t}{B_t + E_t} \geq \mu,$$

where the total asset balance is the sum of the mortgage balance and the equity balance, $B_t + E_t$, and $M_t = \sum_{\ell} M_t^{\ell}$ is the total liability balance.

At time periods $t = 0, \ldots, T$, given the mortgage balance B_t, and the liability balance M_t, we compute the equity balance that fulfills the leverage ratio constraint with equality as

$$E_t = \frac{M_t - B_t(1 - \mu)}{1 - \mu}, \quad t = 0, \ldots, T.$$

We assume that the equity account accrues interest according to the short rate i_t(short), the interest rate of a 3-month zero coupon bond. Thus, we have the following balance equation for the equity account:

$$E_t = E_{t-1}(1 + i_{t-1}(\text{short})) + e_{t-1},$$

where e_t are payments into the equity account (positive) or payments out of the equity account (negative). Using this equation we compute the payments e_t to and from the equity account necessary to maintain the equity balance E_t computed for holding the leverage ratio μ.

7.2.5 Simulation

Using the above specification we may perform a (Monte Carlo) simulation in order to obtain an observation of all cash flows resulting from the mortgage pool and from financing the pool through various bonds. In order to determine in advance how the funding is carried out, we need to specify certain decision rules defining what to do when a bond matures, when to call a callable bond, at what level to fund, and how to manage profits and losses. For the experiment we employed the following six rules.

(i) Initial funding is obtained at the level of the initial balance of the mortgage pool, $M_0 = B_0$.

(ii) Since at time $t = 0$, $M_0 = B_0$, it follows that $E_0 = [\mu/(1 - \mu)]B_0$, an amount that we assume to be an endowed initial equity balance.

(iii) When a bond matures, refunding is carried out using short-term debt (non-callable 3-month bullet bond) until the end of the planning horizon, each time at the level of the balance of the mortgage pool.

(iv) Callable bonds are called according to the call rule specification in Freddie Mac's document (Luytjes 1996). Upon calling, refunding is carried out using short-term debt until the end of the planning horizon, each time at the level of the balance of the mortgage pool.

(v) The leverage ratio (ratio of the difference of all assets minus all liabilities to all assets) is $\mu = 0.025$.

(vi) At each time period t, after maintaining the leverage ratio, we consider a positive sum of all payments as profits and a negative sum as losses.

According to the decision rules, when funding a mortgage pool using a single bond ℓ, we assume at time $t = 0$ that $M_0^\ell = B_0$, i.e. that exactly the amount of the initial mortgage balance is funded using bond ℓ. After bond matures refunding takes place using another bond (according to the decision rules, short-term debt, say, bond $\hat{\ell}$), based on the interest rate and the level of the mortgage balance at the time it is issued. If the initial bond ℓ is callable, it may be called, and then funding carried out through another bond (say, short-term debt $\hat{\ell}$). Financing based on bond $\hat{\ell}$ is continued until the end of the planning horizon, i.e. until $T - \tau < m_{\hat{\ell}}$, and no more bond is issued. Given the type of bond being used for refunding, and given an appropriate calling rule, all finance payments for the initial funding using bond ℓ and the subsequent refunding using bond $\hat{\ell}$ can be determined. We denote the finance payments accruing from the initial funding based on bond ℓ and its consequent refunding based on bond $\hat{\ell}$ as

$$d_t^\ell, \quad t = 1, \ldots, T.$$

Once the funding and the corresponding liability balance M_t^ℓ is determined, the required equity balance $E_t = E_t^\ell$ and the payments $e_t = e_t^\ell$ are computed.

7.2.6 The Net Present Value of the Payment Stream

Finally, we define as

$$P_t^\ell = \beta_t + \alpha_t + d_t^\ell - e_t^\ell, \quad t = 1, \ldots, T, \; P_0 = M_0 - B_0 = 0$$

the sum of all payments in period t, $t = 0, \ldots, T$, resulting from funding a pool of mortgages (initially) using bond ℓ.

Let I_t be the discount factor for period t, i.e.

$$I_t = \prod_{k=1}^{t} (1 + i_k(\text{short})), \quad t = 1, \ldots, T, \; I_0 = 1,$$

where we use the short rate at time t, $i_t(\text{short})$, for discounting. The net present value (NPV) of the payment stream is then calculated as

$$r_\ell = \sum_{t=0}^{T} \frac{P_t^\ell}{I_t}.$$

So far, we consider all quantities that depend on interest rates as random parameters. In particular, P_t^ℓ is a random parameter, since β_t, α_t, d_t^ℓ, and e_t^ℓ are random parameters depending on random interest rates. Therefore, the net present value r_ℓ is a random parameter as well. In order to simplify the notation we do not label any specific outcomes of the random parameters. A particular run of the interest rate model requires $2T$ random outcomes of unit normal random shocks. *We now label a particular path of the interest rates obtained from one run of the interest rate model and all corresponding quantities with ω.* In particular, we label a realization of the net present value based on a particular interest rate path as r_ℓ^ω.

7.2.7 Estimating the Expected NPV of the Payment Stream

We use Monte Carlo sampling to estimate the expected value of the NPV of a payment stream. Under a crude Monte Carlo approach to the NPV estimation, we sample N paths $\omega \in S$, $N = |S|$, using different observations of the distributions of the $2T$ random parameters as input to the interest rate model, and we compute r_ℓ^ω for each $\omega \in S$. Then, an estimate for the expected net present value (NPV) of the cash flow stream based on initial funding using bond ℓ is

$$\bar{r}_\ell = \frac{1}{N} \sum_{\omega \in S} r_\ell^\omega.$$

We do not describe in this document how we use advanced variance reduction techniques (e.g. importance sampling) for the estimation of the expected net present value of a payment stream. We refer to Prindiville (1996) for how importance sampling could be applied.

7.2.8 The Expected NPV of a Funding Mix

Using simulation (as described above) we compute the net present value of the payment stream r_ℓ^ω for each realization $\omega \in S$ and each possible initial funding $\ell \in L$. The net present value of a funding mix is given by the corresponding convex combination of the net present values of the components $\ell \in L$, i.e.

$$r^\omega = \sum_{\ell \in L} r_\ell^\omega x_\ell, \quad \sum_{\ell \in L} x_\ell = 1, \quad x_l \geq 0,$$

where x_ℓ are non-negative weights summing to one. The expected net present value of a funding mix,

$$\bar{r} = \frac{1}{N} \sum_{\omega \in S} r^\omega,$$

is also represented as the convex combination of the expected net present values of the components $\ell \in L$, i.e.

$$\bar{r} = \sum_{\ell \in L} \bar{r}_\ell x_\ell, \quad \sum_{\ell \in L} x_\ell = 1, \quad x_\ell \geq 0.$$

7.2.9 Risk of a Funding Mix

In order to measure risk, we use as an appropriate asymmetric penalty function the negative part of the deviation of the NPV of a funding portfolio from a pre-specified target u, i.e.

$$v^\omega = \left(\sum_{\ell \in L} r_\ell^\omega x_\ell - u \right)^-,$$

and consider risk as the expected value of v^ω, estimated as

$$\bar{v} = \frac{1}{N} \sum_{\omega \in S} v^\omega.$$

A detailed discussion of this particular risk measure is given in Infanger (1996). The efficient frontier with risk as the first lower partial moment is also referred to as the 'put–call efficient frontier;' see, for example, Dembo and Mausser (2000).

7.3   SINGLE-STAGE STOCHASTIC PROGRAMMING

Having computed the NPVs r_ℓ^ω for all initial funding options $\ell \in L$ and for all paths $\omega \in S$, we optimize the funding mix with respect to expected returns and risk by solving the (stochastic) linear program

$$\min \quad \frac{1}{N} \sum v^\omega = \bar{v},$$

$$\text{s.t.} \quad \sum_\ell r_\ell^\omega x_\ell + v^\omega \geq u, \quad \omega \in S,$$

$$\sum_\ell \bar{r}_\ell x_\ell \geq \rho,$$

$$\sum_\ell x_\ell = 1, \quad x_\ell \geq 0, \quad v^\omega \geq 0.$$

The parameter ρ is a pre-specified value that the expected net present value of the portfolio should exceed or be equal to. Clearly, $\rho \leq \rho^{\max} = \max_\ell \{\bar{r}_\ell\}$. Using the model we trace out an efficient frontier starting with $\rho = \rho^{\max}$ and successively reducing ρ until $\rho = 0$, each time solving the linear program to obtain the portfolio with the minimum risk $\bar{v}$ corresponding to a given value of ρ.

The single-stage stochastic programming model optimizes funding strategies based on decision rules defined over the entire planning horizon of $T = 360$ periods, where the net present value of each funding strategy using initially bond ℓ and applying the decision rules is estimated using simulation.

A variant of the model arises by trading off expected NPV and risk in the objective, with λ denoting the risk-aversion coefficient:

$$\min \quad -\sum_\ell \bar{r}_\ell x_\ell + \lambda \frac{1}{N} \sum v^\omega,$$

$$\text{s.t.} \quad \sum_\ell r_\ell^\omega x_\ell + v^\omega \geq u, \quad \omega \in S,$$

$$\sum_\ell x_\ell = 1, \quad x_\ell \geq 0, \quad v^\omega \geq 0.$$

For a risk aversion of $\lambda = 0$, risk is not part of the objective and expected NPV is maximized. The efficient frontier can be traced out by increasing the risk aversion λ successively from zero to very large values, where the risk term in the objective entirely dominates.

This approach is very different to Markowitz's (1952) mean variance analysis in that the distribution of the NPV is represented through scenarios (obtained through simulations over a long time horizon, considering the application of decision rules) and a downside risk measure is used for representing risk.

7.4 MULTI-STAGE STOCHASTIC PROGRAMMING

In the following we relax the application of decision rules at certain decision points within the planning horizon, and optimize the funding decisions at these points. This leads to a multi-stage stochastic programming formulation.

We partition the planning horizon $\langle 0, T \rangle$ into n sub-horizons $\langle T_1, T_2 \rangle, \langle T_2, T_3 \rangle, \ldots,$ $\langle T_n, T_{n+1} \rangle$, where $T_1 = 0$, and $T_{n+1} = T$. For the experiment, we consider $n = 4$, and partition at $T_1 = 0$, $T_2 = 12$, $T_3 = 60$, and $T_4 = 360$. We label the decision points at time $t = T_1$ as stage 1, at time $t = T_2$ as stage 2, and at time $t = T_3$ as stage 3 decisions. Funding obtained at the decision stages is labeled as $\ell_1 \in L_1$, $\ell_2 \in L_2$, and $\ell_3 \in L_3$ according to the decision stages. At time $t = T_4$, at the end of the planning horizon, the (stage 4) decision involves merely evaluating the net present value of each end point for calculating the expected NPV and risk. In between the explicit decision points, at which funding is subject to optimization, we apply the decision rules defined above.

Instead of interest rate paths as used in the single-stage model, we now use an interest rate tree with nodes at each stage. We consider $|S_2|$ paths $\omega_2 \in S_2$ between $t = T_1$ and $t = T_2$; for each node $\omega_2 \in S_2$ we consider $|S_3|$ paths $\omega_3 \in S_3$ between $t = T_2$ and $t = T_3$; for each node $(\omega_2, \omega_3) \in \{S_2 \times S_3\}$ we consider $|S_4|$ paths $\omega_4 \in S_4$ between $t = T_3$ and $t = T_4$. Thus, the tree has $|S_2 \times S_3 \times S_4|$ end points. We may denote $S = \{S_2 \times S_3 \times S_4\}$ and $\omega = (\omega_2, \omega_3, \omega_4)$. Thus a particular path through the tree is now labeled as $\omega = (\omega_2, \omega_3, \omega_4)$ using an index for each partition. Figure 7.1 presents the decision tree of the multi-stage model for only two paths for each period.

The simulation runs for each partition of the planning horizon are carried out in such a way that the dynamics of the interest rate process and the prepayment function are fully carried forward from one partition to the next. Since the interest rate model and the

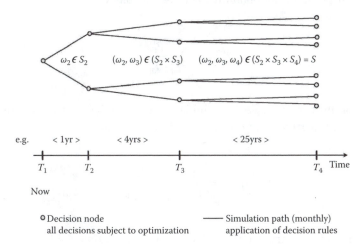

FIGURE 7.1 Multi-stage model setup, decision tree.

prepayment function include many lagged terms and require the storing of 64 state variables, the application of dynamic programming for solving the multi-stage program is not tractable.

Let $I_{\tau t}$ be the discount factor of period t, discounted to period τ, i.e.

$$I_{\tau t} = \prod_{k=\tau+1}^{t} (1 + i_k(\text{short})), \quad t > \tau, \ I_{\tau\tau} = 1,$$

where we use the short rate at time t, $i_t(\text{short})$, for discounting.

Let L_1 be the set of funding instruments available at time T_1. Funding obtained at time T_1 may mature or be called during the first partition (i.e. before or at time T_2), during the second partition (i.e. after time T_2 and before or at time T_3), or during the third partition (i.e. after time T_3 and before or at time T_4). We denote the set of funding instruments issued at time T_1 and matured or called during the first partition of the planning horizon as $L_{11}^{\omega_2}$, the set of funding instruments issued at time T_1 and matured or called during the second partition of the planning horizon as $L_{12}^{\omega_2\omega_3}$, and the set of funding instruments issued at time T_1 and matured or called during the third partition of the planning horizon as $L_{13}^{\omega_2\omega_3\omega_4}$. Clearly, $L_1 = L_{11}^{\omega_2} \cup L_{12}^{\omega_2\omega_3} \cup L_{13}^{\omega_2\omega_3\omega_4}$, for each $(\omega_2, \omega_3, \omega_4) \in \{S_2 \times S_3 \times S_4\}$. Similarly, we denote the set of funding instruments issued at time T_2 and matured or called during the second partition of the planning horizon as $L_{22}^{\omega_2\omega_3}$, and the set of funding instruments issued at time T_2 and matured or called during the third partition of the planning horizon as $L_{23}^{\omega_2\omega_3\omega_4}$. Clearly, $L_2 = L_{22}^{\omega_2\omega_3} \cup L_{23}^{\omega_2\omega_3\omega_4}$, for each $(\omega_2, \omega_3, \omega_4) \in \{S_2 \times S_3 \times S_4\}$. Finally, we denote the set of funding instruments issued at time T_3 and matured or called during the third partition of the planning horizon as L_{33}. Clearly, $L_3 = L_{33} = L_{33}^{\omega_2\omega_3\omega_4}$, for each $(\omega_2, \omega_3, \omega_4) \in \{S_2 \times S_3 \times S_4\}$.

For all funding instruments $\ell_1 \in L_1^{\omega_2}$ initiated at time $t = 0$ that mature or are called during the first partition, we obtain the net present values

$$r_{\ell_1(11)}^{\omega_2} = \sum_{t=0}^{T_2} \frac{P_t^{\ell_1\omega_2}}{I_{0t}^{\omega_2}};$$

for all funding instruments $\ell_1 \in L_{12}^{\omega_2\omega_3}$ initiated at time $t = 0$ that mature or are called during the second partition, we obtain the net present values

$$r_{\ell_1(12)}^{\omega_2,\omega_3} = \frac{1}{I_{0T_2}^{\omega_2}} \sum_{t=T_2+1}^{T_3} \frac{P_t^{\ell_1\omega_3}}{I_{T_2 t}^{\omega_3}};$$

and all initial funding instruments $\ell_1 \in L_{13}^{\omega_2\omega_3\omega_4}$, initiated at time $t = 0$ that mature or are called during the third partition, we obtain the net present values

$$r^{\omega_2,\omega_3,\omega_4}_{\ell_1(13)} = \frac{1}{I^{\omega_2}_{0T_2}} \frac{1}{I^{\omega_3}_{T_2T_3}} \sum_{t=T_3+1}^{T_4} \frac{P^{\ell_1\omega_4}_t}{I^{\omega_4}_{T_3t}}.$$

For all funding instruments $\ell_2 \in L^{\omega_2\omega_3}_{22}$, initiated at time $t = T_2$, that mature or are called during the second partition, we obtain the net present values

$$r^{\omega_2,\omega_3}_{\ell_2(22)} = \frac{1}{I^{\omega_2}_{0T_2}} \sum_{t=T_2+1}^{T_3} \frac{P^{\ell_2\omega_3}_t}{I^{\omega_3}_{T_2t}},$$

and for all funding instruments $\ell_2 \in L^{\omega_2\omega_3}_{23}$, initiated at time $t = T_2$, that mature or are called during the third partition, we obtain the net present values

$$r^{\omega_2,\omega_3,\omega_4}_{\ell_2(23)} = \frac{1}{I^{\omega_2}_{0T_2}} \frac{1}{I^{\omega_3}_{T_2T_3}} \sum_{t=T_3+1}^{T_4} \frac{P^{\ell_2\omega_4}_t}{I^{\omega_4}_{T_3t}}.$$

We obtain for all initial funding $\ell \in L_{33}$ initiated at time $t = T_3$ the net present values

$$r^{\omega_2,\omega_3,\omega_4}_{\ell_3(33)} = \frac{1}{I^{\omega_2}_{0T_2}} \frac{1}{I^{\omega_3}_{T_2T_3}} \sum_{t=T_3+1}^{T_4} \frac{P^{\ell_3\omega_4}_t}{I^{\omega_4}_{T_3t}}.$$

Let $N = |S|$. Let $R^{\omega}_{\ell_1} = r^{\omega_2}_{\ell_1(11)} + r^{\omega_2\omega_3}_{\ell_1(12)} + r^{\omega_2\omega_3\omega_4}_{\ell_1(13)}$, $R^{\omega}_{\ell_2} = r^{\omega_2\omega_3}_{\ell_2(22)} + r^{\omega_2\omega_3\omega_4}_{\ell_2(23)}$, and $R^{\omega}_{\ell_3} = r^{\omega_2\omega_3\omega_4}_{\ell_3(33)}$. Let x_{ℓ_1} be the amount of funding in instrument $\ell_1 \in L_1$ issued at time $t = T_1$, x_{ℓ_2} be the amount of funding in instrument $\ell_2 \in L_2$ issued at time $t = T_2$, and x_{ℓ_3} be the amount of funding in instrument $\ell_3 \in L_3$ issued at time $t = T_3$. Based on the computation of the net present values, we optimize the funding mix solving the multi-stage (stochastic) linear program:

$$\min \quad E\, v^{\omega} = \bar{v},$$

$$\text{s.t.} \quad \sum_{\ell_1 \in L_1} x_{\ell_1} = 1,$$

$$-\sum_{\ell_1 \in L^{\omega_2}_{11}} x_{\ell_1} + \sum_{\ell_2 \in L_2} x^{\omega_2}_{\ell_2} = 0,$$

$$-\sum_{\ell_1 \in L^{\omega_2\omega_3}_{12}} x_{\ell_1} - \sum_{\ell_2 \in L^{\omega_2\omega_3}_{22}} x^{\omega_2}_{\ell_2} + \sum_{\ell_3 \in L_3} x^{\omega_2\omega_3}_{\ell_3} = 0,$$

$$\sum_{\ell_1 \in L_1} R^{\omega}_{\ell_1} x_{\ell_1} + \sum_{\ell_2 \in L_2} R^{\omega}_{\ell_2} x^{\omega_2}_{\ell_2} + \sum_{\ell_3 \in L_3} R^{\omega}_{\ell_3} x^{\omega_2\omega_3}_{\ell_3} - w^{\omega} = 0,$$

$$v^{\omega} + w^{\omega} \geq u, \quad E\, w^{\omega} \geq \rho, \quad x_{\ell_1}, x^{\omega_2}_{\ell_2}, x^{\omega_2\omega_3}_{\ell_3}, \quad v^{\omega} \geq 0,$$

where $E\, w^{\omega} = (1/N) \sum w^{\omega}$ is the estimate of the expected net present value and $E\, v^{\omega} = (1/N) \sum v^{\omega}$ is the estimate of the risk. As in the single-stage model before, the parameter ρ is a pre-specified value for the expected net present value of the portfolio. Starting with $\rho = \rho^{\max}$, the maximum value of ρ that can be assumed without the linear program becoming infeasible, we trace out an efficient frontier by successively reducing ρ from $\rho = \rho^{\max}$ to $\rho = 0$ and computing for each level of ρ the corresponding value of risk $\bar{v}$ by solving the multi-stage stochastic linear program. The quantity ρ, the maximum expected net present value without considering risk, can be obtained by solving the linear program

$$\max E\, w^{\omega} = \rho^{\max},$$

$$\text{s.t.} \quad \sum_{\ell_1 \in L_1} x_{\ell_1} = 1,$$

$$-\sum_{\ell_1 \in L_{11}^{\omega_2}} x_{\ell_1} + \sum_{\ell_2 \in L_2} x_{\ell_2}^{\omega_2} = 0,$$

$$-\sum_{\ell_1 \in L_{12}^{\omega_2 \omega_3}} x_{\ell_1} - \sum_{\ell_2 \in L_{22}^{\omega_2 \omega_3}} x_{\ell_2}^{\omega_2} + \sum_{\ell_3 \in L_3} x_{\ell_3}^{\omega_2 \omega_3} = 0,$$

$$\sum_{\ell_1 \in L_1} R_{\ell_1}^{\omega} x_{\ell_1} + \sum_{\ell_2 \in L_2} R_{\ell_2}^{\omega} x_{\ell_2}^{\omega_2} + \sum_{\ell_3 \in L_3} R_{\ell_3}^{\omega} x_{\ell_3}^{\omega_2 \omega_3} - w^{\omega} = 0, \quad x_{\ell_1}, x_{\ell_2}^{\omega_2}, x_{\ell_3}^{\omega_2 \omega_3} \geq 0.$$

Note that the model formulation presented above does not consider the calling of callable bonds as subject to optimization at the decision stages; rather the calling of callable bonds is handled through the calling rule as part of the simulation. Optimizing also the calling of callable bonds at the decision stages requires only a minor extension to the model formulation, but this is not discussed here.

A variant of the multi-stage model arises by trading off expected net present value and risk in the objective with λ as the risk-aversion coefficient:

$$\min \lambda E\, v^{\omega} - E\, w^{\omega},$$

$$\text{s.t.} \quad \sum_{\ell_1 \in L_1} x_{\ell_1} = 1,$$

$$-\sum_{\ell_1 \in L_{11}^{\omega_2}} x_{\ell_1} + \sum_{\ell_2 \in L_2} x_{\ell_2}^{\omega_2} = 0,$$

$$-\sum_{\ell_1 \in L_{12}^{\omega_2 \omega_3}} x_{\ell_1} - \sum_{\ell_2 \in L_{22}^{\omega_2 \omega_3}} x_{\ell_2}^{\omega_2} + \sum_{\ell_3 \in L_3} x_{\ell_3}^{\omega_2 \omega_3} = 0,$$

$$\sum_{\ell_1 \in L_1} R_{\ell_1}^{\omega} x_{\ell_1} + \sum_{\ell_2 \in L_2} R_{\ell_2}^{\omega} x_{\ell_2}^{\omega_2} + \sum_{\ell_3 \in L_3} R_{\ell_3}^{\omega} x_{\ell_3}^{\omega_2 \omega_3} - w^{\omega} = 0,$$

$$v^{\omega} + w^{\omega} \geq u, \quad x_{\ell_1}, x_{\ell_2}^{\omega_2}, x_{\ell_3}^{\omega_2 \omega_3}, v^{\omega} \geq 0.$$

7.5 DURATION AND CONVEXITY

Since the payments from a mortgage pool are not constant, indeed the prepayments depend on the interest rate term structure and its history since the inception of the pool, an important issue arises as to how the net present value (price) of the mortgage pool changes as a result of a small change in interest rates. The same issue arises for all funding instruments, namely, how bond prices change as a result of small changes in yield. This is especially interesting in the case of callable bonds. In order to calculate the changes in expected net present value due to changes in interest rates, one usually resorts to first- and second-order approximations, where the first-order (linear, or delta) approximation is called the duration and the second-order (quadratic, or gamma) approximation is called the convexity. While the duration and convexity of non-callable bonds could be calculated analytically, the duration and convexity of a mortgage pool and callable bonds can only be estimated through simulation. We use the terms effective duration and effective convexity to refer to magnitudes estimated through simulation.

7.5.1 Effective Duration and Convexity

Let

$$p = \sum_{t=0}^{T} \frac{P_t^{\text{pool}}}{I_t}$$

be the net present value of the payments from the mortgage pool, where $P_t^{\text{pool}} = \alpha_t + \beta_t$ and I_t is the discount factor using the short rate for discounting. We compute p^{ω}, for scenarios $\omega \in S$, using Monte Carlo simulation, and we calculate the expected net present value (price) of the payments of the mortgage pool as

$$\bar{p} = \frac{1}{N} \sum_{\omega \in S} p^{\omega}.$$

Note that the payments $P_t^{\text{pool}} = P_t^{\text{pool}}(i_k, \ k = 0, \ldots, t)$ depend on the interest rate term structure and its history up to period t, where i_t denotes the vector of interest rates for different maturities at time t. Writing explicitly the dependency,

$$\bar{p} = \bar{p}(i_t, \ t = 0, \ldots, T).$$

We now define

$$\bar{p}_+ = \bar{p}(i_t + \Delta, \ t = 0, \ldots, T)$$

as the net present value of the payments of the mortgage pool for an upward shift of all interest rates by $\Delta\%$, and

$$\bar{p}_- = \bar{p}(i_t - \Delta, \ t = 0, \ldots, T)$$

as the net present value of the payments of the mortgage pool for a downward shift of all interest rates by $\Delta\%$, where Δ is a shift of, say, one percentage point in the entire term structure at all periods $t = 1, \ldots, T$.

Using the three points $\bar{p}_-$, $\bar{p}$, $\bar{p}_+$, and the corresponding interest rate shifts $-\Delta$, 0, $+\Delta$, we compute the effective duration of the mortgage pool, as

$$\text{dur} = \frac{\bar{p}_- - \bar{p}_+}{2\Delta\bar{p}},$$

and the effective convexity of the mortgage pool,

$$\text{con} = \frac{\bar{p}_- + \bar{p}_+ - 2\bar{p}}{100\Delta^2\bar{p}}.$$

The quantities of the effective duration and effective convexity represent a local first-order (duration) and second-order (duration and convexity) Taylor approximation of the net present value of the payments of the mortgage pool as a function of interest. The approximation considers the effects on a constant shift of the entire yield curve across all points t, $t = 1, \ldots, T$. The number 100 in the denominator of the convexity simply scales the resulting numbers. The way it is computed, we expect a positive value for the duration, meaning that decreasing interest rates result in a larger expected net present value and increasing interest rates result in a smaller expected net present value. We also expect a negative value for the convexity, meaning that the function of price versus yield is locally concave.

In an analogous fashion we compute the duration dur_ℓ and the convexity con_ℓ for all funding instruments ℓ. Let

$$p_\ell = \sum_{t=0}^{m_\ell} \frac{d_t^\ell}{I_t}$$

be the net present value of the payments of the bond ℓ, where d_t^ℓ represents the payments of bond ℓ until maturity or until it is called. We compute p_ℓ^ω using Monte Carlo simulation over $\omega \in S$, and we calculate the expected net present value (price) of the payments for the bond ℓ as

$$\bar{p}_\ell = \frac{1}{N} \sum_{\omega \in S} p_\ell^\omega.$$

We calculate

$$\bar{p}_{\ell+} = \bar{p}_\ell(i_t + \Delta, \; t = 0, \ldots, T),$$

and

$$\bar{p}_{\ell-} = \bar{p}_\ell(i_t - \Delta, \; t = 0, \ldots, T),$$

the expected net present values for an upwards and downwards shift of interest rates, respectively. Analogously to the mortgage pool, we obtain the duration of bond ℓ as

$$\text{dur}_\ell = \frac{\bar{p}_{\ell-} - \bar{p}_{\ell+}}{2\Delta\bar{p}_\ell},$$

and its convexity as

$$\text{con}_\ell = \frac{\bar{p}_{\ell-} + \bar{p}_{\ell+} - 2\bar{p}_\ell}{100\Delta^2\bar{p}_\ell}.$$

Since in the case of non-callable bonds the payments d_t^ℓ are fixed, changes in expected net present value due to changes in interest rates are influenced by the discount factor only. For non-callable bonds we expect a positive value for duration, meaning that increasing interest rates imply a smaller bond value and decreasing interest rates imply a larger bond value. We expect a positive value for convexity, meaning that the function of expected net present value versus interest rates is locally convex. In the case of callable bonds the behavior of the function of expected net present value versus interest rates is influenced not only by the discount rate but also by the calling rule. If interest rates decrease, the bond may be called and the principal returned. The behavior of callable bonds is similar to that of mortgage pools in that we expect a positive value for duration and a negative value for convexity.

7.5.2   Traditional Finance: Matching Duration and Convexity

Applying methods of traditional finance, one would hedge interest rate risk by constructing a portfolio with a duration and a convexity of close to zero, respectively, thus achieving that the portfolio would exhibit no change in expected net present value (price) due to a small shift in the entire yield curve. Duration and convexity matching is also referred to as immunization (see, for example, Luenberger (1998) or as delta and gamma hedging).

In the situation of funding mortgage pools, hedging is carried out in such a way that a change in the price of the mortgage pool is closely matched by the negative change in the price of the funding portfolio, such that the change of the total portfolio (mortgage pool and funding) is close to zero. Thus, the duration of the total portfolio is close to zero. In addition, the convexity of the mortgage pool is matched by the (negative) convexity of the funding portfolio, such that the convexity of the total portfolio (mortgage pool and funding) is close to zero. We write the corresponding duration and convexity hedging model as

$$\max \sum_\ell r_\ell x_\ell$$

$$\sum_\ell \text{dur}_\ell x_\ell - \text{dg} = \text{dur},$$

$$\sum_\ell \text{con}_\ell x_\ell - \text{cg} = \text{con},$$

$$\sum_\ell x_\ell = 1, \quad x_\ell \geq 0,$$

and

$$-dg^{max} \leq dg \leq dg^{max}, \quad -cg^{max} \leq cg \leq cg^{max},$$

and expected net present value is maximized in the objective. The variable dg accounts for the duration gap, cg accounts for the convexity gap, dg^{max} represents a predefined upper bound on the absolute value of the duration gap, and cg^{max} a predefined upper bound on the absolute value of the convexity gap. Usually, when a duration and convexity hedged strategy is implemented, the model needs to be revised over time as the mortgage pool and the yield curve changes. In practice, updating the funding portfolio may be done on a daily or monthly basis to reflect changes in the mortgage pool due to prepayments and interest rate variations.

The duration and convexity hedging model is a deterministic model, uncertainty is considered as a shift of the entire yield curve, and hedged to the extent of the effect of the remaining duration and convexity gap.

7.5.3 Duration and Convexity in the Single-Stage Model

In the single-stage case we add the duration and convexity constraints

$$\sum_{\ell} dur_{\ell}x_{\ell} - dg = dur,$$

$$\sum_{\ell} con_{\ell}x_{\ell} - cg = con,$$

and

$$-dg^{max} \leq dg \leq dg^{max}, \quad -cg^{max} \leq cg \leq cg$$

to the single-stage linear program using the formulation in which expected net present value and risk are traded off in the objective. Setting the risk aversion λ to zero, only the expected net present value is considered in the objective, and the resulting single-stage stochastic program is identical to the duration and convexity hedging formulation from traditional finance as discussed in the previous section. This formulation allows one to constrain the absolute value of the duration and convexity gap to any specified level, to the extent that the single-stage stochastic program remains feasible. By varying the duration and convexity gap, we may study the effect of the resulting funding strategy on expected net present value and risk.

7.5.4 Duration and Convexity in the Multi-Stage Model

In the multi-stage model we wish to constrain the duration and convexity gap not only in the first stage, but at any decision stage and in any scenario. Thus, in the four-stage model discussed above we have one pair of constraint for the first stage, $|S_2|$ pairs of constraints in the second stage, and $|S_2 \times S_3|$ pairs of constraints in the third stage. Accordingly, we need to compute the duration and the convexity for the mortgage pool and all funding instruments at any decision point in all stages from one to three.

In order to simplify the presentation, we omit the scenario indices $\omega_2 \in S_2$ and $(\omega_2, \omega_3) \in \{S_2 \times S_3\}$. Let dur_t and con_t be the duration and convexity of the mortgage pool in each stage t, and dur_{ℓ_t} and con_{ℓ_t} be the duration and convexity of the funding instruments issued at each period t.

Let dur_t^F and con_t^F be symbols used to conveniently present the duration and convexity of the funding portfolio at each period t. In the first stage, $\mathrm{dur}_1^F = \sum_{\ell \in L_1} \mathrm{dur}_\ell x_{\ell_1}$. In the second stage, $\mathrm{dur}_2^F = \sum_{\ell_2 \in L_2} \mathrm{dur}_{\ell_2} x_{\ell_2} + \sum_{\ell_1 \in L_{12}} \mathrm{dur}_{\ell_1(12)} x_{\ell_1}$, where $\mathrm{dur}_{\ell_1(12)}$ represents the duration of the funding instruments from the first stage that are still available in the second stage. In the third stage, $\mathrm{dur}_3^F = \sum_{\ell_3 \in L_3} \mathrm{dur}_{\ell_3} x_{\ell_3} + \sum_{\ell_2 \in L_{23}} \mathrm{dur}_{\ell_2(23)} x_{\ell_2} + \sum_{\ell_1 \in L_{13}} \mathrm{dur}_{\ell_1(13)} x_{\ell_1}$, where $\mathrm{dur}_{\ell_2(23)}$ represents the duration of the funding instruments issued in the second stage still available in the third stage, and $\mathrm{dur}_{\ell_1(13)}$ represents the duration of the funding instruments from the first stage still available in the third stage.

Analogously, in the first stage, $\mathrm{con}_1^F = \sum_{\ell \in L_1} \mathrm{con}_{\ell_1} x_{\ell_1}$. In the second stage, $\mathrm{con}_2^F = \sum_{\ell_2 \in L_2} \mathrm{con}_{\ell_2} x_{\ell_2} + \sum_{\ell_1 \in L_{12}} \mathrm{con}_{\ell_1(12)} x_{\ell_1}$, where $\mathrm{con}_{\ell_1(12)}$ is the convexity of the funding instruments from the first stage still available in the second stage. In the third stage, $\mathrm{con}_3^F = \sum_{\ell_3 \in L_3} \mathrm{con}_{\ell_3} x_{\ell_3} + \sum_{\ell_2 \in L_{23}} \mathrm{con}_{\ell_2(23)} x_{\ell_2} + \sum_{\ell_1 \in L_{13}} \mathrm{con}_{\ell_1(13)} x_{\ell_1}$, where $\mathrm{con}_{\ell_2(23)}$ is the convexity of the funding instruments issued in the second stage still available in the third stage, and $\mathrm{con}_{\ell_1(13)}$ is the convexity of the funding instruments from the first stage still available in the third stage.

Thus, in the multi-stage case, we add in each stage $t = 1, \ldots, 3$ and in each scenario $\omega_2 \in S_2$ and $(\omega_2, \omega_3) \in \{S_2 \times S_3\}$ the constraints

$$\mathrm{dur}_t^F - \mathrm{dg}_t = \mathrm{dur}_t,$$
$$\mathrm{con}_t^F - \mathrm{cg}_t = \mathrm{con}_t,$$

and

$$-\mathrm{dg}_t^{\max} \le \mathrm{dg}_t \le \mathrm{dg}_t^{\max}, \quad -\mathrm{cg}_t^{\max} \le \mathrm{cg}_t \le \mathrm{cg}_t^{\max}.$$

The variables dg_t, cg_t account for the duration and convexity gap, respectively, in each decisions stage and in each of the scenarios $\omega_2 \in S_2$ and $(\omega_2, \omega_3) \in \{S_2 \times S_3\}$. At each decision node the absolute values of the duration and convexity gap are constrained by dg_t and cg_t, respectively. This formulation allows one to constrain the duration and convexity gap to any specified level, even at different levels at each stage, to the extent that the multi-stage stochastic linear program remains feasible.

Using the formulation of the multi-stage model, in which expected net present value and risk are traded off in the objective, and setting the risk aversion coefficient λ to zero, we obtain a model in which the duration and convexity are constrained at each node of the scenario tree and expected net present value is maximized. Since the scenario tree represents simulations of possible events of the future, the model results in a duration and convexity hedged funding strategy, where duration and convexity are constrained at the decision points of the scenario tree, but not constrained at points in between, where the funding portfolios are carried forward using the decision rules. We use this as an approximation for other duration and convexity hedged strategies, in which duration and convexity are matched, for example, at every month during the planning horizon. One could, in addition, apply funding rules in the simulation that result in duration and convexity hedged portfolios at every month, but such rules have not been applied in the computations used for this chapter.

We are now in the position to compare the results from the multi-stage stochastic model trading off expected net present value and downside risk with the deterministic duration and convexity hedged model on the same scenario tree. The comparison looks at expected net present value and risk (represented by various measures), as well as underlying funding strategies.

7.6 COMPUTATIONAL RESULTS

7.6.1 Data Assumptions

For the experiment we used three data sets, based on different initial yield curves, labeled 'Normal,' 'Flat' and 'Steep.' The data represent assumptions about the initial yield curve, the parameters of the interest rate model, and the prepayment function, assumptions about the funding instruments, assumptions about refinancing and the calling rule, and the planning horizon and its partitioning.

Table 7.1 presents the initial yield curve (corresponding to a zero coupon bond) for each data set. For each data set the mortgage contract rate is assumed to be one percentage point above the 10-year rate (labeled 'y10').

For the experiment we consider 16 different funding instruments. Table 7.2 presents the maturity, the time after which the instrument may be called, and the initial spread over the corresponding zero coupon bond (of the same maturity) for each instrument and for each of the data sets. For example, 'y03nc1' refers to a callable bond with a maturity of 3 years (36 months) and callable after 1 year (12 months). Corresponding to the data set 'Normal,' it could be issued initially (at time $t = 0$) at a rate of

TABLE 7.1 Initial Yield Curves

		Interest rate (%)		
Label	Maturity (months)	Normal	Flat	Steep
m03	3	5.18	5.88	2.97
m06	6	5.31	6.38	3.16
y01	12	5.55	6.76	3.36
y02	24	5.97	7.06	4.18
y03	36	6.12	7.36	4.58
y05	60	6.33	7.56	5.56
y07	84	6.43	7.59	5.97
y10	120	6.59	7.64	6.36
y30	360	6.87	7.76	7.20

TABLE 7.2 Spreads for Different Funding Instruments

Label	Maturity (months)	Callable after (months)	Spread (%)		
			Normal	Flat	Steep
m03n	3		0.17	0.15	0.19
m06n	6		0.13	0.10	0.13
y01n	12		0.02	0.08	0.08
y02n	24		0.00	0.11	0.1
y03n	36		0.04	0.18	0.12
y03nc1	36	12	0.36	0.61	0.22
y05n	60		0.08	0.21	0.13
y05nc1	60	12	0.65	0.84	0.22
y05nc3	60	36	0.32	0.41	0.18
y07n	84		0.17	0.22	0.15
y07nc1	84	12	0.90	0.95	0.45
y07nc3	84	36	0.60	0.73	0.35
y10n	120		0.22	0.29	0.22
y10nc1	120	12	1.10	1.28	0.57
y10nc3	120	36	0.87	0.94	0.48
y30n	360		0.30	0.32	0.27

$6.12\% + 0.36\% = 6.48\%$, where 6.12% is the interest rate from Table 7.1 and 0.36% is the spread from Table 7.2.

We computed the results for a pool of \$100M. As the target for risk, u, we used the maximum expected net present value obtained using single-stage optimization, i.e. we considered risk as the expected net present value below this target, defined for each data set. In particular, the target for risk equals $u = 10.5M$ for the 'Normal' data set, $u = 11.3M$ for the 'Flat' data set and $u = 18.6M$ for the 'Steep' data set.

We first used single-stage optimization using $N = 300$ interest rate paths. These results are not presented here. Then, in order to more accurately compare single-stage and multi-stage optimization, we used a tree with $N = 4000$ paths, where the sample sizes in each stage are $|S_2| = 10$, $|S_3| = 20$ and $|S_4| = 20$. For this tree the multi-stage linear program has 8213 rows, 11 377 columns and 218 512 non-zero elements. The program can easily be solved on a modern personal computer in a very small amount of (elapsed) time. Also, the simulation runs to obtain the coefficients for the linear program can easily be carried out on a modern personal computer.

7.6.2 Results for the Single-Stage Model

As a base case for the experiment we computed the efficient frontier for each of the data sets using the single-stage model. Figure 7.2 presents the result for the 'Normal' data set in comparison with the efficient frontiers obtained from the multi-stage model. (The single-stage results obtained from optimizing on the tree closely resemble the efficient frontiers obtained from optimizing on 300 interest rate paths.) The efficient frontiers for the 'Flat' and 'Steep' data sets are similar in shape to that of the 'Normal' data set. While 'Normal' and 'Flat' have the typical shape one would expect, i.e. steep at low levels of risk and

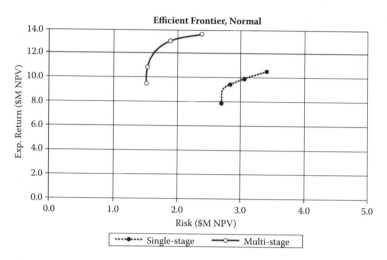

FIGURE 7.2 Efficient frontier for the Normal data set, single- versus multi-stage.

TABLE 7.3 Funding Strategies from the Single-Stage Model, Case 95% of Maximum Expected Return, for All Three Data Sets

Model	Normal		Flat		Steep
Stage 1			Allocation		
	j1m06n	j1y07n	j1m03n	j1y07n	j1m03n
	0.853	0.147	0.911	0.089	1.000
Stage 2			Allocation		
Scenario	j2m03n		j2m03n		j2m03n
1	0.853		0.911		1.000
2	0.853		0.911		1.000
3	0.853		0.911		1.000
4	0.853		0.911		1.000
5	0.853		0.911		1.000
6	0.853		0.911		1.000
7	0.853		0.911		1.000
8	0.853		0.911		1.000
9	0.853		0.911		1.000
10	0.853		0.911		1.000

bending more flat with increasing risk, it is interesting to note that the efficient frontier for data set 'Steep' is very steep at all levels of risk.

As a base case we look at the optimal funding strategy at the point of 95% of the maximum expected net present value. In the graphs of the efficient frontiers this is the second point down from the point of the maximum expected net present value point.

The funding strategies for all three data sets are presented in Table 7.3. The label 'j1' in front of the funding acronyms means that the instrument is issued in stage 1, and the label 'j2' refers to the instrument's issuance in stage 2. For example, a funding instrument called 'j1y07n' is a non-callable bond with a maturity of 7 years issued at stage 1. While the single-stage model naturally does not have a second stage, variables with the prefix 'j2' for the second stage represent the amount of each instrument to be held at stage 2 according to the decision rules. This is to facilitate easy comparisons between the single-stage and multi-stage funding strategies.

In the 'Normal' case, the optimal initial funding mix consists of 85.3% six-month non-callable debt and 14.7% seven-year non-callable debt. The risk level of this strategy is 3.1M NPV. In the 'Flat' case, the optimal initial funding mix consists of 91.1% three-month non-callable debt and 8.9% seven-year non-callable debt. The risk level of this strategy is 3.1M NPV. In the 'Steep' case, the initial funding mix consists of 100.0% three-month non-callable debt (short-term debt). The risk level of this strategy is 2.6M NPV.

7.6.3   Results for the Multi-Stage Model

Figure 7.2 presents the efficient frontier for the 'Normal' data set. In addition to the multi-stage efficient frontier, the graph also contains the corresponding single-stage efficient frontier for better comparison. The results for the 'Flat' and 'Steep' data sets are very similar in shape to that of the 'Normal' data set and therefore are not presented graphically. The results show substantial differences in the risk and expected net present value profile of multi-stage versus single-stage funding strategies. For any of the three data sets, the efficient frontier obtained from the multi-stage model is significantly north-west of that from the single-stage model, i.e. multi-stage optimization yields a larger expected net present value at the same or smaller level of risk.

In all three data cases, we cannot compare the expected net present values from the single-stage and multi-stage model at the same level of risk, because the risk at the minimum risk point of the single-stage model is larger than the risk at the maximum risk point of the multi-stage model. In the 'Normal' case, the minimum risk of the single-stage curve is about 2.7M NPV. Since the efficient frontier is very steep at low levels of risk, we use the point of 85% of the maximum risk as the 'lowest risk' point, even if the risk could be further decreased by an insignificant amount. The maximum expected net present value point of the multi-stage curve has a risk of about 2.4M NPV. At this level of risk, the expected net present value on the single-stage efficient frontier is about 8.9M NPV, versus the expected net present value on the multi-stage curve of about 13.6M NPV, which represents an improvement of 52.8%. In the 'Flat' case, we compare the point with the smallest risk of 3.0M NPV on the single-stage efficient frontier with that with the largest risk of 2.5M NPV on the multi-stage efficient frontier. The expected net present value at the two points is 13.6M NPV (multi-stage) versus 9.6M NPV (single-stage), which represents an improvement of 41.7%. In the 'Steep' case, we compare the point with the smallest risk of 2.6M NPV on the single-stage efficient frontier with that with the largest risk of 1.9M NPV on the multi-stage efficient frontier. The expected net present value at the two points is 20.4M NPV (multi-stage) versus 18.6M NPV (single-stage), which represents an improvement of 9.7%.

As for the single-stage model, we look at the funding strategies at the point of 95% of the maximum expected return (on the efficient frontier the second point down from the

TABLE 7.4 Funding Strategy from the Multi-Stage Model, Case 95% of Maximum Expected Return, Normal Data Set

Stage 1	Allocation							
	j1m06n	**j1y05nc1**						
	0.785	0.215						
Stage 2	Allocation							
Scenario	**j2m03n**	**j2y03n**	**j2y03nc1**	**j2y05nc1**	**j2y10n**	**j2y10nc1**	**j2y10nc3**	**j2y30n**
1								1.000
2						0.597	0.188	
3		0.125	0.420		0.240			
4	0.785							
5			0.508		0.423	0.069		
6	0.785							
7						0.785		
8	0.251				0.749			
9	0.301			0.035	0.261	0.188		
10	0.785							

maximum expected return point). The funding strategies are presented in Table 7.4 for the 'Normal' data set, and in Table 7.A1 and Table 7.A2 of Appendix 7.A for the 'Steep' and 'Flat' data sets, respectively.

In the 'Normal' case, the initial funding mix consists of 78.5% six-month non-callable debt, and 21.5% five-year debt callable after one year. After one year the 78.5% six-month debt (that according to the decision rules is refinanced through short-term debt and is for disposition in the second stage) and, if called in certain scenarios, also the five-year callable debt are refunded through various mixes of short-term debt: three-year, five-year and ten-year callable and non-callable debt. In one scenario, labeled '1,' in which interest rates fall to a very low level, the multi-stage model resorts to funding with 30-year non-callable debt in order to secure the very low rate for the future. The risk associated with this strategy is 1.85M NPV and the expected net present value is 12.9M NPV. The corresponding (95% of maximum net present value) strategy of the single-stage model, discussed above, has a risk of 3.1M NPV and a net present value of 10.0M NPV. Thus, the multi-stage strategy exhibits 57.4% of the risk of the single-stage strategy and a 29% larger expected net present value.

In the 'Flat' case, the initial funding mix consists of 50.9% three-month non-callable debt and 49.1% six-month non-callable debt. After one year the entire portfolio is refunded through various mixes of short-term, three-year, and ten-year callable and non-callable debt. The multi-stage strategy exhibits 68% of the risk of the single-stage strategy and a 20.6% larger expected net present value. In the 'Steep' case, the initial funding consists of 100% three-month non-callable debt. After one year the portfolio is refunded through various mixes of short-term, three-year and five-year non-callable, and ten-year callable and non-callable debt. The multi-stage strategy exhibits 62% of the risk of the single-stage strategy, and a 4.3% larger expected net present value.

Summarizing, the results demonstrate that multi-stage stochastic programming potentially yields significantly larger net present values at the same or even lower levels of risk, with significantly different funding strategies, compared with single-stage optimization. Using multi-stage stochastic programming for determining the funding of mortgage pools promises to lead, in the average, to significant profits compared with using single-stage funding strategies.

7.6.4 Results for Duration and Convexity

Funding a mortgage pool by a portfolio of bonds that matches the (negative) value of duration and convexity, the expected net present value of the total of mortgage pool and bonds is invariant to small changes in interest rates. However, duration and convexity give only a local approximation, and the portfolio needs to be updated as time goes on and interest rates change. The duration and convexity hedge is one-dimensional, since it considers only changes of the whole yield curve by the same amount and does not consider different shifts for different maturities. The multi-stage stochastic programming model takes into account multi-dimensional changes of interest rates and considers (via sampling) the entire distribution of possible yield curve developments. In this section we quantify the difference between duration and convexity hedging versus hedging using the single- and multi-stage stochastic programming models. Table 7.5 gives the initial (first stage) values for duration and convexity (as obtained from the simulation runs) for the mortgage pool and for all funding instruments for each of the three yield curve cases 'Normal,' 'Flat' and 'Steep.' In each of the three yield curve cases the mortgage pool exhibits a positive value for duration and a negative value for convexity. All funding instruments have positive values for duration, non-callable bonds exhibit a positive value

TABLE 7.5 Initial Duration and Convexity

Label	Normal		Flat		Steep	
	Dur.	Conv.	Dur.	Conv.	Dur.	Conv.
mortg.	3.442	−1.880	2.665	−1.548	2.121	−2.601
m03n	0.248	0.001	0.247	0.001	0.249	0.001
m06n	0.492	0.003	0.491	0.003	0.495	0.003
y01n	0.971	0.010	0.964	0.010	0.982	0.011
y02n	1.882	0.038	1.861	0.038	1.917	0.039
y03n	2.737	0.081	2.686	0.079	2.801	0.084
y03nc1	1.596	−0.384	1.376	−0.186	1.338	−0.235
y05n	4.278	0.205	4.160	0.197	4.378	0.212
y05nc1	1.914	−0.810	1.613	−0.700	1.478	−0.632
y05nc3	3.274	−0.012	3.068	−0.111	3.114	−0.241
y07n	5.622	0.367	5.448	0.351	5.769	0.380
y07nc1	2.378	0.128	1.671	0.959	0.898	0.032
y07nc3	3.406	0.269	3.117	−0.298	3.122	−0.230
y10n	7.335	0.656	7.083	0.624	7.538	0.681
y10nc1	1.555	0.246	1.397	−0.052	0.593	−0.119
y10nc3	3.409	−0.160	3.265	−0.346	3.086	−0.446
y30n	13.711	2.880	13.306	2.743	14.204	3.011

for convexity, and callable bonds show a negative value for convexity. Note as an exception the positive value for convexity of bond 'y07ncl.'

To both the single-stage and multi-stage stochastic programming models we added constraints that, at any decision point, the duration and convexity of the mortgage pool and the funding portfolio are as close as possible. Maximizing expected return by setting the risk aversion λ to zero, we successively reduced the gap in duration and convexity between the mortgage pool and the funding portfolio. We started from the unconstrained case (the maximum expected return–maximum risk case from the efficient frontiers discussed above) and reduced first the duration gap and subsequently the convexity gap, where we understand as duration gap the absolute value of the difference in duration between the funding portfolio and the mortgage pool, and as convexity gap the absolute value of the difference in convexity between the funding portfolio and the mortgage pool. We will discuss the results with respect to the downside risk measure (expected value of returns below a certain target), as discussed in Section 7.2.9, and also with respect to the standard deviation of the returns. We do not discuss the results for duration and convexity obtained from the single-stage model and focus on the more interesting multi-stage case.

7.6.5 Duration and Convexity, Multi-Stage Model

Figures 7.3 and 7.4 give the risk–return profile for the 'Normal' case with respect to downside risk and standard deviation, respectively. We compare the efficient frontier (already depicted in Figure 7.2) obtained from minimizing downside risk for different levels of expected return (labeled 'Downside') with the risk–return profile obtained from restricting the duration and convexity gap (labeled 'Delta–Gamma'). For different levels of duration and convexity gap, we maximized expected return. The unconstrained case with respect to duration and convexity is identical to the point on the efficient frontier with the maximum expected return. Figure 7.3 shows that the downside risk eventually increases

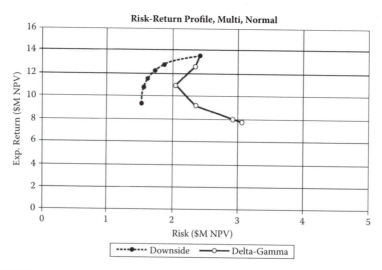

FIGURE 7.3 Risk-return profile, multi-stage model, Normal data set, risk as downside risk.

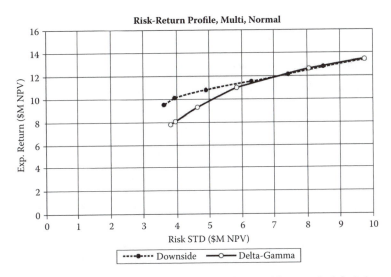

FIGURE 7.4 Risk-return profile, multi-stage model, Normal data set, risk as standard deviation.

when decreasing the duration and convexity gap and was significantly larger than the minimized downside risk on the efficient frontier. We are especially interested in comparing the point with the smallest duration and convexity gap with the point of minimum downside risk on the efficient frontier. The point with the smallest downside risk on the efficient frontier exhibits expected return of 9.5M NPV and a downside risk of 1.5M NPV. The point with the smallest duration and convexity gap has expected return of 7.9M NPV and a downside risk of 3.1M NPV. The latter point is characterized by a maximum duration gap in the first and second stage of 0.5 and of 1.0 in the third stage, and by a maximum convexity gap of 2.0 in the first and second stage and of 4.0 in the third stage. A further decrease of the convexity gap was not possible as it led to an infeasible problem. The actual duration gap in the first stage was −0.5, where the negative value indicates that the duration of the funding portfolio was smaller than that of the mortgage pool, and the actual convexity gap in the first stage was 1.57, where the positive value indicates that the convexity of the funding portfolio was larger than that of the mortgage pool. Comparing the points with regard to their performance, restricting the duration and convexity gap led to a decrease in expected return of 15% and an increase of downside risk by a factor of 2. It is interesting to note that, for the point on the efficient frontier with the smallest risk, the first-stage duration gap was −0.97 and the first-stage convexity gap was 1.75. Looking at the risk in terms of standard deviation of returns, both minimizing downside risk and controlling the duration and convexity gap led to smaller values of standard deviation. In the unconstrained case, the smallest standard deviation was 3.3M NPV obtained at the minimum downside risk point, and the smallest standard deviation in the constrained case (3.8M NPV) was obtained when the duration and convexity gap was smallest.

Table 7.6 gives the funding strategy for the minimum downside risk portfolio and Table 7.7 the funding strategy for the duration and convexity constrained case at

TABLE 7.6 Funding Strategy, Case Minimum Downside Risk, Multi-Stage Model, Normal Data Set

Stage 1	Allocation							
	j1m06n	j1y05n	j1y05nc1	j1y05nc3				
	0.266	0.235	0.218	0.281				
Stage 2	Allocation							
Scenario	j2m03n	j2y01n	j2y03nc1	j2y05n	j2y10n	j2y10nc1	j2y10nc3	j2y30n
1		0.017		0.148	0.240			0.079
2						0.102	0.165	
3			0.117		0.149			
4	0.266							
5	0.031		0.032		0.237	0.185		
6	0.266							
7					0.014	0.252		
8	0.118				0.300	0.019		0.047
9	0.012				0.201	0.053		
10	0.266							

minimum gap. Both strategies use six-month non-callable debt, five-year non-callable debt and five-year debt callable after one year for the initial funding. The minimum downside risk portfolio used in addition five-year debt callable after three years, while the duration and convexity constrained case used seven-year non-callable debt and ten-year non-callable debt. In the second stage, funding differed significantly across the different scenarios for both funding strategies. In the duration and convexity constrained case, the multi-stage optimization model resorted to calling five-year debt callable after one year in scenario '4' at a fraction and in scenario '10' at the whole amount. In the minimum risk case, calling of debt other than by applying the decision rules did not happen. In the second stage, interest rates were low in scenarios '1' and '8' and were high in scenarios '4', '6' and '10'. The minimum downside risk strategy tended towards more long-term debt when interest rates were low and towards more short-term debt when interest rates were high. The amounts for each of the scenarios depended on the dynamics of the process and the interest rate distributions. The duration and convexity constrained strategy was, of course, not in the position to take advantage of the level of interest rates, and funding was balanced to match the duration and convexity of the mortgage pool.

The results for the case of the 'Flat' and 'Steep' yield curves are very similar. For the 'Flat' case, the maximum duration gap in the first and second stage was set to 0.5 and to 1.0 in the third stage, and the maximum convexity gap was set to 2.0 in the first and second stage and to 4.0 in the third stage. The duration and convexity gap could not be decreased further since the problem became infeasible. The actual duration gap in the first stage was −0.5, and the actual convexity gap in the first stage was 1.5. In Appendix 7.A, Table 7.A3 gives the initial funding and the second-stage updates for the minimum downside risk portfolio, and Table 7.A4 gives the funding strategy for the duration and convexity constrained case. For the 'Steep' case, the maximum duration gap in the first and second stage was set to 0.5 and was set to 1.0 in the third stage, and the maximum

TABLE 7.7 Funding Strategy, Case Duration and Convexity Constrained, Multi-Stage Model, Normal Data Set

Stage 1	Allocation						
	j1m06n	j1y05n	j1y05ncl	j1y07n	j1y10n		
	0.122	0.083	0.534	0.239	0.022		
Stage 2	Allocation						
Scenario	j2m03n	j2m06n	j2y01n	j2y03ncl	j2y05n	j2y05ncl	j2y07ncl
1	0.466						0.117
2							
3				0.064			
4							
5				0.506	0.043	0.044	
6			0.064				
7							
8	0.567						
9	0.072						
10		0.316	0.340				

Stage 2	Allocation			Call
]Scenario	]j2y10n	]j2y10ncl	]j2y10nc3	j1y05ncl
1			0.073	
2	0.011	0.112		
3	0.058			
4	0.152			0.029
5	0.058	0.004		
6	0.059			
7	0.122			
8	0.089			
9			0.051	
10				0.534

convexity gap was set to 3.0 in the first and second stage and to 6.0 in the third stage. Further decrease made the problem become infeasible. The actual duration gap in the first stage was −0.5, and the actual convexity gap in the first stage was 1.6. In Appendix 7.A, Table 7.A5 gives the initial funding and the second-stage updates for the minimum downside risk portfolio, and Table 7.A6 gives the funding strategy for the duration and convexity constrained case. Again, in both the 'Flat' and 'Steep' case, one could see in the minimum downside risk case a tendency to use longer-term debt when interest rates were low and shorter-term debt when interest rates were high, and in the duration and convexity constrained case, funding was balanced to match the duration and convexity of the mortgage pool.

Summarizing, in each case reducing the duration and convexity gap helped control the standard deviation of net present value but did nothing to reduce downside risk. Multi-stage stochastic programming led to larger than or equal expected net present value at each level of risk (both downside risk and standard deviation of net present value). There may be reasons for controlling the duration and convexity gap in addition to controlling

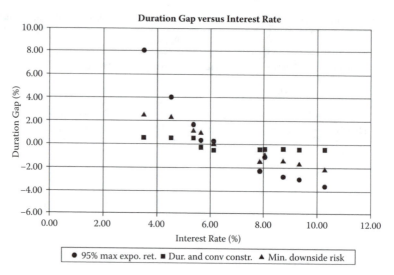

FIGURE 7.5 Duration gap versus interest rate, multi-stage, Normal data set.

downside risk via the multi-stage stochastic programming model. For example, a conduit might like the results of the stochastic programming model, but not wish to take on too much exposure regarding duration and convexity gap. In the following we will therefore compare runs of the multi-stage model with and without duration and convexity constraints.

Figure 7.5 sheds more light on the cause of the performance gain of the multi-stage model versus the duration and convexity hedged strategy. The figure presents the duration gap (the difference of the duration of the funding portfolio and the mortgage pool) versus the interest rate (calculated as the average of the yield curve) for each of the second-stage scenarios of the data set 'Normal.' When interest rates are very low, the 95% maximum expected return strategy takes on a significant positive duration gap to lock in the low rates for a long time. It takes on a negative duration gap when interest rates are high, in order to remain flexible, should interest rates fall in the future. The minimum downside risk strategy exhibits a similar pattern, but less extreme. Thus, the multi-stage model makes a bet on the direction interest rates are likely to move, based on the information about the interest rate process. In contrast, the duration and convexity constrained strategy cannot take on any duration gap (represented by the absolute value of 0.5 at which the gap was constrained) and therefore must forsake any gain from betting on the likely direction of interest rates.

7.6.6 Out-of-Sample Simulations

In order to evaluate the performance of the different strategies in an unbiased way, true out-of-sample evaluation runs need to be performed. Any solution at any node in the tree obtained by optimization must be evaluated using an independently sampled set of observations.

For the single-stage model this evaluation is rather straightforward. Having obtained an optimal solution from the single-stage model (using sample data set one), we run simulations again with a different seed. Using the new independent sample (data set two), we start the optimizer again, however now with the optimal solution fixed at the values as obtained from the optimization based on sample data set one. Using the second set of observations of data set two we calculate risk and expected returns.

To independently evaluate the results obtained from a K-stage model (having $n = K-1$ sub-periods), we need K independent sets of K-stage trees of observations. We describe the procedure for $K = 4$. Using data set one we solve the multi-stage optimization problem and obtain an optimal first-stage solution. We simulate again (using a different seed) over all stages to obtain data set two. Fixing the optimal first-stage solution at the value obtained from the optimization based on data set one, we optimize based on data set two and obtain a set of optimal second-stage solutions. We simulate again (with a different seed) to obtain independent realizations for stages three and four, thereby keeping the observations for stage two the same as in data set two, and obtain data set three. Fixing the first-stage decision at the level obtained from the optimization using data set one and all second-stage decisions at the level obtained from the optimization based on data set two, we optimize again to obtain a set of optimal third-stage decisions. We simulate again (using a different seed) to obtain independent outcomes for stage four, thereby keeping the observations for stage two and three the same as in data sets two and three, respectively, and obtain data set four. Fixing the first-stage decision, all second-stage decisions, and all third-stage decisions at the levels obtained from the optimization based on data sets one, two and three, respectively, we finally calculate risk and returns based on data set four.

The out-of-sample evaluations resemble how the model could be used in practice. Solving the multi-stage model (based on data set one), an optimal first-stage solution (initial portfolio) would be obtained and implemented. Then one would follow the strategy (applying the decision rules) for 12 months until decision stage two arrives. At this point, one would re-optimize, given that the initial portfolio had been implemented and that particular interest rates and prepayments had occurred (according to data set two). The optimal solution for the second stage would be implemented, and one would follow the strategy for four years until decision stage three arrives. At this point, one would re-optimize, given that the initial portfolio and a second stage update had been implemented and that particular interest rates and prepayments had occurred (according to data set three). The optimal solution for the third stage would be implemented, and one would follow the strategy (applying the decision rules) until the end of the horizon (according to data set four). Now one possible path of using the model has been evaluated. Decisions had no information about particular outcomes of future interest rates and prepayments, and were computed based on model runs using data independent from the observed realization of the evaluation simulation. Alternatively, one could simulate a strategy involving re-optimization every month, but this would take significantly more computational effort with likely only little to be gained.

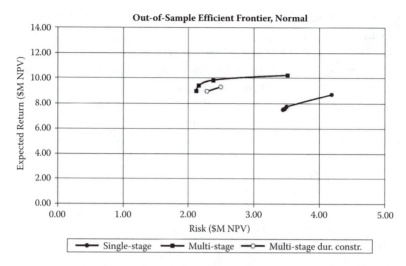

FIGURE 7.6 Out-of-sample risk–return profile, multi-stage model, Normal data set, risk as downside risk.

Using the out-of-sample evaluation procedure, we obtain $N = |S|$ out-of-sample simulations of using the model as discussed in the above paragraph, and we are now in the position to derive statistics about out-of-sample expected returns and risk.

Figure 7.6 presents the out-of-sample efficient frontiers for the 'Normal' data set and downside risk. The figure gives the out-of-sample efficient frontier for the multi-stage model without duration and convexity constraints, the out-of-sample efficient frontier when the duration gap was constrained to be less than or equal to 1.5%, and the out-of-sample efficient frontier for the single-stage model. The out-of-sample evaluations demonstrate clearly that the multi-stage model gives significantly better results than the single-stage model. For example, the point with the maximum expected returns of the multi-stage model gave expected returns of 10.2M NPV and a downside risk of 3.5M NPV. The minimum risk point on the single-stage out-of-sample efficient frontier gave expected returns of 7.6M NPV, and a downside risk of also 3.5M NPV. Thus, at the same level of downside risk the multi-stage model gave 34% higher expected returns. The minimum downside risk point of the multi-stage model gave expected returns of 9.0M NPV and a downside risk of 2.1M NPV. Comparing the minimum downside risk point of the multi-stage model with that of the single-stage model, the multi-stage model had 19.2% larger expected returns at 61% of the downside risk of the single-stage model. The efficient frontier of the duration-constrained strategy was slightly below that without duration and convexity constraints.

Figure 7.7 gives the out-of-sample risk–return profiles when measuring risk in terms of standard deviation. In this setting the multi-stage strategy performed significantly better than the single-stage strategy at every level of risk, where the difference was between 17.6% and 23.6%. The duration-constrained strategy did not span as wide a range in risk as the unconstrained strategy. But for the risk points attained by the constrained strategy, the unconstrained strategy achieved a somewhat higher expected return.

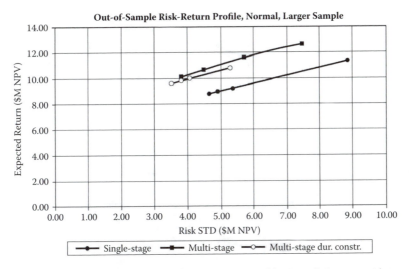

FIGURE 7.7 Out-of-sample risk–return profile, multi-stage model, Normal data set, risk as standard deviation.

The out-of-sample evaluations for the 'Flat' and 'Steep' data sets gave very similar results qualitatively. The results are included in Appendix 7.A. For the 'Flat' data set, Figure 7.A1 presents out-of-sample efficient frontiers for downside risk and Figure 7.A2 the risk–return profile for risk in standard deviations. For the 'Steep' data set, Figure 7.A3 presents out-of-sample efficient frontiers for downside risk and Figure 7.A4 the risk–return profile for risk in standard deviations for the data set 'Steep.' For both data sets, the multi-stage model gave significantly better results; the downside risk was smaller and expected returns were larger.

7.6.7   Larger Sample Size

All results discussed so far were obtained from solving a model with a relatively small number of scenarios at each stage, i.e. $|S_2| = 10$, $|S_3| = 20$ and $|S_4| = 20$, with a total of 4000 scenarios at the end of the forth stage. This served well for analysing and understanding the behavior of the multi-stage model in comparison with the single-stage model and with Gamma and Delta hedging. Choosing a larger sample size will improve the obtained strategies (initial portfolio and future revisions) and therefore result in better (out-of-sample simulation) results. Of course, the accuracy of prediction of the models will be improved also. In order to show the effect of using a larger sample size, we solved and evaluated the models using a sample size of 24 000 scenarios, i.e. $|S_2| = 40$, $|S_3| = 30$ and $|S_4| = 20$. The results for the data set 'Normal' are presented in Figure 7.8 for downside risk and in Figure 7.9 for risk as standard deviation. Indeed, one can see improved performance in both smaller risk and larger expected NPV compared with the smaller sample size (compare with Figures 7.6 and 7.7).

The point with the maximum expected returns for the multi-stage model gave expected returns of 12.9M NPV and a downside risk of 2.4M NPV. The minimum risk point for the

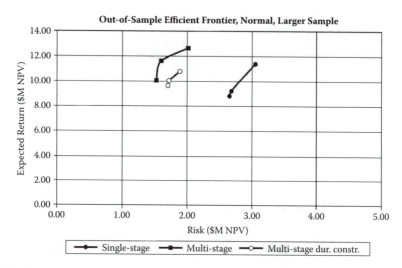

FIGURE 7.8 Larger sample, out-of-sample risk–return profile for the multi-stage model, Normal data set, risk as downside risk.

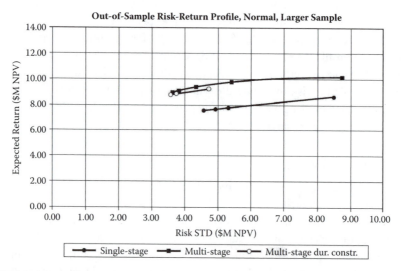

FIGURE 7.9 Larger sample, out-of-sample risk–return profile for the multi-stage model, Normal data set, risk as standard deviation.

single-stage model had expected returns of 9.0M NPV and a downside risk of 2.7M NPV. So, at slightly smaller downside risk the multi-stage model gave expected returns that were 43% higher. The minimum downside risk point of the multi-stage model had expected returns of 10.9M NPV and a downside risk of 1.5M NPV. Comparing the minimum downside risk points of the multi-stage and single-stage models, the multi-stage model has 18.4% larger expected returns at 57% of the downside risk of the single-stage model. Again, the efficient frontier of the duration-constrained strategy was slightly below that

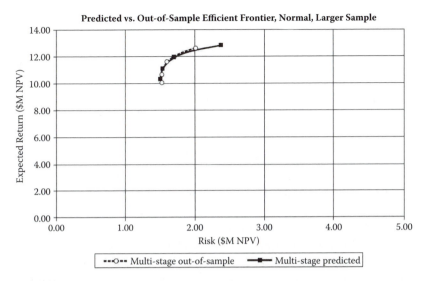

FIGURE 7.10 Larger sample, predicted versus out-of-sample efficient frontier for the multi-stage model, Normal data set.

without duration and convexity constraints. Measuring risk in terms of standard deviation, results are similar to those when using the small sample size: the multi-stage strategy performed significantly better than the single-stage strategy at every level of risk, with a difference in expected returns between 23.3% and 24.0%. Again, the duration-constrained strategy exhibited smaller risk (in standard deviations) at the price of slightly smaller expected returns.

Figure 7.10 gives a comparison of the multi-stage efficient frontiers predicted (in-sample) versus evaluated out-of-sample. One can see that when using the larger sample size of 24,000, the predicted and the out-of-sample evaluated efficient frontier look almost identical, thus validating the model. It is evident that using larger sample sizes will result in both better performance and a more accurate prediction.

7.7   LARGE-SCALE RESULTS

For the actual practical application of the proposed model, we need to consider a large number of scenarios in order to obtain small estimation errors regarding expected returns and risk and accordingly stable results. We now explore the feasibility of solving large-scale models. Table 7.8 gives measures of size for models with larger numbers of scenarios. For example, model L4 has 80,000 scenarios at the fourth stage, composed of $|S_2| = 40$, $|S| = 40$ and $|S_4| = 50$, and model L5 has 100,000 scenarios at the fourth stage composed of $|S_2| = 50$, $|S_3| = 40$ and $|S_4| = 50$; both models have a sufficiently large sample size in each stage.

In the case of problem L5 with 100,000 scenarios the corresponding linear program had 270,154 constraints, 249,277 variables, and 6,291,014 non-zero coefficients. Table 7.9 gives

TABLE 7.8 Large-Scale Models, Dimensions

Model	Scenarios				Problem Size		
	Stage 2	Stage 3	Stage 4	Total	Rows	Columns	Non-zeros
L1	10	40	50	20,000	54,034	49,918	1,249,215
L2	20	40	50	40,000	108,064	99,783	2,496,172
L3	30	40	50	60,000	162,094	149,623	3,756,498
L4	40	40	50	80,000	216,124	199,497	5,021,486
L5	50	40	50	100,000	270,154	249,277	6,291,014
L20	200	40	50	400,000	808,201	931,271	14,259,216

the elapsed times for simulation and optimization, obtained on a Silicon Graphics Origin 2000 workstation.

While the Origin 2000 at our disposition is a multi-processor machine with 32 processors, we did not use the parallel feature, and all results were obtained using single processor computations. For the direct solution of the linear programs, we used CPLEX (1989–1997) as the linear programming optimizer. We contrast the results to using DECIS (Infanger 1989–1999), a system for solving stochastic programs, developed by the author. Both CPLEX and DECIS were accessed through GAMS (Brooke et al. 1988). DECIS exploited the special structure of the problems and used dual decomposition for their solution. The problems were decomposed into two stages, breaking between the first and the second stage. The simulation runs for model L5 took less than an hour. The elapsed solution time solving the problem directly was 13 hours and 28 minutes. Solving the problem using DECIS took significantly less time, 2 hours and 33 minutes. Encouraged by the quick solution time using DECIS, we generated problem L20 with 400,000 scenarios (composed of $|S_2| = 200$, $|S_3| = 40$ and $|S_4| = 50$) and solved it in 11 hours and 34 minutes using DECIS. Problem L20 had 808,201 constraints, 931,271 variables, and 14,259,216 non-zero coefficients.

Using parallel processing, the simulation times and the solution times could be reduced significantly. Based on our experiences with parallel DECIS, using six processors we would expect the solution time for the model L5 with 100,000 scenarios to be less than 40 minutes, and using 16 processors one could solve model L20 with 400,000 scenarios in about one hour. We actually solved a version of the L3 model with 60,000 scenarios, composed of $|S_2| = 50$, $|S_3| = 40$ and $|S_4| = 30$, in less than 5 minutes using parallel DECIS on 16 processors.

TABLE 7.9 Large-Scale Models, Solution Times

Model	Scenarios	Simul. time (s)	Direct sol. time (s)	Decomp. sol. time (s)
L1	20,000	688	1389.75	1548.91
L2	40,000	1390	6144.56	
L3	60,000	2060	14 860.54	5337.76
L4	80,000	2740	28 920.69	
L5	100,000	3420	48 460.02	9167.47
L20	400,000			41 652.28

7.8   SUMMARY

The problem of funding mortgage pools represents an important problem in finance faced by conduits in the secondary mortgage market. The problem concerns how to best fund a pool of similar mortgages through issuing a portfolio of bonds of various maturities, callable and non-callable, and how to refinance the debt over time as interest rates change, prepayments are made and bonds mature. This chapter presents the application of stochastic programming in combination with Monte Carlo simulation for effectively and efficiently solving the problem.

Monte Carlo simulation was used to estimate multiple realizations of the net present value of all payments when a pool of mortgages is funded initially with a single bond, where pre-defined decision rules were applied for making decisions not subject to optimization. The simulations were carried out in monthly time steps over a 30-year horizon. Based on a scenario tree derived from the simulation results, a single-stage stochastic programming model was formulated as a benchmark. A multi-stage stochastic programming model was formulated by splitting up the planning horizon into multiple sub-periods, representing the funding decisions (the portfolio weights and any calling of previously issued callable bonds) for each sub-period, and applying the pre-defined decision rules between decision points.

In order to compare the results of the multi-stage stochastic programming model with hedging methods typically used in finance, the effective duration and convexity of the mortgage pool and of each funding instrument was estimated at each decision node, and constraints bounding the duration and convexity gap to close to zero were added (at each node) to the multi-stage model to approximate a duration and convexity hedged strategy.

An efficient method for obtaining an out-of-sample evaluation of an optimal strategy obtained from solving a K-stage stochastic programming model was presented, using K independent (sub-)trees for the evaluation.

For different data assumptions, the efficient frontier of expected net present value versus (downside) risk obtained from the multi-stage model was compared with that from the single-stage model. Under all data assumptions, the multi-stage model resulted in significantly better funding strategies, dominating the single-stage model at every level of risk, both in-sample and by evaluating the obtained strategies via out-of-sample simulations. Also, for all data assumptions, the out-of-sample simulations demonstrated that the multi-stage stochastic programming model dominated the duration and convexity hedged strategies at every level of risk. Constraining the duration and convexity gap reduced risk in terms of the standard deviation of net present value at the cost of a smaller net present value, but failed in reducing the downside risk.

The results demonstrate clearly that using multi-stage stochastic programming results in significantly larger profits, both compared with using single-stage optimization models and with using duration and convexity hedged strategies. The multi-stage model is better than the single-stage model because it has the option to revise the funding portfolio according to changes in interest rates and pre-payments, therefore reflecting a more realistic representation of the decision problem. It is better than the duration and convexity hedged strategies because it considers the entire distribution of the yield curve represented by the stochastic process of interest rates, compared with the much simpler hedge against a small shift of the entire yield curve as used in the duration and convexity hedged strategies.

Small models with 4000 scenarios and larger ones with 24,000 scenarios were used for determining the funding strategies and the out-of-sample evaluations. The out-of-sample efficient frontiers of the larger models were shown to be very similar to the (in-sample) predictions, indicating a small optimization bias. The larger models were solved in a very short (elapsed) time of a few minutes. Large-scale models with up to 100,000 scenarios were shown to solve in a reasonable elapsed time using decomposition on a serial computer, and in a few minutes on a parallel computer.

ACKNOWLEDGEMENTS

Funding for research related to this chapter was provided by Freddie Mac. The author is grateful to Jan Luytjes for many valuable discussions on the subject. Editorial comments by three anonymous referees and by John Stone on previous versions of this paper helped enhance the presentation.

REFERENCES

Bellman, R., *Dynamic Programming*, 1957 (Princeton University Press: Princeton, NJ).

Brooke, A., Kendrik, D. and Meeraus, A., *GAMS, A Users Guide*, 1988 (Scientific Press: South San Francisco, CA).

Carino, D.R., Kent, T., Myers, D.H., Stacy, C., Sylvanus, M., Turner, A.L., Watanabe, K. and Ziemba, W.Y., The Russel–Yasuda Kasai model: An asset-liability model for a Japanese insurance company using multistage stochastic programming. *Interfaces*, 1994, **24**, 29–49.

Cox, J.C., Ingersoll, J.E. and Ross, S.A., A theory of the term structure of interest rates. *Econometrica*, 1985, **53**, 385–407.

CPLEX Optimization, Inc., *Using the CPLEX Callable Library*, 1989–1997 (CPLEX Optimization, Inc.: Incline Village, NV).

Dantzig, G.B. and Glynn, P.W., Parallel processors for planning under uncertainty. *Ann. Oper. Res.*, 1990, **22**, 1–21.

Dembo, R. and Mausser, H., The put/call efficient frontier. *Algo. Res. Q.*, 2000, **3**, 13–25.

Dempster, M.A.H. and Thorlacius, A.F., Stochastic simulation of international economic variables and asset returns, the falcon asset model in *8th International AFIR Colloquium*, 1998, pp. 29–45.

Di Domenica, N., Valente, P., Lucas, C.A. and Mitra, G., A review of scenario generation for stochastic programming and simulation, CARISMA Research Report CTR 31, to appear in *The State of the Art in Planning under Uncertainty (Stochastic Programming)*, edited by G.B. Dantzig and G. Infanger, 2006 (Kluwer: Dordrecht).

Frauendorfer, K. and Schürle, M., Refinancing mortgages in Switzerland. *Applications of Stochastic Programming* (MPS-SIAM Series on Optimization), edited by S.W. Wallace and W.T. Ziemba, 2005 (SIAM: Philadelphia).

Golub, B., Holmer, M., McKendal, R., Pohlman, L. and Zenios, S.A., A stochastic programming model for money management. *Eur. J. Oper. Res.*, 1995, **85**, 282–296.

Hiller, R.S. and Eckstein, J., Stochastic dedication: Designing fixed income portfolios using massively parallel benders decomposition. *Mgmt Sci.*, 1993, **39**, 1422–1438.

Ho, T.S.Y. and Lee, S.-B., Term structure movements and pricing interest rate contingent claims. *J. Finan.*, 1986, **XLI**, 1011–1029.

Hull, J. and White, A., Pricing interest rate securities. *Rev. Finan. Stud.*, 1990, **3**, 573–592.

Infanger, G., Monte Carlo (importance) sampling within a benders decomposition algorithm for stochastic linear programs. *Ann. Oper. Res.*, 1992, **39**, 65–95.

Infanger, G., *Planning Under Uncertainty—Solving Large-Scale Stochastic Linear Programs*, 1994 (Boyd and Fraser) (The Scientific Press Series: Danvers, MA).

Infanger, G., Managing risk using stochastic programming, In *INFORMS National Meeting*, 1996.

Infanger, G., *DECIS User's Guide*, 1989–1999 (Infanger Investment Technology: Mountain View, CA).

Infanger, G., Dynamic asset allocation strategies using a stochastic dynamic programming approach. In *Handbook of Asset and Liability Management*, edited by S. Zenios and W.T. Ziemba, 2006 (Elsevier: Amsterdam).

Infanger, G. and Morton, D., Cut sharing for multistage stochastic linear programs with interstage dependency. *Math. Program.*, 1996, **75**, 241–256.

Kang, P. and Zenios, S., Complete prepayment models for mortgage-backed securities. *Mgmt Sci.*, 1992, **38**, 1665–1685.

Kusy, M.I. and Ziemba, W.T., A bank asset and liability management model. *Oper. Res.*, 1986, **35**, 356–376.

Lekkas, V. and Luytjes, J., Estimating a prepayment function for 15- and 30-year fixed rate mortgages. FHLMC Technical Report, 1993; Rates model enhancement summary. Freddie Mac Internal Document.

Longstaff, F.A. and Schwarz, E.S., Interest rate volatility and the term structure: A two-factor general equilibrium model. *J. Finan.*, 1992, **XLVII**, 1259–1282.

Luenberger, D.G., *Investment Science*, 1998 (Oxford University Press: New York).

Luytjes, J., Estimating a yield curve/inflation process. FHLMC Technical Report, 1993.

Luytjes, J., Memorandum FR call rule—2Q96 release. FHLMC Technical Report, 1996.

Markowitz, H., Portfolio selection. *J. Finan.*, 1952, **7**, 77–91.

Mulvey, J.M. and Vladimirou, H., Stochastic network programming for financial planning problems. *Mgmt Sci.*, 1992, **38**, 1642–1664.

Nielsen, S. and Zenios, S.A., A stochastic programming model for funding single-premium deferred annuities. *Math. Program.*, 1996, **75**, 177–200.

Prindiville, M., Advances in the application of stochastic programming to mortgage finance. PhD thesis, Department of Engineering—Economic Systems and Operations Research, Stanford University, CA, 1996.

Vasicek, O.A., An equilibrium characterization of the term structure. *J. Finan. Econ.*, 1977, **5**, 177–188.

Wallace, S.W. and Ziemba, W.T., *Applications of Stochastic Programming*, (MPS-SIAM Series on Optimization), 2005 (SIAM: Philadelphia).

Zenios, S.A., A model for portfolio management with mortgage-backed securities. *Ann. Oper. Res.*, 1993, **43**, 337–356.

Ziemba, W.T., *The Stochastic Programming Approach to Asset, Liability, and Wealth Management*, 2003 (The Research Foundation of AIMR: Charlottesville, VA).

APPENDIX 7.A: TABLES AND GRAPHS FOR DATA SETS 'FLAT' AND 'STEEP'

TABLE 7.A1 Funding Strategy from the Multi-Stage Model, Case 95% of Maximum Expected Return, Flat Data Set

Stage 1	Allocation						
	j1m03n	j1m06n					
	0.509	0.491					
Stage 2	**Allocation**						
Scenario	j2m03n	j2y03n	j2y03nc1	j2y10n	j2y10nc1	j2y10nc3	j2y30n
1				0.397			0.603
2					1.000		
3	0.251	0.590		0.011		0.148	
4	1.000						
5			0.514	0.248	0.238		
6	1.000						
7	0.984				0.016		
8	0.620			0.380			
9	0.521			0.151	0.328		
10	1.000						

TABLE 7.A2 Funding Strategy from the Multi-Stage Model, Case 95% of Maximum Expected Return, Steep Data Set

Stage 1	Allocation						
	j1m03n						
	1.000						
Stage 2	**Allocation**						
Scenario	j2m03n	j2y03n	j2y03nc1	j2y05n	j2y10n	j2y10nc1	j2y30n
1		0.038		0.296	0.544		0.122
2	0.717		0.283				
3			1.000				
4	1.000						
5			0.745		0.085	0.170	
6	1.000						
7	1.000						
8	0.609				0.272	0.119	
9	0.973				0.027		
10	1.000						

TABLE 7.A3 Funding Strategy from the Multi-Stage Model, Case Minimum Downside Risk, Flat Data Set

Stage 1	Allocation					
	j1m06n	j1y05nc3				
	0.698	0.302				

Stage 2	Allocation					
Scenario	j2m03n	j2m06n	j2y03n	j2y03nc1	j2y05n	j2y05nc3
1		0.046			0.418	0.052
2						
3	0.201		0.292			
4	0.698					
5	0.119			0.145		
6	0.698					
7	0.603					
8	0.333					
9	0.302					
10	0.698					

Stage 2	Allocation			
Scenario	j2y10n	j2y10nc1	j2y10nc3	j2y30n
1	0.129			0.053
2		0.698		
3	0.039	0.039	0.127	
4				
5	0.221	0.214		
6				
7		0.095		
8	0.280	0.085		
9	0.095	0.301		
10				

TABLE 7.A4 Funding Strategy from the Multi-Stage Model, Case Duration and Convexity Constrained, Flat Data Set

Stage 1	Allocation							
	j1m06n	j1y05n	j1y05nc1	j1y07n				
	0.448	0.225	0.201	0.125				
Stage 2	Allocation							Call
Scenario	j2m03n	j2m06n	j2y02n	j2y03n	j2y03nc1	j2y10n	j2y10nc1	j1y05nc1
1	0.567					0.082		
2	0.102						0.346	
3	0.370			0.078				
4		0.251	0.033			0.164		
5	0.018				0.631			
6		0.397				0.051		
7						0.036	0.412	
8	0.634					0.016		
9	0.448							
10	0.277	0.372						0.201

TABLE 7.A5 Funding Strategy from the Multi-Stage Model, Case Minimum Downside Risk, Steep Data Set

Stage 1	Allocation							
	j1m03n							
	1.000							
Stage 2	Allocation							
Scenario	j2m03n	j2y01n	j2y03n	j2y03nc1	j2y05n	j2y10n	j2y10nc1	j2y30n
1		0.299			0.389	0.215		0.097
2	0.644			0.356				
3				1.000				
4	1.000							
5				0.734		0.088	0.178	
6	1.000							
7	1.000							
8	0.600		0.005			0.246	0.149	5.126743E-4
9	0.958					0.042	8.23454E-4	
10	1.000							

TABLE 7.A6 Funding Strategy from the Multi-Stage Model, Case Duration and Convexity Constrained, Steep Data Set

Stage 1	Allocation		
	j1y01n	j1y03n	j1y10n
	0.682	0.305	0.013

Stage 2	Allocation						
Scenario	j2m03n	j2y01n	j2y02n	j2y03n	j2y03nc1	j2y10n	j2y30n
1	0.598						0.084
2		0.053		0.629			
3	0.049	0.633					
4				0.531		0.151	
5					0.682		
6		0.158		0.524			
7		0.500				0.182	
8	0.682						
9	0.682						
10		0.392	0.289				

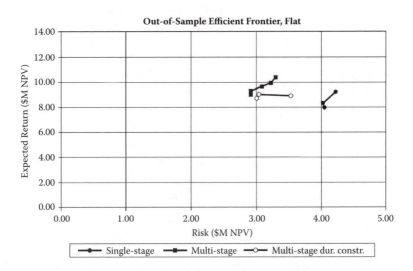

FIGURE 7.A1 Out-of-sample risk–return profile for the multi-stage model, Flat data set, risk as downside risk.

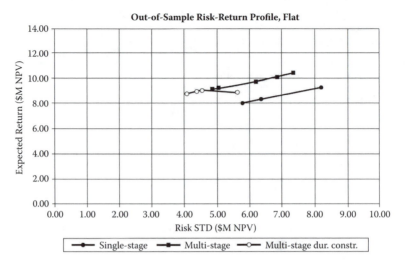

FIGURE 7.A2 Out-of-sample risk–return profile for the multi-stage model, Flat data set, risk as standard deviation.

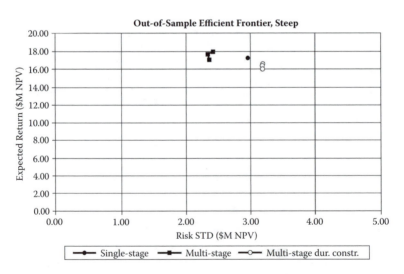

FIGURE 7.A3 Out-of-sample risk–return profile for the multi-stage model, Steep data set, risk as downside risk.

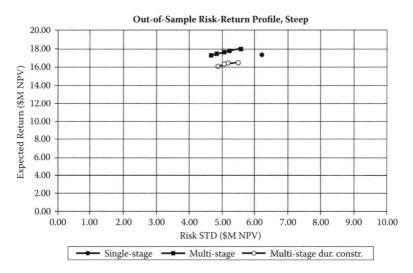

FIGURE 7.A4 Out-of-sample risk–return profile for the multi-stage model, Steep data set, risk as standard deviation.

CHAPTER 8

Scenario-Generation Methods for an Optimal Public Debt Strategy

MASSIMO BERNASCHI, MAYA BRIANI, MARCO PAPI and DAVIDE VERGNI

CONTENTS

8.1 INTRODUCTION

T HE MANAGEMENT OF THE PUBLIC DEBT is of paramount importance for any country. Mathematically speaking, this is a stochastic optimal control problem with a number of constraints imposed by national and supranational regulations and by market practices.

The Public Debt Management Division of the Italian Ministry of Economy decided to establish a partnership with the Institute for Applied Computing in order to examine this problem from a quantitative viewpoint. The goal is to determine the composition of the portfolio issued every month that minimizes a predefined *objective function* (Adamo *et al.* 2004), which can be described as an optimal combination between cost and risk of the public debt service.

Since the main stochastic component of the problem is represented by the evolution of interest rates, a key point is to determine how various issuance strategies perform under different scenarios of interest rate evolution. In other words, an optimal strategy for the management of the public debt requires a suitable modeling of the stochastic nature of the term structure of interest rates.

Note that hereafter we do not report how to *forecast* the *actual* future term structure, but how to generate a set of realistic possible scenarios. For our purposes, the scenarios should cover a wide range of possible outcomes of future term structures in order to provide a reliable estimate of the possible distribution of future debt charges. This distribution is useful in a risk-management setting to estimate, for instance, a sort of *Cost at Risk* (CaR) of the selected issuance policy (Risbjerg and Holmund 2005).

The chapter is organized as follows. Section 8.2 describes the problem. Section 8.3 describes the models employed for generating future interest rate scenarios. Section 8.4 introduces criteria to validate a scenario. Section 8.5 concludes with future perspectives of this work.

8.2   PROBLEM DESCRIPTION AND BASIC GUIDELINES

It is widely accepted that stock prices, exchange rates and most other interesting observables in finance and economics cannot be forecast with certainty. Interest rates are an even more complex issue because it is necessary to consider the term structure, which is a multi-value observable. Despite this difficulty, there are a number of studies that, from both the academic and practitioner viewpoint, deal with interest rate modeling (for a comprehensive survey of interest rate modeling, see James and Webber (2000)).

The most common application of existing term structure models is the evaluation of interest-rate-contingent claims. However, our purpose is different, since we aim to find an optimal strategy for the issuance of Italian public debt securities. In a very simplified form the problem can be described as follows. The Italian Treasury Department issues a number of different types of securities. Securities differ in the maturity (or expiration date) and in the rules for the payment of interest. Short-term securities (those having maturity up to two years) do not have coupons. Medium- and long-term securities (up to 30 years) pay cash dividends, every 6 months, by means of coupons. The problem is to find a strategy for the selection of public debt securities that minimizes the expenditure for interest payment (according to the ESA95 criteria (Jackson 2000)) and satisfies, at the same time, the constraints on debt management. The cost of future interest payments depends on the future value of the term structure (roughly speaking, when a security expires or a coupon is paid, there is the need to issue a new security whose cost depends on the term structure at expiration time). That is the reason why we need to generate scenarios of future interest rates. Adamo *et al.* (2004) show that, for a set of term structure evolutions and Primary Budget Surplus (PBS) realizations, such an optimization problem can be formulated as a linear programming problem with linear constraints.

Broadly speaking, this is a typical risk-management problem: once we find an optimal strategy for a specific realization of the term structure evolution, we need to determine the expenditure for interest if a different scenario takes place. As a consequence, we need to simulate the term structure under the objective measure dynamics. This requirement entails an implicit evaluation of market-price-of-risk dynamics.

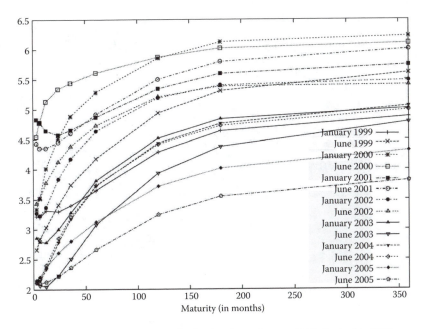

FIGURE 8.1 Evolution of the term structure in the time frame January 1999–September 2005.

The first issue to consider in the model selection process is the time frame. For the purposes of the Ministry, a reasonable planning period is 3–5 years. Within such a long period, the term structure may change significantly, as shown in Figure 8.1.

Figure 8.2 reports the monthly evolution of the swap interest rates for the following maturities: 3, 6, 12, 24, 36, 60, 120, 180, 360 months, along with the value of the European Central Bank (ECB) official rate and a simple interpolation of such rate in the same period (January 1999–September 2005). The interpolation is obtained by joining two successive jumps of the ECB official rate by means of a line. Such an interpolation mimics the evolution of interest rates, especially for short maturities, and we use it as a simple approximation of the ECB trend.

This is our basic data set for the analysis and generation of new scenarios of term structure evolution. We selected swap rates due to their availability for any maturity regardless of the actual issuance of a corresponding security.

8.3   MODELS FOR THE GENERATION OF INTEREST RATE SCENARIOS

From Figure 8.2 it is apparent that any rate (regardless of its maturity) has a strong correlation with the ECB rate. This is not surprising and we use this observation to develop an original approach to the generation of future term-structure scenarios that can be described as a multi-step process:

  i. generation of a scenario for the future ECB official rate;
 ii. generation of the fluctuations of each rate with respect to the ECB official rate;
iii. validation of the resulting scenario to determine whether it is acceptable or not.

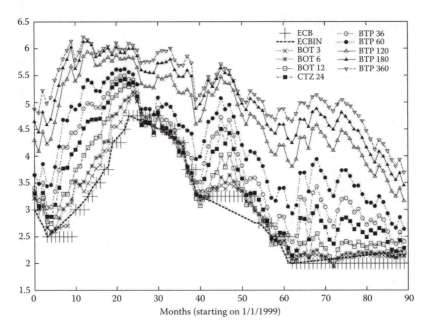

FIGURE 8.2 Evolution of interest rates with maturity from three months up to 30 years in the time frame January 1999–September 2005. The thick line is a linear interpolation of the ECB official rate (represented by +).

The basic idea is that the ECB official rate represents a 'reference rate' and all the other rates are generated by the composition of this 'reference rate' plus a characteristic, maturity-dependent, fluctuation. In other words, each rate is determined by the sum of two components: the first component is proportional to the (linearly interpolated) ECB official rate; the second component is the orthogonal fluctuation of that rate with respect to the ECB. In mathematical terms, each rate r^h is decomposed as

$$r_t^h = \alpha_h r_t^{\text{ecb}} + p_t^{h,\perp}, \tag{8.1}$$

where r_t^{ecb} is the linear interpolation of the ECB official rate and α_h is given by the expression

$$\alpha_h = \frac{\langle r^h \cdot r^{\text{ecb}} \rangle - \langle r^h \rangle \cdot \langle r^{\text{ecb}} \rangle}{\langle (r^{\text{ecb}})^2 \rangle - \langle r^{\text{ecb}} \rangle^2}, \tag{8.2}$$

where $\langle \cdot \rangle$ denotes the average value of the enclosed time series. By construction, the time series $p_t^{h,\perp}$ has null correlation with r^{ecb}. The factors α^h are different for each maturity and a larger value of α^h indicates a stronger correlation with the ECB official rate. Table 8.1 reports the value of α^h for each maturity considered. As expected, longer maturities are less correlated with the ECB official rate.

TABLE 8.1 Value of α for Different Maturities

Maturity	α_h	Maturity	α_h	Maturity	α_h
3 months	0.9833	2 years	0.7693	10 years	0.3949
6 months	0.9514	3 years	0.6695	15 years	0.3545
12 months	0.9094	5 years	0.5425	30 years	0.3237

8.3.1   Simulation of the ECB Official Rate

For the simulation of the future ECB official rate we do not employ a macroeconomic model since we assume that the interventions of the ECB can be represented as a stochastic jump process (we return to this point in Section 8.5). Some features of this process are readily apparent by looking at the evolution of the ECB official rate since January 1999 (see Figure 8.2) when the ECB official rate replaced national official rates for the Euro 'zone' countries:

- there are, on average, three ECB interventions per year;
- the ECB rate jumps (with approximately the same probability) by either 25 or 50 basis points; and
- there is a strong 'persistence' in the direction of the jumps. That is, there is a high probability that a rise will be followed by another rise and that a cut will be followed by another cut.

We model the ECB interventions as a combination of two processes: (i) a Poisson process that describes when the ECB changes the official rate; and (ii) a Markov process that describes the sign of the change. Then we resort to an exponential distribution to simulate the waiting time between two changes of the ECB official rate, and to a Markov Chain for simulation of the direction (positive or negative) of the change.

The parameter of the exponential distribution can easily be estimated from available data (that is from the waiting times, in months, between the jumps occurring in the past) by means of the Maximum Likelihood Estimation (MLE) method. It turns out to be approximately equal to three months.

Since there are two possible states (positive and negative) in the Markov chain, the corresponding transition matrix has four entries (positive–positive, positive–negative, negative–negative, negative–positive). We estimate the values of each entry by looking at the historical data and, in particular, at the probability that a change is in the same, or in the opposite, direction of the previous change. It is interesting to note that the probability of having a change in the same direction of the previous change is pretty high, approximately 85%. The width of the jump is selected between two possible values (25 or 50 basis point) with the same probability. In summary, the ECB official rate at time t is defined as

$$\text{ECB}_t = \text{ECB}_0 + \sum_{s=1}^{N_t} a_s C_{s-1,s}, \tag{8.3}$$

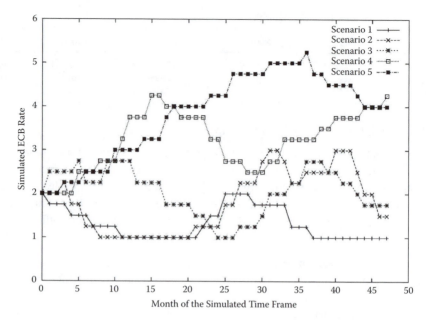

FIGURE 8.3 Simulations of the future ECB official rate for a planning period of 48 months (the initial value is always the September 2005 official rate, that is 2%).

where N_t is the realization at time t of the Poisson process that represents the total number of jumps up to time t, a_s is the random width of jump s, and $C_{s-1,s}$ represents the sign of jump s given the sign of jump $s-1$.

Moreover, in order to prevent the ECB official rate from reaching unrealistic values (e.g. negative values), we impose a lower bound on the simulated rate. This lower bound is set equal to a fixed value (currently 1%). Of course, such a value can easily be modified if the evolution of the *real* ECB rate shows that it is no longer adequate. Any jump of the ECB rate that brings it below the lower bound is discarded. Figure 8.3 shows a few simulations of the future ECB official rate produced using this approach.

8.3.2 Simulation of the Fluctuations

Figure 8.4 shows the result of decomposition (8.1) applied to the data of Figure 8.2 (only the component having null correlation with the ECB rate is shown in the figure).

First, we highlight that the historical fluctuations are correlated. This is apparent from the correlation matrix reported in Table 8.2. As a consequence, simulation of their evolution requires multivariate models in order to represent this correlation. In other words, it is not possible to model the dynamics of the fluctuations of a single rate without taking into account the dynamics of the fluctuations of all rates. To this end, we decided to follow two approaches having different features that complement each other. The first approach is based on principal component analysis.

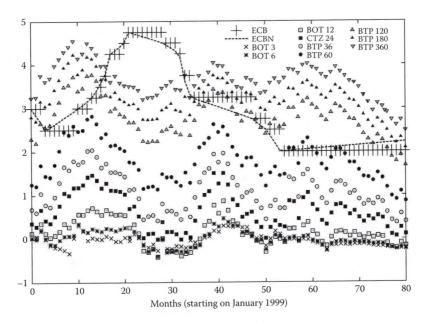

FIGURE 8.4 Evolution of the components of interest rates, $p_t^{h,\perp}$, having null correlation with the linearly interpolated ECB rate in the time frame January 1999–September 2005.

TABLE 8.2 Correlation Matrix of the Components of Each Rate Having Null Correlation with the (Linearized) ECB Official Rate

	$p^{3,\perp}$	$p^{6,\perp}$	$p^{12,\perp}$	$p^{24,\perp}$	$p^{36,\perp}$	$p^{60,\perp}$	$p^{120,\perp}$	$p^{180.\perp}$	$p^{360,\perp}$
$p^{3,\perp}$	1.000	0.822	0.625	0.533	0.505	0.490	0.431	0.384	0.311
$p^{6,\perp}$	0.822	1.000	0.937	0.867	0.827	0.779	0.682	0.612	0.520
$p^{12,\perp}$	0.625	0.937	1.000	0.970	0.934	0.877	0.770	0.697	0.610
$p^{24,\perp}$	0.533	0.867	0.970	1.000	0.991	0.956	0.869	0.803	0.717
$p^{36,\perp}$	0.505	0.827	0.934	0.991	1.000	0.986	0.922	0.866	0.786
$p^{60,\perp}$	0.490	0.779	0.877	0.956	0.986	1.000	0.971	0.931	0.862
$p^{120,\perp}$	0.431	0.682	0.770	0.869	0.922	0.971	1.000	0.990	0.949
$p^{180,\perp}$	0.384	0.612	0.697	0.803	0.866	0.931	0.990	1.000	0.981
$p^{360,\perp}$	0.311	0.520	0.610	0.717	0.786	0.862	0.949	0.981	1.000

8.3.3 Principal Component Analysis

Principal component analysis (PCA) is a well-known technique in time series analysis and has been in use for a number of years in the study of fixed income markets (Litterman and Scheinkman 1991). In general, PCA assumes that the underlying process is a diffusion. The data we employ do not have the jump components produced by the ECB interventions thanks to the decomposition procedure (8.1) described previously. From this viewpoint the data appear suitable for PCA. The procedure we follow is the standard one.

- For each de-trended rate $p_t^{h,\perp}$ we calculate the differences $\delta_t^h = p_t^{h,\perp} - p_{t-1}^{h,\perp}$ (t is the time index).
- We apply PCA to the m time series of the differences, which we indicate by $\bar{\delta}$. This means:
  - calculating the empirical covariance matrix (Λ_{ij}) of $\bar{\delta}$; and
  - finding a diagonal matrix of eigenvalues Λ^d and an orthogonal matrix of eigenvectors $E = [e_1, \ldots, e_m]$ such that the covariance matrix is factored as $\Lambda^d = E\Lambda^{\text{hist}}E^T$.

The eigenvectors with the largest eigenvalues correspond to the components that have the strongest correlation in the data set.

For the data set of the monthly yield changes from January 1999 to September 2005, the first three components represent 98.3% of the total variance in the data. The projections of the original data onto the first three principal components do not show autocorrelation phenomena (the covariance matrix is, as expected in the PCA, very close to the Λ^d matrix). Usually, the first three components are interpreted respectively as (i) a level shift, (ii) a slope change, and (iii) a curvature change. Since the PCA is applied, in our case, to de-trended data (with respect to the ECB official rate) the meaning of the components could be different. However, the plot of the first three components shown in Figure 8.5 does not seem too different from similar studies that consider yield changes directly (Baygun *et al.* 2000).

To create a new scenario for the fluctuations it is necessary to take a linear combination of the principal components. If N is the number of principal components ($N = 3$ in the present case), an easy way is to compute $\bar{\delta} = F\bar{\mu}$, where $F = [e_1, \ldots, e_N]$ is an $m \times N$ matrix composed of the eigenvectors corresponding to the first N eigenvalues, and $\bar{\mu}$ a vector with

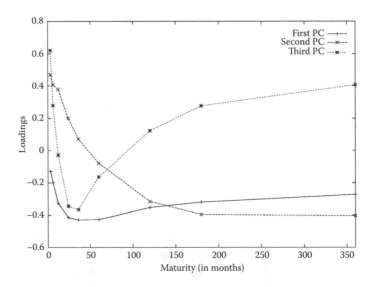

FIGURE 8.5 The three most significant components computed from monthly yield changes, January 1999–September 2005.

N elements. In principle, the value of each element of this vector could be selected at will. However, there are two common choices: the first is to assign a value taking into account the meaning of the corresponding principal component. For instance, if the purpose is to test the effect of a level shift with no slope or curvature change, it is possible to assign a value only to the first element, leaving the other two elements equal to zero. The other choice is to compute $\bar{\delta} = F\sqrt{\Lambda^d}\bar{Z}$, where $\bar{Z}$ is a vector of N independent, normally distributed, variables. Therefore, each element of the vector $\bar{\mu}$ is drawn from a normal distribution with variance equal to the eigenvalue corresponding to the principal component. In the present work, we followed the second approach since we did not want to make *a priori* assumptions. Hence, the evolution of $p_t^{h,\perp}$ is described by the following equation:

$$p_t^{h,\perp} = p_{t-1}^{h,\perp} + \sum_{j=1}^{N} F_{hj}\sqrt{\lambda_j}\, Z_j, \qquad (8.4)$$

where λ_j is the jth eigenvalue. Note that a linear combination of the principal components provides a vector of fluctuations for *one* time period only. Actually, since the planning period is 3–5 years and the time step is one month, we need, for a single simulation, a minimum of 36 up to a maximum of 60 vectors of fluctuations. Obviously, it is possible to repeat the procedure for the generation of the fluctuations, but, as a result of limited sampling, the covariance matrix of the resulting simulated fluctuations may appear quite different with respect to the covariance matrix of the historical fluctuations. Although a different behavior does not necessarily imply that the simulated fluctuations are meaningless, it is clear that we must keep this difference under control. In Section 8.4 we describe how we dealt with this issue.

Figure 8.6 shows the results of a simulation, based on the PCA, of the evolution of interest rates for a planning period of 48 months and a few samples of the corresponding term structures. The results of another simulation reported in Figure 8.7 show how this technique is able to also produce *inverted* term structures in which long-term or medium-term maturities have lower returns with respect to short-term maturities. This is not the most common situation, of course, but it may happen, as shown in Figure 8.1 (see the term structure of January 2001).

The second approach for the simulation of the fluctuations aims at maintaining a closer relationship with the historical fluctuations and assumes that each fluctuation has its own *long-term* level to which it tends to return. For this reason, we simulate the fluctuations as a set of mean-reverting processes.

8.3.4 Multivariate Mean-Reverting Models

8.3.4.1 The basic stochastic process A widely used model for the description of interest rate evolution is based on the following equation proposed by Cox, Ingersoll and Ross (1985) (CIR):

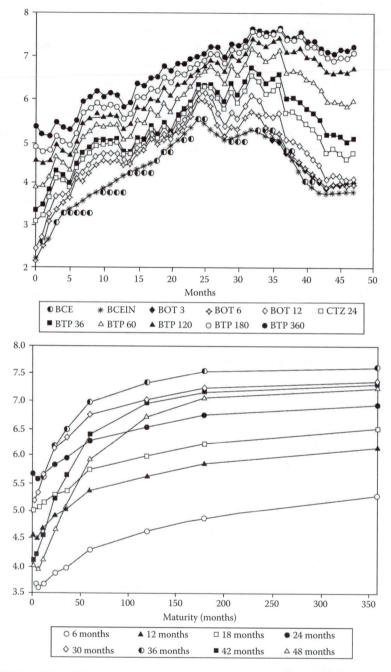

FIGURE 8.6 Simulation of the evolution of interest rates based on the PCA (top panel) and the corresponding term structures (bottom panel).

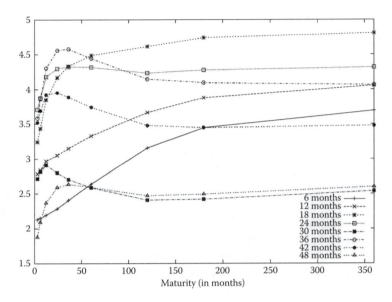

FIGURE 8.7 Another set of term structures produced by the PCA-based technique. Note how those corresponding to 30, 36, 42 and 48 months are *inverted*.

$$dr_t = k(\mu - r_t)\,dt + \sigma\sqrt{r_t}\,dB_t, \qquad (8.5)$$

where k, μ and σ are positive constants and r_t is the short-term interest rate. The CIR model belongs to the class of *mean-reverting* and exponential-affine models (Duffie and Kan 1996). This means that the short rate is elastically pulled to the constant long-term value μ (mean-reversion) and that the yield to maturity on all bonds can be written as a linear function of the state variables (affine models have been carefully analysed by Duffie *et al.* 2000). With respect to other, more simple, models belonging to the same class, such as the Vasicek (1977) model, the CIR model guarantees that the interest rates are represented by continuous paths and remain positive if condition $2k\mu > \sigma^2$ is fulfilled, and $r_0 > 0$. Moreover, the volatility is proportional to the interest rate. A review of some of the estimation methods for these models is reported in Appendix 8.A.

As already mentioned, if we used the simple one-factor model (8.5), we would neglect a fundamental point for the generation of scenarios of the future term structure, that is the correlation among fluctuations corresponding to different maturities. In order to capture this correlation element, in the following we propose a simple multi-dimensional extension of model (5) for the generation of the orthogonal fluctuations with respect to the ECB official rate. We consider the following equation:

$$dp_t^{h,\perp} = k_h(\mu_h - p_t^{h,\perp})\,dt + \sqrt{p_t^{h,\perp}}\sum_{j=1}^{M}\sigma_{hj}\,dB_t^{j}, \qquad \text{for } h = 1, \ldots, M \qquad (8.6)$$

on the interval $[0, T]$, T being the time horizon for the generation of scenarios. Here $B_t = (B_t^1, \ldots, B_t^M)$ is an M-dimensional Brownian motion representing M sources of randomness in the market, and $k_h > 0$, $\mu_h > 0$, and σ_{hj} are constants such that the matrix σ is symmetric and the following conditions hold true:

$$\det(\sigma) \neq 0, \quad k_h\mu_h > \frac{1}{2}(\sigma^2)_{hh}, \quad \forall h = 1, \ldots, M. \tag{8.7}$$

It is easy to show that the inequality (8.7) is a sufficient condition for the existence of a unique solution with positive trajectories to the stochastic differential Equation (8.6) starting at $p_0 \in (0, \infty)^M$; see, in particular, Duffie and Kan (1996). Since the var–cov matrix $\Sigma(p)\Sigma(p)^\top$, with $\Sigma_{i,j}(p) = \sqrt{p_i}\sigma_{ij}$, $i, j = 1, \ldots, M$, is not an affine function of p, model (8.6) does not belong to the affine yield class (Duffie et al. 2000) unless σ is diagonal, in which case the model reduces to a collection of M uncorrelated CIR univariate processes. However, model (8.6) preserves some features of the class studied in Duffie anf Kan (1996) and Duffie et al. (2000), such as analytical tractability and convenient econometric implementation, while the volatility of each component is proportional only to the component itself. Below, we focus our attention on the problem of the estimation of (8.6).

8.3.4.2. *The estimation of model (8.6)* There is a growing literature on estimation methods for term structure models; see, in particular, Ait-Sahalia (2002). A vast literature is specifically devoted to the estimation of affine models (see Balduzzi et al. 1996 and references therein). Here we shall discuss an *ad hoc* method to estimate model (8.6), which is based on a discrete-time maximum likelihood method (MLE). It is now well recognized that discretization of continuous-time stochastic differential equations introduces a bias in the estimation procedure. However, such a bias is negligible when the data have daily frequency (Bergstrom 1988). For this reason we use daily swap data in this case. In discrete time, the process in Equation (8.6) becomes

$$p_{t_{i+1}}^h = p_{t_i}^h + k_h(\mu_h - p_{t_i}^h)\Delta + \sqrt{p_{t_i}^h\Delta} \sum_{j=1}^{M} \sigma_{hj} Z_i^j, \quad i = 1, \ldots, n-1, \tag{8.8}$$

for $h = 1, \ldots, M$, where $t_i = i\Delta$ and $\Delta = T/n$. Here, $Z_i = (Z_i^1, \ldots, Z_i^M)$, $i = 1, \ldots, n-1$, are independent multivariate normal random variables with zero mean and covariance matrix I_M, I_M being the identity matrix of order M. We observe that the distributional properties of the process (8.8) depend only on k, μ and σ^2. Therefore, in order to generate scenarios from model (8.8), it suffices to know an estimate of this matrix. For this, we introduce the matrix $\Gamma = [\sigma^2]^{-1}$. Estimation involves maximizing the log-likelihood function associated with a sequence of observations $\hat{p}_{t_i}^h > 0$:

$$\ell_n = \zeta + \frac{n-1}{2}\log(\det\Gamma) - \frac{1}{2}\sum_{i=1}^{n-1} e_i^\top \Gamma e_i, \tag{8.9}$$

where $\zeta = -[(n-1)M\log(2\pi)]/2$ and

$$e_i^h := \epsilon_i^h + \eta_i^h k_h \mu_h + \psi_i^h k_h, \tag{8.10}$$

ϵ_i^h, η_i^h and ψ_i^h being constants independent of the parameters of the model, given by

$$\epsilon_i^h = \frac{\hat{p}_{t_{i+1}}^h - \hat{p}_{t_i}^h}{\sqrt{\hat{p}_{t_i}^h \Delta}}, \quad \eta_i^h = -\sqrt{\frac{\Delta}{\hat{p}_{t_i}^h}}, \quad \psi_i^h = \sqrt{\hat{p}_{t_i}^h \Delta}, \tag{8.11}$$

for every $i = 1, \dots, n-1$ and $h = 1, \dots, M$. We reduce the number of parameters from $2M + M^2$ to $2M + [M(M+1)/2]$ by introducing the following new variables:

$$\Gamma = C^\top C, \quad \alpha_h = k_h \mu_h, \quad h = 1, \dots, M, \tag{8.12}$$

where C is a lower triangular matrix with strictly positive entries on the main diagonal. (using the Cholesky factorization of $\Gamma^{-1} = A^\top A$, with A upper triangular, $C = [A^{-1}]^\top$). We associate with this matrix a vector $c \in \mathbb{R}^{M(M+1)/2}$ according to the relation

$$C_{i,j} = c_{[i(i-1)/2]+j}, \quad \forall M \geq i \geq j \geq 1. \tag{8.13}$$

Using these variables, we have $\det \Gamma = [\det C]^2 = \Pi_{h=1}^M C_{hh}^2$, while (8.9) can be rewritten as follows:

$$\ell_n(c, \alpha, k) = \zeta + (n-1) \sum_{h=1}^M \log(C_{hh}) - \frac{1}{2} \sum_{i=1}^{n-1} |Ce_i|^2. \tag{8.14}$$

Therefore, the calibration of model (8.6) can be obtained by computing the maximum of ℓ_n for $\alpha, k \in (0, \infty)^M$ and $c \in \mathbb{R}^{M(M+1)/2}$, satisfying $c_{h(h+1)/2} > 0$, for any $h = 1, \dots, M$. We observe that the computation can be reduced to the maximum of

$$U(\alpha, k) = \sup_c \ell_n(c, \alpha, k) \tag{8.15}$$

on $(0, \infty)^M \times (0, \infty)^M$. Since ℓ_n is concave compared with c, it is easy to show that U is well defined and, for every $(\alpha, k) \in (0, \infty)^{2M}$, there is a $c^* = c^*(\alpha, k)$ such that

$$U(\alpha, k) = \ell_n(c^*, \alpha, k) \tag{8.16}$$

holds true. The optimizer c^* is related, via relation (8.13), to the lower triangular matrix $(R_{ij}^*/\sqrt{R_{ii}^*})_{i \geq j}$ defined by

$$\begin{pmatrix} R_{i1}^* \\ R_{i2}^* \\ \vdots \\ R_{ii}^* \end{pmatrix} = V_i^{-1} \begin{pmatrix} 0 \\ 0 \\ \vdots \\ n-1 \end{pmatrix}, \quad i = 1, \dots M, \tag{8.17}$$

where

$$(V_i)_{h_1 h_2} = (V_i(\alpha, k))_{h_1 h_2} := [e^{h_1}]^\top e^{h_2}, \quad h_1, h_2 = 1, \dots, i. \tag{8.18}$$

For details of this result, see Papi (2004).

Summarizing, it is possible to select an optimal value of $C = C(\alpha, k)$ and then calculate the maximum of the function U which depends only on $2M$ variables, thus reducing the computational burden. Since we do not know whether the function U is concave or not, it is not possible to resort to gradient methods in order to find a global maximum. For this reason, we employ a stochastic algorithm based on Adaptive Simulated Annealing (ASA) combined with Newton's method. Table 8.3 reports representative results of this reduction method applied to the estimation of the parameters of model (8.6).

Note that Equation (8.6) may be generalized as follows:

$$dp_t^{h,\perp} = k_h(\mu_h - p_t^{h,\perp})\,dt + [p_t^{h,\perp}]^\nu \sum_{j=1}^M \sigma_{hj}\,dB_t^j, \quad \text{for } h = 1, \dots, M, \tag{8.19}$$

where the exponent ν belongs to the interval $[0, 1.5]$ (Duffie 1996) (this restriction is needed for the existence and uniqueness of solutions).

The discrete-time maximum-likelihood method does not involve particular difficulties with respect to the case we have dealt with (i.e. $\nu = 0.5$). More precisely, the reduction method described by (8.12)–(8.18) can be easily adapted to this more general situation. Table 8.4 reports the results of a comparison (for the sake of simplicity, only two maturities are considered) between the estimates provided by model (8.6) and a simpler pure-Gaussian version (i.e. $\nu = 0$) corresponding to a multivariate Vasicek model.

Path generation using model (8.6) is carried out by means of the discrete version (8.8). The generation of a scenario with time frequency $\Delta > 0$ assigns the last observed data to

TABLE 8.3 MLE of the Multivariate Model. Parameters for Some Representative Maturities Provided by the Multivariate Model (8.6) of the Orthogonal Components ($p_t^{h,\perp}$ of Equation (8.1)) of Italian Treasury Rates. The Estimate is Obtained from Daily Data of the 1999–2005 Period by Means of the Techniques Described in Section 8.3.4.2 using Adaptive Simulated Annealing

Maturity	k	μ	σ_6	σ_{36}	σ_{60}	σ_{360}
6 months	4.3452	0.4353	0.7356	0.0421	0.0081	−0.0167
36 months	3.4351	2.0429	0.0421	0.2994	0.1871	0.0874
60 months	3.4195	2.8469	0.0081	0.1871	0.2455	0.1010
360 months	4.3450	4.5369	−0.0167	0.0874	0.1010	0.1993

TABLE 8.4 Comparison of the Parameters of the Multivariate CIR Model (Equation (8.8)) and a Multivariate Vasicek (i.e. Pure Gaussian) Model, Corresponding to Equation (8.19) with $v = 0$. For the Estimation We Resorted to the Discrete Maximum-Likelihood Method Applied to Daily Observations in the Period January 1999–September 2005 of the $p_t^{h,\perp}$ Components of Two Maturities (60 and 120 Months, Indicated, Respectively, as 1 and 2). The Values of the t-statistics in the Multivariate CIR Model and the Larger Log-likelihood Indicate that Model (8.8) provides a Better Fit of the Original Data with Respect to the Pure-Gaussian Model

	Model			
	Multivariate CIR		Multivariate Vasicek	
Parameter	Estimate	t-Stat.	Estimate	t-Stat.
k_1	1.6208	17.1201	1.4542	17.2518
k_2	1.4175	15.0218	1.4243	16.4660
μ_1	1.6959	0.0917	1.6992	0.0950
μ_2	2.6422	0.5381	2.6557	0.5463
σ_{11}	0.5545	2.0263	0.6667	34.190
σ_{12}	0.2788	2.0269	0.4079	24.419
σ_{22}	0.3302	1.4483	0.5686	33.843
Log-likelihood	-1001.925337		-2206.122226	

p_1^h for every $h = 1, \dots, M$. Then, at each time step i, where $1 \leq i \leq n-1$, we generate independent random vectors z_i from the multivariate normal distribution $N(0, I_M)$, and we set

$$p_{i+1}^h = p_i^h + k_h(\mu_h - p_i^h)\Delta + \sqrt{p_i^h \Delta} (C^{-1} \cdot z_i)_h, \tag{8.20}$$

for $i = 1, \dots, n-1$, where k, μ and C are the MLE estimators.

Figure 8.8 shows the results of a simulation, based on the multivariate CIR model described in this section, for a planning period of 48 months.

8.4 VALIDATION OF THE SIMULATED SCENARIOS

The stochastic models presented above allow the generation of a 'realistic' future term structure. However, we need to generate a (pretty long) *temporal sequence* of term structures starting from the present interest rates curve. As mentioned in Section 8.3, this requires control of the evolution in time of the simulated term structure in such a way that, for instance, its behavior is not too different from the behavior observed in the past.

Besides control on the value of the simulated ECB with respect to a predefined lower bound as described in Section 8.3.1, we resort to two techniques to assess the reliability of the sequence of simulated term structures. The first method, which we classify as 'local,' ensures that, at each time step of the planning period, the simulated term structure is 'compatible' with the historical term structure. The second (which we classify as 'global') considers the whole term structure evolution, and ensures that the correlation among the increments of $p_t^{h,\perp}$ are close to the correlation of the increments of the historical fluctuations.

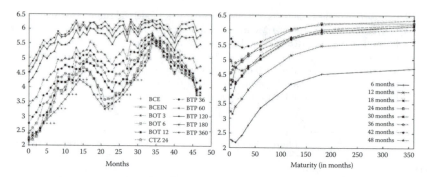

FIGURE 8.8 Simulation of the evolution of interest rates based on the multivariate CIR model (left panel) and the corresponding term structures (right panel).

The local test is based upon the observation that the shape of the term structure does not vary widely over time. Obviously, there is some degree of variability. For instance, it is known that the term structure is usually an increasing function of the maturity, but, at times, it can be inverted for a few maturities. This is the fundamental assumption of Nelson and Siegel's (1987) parsimonious model of the term structure.

In the present case, since we are not interested in a functional representation of the term structure but in the interest rates at fixed maturities, we take as the indicator of the term structure shape the relative increment and the local convexity of the interest rates:

$$d_t^h = r_t^h - r_t^{h-1}, \quad c_t^h = \tilde{r}_t^h - r_t^h, \tag{8.21}$$

where $\tilde{r}_t^h$ is the value at m^h of the linear interpolation between r_t^{h+1} and r_t^{h-1}:

$$\tilde{r}_t^h = r_t^{h-1} + \frac{r_t^{h+1} - r_t^{h-1}}{m^{h+1} - m^{h-1}}(m^h - m^{h-1}).$$

A positive convexity $(c_t^h > 0)$ means that the rate r_t^h at maturity m^h is below the line joining the rates r_t^{h-1} and r_t^{h+1} at maturities m^{h-1} and m^h, respectively, therefore the curvature opens upward. The case of $c_t^h < 0$ can be interpreted along the same lines, but the curvature opens downward.

At any time step, we accept the simulated interest rates if the corresponding values of d_t^h and c_t^h are not too different from the historical values. Briefly, we compute the historical mean and standard deviation (μ_d^h, σ_d^h) and (μ_c^h, σ_h^c) of d_t^h and c_t^h, respectively, and then we check that

$$\sum_{h=1}^{M-1} \frac{1}{(\sigma_d^h)^2} (d_t^h - \mu_d^h)^2 \le (M-1),$$

$$\sum_{h=2}^{M-1} \frac{1}{(\sigma_c^h)^2} (c_t^h - \mu_c^h)^2 \le (M-2). \tag{8.22}$$

The meaning of such tests is straightforward. If the simulated interest rates pass the tests, we are quite confident that their increments and the local convexity have statistical behaviour similar to the historical rates.

The global test controls the correlation among the increments of $p_t^{h,\perp}$, since it can be quite different from the correlation of the historical data. In order to avoid 'pathological' situations like anti-correlated increments or increments that are too correlated with each other, we compute the 'one-norm' of the matrix difference between the correlation matrix of the increments of the historical fluctuations and the correlation matrix of the simulated increments. The 'one-norm' is defined as follows: for each column, the sum of the absolute values of the elements in the different rows of that column is calculated. The 'one-norm' is the maximum of these sums:

$$\|\Sigma_{i,j}^{\text{hist}} - \Sigma_{i,j}\|_1 = \sup_i \sum_{j=1}^{M} |\Sigma_{i,j}^{\text{hist}} - \Sigma_{i,j}|. \tag{8.23}$$

We compare the result of this calculation with a predefined acceptance threshold. If the 'one-norm' is below the threshold, the simulated scenario is accepted, otherwise it is rejected and a new set of fluctuations is generated. Currently, the threshold is set equal to 0.05. Since the correlation is a number within the range $[-1,1]$, this simple mechanism guarantees that the covariance among the increments of the interest rates of different maturities in the simulated scenarios is pretty close to the historical covariance.

Note how the local and global tests are complementary since the first test involves the shape of the term structure, whereas the second test concerns how the rates' increments are correlated in time, that is, how the term structure evolves from one time step of the simulation to the next.

8.5 CONCLUSIONS AND FUTURE PERSPECTIVES

The management of public debt is of paramount importance for any country. Such an issue is especially important for European countries after the definition of compulsory rules by the Maastricht Treaty. Together with the Italian Ministry of Economy and Finance, we have studied the problem of finding an optimal strategy for the issuance of public debt securities. This turned out to be a stochastic optimal control problem where the evolution of the interest rates plays a crucial role.

We have presented the techniques we employ to simulate the future behaviour of interest rates for a wide range of maturities (from 3 months up to 30 years). Since the planning period that we consider is pretty long (up to five years), most existing models need to be modified in order to provide realistic scenarios. In particular, the evolution of the 'leading' rate (which we assume to be the European Central Bank's official rate) is simulated as if it were a stochastic jump process. We are aware that there is a long tradition of studies that try to explain the links between prices, interest rates, monetary policy, output and inflation. Webber and Babbs (1997) were among the first to propose yield-curve models able to capture some aspects of monetary policies. Piazzesi (2001) introduced a class of linear-quadratic jump-diffusion processes in order to develop an arbitrage-free time-series model of yields in continuous time that incorporates the central bank policy and estimates the model with U.S. interest rates and the Federal Reserve's target rate.

A general framework for these models has recently been presented by Das and Chacko (2002), where the authors introduce factors that influence the marginal productivity of capital, and thus the interest rates, in the economy. Their technique is general, since it applies to any multi-factor, exponential-affine term structure model with multiple Wiener and jump processes.

From an empirical point of view, there is some evidence that important central banks, like the U.S. Federal Reserve, conduct a monetary policy (i.e. set the official rate) that is well described by the so-called Taylor's rule (Taylor 1993). Basically, the rule states that the 'real' short-term interest rate (that is, the interest rate adjusted for inflation) should be determined according to three factors: (1) where the actual inflation is relative to the targeted level that the central bank wishes to achieve, (2) how far economic activity is above or below its 'full employment' level, and (3) what the level of the short-term interest rate is that would be consistent with full employment. Although Taylor's rule appears to be more robust than more-complex rules with many other factors, it requires knowledge of inflation and the real Gross Domestic Product (GDP). The simulation of future inflation and GDP is far from easy, so, in some sense, the application of Taylor's rule changes, but does not solve, the problem of generating meaningful scenarios for the evolution of the ECB official rate.

The models proposed in the present chapter are fully integrated into the software prototype in use at the Ministry. The time required by the simulations (a few seconds on a personal computer) is such that we can afford on-line generation of the simulated scenarios even if the validation procedures described in Section 8.4 may require multiple executions before a scenario is accepted.

Open problems and future analysis directions include the following:

- Modeling of the Primary Budget Surplus, Gross Domestic Product and Inflation in order to implement Taylor's rule and possibly other models for the evolution of the European Central Bank's official rate.
- Overcoming the assumption that interest rates are independent of the portfolio of existing securities and independent of the new securities issued every month. To limit the impact on the optimization problem, we should devise a description of these interactions that is compatible with the linear formulation of the problem.

ACKNOWLEDGEMENTS

We thank A. Amadori, D. Guerri and B. Piccoli for useful discussions. This work has been partially supported by a FIRB project of the Italian Ministry of Research.

REFERENCES

Adamo, M., Amadori, A.L., Bernaschi, M., La Chioma, C., Margio, A., Piccoli, B., Sbaraglia, S., Uboldi, A. and Vergni, D., Optimal strategies for the issuances of public debt securites. *Int. J. Theor. Appl. Finan.*, 2004, **7**, 805–822.

Ait-Sahalia, Y., Maximum-likelihood estimation of discretely-sampled diffusions: A closed-form approach. *Econometrica*, 2002, **70**, 223–262.

Balduzzi, P., Das, S., Foresi, S. and Sundaram, R., A simple approach to three factor affne models of the term structure. *J. Fixed Income*, 1996, **3**, 43–53.

Baygun, B., Showers, J. and Cherpelis, G., *Principles of Principal Components*, 2000 (Salomon Smith Barney).

Bergstrom, A.R., The history of continuous time econometric models. *Economet. Theory*, 1988, **4**, 365–383.

Bibby, B.M. and Sorensen, M., Martingale estimation functions for discretely observed diffusion processes. *Bernoulli*, 1995, **1**, 17–39.

Cox, J.C., Ingersoll, J.E. and Ross, S.A., A theory of the term structure of interest rates. *Econometrica*, 1985, **53**, 385–487.

Dacunha-Castelle, D. and Florens-Zmirou, D., Estimation of the coefficients of a diffusion from discrete observations. *Stochastics*, 1986, **19**, 263–284.

Das, S. and Chacko, G., Pricing interest rate derivatives: A general approach. *Rev. Finan. Stud.*, 2002, **15**, 195–241.

Duffie, D., *Dynamic Asset Pricing Theory*, 1996 (Princeton University Press: Pinceton, NJ).

Duffie, D. and Kan, R., A yield-factor model of interest rates. *Math. Finan.*, 1996, **6**, 379–406.

Duffie, D., Pan, J. and Singleton, K., Transform analysis and asset pricing for affine jump-diffusions. *Econometrica*, 2000, **68**, 1343–1376.

Jackson, D., *The New National Accounts: An Introduction to the System of National Accounts 1993 and the European System of Accounts 1995*, 2000 (Edward Elgar: Camberley, U.K.).

James, J. and Webber, N., *Interest Rate Modelling*, 2000 (Wiley: New York).

Litterman, R. and Scheinkman, J., Common factor affecting bond returns. *J. Fixed Income*, 1991, **1**, 54–61.

Nelson, C.R. and Siegel, A.F., Parsimonious modeling of yield curves. *J. Bus.*, 1987, **60**, 473–489.

Papi, M., Term structure model with a central bank policy. In preparation, 2004.

Piazzesi, M., An econometric model of the yield curve with macroeconomic jump effects. NBER Working Paper Series, 2001.

Risbjerg, L. and Holmlund, A., Analytical framework for debt and risk management. *Advances in Risk Management of Government Debt*, 2005 (Organization for Economic Cooperation and Development).

Taylor, J.B., Discretion versus policy rules in practice. *Carnegie–Rochester Conf. Ser. Public Policy*, 1993, **39**, 195–214.

Vasicek, O., An equilibrium characterization of the term structure. *J. Finan. Econ.*, 1977, **5**, 177–188.

Webber, N. and Babbs, S., *Mathematics of Derivative Securities*, 1997 (Cambridge University Press: Cambridge).

APPENDIX 8.A: ESTIMATION WITH DISCRETE OBSERVATIONS

We deal with the case of a diffusion process X_t (i.e. the interest rate r_t) that is observed at discrete times $0 = t_0 < t_1 < \cdots < t_n$, not necessarily equally spaced.

If the transition density of X_t from y at time s to x at time t is $p(x,t,y,s; \theta)$, where θ is a collection of parameters to be estimated, we can resort to the Maximum Likelihood Estimator (MLE) $\hat{\theta}_n$ which maximizes the likelihood function

$$L_n(\theta) = \prod_{i=1}^{n} p(X_{t_i}, t_i, X_{t_{i-1}}, t_{i-1}, \theta), \qquad (8.A1)$$

or, equivalently, the log-likelihood function

$$\ell_n(\theta) = \log(L_n(\theta)) = \sum_{i=1}^{n} \log(p(X_{t_i}, t_i, X_{t_{i-1}}, t_{i-1}; \theta)). \qquad (8.A2)$$

In the case of observations equally spaced in time, the consistency and asymptotic normality of $\hat{\theta}_n$ as $n \to \infty$ can be proved (Dacunha-Castelle and Florens-Zmirou 1986). In general, the transition density of X_t is not available. In this case the classical alternative estimator is obtained by means of an approximation of the log-likelihood function for θ based on continuous observations of X_t. Unfortunately, this approach has the undesirable property that the estimators are biased, unless $\pi = \max_i |t_i - t_{i-1}|$ is small. To overcome the difficulties due to the dependence of the parameters on π, different solutions have been proposed. One of the most efficient methods resorts to martingale estimating functions (Bibby and Sorensen 1995). This method is based on the construction of estimating equations having the following form:

$$G_n(\theta) = 0, \qquad (8.A3)$$

where

$$G_n(\theta) = \sum_{i=1}^{n} g_{i-1}(X_{t_{i-1}}; \theta)(X_{t_i} - \mathbb{E}(X_{t_i} \mid X_{t_{i-1}})), \qquad (8.A4)$$

g_{i-1} being continuously differentiable in θ, for $i = 1, \ldots, n$. Bibby and Sorensen (1995) proved, under technical conditions on g_{i-1}, that an estimator that solves Equation (8.A3) exists with probability tending to one as $n \to \infty$, and this estimator is consistent and asymptotically normal.

A simpler approach, which can be used when π is sufficiently small, is based on the Euler discretization of the diffusion equation associated with X_t. In this case, one can use the log-likelihood approach since the transition density can be easily computed. We discuss the application of this method to (8.5). Let

$$r_{t_{i+1}} = r_{t_i} + k(\mu - r_{t_i})\Delta + \sigma\sqrt{r_{t_i}}\Delta Z_i, \quad i = 1, \ldots, n-1, \tag{8.A5}$$

be the first-order discretization of Equation (8.5) on time interval $[0, T]$, where $\Delta = T/n$ and with Z_i the increment ΔB_i of the Brownian motion between $t_i = i\Delta$ and $t_{i+1} = (i+1)\Delta$. Since these increments are independent $N(0, \Delta)$ distributed random variables, the transition density from $y > 0$ to x during an interval of length Δ is

$$p(x \mid y; \theta) = \frac{1}{\sqrt{2\pi y\sigma^2\Delta}} \exp\left(-\frac{1}{2y\sigma^2\Delta}(x - k(\mu - y)\Delta)^2\right), \tag{8.A6}$$

where $\theta = (k, \mu, \sigma)$. Therefore, given a sequence of observations $\{r_i\}_i$, an estimator of (8.32) is obtained by maximizing the log-likelihood function

$$\max_\theta \sum_{i=1}^{n-1} \log(p(r_{t_{i+1}} \mid r_{t_i}; \theta)). \tag{8.A7}$$

If one expands the function in (8.34) and sets equal to zero its partial derivatives with respect to (k, μ, σ^2), it is easy to show that the corresponding equations admit a unique solution that is a maximum point of the log-likelihood function. The following relations represent this *ML* estimator:

$$\mu = \frac{EF - 2CD}{2EB - FD}, \quad k = \frac{D}{2\mu B - F}, \tag{8.A8}$$

$$\sigma^2 = \frac{1}{n-1}(A + k^2\mu^2 B + k^2 C - k\mu D + kE - k^2\mu F), \tag{8.A9}$$

TABLE 8.A1 Assessment of the Discrete Estimator (1D case). Results for a Simulation Evaluation of the Log-likelihood Estimator to the Univariate CIR Model. Using the True Values of the Parameters We Simulated 500 Sample Paths of Length 2610 Daily Observations Each. For Each Sample Path We Undertook Discrete MLE Estimation Via Euler Discretization. The Table Presents Summary Statistics of the Simulated Estimations. We Computed the Mean and Standard Error for the Estimator $(\hat{k}, \hat{\mu}, \hat{\sigma})$, and Computed t-Statistics for the Difference between the Simulated Parameter Estimate and the True Parameter. The Null Hypothesis Cannot be Rejected at the 5% Significance Level

	Parameter		
Statistics ($N=500$)	k	μ	σ
True parameter	0.8	3.24	0.03
Estimated parameter	0.79928	3.2405	0.030009
Standard error	0.00081491	0.0011	0.00001936
t-Statistics	0.8835	0.4545	0.4649

TABLE 8.A2 MLE of the Univariate CIR Model. Results for the Estimation of the Univariate CIR Model of $p_t^{h,\perp}$ of Equation (8.1). Estimation was Carried Out Using Discrete Maximum-likelihood (8.34). We Report the Value of the Log-likelihood (Log (L)), the Norm of its Gradient (|∇Log (L)|) and the Maximum Eigenvalue (l_L) of its Hessian Matrix at the Maximum Point

Parameter	BOT3	BOT6	BOT12	CTZ	
k	5.4611	4.0217	2.1634	1.1265	
μ	0.32712	0.43815	0.72852	1.457	
σ	0.78513	0.76817	0.74801	0.57573	
log (L)	1866.541	2174.4863	1920.0013	1791.6797	
|∇log (L)|	1.2268×10^{-13}	3.0103×10^{-15}	6.3497×10^{-15}	2.2739×10^{-13}	
λ_L	-0.54044	-1.1819	-1.5844	-1.7826	
	BTP36	BTP60	BTP120	BTP180	BTP360
k	1.1488	1.4039	1.7635	1.9973	2.4133
μ	2.0531	2.8686	3.9144	4.2957	4.5355
σ	0.50163	0.4293	0.33892	0.31716	0.30738
log (L)	1756.6086	1743.7829	1822.044	1839.9547	1840.362
|∇log (L)|	1.4118×10^{-14}	0	9.0949×10^{-13}	0	1.504×10^{-13}
λ_L	-1.8906	-1.8071	-1.5539	-1.3116	-0.93247

where

$$A = \sum_{i=0}^{n-1} \frac{(r_{i+1} - r_i)^2}{r_i \Delta}, \quad B = \Delta \sum_{i=0}^{n-1} \frac{1}{r_i}, \quad C = \Delta \sum_{i=0}^{n-1} r_i,$$

$$D = 2 \sum_{i=0}^{n-1} \frac{r_{i+1} - r_i}{r_i}, \quad E = 2(r_n - r_0), \quad F = 2\Delta(n-1). \tag{8.A10}$$

Here, r_i is the interest rate at time t_i. We report the results of this method in Table 8.A1. In Table 8.A2 the method is applied to each orthogonal component $p_t^{h,\perp}$.

CHAPTER 9

Solving ALM Problems via Sequential Stochastic Programming

FLORIAN HERZOG, GABRIEL DONDI, SIMON KEEL,
LORENZ M. SCHUMANN and HANS P. GEERING

CONTENTS

9.1 INTRODUCTION

I N RECENT YEARS, A GROWING NUMBER of real-world applications of asset liability management (ALM) with discrete-time models have emerged. Insurance companies and pension funds pioneered these applications, which include the Russell–Yasuada investment system (Carino and Ziemba 1998), the Towers–Perrin System (Mulvey 1995), the Siemens Austria Pension Fund (Ziemba 2003; Geyer *et al.* 2004), and Pioneer Investment guaranteed funds (Dempster *et al.* 2006). In each of the applications, the investment decisions are linked to liability choices, and the funds are maximized over time using multi-stage stochastic programming methods. Other examples of the use of stochastic programming to solve dynamic ALM problems are given by Dempster and Consigli (1998) and Dondi (2005).

All authors propose stochastic programming as the most suitable solution framework for ALM problems. However, since most of the asset or liability models are dynamic stochastic models, the ALM problem is one of dynamic optimization which can be solved by applying the continuous state dynamic programming (DP) algorithm. In this chapter, we show that the DP algorithm can be approximated locally by stochastic programming (SP) methods. By using a sufficient number of scenarios, the difference between the exact solution and the approximation can be made arbitrarily small. The SP optimization is re-solved for every time-step based on a new set of stochastic scenarios that is computed according to the latest conditional information. In this way, a feedback from the actual observed state is introduced which is not from the coarse scenario approximation from an earlier time-step. This procedure, often called rolling-horizon planning, is frequently used as heuristic, but we show here that by carefully posing and computing the SP optimization, the continuous state DP algorithm is being applied approximately.

This DP approximation is applied to the problem of a fund that guarantees a minimal return on investments and faces transaction costs when investing. The situation resembles the problem faced by Swiss pension funds or German life insurance policies. The minimal return guarantee changes the fund problem from a pure asset allocation problem to an ALM problem, see Dempster *et al.* (2007). First, models of the asset returns and portfolio dynamics with transaction costs are introduced. Then we propose that the most suitable risk measure for such a situation is a shortfall risk measure that penalizes all possible scenarios for the future in which the minimal return is not achieved. The optimization problem is solved with the aim of maximizing the return above the guarantee over the planning horizon, while limiting the shortfall risk. The problem is tested in an eight year out-of-sample backtest with a quarterly trading frequency from the perspective of a Swiss Fund that invests domestically and in the EU markets and faces transaction costs.

The chapter is organized as follows: in Section 9.2, the approximation of dynamic programming is introduced. In Section 9.3, we present the application of the stochastic programming approximation to a portfolio problem with transaction costs. Additionally, we introduce a dynamically coherent risk measure for asset liability situations. In Section 9.4, a case study is examined from the perspective of a Swiss fund that invests domestically in stocks, bonds and cash, as well as in the EU stock and bond markets. Section 9.5 concludes the chapter.

9.2 STOCHASTIC PROGRAMMING APPROXIMATION OF THE CONTINUOUS STATE DYNAMIC PROGRAMMING ALGORITHM

We discuss how to use dynamic stochastic programming as one possible approximation of dynamic programming (DP). It is well known that any dynamic optimization problem (DOP) in discrete-time may be solved in principle by employing the DP algorithm. However for situations with realistic assumptions, such as control constraints or non-Gaussian white noise processes, it is very difficult to obtain closed-form solutions. In these cases it is necessary to resort to numerical solution methods.

9.2.1 Basic Problem and Description of the Algorithm

In this chapter, we approximate the continuous state dynamic programming method by discretizing not the state space but the possible outcomes of the white noise process. Starting from the current state value the DP approximation solves a SP problem over a finite horizon at each sampling time. The optimization procedure yields an optimal control sequence over the planning horizon but only the first control decision of the sequence is applied to the system and the rest are disregarded. At the next time-step, the calculation is repeated based on the new value of the state variable and over a shifted horizon, which leads to a receding horizon policy. The receding horizon policy means that we solve the multi-period decision problem always with the same number of periods to the horizon. Other authors have investigated the same idea in a portfolio optimization context, see Brennan *et al.* (1997), but based on numerical solutions of the Hamilton–Jacobi–Bellman PDE to solve the optimal control problem. A useful advantage of the present DP approximation is the capability to solve DP problems using established methods from stochastic programming and their known properties and algorithms.

Other techniques, such as discrete DP solutions, often lead to computationally overwhelming tasks which often prevent their application. The technique proposed here however, solves the problem only for the current state and the approximate trajectory of the underlying dynamical system and thus avoids the curse of dimensionality. In Figure 9.1, the approximation algorithm is summarized; in Table 9.1 it is described in detail.

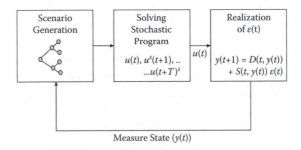

FIGURE 9.1 Graphical description of the approximation algorithm.

TABLE 9.1 Algorithm to Compute the Stochastic Programming Approximation

---

1. Based on the information at time t, determine $y(t)$. Set an accuracy parameter η, set $\underline{J} = \infty$, and define the number of scenarios per period. Start with a relatively low number of scenarios.
2. Compute the scenario approximation of $\epsilon(\tau)$ for $\tau = t, t+1, \ldots, T-1$ based on the number of samples per time period.
3. Solve the multi-stage stochastic program as outlined in Proposition 9.1.
4. In the case that $|J^s(t, y(t)) - \underline{J}| < \eta$ stop and go to step 5. Otherwise set $\underline{J} := J^s(t, y(t))$ and increase the number of scenarios to refine the approximation, see Section 9.4.2.1. Go to step 2.
5. Apply only the first control decision $\overline{u}(t)$ and disregard all future control decisions for the next time-step. Until the optimization has reached its fixed end date (horizon) go to step 1, else stop.

A general DOP can be stated as

$$J(t, y(t)) := \max_{\overline{u} \in \mathcal{U}} \left\{ \mathbb{E} \left[\sum_{\tau=t}^{T-1} L(\tau, y(\tau), \overline{u}(\tau)) + M(T, y(T)) \right] \right\} \tag{9.1}$$
$$\text{s.t. } y(\tau+1) = D(\tau, y(\tau), \overline{u}(\tau)) + S(\tau, y(\tau), \overline{u}(\tau))\epsilon(\tau),$$

where $\tau = t, t+1, \ldots, T-1$, $L(\cdot)$ and $M(\cdot)$ are the strictly concave value functionals, $D(\cdot)$ and $S(\cdot)$ define the state dynamics and are assumed to be continuously differentiable functions, $\overline{u}(\tau)$ is a bounded control vector constrained to the convex set $\mathcal{U}$ and $\epsilon(\tau)$ is a strictly covariance stationary white noise process. It is assumed that all functionals $L(\cdot)$, $M(\cdot)$, $D(\cdot)$ and $S(\cdot)$ are Lipschitz continuous and fulfil the necessary conditions for this dynamic optimization problem to be well defined and possesses a unique solution. For details refer to Bertsekas and Shreve (1978). Note that the white noise process is stationary, but the dynamics of $y(\tau)$ are both state and time dependent since the functional $S(\tau, y(\tau), \overline{u}(\tau))$ depends on both $y(\tau)$ and τ.

In order to obtain a feedback solution to the DOP problem (9.1), the dynamic programming (DP) algorithm is given by

$$J(T, y(T)) = M(T, y(T)),$$
$$J(\tau, y(\tau)) = \max_{\overline{u}(\tau) \in \mathcal{U}} \left\{ \mathbb{E}[L(\tau, y(\tau)) + J(\tau+1, D(\tau, y(\tau), \overline{u}(\tau)) + S(\tau, y(\tau), \overline{u}(\tau)), \epsilon(\tau))] \right\}.$$
$$\tag{9.2}$$

This condition for optimality can be found in Bertsekas (1995, Chapter 1). Instead of solving the DP algorithm for J to yield the true stochastic dynamics, we locally approximate the stochastic dynamics by a finite number of scenarios at the current decision time and solve the problem repeatedly at each decision time-step.

The standard DP procedure discretizes the state space for each dimension of $y(t)$ and each time-step until the horizon T. Then, beginning at time T, the optimization problem of (9.2) is solved for each discretized state of $y(t)$. Based on the optimal solution for each state discretization of $y(t)$ the optimal value of $J(T, y(T))$ is known. The optimal control decision for time $T-1$ is solved by maximizing the backward recursion in (9.2). This procedure is repeated until we reach the current time t. In this way, the DP algorithm solves $(T - t + 1)p^k$ optimization problems, where p is the number of states, k is the number of discretizations for each state and $T - t + 1$ is the number of remaining time-steps. For example, for a problem with five states, 10 discretizations and five remaining time-steps, the DP procedure involves 48 millon optimizations, which is already an enormous amount for this rather coarse discretization. This fact is often called the 'curse of dimensionality' since the number of computations increases exponentially with the number of states.

9.2.2 Scenario Approximations

If a regular grid is used to discretize the state space of (1), the optimization is computed at many points in the state space which are reached with low probability. A standard approach to overcome these drawbacks is to use a scenario approach and a sampling approximation of the true expectation. The difficulty with solving the DP problem directly is the computation of the recursion under realistic assumptions. Instead of computing the exact optimal control policy, we solve an approximate problem where we replace the expectation by the sample mean and the true dynamics by a finite number of representative scenarios. To calculate the scenario and sample mean, a number of samples has to be drawn from the underlying path probability distribution. This procedure is repeated at each time-step so that a new control decision is based on the current time and state. Instead of solving the DP problem for all possible states and time, the DP problem is approximated at the current state and time by an SP problem. At each time-step t, we replace the probability distribution of $\epsilon(\tau)$ by $\kappa(\tau)$ scenarios, denoted by $\epsilon^s(\tau)$. We denote by $\tau = t, t + 1, \ldots, T - 1$ time in the scenario tree and thus in the SP problem. The 'physical' time is denoted by t. At time $\tau + 1$ conditional on the scenario $\epsilon^s(\tau)$, we generate $\kappa(\tau + 1)$ scenarios for each previous scenario of $\epsilon^s(\tau)$ as shown in Figure 9.2. Since we assume that the white noise process is stationary, at each node $\epsilon(\tau)$ is replaced by the same scenario values. The system dynamics however are different at each node, since both $D(\cdot)$ and $S(\cdot)$ are time and state dependent. The scenarios are defined as the set S

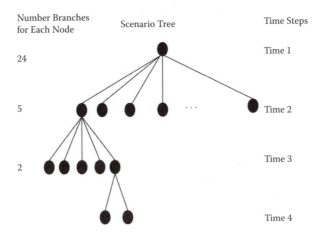

FIGURE 9.2 Graphical description of a scenario tree.

that represents a reasonable description of the future uncertainties. A scenario $s \in S$ describes a unique path through consecutive nodes of the scenario tree, as depicted in Figure 9.2. In this way, we generate a tree of scenarios that grows exponentially with the time horizon and the number of scenarios is given by $N = \prod_{i=t}^{T} \kappa(i)$. By this procedure we generate an irregular grid of the state space along a highly probable evolution of the system. The optimization is then computed for the approximate dynamics in terms of the sample expectation. For the approximate problem we still effectively execute the DP algorithm, but only for the approximate paths of the state dynamics and the current state and time.

9.2.3 Approximate of Dymanic Programming

When we replace the 'true' stochastic dynamics of the state equation by its sample approximation, we need to compute the sample mean of the objective function instead of its expectation. The objective function at time τ on one scenario s with feedback mapping (policy) $\Pi^s(\tau) := [\pi^s(\tau), \pi^s(\tau+1), \ldots, \pi^s(T-1)]$ is given by

$$V^s(\tau, y^s(\tau), \Pi^s(\tau)) = \sum_{i=\tau}^{T-1} L(i, y^s(i), \pi^s(i)) + M(T, y^s(T)). \tag{9.3}$$

The control decisions $\Pi^s(\tau)$ define a feedback mapping, since for each scenario s and time τ a predetermined control decision based on a feedback rule $\pi^s(\tau): = u(\tau, y^s(\tau))$ is used. The sample mean of the objective function is given by

$$\hat{E}\left[V^s(\tau, y^s(\tau), \Pi^s(\tau))\right] = \frac{1}{S}\sum_{s=1}^{N} V^s(\tau, y^s(\tau), \Pi^s(\tau)), \tag{9.4}$$

where $\hat{E}[\cdot] = (1/S)\sum_{s=1}^{S}[\cdot]$ denotes the sample mean. Using (9.4) we define the sample approximation of the dynamic optimization problem as

$$\hat{J}^s(\tau, y(\tau)) := \max_{\Pi^s(\tau) \in \mathcal{U}} \left\{ \hat{E}\left[V^s(\tau, y^s(\tau), \Pi^s(\tau))\right] \right\}$$
$$\text{s.t. } y^s(\tau+1) = D(\tau, y^s(\tau), \pi^s(\tau)) + S(\tau, y^s(\tau), \pi^s(\tau))\epsilon^s(\tau), \tag{9.5}$$

where $\hat{J}^s(\tau, y(\tau))$ is the value of the objective function at time t under scenario s and we must impose non-anticipativity constraints, i.e. $y(\tau)^s = y(\tau)^{s'}$ when s and s' have the same past until time τ. At the root of the scenario tree $\hat{J}^s(\cdot)$ and $\pi^s(\tau)$ are the same for all scenarios ($s = 1, \ldots, N'$).

Theorem 9.1: *The sample approximation of the optimization problem given in (9.5) can be recursively computed by the following dynamic program:*

$$\hat{J}^s(\tau, y^s(\tau)) = \max_{u(\tau)\in\mathcal{U}} \left\{ \hat{E}\Big[L(\tau, y^s(\tau), u^s(\tau)) + \hat{J}^s(\tau+1, D(\tau, y^s(\tau), \right.$$
$$\left. u^s(\tau)) + S(\tau, y^s(\tau), u^s(\tau))\epsilon^s(\tau)))\Big] \right\} \tag{9.6}$$

with terminal condition $\hat{J}^s(T\, y^s(T)) = M(T, y^s(T))$.

The proof of Theorem 9.1 is given in Appendix 9.B. Note that using (9.6) to solve (9.5), we use a backward recursion that automatically takes into account the non-anticipative constraints which prevent exact knowledge of future scenarios.

Proposition 9.1: *The dynamic programming formulation of the approximated dynamic optimization problem can be written as a multi-stage stochastic program.*

This proposition follows directly from Louveaux and Birge (1997, Chapter 3, p. 128). In Table 9.1 we state the algorithm to compute the stochastic programming approximation of the dynamic programming approach. By starting with a low number of scenarios, we ensure that the multi-stage stochastic program is solved rather quickly. The algorithm cycles between step 2 and step 4 until the desired accuracy has been reached. The relation of the approximation algorithm and the true problem defined in (9.1) is given in Proposition 9.2.

Proposition 9.2: *Under the assumptions of Section 9.2.1, the sample approximation of the objective function $\hat{J}^s(\tau, y(\tau))$ defined in (9.5) converges with probability 1 to the true objective function $J(\tau, y(\tau))$ as defined in (9.1) for $N \to \infty$. Especially $\hat{J}^s(t, y(t))$ converges with probability 9.1 to $J(t, y(t))$ for $N \to \infty$.*

Proof: Given a predetermined feedback mapping $\Pi^s(\tau)$ as defined above and using the Tchebychev inequality (Casella and Berger 2002) the following holds:

$$\lim_{N\to\infty} \mathbb{P}\left(\left| \hat{E}[V(\tau, y^s(\tau), \Pi^s(\tau))] - \mathbb{E}[V(\tau, y(\tau), \Pi(\tau))] \right| < \eta \right) = 1, \qquad \forall \eta > 0, \tag{9.7}$$

where $\tau = t, t+1, \ldots, T-1$. We know that exchanging the expectation with the sample mean has a negligible effect with arbitrarily large probability, since $\hat{E}[V(\tau, y^s(\tau), \Pi^s(\tau))]$ converges to $E[V(\tau, y(\tau), \Pi(\tau))]$ with probability 1 and using

$$\hat{J}^s(\tau, y^s(\tau)) = \max_{\Pi^s(\tau)\in\mathcal{U}} \left\{ \hat{E}\Big[V^s(\tau, y^s(\tau), \Pi^s(\tau)) \Big] \right\},$$

it follows under suitable assumptions that

$$\lim_{N\to\infty} \mathbb{P}\left(\left| \hat{E}[\hat{J}(\tau, y^s(\tau))] - \mathbb{E}[J(\tau, y(\tau))] \right| < \eta \right) = 1, \qquad \forall \eta > 0. \tag{9.8}$$

□

The proof of Proposition 9.2 holds only under the restrictive assumptions for the problem stated in Section 9.2.1. The more general case is discussed in detail in Römisch (2003) and Dempster (2004). The result of Proposition 9.2 states that the approximate objective function converges in probability to the true value of the objective function. However this does not imply that the control law computed by the approximation converges in probability to the true control law. The approximation by stochastic programming techniques determines only the control law for the scenarios used. For other values of the state variables, which are not covered by the scenario approximation, the control law is not defined. Since we only apply the first control decision for the fixed (measured) state $y(t)$ the issue of feedback for uncovered state variables may not constitute a problem in the absence of nonlinearities. By using the approximation procedure at every time-step, however, the control decisions are always based on current information, but remain scenario dependent. A similar analysis for linear stochastic systems with quadratic performance criteria can be found in Batina *et al.* (2002), however without the explicit connection to stochastic programming.

The scenario approximation does not depend on the dimension of the state variables but on the number of scenarios used. The algorithm's complexity is thus independent of the state space dimension. However, to obtain results with a desired accuracy we need a sufficiently large number of scenarios. By solving the stochastic programming problem at every time-step we introduce feedback into our system. This method requires the solution of $T - t + 1$ stochastic programming problems. The approach is very suitable for historical backtesting, but is less suitable for simulation studies, since for each time-step in the simulation a stochastic program must be solved. Furthermore, this approach is limited by the exponential growth of the scenario tree. If a very large number of scenarios is needed, it becomes very slow and computationally intractable.

The convergence speed of the proposed method relies on the convergence speed of the stochastic program and the scenario generation. As shown in Koivu (2005), the scenario generation method determines the convergence speed and the accuracy of the solution. For this reason we use the method proposed in Koivu (2005) for scenario generation, since it outperforms standard Monte Carlo techniques (see Section 9.4.2.1).

9.3 PORTFOLIO OPTIMIZATION

In this section, the general asset return statistical model and the portfolio model with transaction costs are given. Further, the objective function and the problem of portfolio optimization under transaction costs is stated.

9.3.1 Asset Return Models

The returns of assets (or asset classes) in which we are able to invest are described by

$$r(t+1) = Gx(t) + g + \epsilon^r(t), \tag{9.9}$$

where $r(t) = (r_1(t), r_2(t), \ldots, r_n(t))^T \in \mathbb{R}^n$ is the vector of asset returns, $\epsilon^r(t) \in \mathbb{R}^n$ is a white noise process with $E[\epsilon^r(t)] = 0$ and $E[\epsilon^r(t)\epsilon^{rT}(t)] := \Sigma(t) \in \mathbb{R}^{n \times n}$ is the covariance matrix, $Gx(t) + g \in \mathbb{R}^n$ is the local expected return, $x(t) \in \mathbb{R}^m$ is the vector of factors, $G \in \mathbb{R}^{n \times m}$ is the factor loading matrix, and $g \in \mathbb{R}^n$ is a constant vector. We assume that the conditional expectation is time-varying and stochastic, since $G\,x(t) + g$ is a function of the factor levels $x(t)$ which themselves are governed by a stochastic process. The white noise process $\epsilon^r(t)$ is assumed to be strictly covariance stationary. The prices of the risky assets evolve according to

$$P_i(t+1) = P_i(t)\big(1 + r_i(t)\big), \quad P_i(0) = p_{i0} > 0, \tag{9.10}$$

where $P(t) = (P_1(t), P_2(t), \ldots, P_n(t))$ denotes the prices of the risky assets. A locally risk-free bank account with interest rate $r_0(t, x(t))$ is given as

$$P_0(t+1) = P_0(t)\big(1 + F_0 x(t) + f_0\big), \quad P_0(0) = p_{00} > 0, \tag{9.11}$$

where $P_0(t)$ denotes the bank account. The interest rate of the bank account, described by (9.11), is modelled by

$$r_0(t) = F_0 x(t) + f_0, \tag{9.12}$$

where $F_0 \in \mathbb{R}^{1 \times m}$ and $f_0 \in \mathbb{R}$.

The factor process affecting the expected return of the risky assets and the interest rate of the bank account is described by the following linear stochastic process difference equation

$$x(t+1) = Ax(t) + a + v\epsilon^x(t), \tag{9.13}$$

where $A \in \mathbb{R}^{m \times m}$, $a \in \mathbb{R}^m$, $v \in \mathbb{R}^{m \times m}$ and $\epsilon^x(t) \in \mathbb{R}^m$ is a strictly covariance stationary white noise process. The white noise process of the risky asset dynamics $\epsilon^r(t)$ is not restricted to have a Gaussian distribution. We also assume that $\epsilon^r(t)$ and $\epsilon^x(t)$ are correlated. The stochastic process of the asset returns has a Markov structure and therefore we can apply DP techniques to solve the corresponding portfolio optimization problem.

9.3.2 Portfolio Dynamics with Transaction Costs

For the case of transaction costs, we limit our description of the wealth dynamics to linear transaction costs and use the scenario approach to describe multi-period asset prices. Many different formulations of multi-period investment problems can be found in the literature. Here, we adopt the basic model formulation presented in Mulvey and Shetty (2004). The portfolio optimization horizon consists of T time-steps represented by $t = \{1, 2, \ldots, T\}$. At every time-step t, the investors are able to make a decision regarding their investments and face inflows and outflows from and to their portfolio. The investment classes belong to the set $I = \{1, 2, \ldots, n\}$.

Let $z_i^s(t)$ be the amount of wealth invested in instrument i at the beginning of the time-step t under scenario s. The units we use are the investor's home currency (e.g. Swiss francs). Foreign assets, hedged or un-hedged against exchange rate fluctuations, are also denoted in the portfolio's home currency. At time t the total wealth of the portfolio is

$$W^s(t) = \sum_{i=1}^{n} z_i^s(t) \qquad \forall s \in S, \tag{9.14}$$

where $W^s(t)$ denotes the total wealth under scenario s. Given the returns of each investment class, the asset values at the end of the time period are

$$z_i^s(t)(1 + r_i^s(t)) = \tilde{z}_i^s(t) \qquad \forall s \in S, \quad \forall i \in I, \tag{9.15}$$

where $r_i^s(t)$ is the return of investment class i at time t under scenario s. The returns are obtained from the scenario generation system. Therefore, $\tilde{z}_i^s(t)$ is the ith asset value at the end of the time period t under scenario s. The sales or purchases of assets occur at the beginning of the time period, where $d_i^s(t) \geq 0$ denotes the amount of asset i sold at time t under scenario s, and $p_i^s(t) \geq 0$ denotes the purchase of asset i at time t under scenario s. The asset balance equation for each asset is

$$z_i^s(t) = \tilde{z}_i^s(t-1) + p_i^s(t)(1 - \delta_i) - d_i^s(t) \quad \forall s \in S, \quad \forall i \in I \setminus \{1\}, \tag{9.16}$$

where δ_i is the proportional (linear) transaction cost of asset i. We make the assumption that the transaction costs are not a function of time, but depend only on the investment class involved.

We treat the cash component of our investments as a special asset. The balance equation for cash is

$$z_1^s(t) = \tilde{z}_1^s(t-1) + \sum_{i=2}^{n} d_i^s(t)(1 - \delta_i) - \sum_{i=2}^{n} p_i^s(t) + q^s(t) \quad \forall s \in S, \tag{9.17}$$

where $z_1^s(t)$ is the cash account at time t under scenario s and $q^s(t)$ is the inflow or outflow of funds at time t under scenario s, respectively.

All of the variables in Equations (9.14)–(9.17) are dependent on the actual scenario s. These equations could be decomposed into subproblems for each scenario in which we anticipate that this scenario will evolve. To model reality, we must, however, impose non-anticipativity constraints. All scenarios which inherit the same past up to a certain time period must evoke the same decisions in that time period, otherwise the non-anticipativity requirement would be violated. So $z_i^s(t) = z_i^{s'}(t)$ when s and s' have same past until time t.

9.3.3 Risk Measure and Objective Function

We introduce a linear risk measure that is well suited for problems with assets and liabilities. Liabilities can be explicit payments promised at future dates, as well as capital guarantees (promises) to investors or investment goals.

We define our risk measure as a penalty function for net-wealth, i.e. wealth minus liabilities. We want to penalize small 'non-achievement' of the goal differently from large 'non-achievement.' Therefore, the penalty function should have an increasing slope with increasing 'non-achievement.' The risk of the portfolio is measured as one-sided downside risk based on non-achievement of the goals. As penalty function $P_f(\cdot)$ we choose the expectation of a piecewise linear function of the net wealth as shown in Figure 9.3. The penalty function is convex, but not a coherent risk measure in the sense of Artzner *et al.* (1999) However, it fulfills the modified properties for coherent risk measures defined by Ziemba and Rockafellar (2000). Furthermore, we can formulate a backward recursion to compute the risk measure and, thus, it is dynamically coherent in the sense of Riedel (2004). The same approach is discussed in detail by Dondi *et al.* (2007) where this method is applied to the management of a Swiss pension fund. Furthermore, for the case of multi-period capital guarantees, a suitable risk measure that is linear is given by

$$\Gamma = \sum_{s=1}^{N}\left[\sum_{\tau=t}^{T}P_f\big(W^s(\tau) - G(\tau)\big)\right],\tag{9.18}$$

where P_f denotes the piecewise linear penalty function, $G(\tau)$ denotes the capital guarantee at time τ and $W(\tau)$ the portfolio value at time τ. By multi-period capital guarantee we mean a capital guarantee not only for the final period but for all intermediate periods. This risk measure is convex and fulfills the properties of a dynamic risk measure.

Standard coherent risk measures (CVaR, maximum loss) or traditional risk measures (utility functions, variance, VaR) can be used in ALM situations, when applied to the fund's net wealth, e.g. the sum of the assets minus all the present value of the remaining liabilities. For example, CVaR penalizes linearly all events which are below the VaR limit for a given confidence level. The inherent VaR limit is a result of the CVaR optimization, see Rockafellar and Uryasev (2000, 2002). The VaR limit therefore depends on the confidence level chosen and the shape of the distribution. The VaR limit (quantile) may be a negative number, i.e. a negative net wealth may result. For a pension fund we do not only want to penalize scenarios that are smaller than a given quantile, but all scenarios

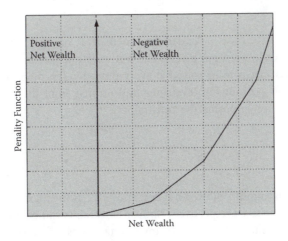

FIGURE 9.3 Depiction of the penalty function.

where the net wealth is non-positive. Similarly, traditional risk measures do not measure the shortfall, but the inherent 'randomness' of the investment policy. We have therefore chosen the approach described in the previous paragraph.

The objective of our ALM problem is to maximize the expected wealth at the end of the planning horizon, while keeping the penalty function under a certain level. Therefore, we propose the following objective function

$$\max \left\{ \sum_{s=1}^{N} \left[W^s(T) + \lambda \sum_{\tau=t}^{T} P_f\big(\tau, W^s(\tau) - G(\tau)\big) \Big| \mathcal{F}(t) \right] \right\}, \qquad (9.19)$$

where $\tau = t, t+1, \ldots, T$, and $\lambda < 0$ denotes the coefficient of risk aversion.

9.3.4 Multi-Period Portfolio Optimization with Transaction Costs

For the case of portfolio optimization problems with long-term goals and transaction costs we use the approach presented in Section 9.2. The optimization problem for the minimal guaranteed return fund is

$$\max_{p_i^s(\tau), d_i^s(\tau)} \left\{ \sum_{s=1}^{S} \left[W^s(T) + \lambda \sum_{\tau=t}^{T} P_f\big(\tau, W(\tau) - G(\tau)\big) \Big| \mathcal{F}(t) \right] \right\},$$

$$W^s(\tau) = \sum_{i=1}^{n} z_i^s(\tau) \qquad \forall s \in S,$$

$$z_i^s(\tau)(1 + r_i^s(\tau)) = \tilde{z}_i^s(\tau) \qquad \forall s \in S, \quad \forall i \in I,$$

$$z_i^s(\tau) = \tilde{z}_i^s(\tau - 1) + p_i^s(\tau)(1 - \delta_i) - d_i^s(\tau) \qquad \forall s \in S, \quad \forall i \in I \backslash \{1\}, \qquad (9.20)$$

$$z_1^s(\tau) = \tilde{z}_1^s(\tau - 1) + \sum_{i=2}^{n} d_i^s(\tau)(1 - \delta_i) - \sum_{i=2}^{n} p_i^s(\tau) + q^s(\tau) \qquad \forall s \in S,$$

where $\tau = t, t+1, \ldots, T$, $\lambda < 0$, denotes coefficient of risk aversion, and initial conditions $z_i(t)$, $i = 1, \ldots, n$ are given. The optimization problem is given directly as the SP problem that needs to be solved at every time-step. Furthermore, we may impose constraints for short positions or leveraging and we may limit the position in a specific asset (or asset class) by imposing the constraints

$$\bar{u}_i^{(\min)} \leq \frac{z_i^s(\tau)}{W^s(\tau)} \leq \bar{u}_i^{(\max)},$$

where $\bar{u}_i^{(\min)}$ is the minimum allowed fraction and $\bar{u}_i^{(\max)}$ is the maximum allowed fraction of wealth invested in asset i respectively. Similar constraints can be introduced to enforce minimum and maximum investments into certain sets of assets, e.g. international investments, or stocks. This kind of constraint can be formulated as linear and thus, the optimization given in (9.20) is still a large-scale LP. Many specialized algorithms exist to solve this special form of an LP, such as the L-shaped algorithm, see e.g. Birge *et al.* (1994).

The portfolio optimization problem is solved by employing the algorithm outlined in Section 9.2. At every time-step, we generate the scenario approximation of the return dynamics used for the specific investment universe based on the current information at time τ. Then, we solve the optimization problem given by (9.20) with all of the corresponding constraints. Next we then check the accuracy and refine the scenario tree. When we reach the predetermined accuracy, we terminate the tree generation and optimization and apply the investment decisions $p_i(\tau)$ and $p_i(\tau)$. The procedure then moves one time-step ahead.

9.4 CASE STUDY: ASSET AND LIABILITY MANAGEMENT WITH TRANSACTION COSTS FOR A SWISS FUND

In this case study, we take the view of an investment fund that resides in Switzerland which invests domestically and abroad. The fund is assumed to be large and we therefore cannot neglect the market impacts of its trading activities which result in trading costs. The problem of portfolio optimization is solved with the framework presented in Section 9.3.2. Moreover, we assume that the fund gives a capital (performance) guarantee which introduces a liability. The situation resembles the situation of a Swiss pension fund and, thus, we impose similar restrictions on the case study. Designing funds with performance guarantees is also discussed in Dempster *et al.* (2006, 2007).

The critical connection between the assets and liabilities is modelled through the capital guarantee $G(t)$. In the case of fixed discount rates, such as in the Swiss or German case, the capital guarantee increases with the discount rate. For the case in which the capital guarantee is linked to the current term structure of interest rates, the capital guarantee would increase along the term structure. The guarantee is not necessarily deterministic, it could be stochastic, e.g. the guarantee could increase with a LIBOR rate that changes in the future. Then the guarantee must be part of the stochastic scenario generation. In this case study, we assume that the discount rate is fixed to 4% and the nominal liabilities increases accordingly. Therefore, we do not include the capital guarantee in the scenario generation. Other approaches that feature a two-step method to calculate the optimal ALM are described in Mulvey *et al.* (2000), where the liabilities are described by a detailed interest rate model. A three factor yield curve model is also used in Dempster *et al.* (2006, 2007). Classical techniques are the immunization of liabilities by bonds, as described in Fabozzi (2005). A detailed overview of different approaches to ALM modelling is given in Dondi *et al.* (2007) and Zenios and Ziemba (2007).

9.4.1 Data Sets and Data Analysis

The data sets consists of the Datastream (DS) Swiss total stock market index, the DS Swiss government benchmark bond index, the DS European Union (EU) total stock market, and the DS EU government benchmark bond index. For the money market account, we use the 3-month LIBOR (SNB) interest rate. The data set starts on 1 January 1988 and finishes on 1 January 2005 with quarterly frequency. The two international indices are used in two different ways in our case study. In the first the indices are simply recalculated in Swiss francs (CHF) and the in other the currency risk is eliminated by completely hedging the currency risk. The risk-return profile is changed in the latter case since hedging introduces costs that reduce the performance but eliminate the currency risk. The hedging costs are computed on the basis of the difference of the 3-month forward rates between the Swiss franc and euro. Before 1999, we use the difference between forward rates of the Swiss franc and German mark as an approximation for the Euro–Swiss France hedging costs. Also, before 1999 the EU stock and bond indices are calculated in German marks. We simply substitute the euro by German mark (with euro reference conversion of 1.95583), since the German mark was the major currency of the euro predecessor the ECU (the German mark and linked currencies such as the Dutch gilder made up more than 50% of the ECU). The correlation of the ECU and the German mark were extremely high, especially after 1996 where the out-of-sample backtesting starts. This substitution is of course a simplification, but does not change the return distributions significantly. In Figure 9.4, the return histograms for the Swiss stock market, the Swiss bond market, the EU stock market in CHF, and the EU stock market hedged are shown. The figure shows the histograms and the best fits of a normal distribution. Except for the bond market index, we can clearly reject the assumption that the stock market data are normally distributed. When we fit a multivariate student-t distribution on a rolling basis to the stock market data, we get degrees of freedom between 6.7 and 9.6, which indicate a very clear deviation from normality. These results are supported by two tests for normality which reject the assumption of normality at 5% confidence level. The two tests are the Jarque–Bera test

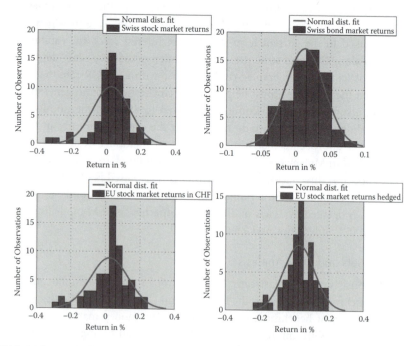

FIGURE 9.4 Histogram of the quarterly returns for different assets of the Swiss case study.

(J–B) and the Lilliefors test (LF) (Alexander 2001, Chapter 10). For these reasons, we model the returns of the stock market data by a non-normal distribution. The results of the normality tests for the out-of-sample period are given in Table 9.4, Section 9.4.4. We obtain similar results for the in-sample time period from 1 January 1988 to 1 June 1996.

9.4.2 Implementation

The implementation of the out-of-sample test for the portfolio allocation method consists of three main steps: the scenario generation, the parameter estimation and the computation of the asset allocation. By using the methods discussed in Section 9.3.2 one crucial step of the implementation is the generation of scenarios. The scenarios describe the future stochastic evolution of the asset returns and must reasonably well approximate the underlying stochastic model.

9.4.2.1. Scenario generation The most common techniques to generate scenarios for multistage stochastic programs are discussed in Dupacova *et al.* (2000). Among the most important methods of scenario generation are moment matching (Hoyland *et al.* 2003), importance sampling (Dempster and Thompson 1999, Dempster 2004) or discretizations via integration quadratures (Pennanen 2005, Pennanen and Koivu 2005).

We use the method of discretization via integration quadratures, because we believe that this method is superior to Monte Carlo methods, especially for high dimensional problems. Furthermore, numerical tests validate the stability of the optimization results, as shown in Pennanen and Koivu (2005). The method is used by approximating the white noise process at every stage of the dynamic model. Since we have assumed that the white noise process is stationary and identically distributed, we can use the same scenario generation method for $\epsilon^r(t)$ and $\epsilon^x(t)$ for each time-step and scenario. The resulting scenarios of the dynamic stochastic evolution of the system are different for each scenario, since the asset return evolution depends on the evolution of the factors and the correlation of asset return dynamics and factors. The scenarios for the asset and portfolio evolution therefore become stochastic and dynamic. For the generation of the low discrepancy sequences, which are essential for the discretization we use the Sobol sequence, see Bratley and Fox (1988).

We discuss the implementation of scenario generation for the student-t distribution which we assume for the white noise process of the asset returns. The multivariate student-t distribution possesses the following parameters: mean vector μ, degrees of freedom v and diffusion matrix Σ. The algorithm to compute an s-sample from a multivariate student-t distribution is based on the following result (Glasserman 2004, Chapter 9, p. 510):

$$ X \sim \mathcal{N}(\mu, \Sigma), \quad z \sim \chi_v^2, \quad Y = \sqrt{v}\,\frac{X}{\sqrt{z}} \sim t_v^n, $$

where $\mathcal{N}$ denotes the normal distribution in $\mathbb{R}^n$, χ_v^2 the standard chi-square distribution with v degrees of freedom, and t_v^n the student-t distribution in $\mathbb{R}^n$ with v degrees of freedom. The covariance of the student-t distribution is given by $v/(v-2)\Sigma$ and exists

TABLE 9.2 Scenario Generation for a Multivariate Student-t Distribution

---

1. Use the Sobol sequence to generate an s-sample from the marginally independent uniform distribution in the unit cube $[0,1]^{n+1}$ distributions, denoted by $S \in [0,1]^{(n+1) \times s}$.
2. Use the tables of the inverse standard normal distribution to transform the uniformly distributed realizations $(S_i, i = 1,\ldots, n)$ to standard normal distributed random variables $(\tilde{\xi}_i, i = 1,\ldots, n)$, i.e. $\tilde{\xi}_i(j) = \Phi^{-1}(S_i(j))$, $i = 1,\ldots, n, j = 1,\ldots, s$. The realizations of the n normal distribution are summarized in the vector $\tilde{\xi}(j) = [\tilde{\xi}_1(j),\ldots,\tilde{\xi}_n(j)], j = 1,\ldots, s$. Use tables of the inverse of the chi-square distribution to transform $S_{n+1}(j)$ to chi-square distributed realizations which we denote by $\phi(j), j + 1,\ldots, s$.
3. Compute the covariance matrix $\tilde{\Sigma} := \text{cov}(\tilde{\xi})$, the covariance matrix of the realizations of the normal distribution of step 2, and calculate the normalized sequence of standard random variables $\bar{\xi}(j) = \tilde{\Sigma}^{-1}\tilde{\xi}(j)$, $j = 1,\ldots, s$. In this way, we ensure that $\bar{\xi}(s)$ possesses unit variance.
4. Calculate the realizations of the white noise by $\epsilon(j) = \sqrt{v}(\mu + \sigma\bar{\xi}(j))/\sqrt{\phi(j)}, j = 1,\ldots, s$ where $\Sigma = \sigma\sigma^T$, μ and v are the parameters of the multivariate student-t distribution.

---

only for $v > 2$. The scenario generation algorithm for the student-t distribution is given in Table 9.2. With similar algorithms for scenario generation, any kind of normal (variance) mixture distributions as defined in McNeil *et al.* (2005, Chapter 3, p. 78) can be approximated, as long as the method of inverses (Glasserman 2004, Chapter 2, p. 54) can be applied. Distributions such as the generalized hyperbolic distribution which belong to the family of normal (variance) mixture distributions can be expressed as a function of the multivariate normal distribution and a mixing univariate distribution. The normal distribution can again be simulated by generating n independent realizations of the uniform $[0, 1]$ distribution and using the tables of the inverse univariate standard normal distribution. This is possible since we can generate the multivariate normal distribution from its univariate marginal distributions and the Cholesky factorization of its covariance matrix. For the univariate mixing distribution also realizations of the uniform $[0, 1]$ distribution are generated and realizations of the mixing distribution are calculated from tables of the inverse. The resulting realizations of the non-normal distributions are generated by using the functional relationship of the normal (variance) mixture distributions.

9.4.3 Factor Selection, Parameter Estimation and Asset Allocation Strategy

The factor selection determines which of the factors best explain the expected returns of the risky assets. Factor selection is used for the regression problem between the expectation of the risky asset returns $Gx(t) + g$ and the factors $x(t)$. Often, this selection is predetermined using literature recommendations or economic logic. However, when we use a very large set of factors, it is difficult to decide which factors explain the expected returns best. Moreover, by using a predefined set of factors a bias is introduced into the out-of-sample test, since from knowing the history we include factors where it is known that they have worked as predictors. In order to solve this problem, we employ a heuristic as described in Givens and Hoeting (2005, p. 55) known as stepwise regression. When we want to choose the best possible subset of factors, we face a combinatorial number of possible subsets. For this reason we use a 'greedy' strategy that reduces the number of factors by the factor with the lowest impact. Given m factors, the heuristic creates only m subsets and we choose the best subset by a so-called information criterion.

The information criterion is a trade-off between the number of regressors and the regression quality. For factor selection we either use the modified Akaike criterion or the Schwartz–Bayesian criterion. The information criterion here is used to select the best regression model for the expected return, but is not applied to select the best distribution. This factor selection procedure is used for all risky assets independently. Different expected returns of the risky assets are regressed on different sets of factors. The factor selection is recomputed every 12 months, i.e. the factors are selected at the beginning of every year. In this way, we remove of the factor selection bias and the model selects only the factors that have worked in the past without any information about the future. This heuristic and the two information criteria are discussed in detail in Illien (2005).

The estimation of parameters for the risky return is computed on a rolling basis with the last 8 years of data used. See Table 9.A1 in Appendix 9.A for a list of potential factors. The factor selection determines which time series of factors are used to predict the expected returns of the four risky assets.

We assume that the fund faces different transaction costs for domestic and international assets. The costs (due to market impact and international brokerage cost) for the domestic market are 1% for stocks and 0.5% for bonds. For the European stock market, the transaction costs are assumed to be 2% and for the bond market 1%. The transaction costs for European assets are independent of the hedging, since we calculate the hedging cost as part of the realized returns in Swiss francs.

The asset allocation decisions are calculated with the optimization algorithm proposed in (9.20). We assume that the fund possess a two-year moving investment horizon and we use a tree structure with 50, 20 and 5 branches which results in 5000 scenarios which we denote by 5000 (50,20,5). For the first branching, we use a one quarter time-step, for the second branching we use two quarters, and for the third branching we use five quarters. The algorithm to approximate (locally) the DP algorithm given in Section 9.2 is implemented by first computing 500 (10,10,5), 1000 (20,10,5), 3000 (20,15,10), 4000 (40,20,5) and 5000 (50,20,5) scenarios. The relative error between using 4000 and 5000 scenarios was smaller than 1% (measured by the objective function value obtained with 5000 scenarios). This test was done at the first time-step of the out-of-sample test and repeated every 12 quarters. In all tests, the difference was smaller than the 1%.

The constraints for the optimization are similar to the constraints which Swiss pension funds face. In Table 9.3 the maximum limits for investments in the different asset classes are given.

TABLE 9.3 Investment Constraints

Swiss stock market	50%
Swiss bond market	100%
EU stock market	25%
EU bond market	70%
EU stock market in CHF (hedged)	40%
EU bond market in CHF (hedged)	70%
All international assets	70%
All stock market investments	50%

As mentioned before, we assume that the fund gives the same capital guarantee of 4% as a Swiss pension fund (until 2002). Since 2002, the minimum guarantee has been reduced to the region of 2.5% and, therefore, we calculate a second portfolio with minimal guarantee of 2.5%. The capital guarantee function is given as $G(q) = 100(1 + r/4)^q$ where r is either 4% or 2.5% and q is the number of quarters in the backtesting.

When we solve the optimization problem at every time-step, the current asset allocation is taken as the initial asset allocation. In this way, the transaction costs for every rebalancing of the portfolios are correctly considered.

The risk measure used for this case study is an expected shortfall measure, computed as the expected shortfall of the portfolio wealth minus the capital guarantees. Therefore we use the piecewise linear risk measure given in (9.18) but with only one linear function and a slope of 5. The expected shortfall is not only used at the terminal date of the optimization but at all time-steps in between. In our strategy we compute the expected shortfall for one quarter, three quarters and eight quarters in advance.

9.4.4 Results of the Historical Out-of-Sample Backtest

The out-of-sample test starts on 1 June 1996 and ends on 1 January 2005 with a quarterly frequency. The results are simulated results with all the inherent weaknesses and deviations from actual policies of large financial institutions. The statistics of the out-of-sample test for the portfolios and the assets are shown in Table 9.4. The asset and the portfolio evolutions throughout the historical out-of-sample test are shown in Figure 9.5. The graph shows that both portfolios would have a relatively steady evolution throughout the historical backtest with only one longer drawdown period between the third quarter in 2000 until the third quarter in 2002. The largest loss occurs in the third quarter of 1998, where the portfolio (9.1) loss is 16.3%. The initial investments are mostly into the EU bond market and the subsequent portfolio gain allows the system to invest more into the stock market between 1996 and 1998. In the bull market phase, the portfolio with lower return guarantee would have had a higher return, since it could invest more into the risky assets. This higher allocation into the stock and bond market arises from the higher

TABLE 9.4 Summary Statistics of the 4 Indices from 1 June 1996 to 1 January 2005 and Results of the Normality Tests for the Standard Residuals. A 0 Indicates that We Cannot Reject the Normality Assumption and a 1 Indicates the Rejection of Normality

Time series	$r(\%)$	$\sigma(\%)$	SR	krt	skw	J–B (5%)	LF (5%)
Swiss stock market	7.0	24.6	0.22	1.5	−0.9	1	1
Swiss bond market	5.5	4.5	0.88	−0.9	−0.3	0	0
EU stock market in CHF	9.1	26.2	0.28	0.5	−0.8	1	1
EU bond market in CHF	9.2	7.4	1.04	−0.7	0.0	0	0
EU stock market hedged	10.0	22.1	0.39	0.0	−0.5	1	1
EU bond market hedged	9.4	6.1	1.29	−0.3	−0.3	0	0
3-month LIBOR (SNB)	1.41	—	—	—	—	0	0
Portfolio (1) (4% guarantee)	7.2	9.5	0.6	3.4	−1.4	1	1
Portfolio (2) (2.5% guarantee)	5.7	11.3	0.4	4.5	−0.9	1	1

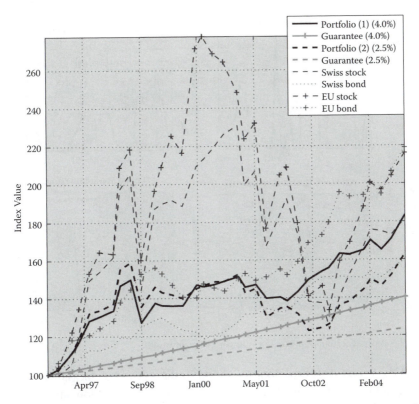

FIGURE 9.5 Results of the out-of-sample test for the Swiss case study and comparison to the asset classes.

distance to the minimum guarantee barrier and thus constrains the allocation less in this market phase. The evolution of the asset allocations of portfolio (1) are shown in Figure 9.6 as percentages of portfolio value. The large loss in the third quarter of 1998 leads to a dramatic increase of the money market investments and a sharp decrease of the investments into the Swiss stock market. The capacity to incur losses is reduced at this moment and the risky investments are consequently reduced. A similar behaviour can be seen during the drawdown from 2000 to early 2003, where the fund invests mostly into Swiss bonds and non-hedged EU bonds. Large changes in the asset allocation happen usually after significant changes in the portfolio value or after significant changes in the risk-return perception of the assets. In this phase (2000–2003) the portfolio (2) has a longer and stronger drawdown than portfolio (1), since portfolio (1) is more constrained by the distance to the minimal return guarantee. Also, the higher return guarantee forces the allocation to be more conservative and, therefore, limits the losses in this phase. When the stock markets are rising again (mid 2003–2005), portfolio (1) increases but not as strongly as portfolio (2). Therefore, we can conclude that the return guarantee acts as a security measure that limits losses in unfavourable times but has significant opportunity costs in rising markets. The dynamic changes in the asset allocation hold the wealth above the guarantee barrier, but would be impossible to be implemented in reality by a large

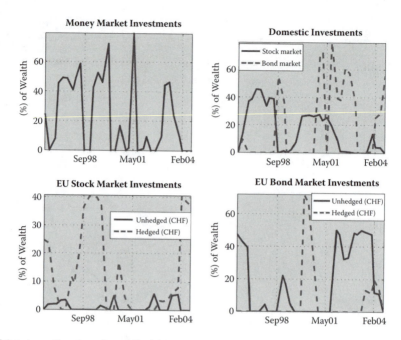

FIGURE 9.6 Asset allocation of portfolio (1) in percentage of wealth.

financial institution. This behaviour is induced by the type of risk measure and the factor model. Institutions stick much closer to their strategic allocation and change their allocation less dramatically. This method could, however, provide a guide for tactical asset allocation.

Most of the stock investments are hedged as shown in Figure 9.6. The risk-return trade-off for stocks seems to be more acceptable with hedging than without hedging. Moreover, most of the investment in stocks occurs before 2001 and in this period the Swiss franc is constantly gaining in value and thus reducing returns from the international investments. The Swiss franc loses value after mid-2002 when most of the international bond investment occurs. Therefore, most of the bond investment takes place without hedging. By introducing the international assets twice as either hedged or not hedged, we use the standard portfolio optimization to make the hedging decision in parallel with the portfolio construction.

However, a historical backtest starting after 2000 would have much more difficulties to remain above the barriers, since the two least risky assets (Swiss money market and bond market) did not yield returns above 4%. For this reason, many Swiss pension funds came into a situation of severe financial stress. This was one of the reasons why the capital guarantee has been strongly reduced and is adjusted biennially according to market expectations.

Despite the relatively high transaction costs, the performance of portfolio (1) is satisfactory with an average return of 7.1%. The performance is similar to the Swiss stock market but markedly higher than that of the Swiss bond market or money market accounts.

The standard deviation is moderate, but much smaller than any of the stock market investment possibilities. Hence, the Sharpe ratio lies between the one for the stock and the bond markets. The performance and the risk numbers for portfolio (2) are not as good, but a backtest that would have started earlier, e.g. 1990, would have shown better results.

Both portfolios show the strengths and weaknesses of performance guarantees for funds. Often guarantees are only judged with respect to the opportunity costs they carry. In this case study, since it includes a strong bull and bear market, the positive sides are also highlighted. However, the very dynamic changes in the allocation would be very difficult to follow by a large financial institution which limits this backtest to be only a demonstration of the proposed method.

9.5 CONCLUSION AND OUTLOOK

The first part of the chapter shows that we can approximately solve dynamic programming problems for dynamic portfolio management problems. We prove that by approximating the true dynamics by a set of scenarios and re-solving the problem at every time-step, we solve the dynamic programming problem with an arbitrarily objective function error. Future work should address the question of computational efficiency versus other established method.

In the second part of the chapter, we describe asset return and portfolio dynamics. We argue that the most suitable risk measures for ALM situations or guaranteed return funds are shortfall risk measures. A penalty function classifies the shortfall of the assets with respect to the liabilities where large 'non-achievements' are more severely penalized than small 'non-achievements' on future scenarios where the minimal return is not achieved. The optimization problem is solved with the aim of maximizing the return above the guarantee over the planning horizon, while keeping the shortfall risk below a predetermined limit.

In this case study, the risk aversion varies throughout the historical backtest and depends on the distance to the barrier. The optimization always reduces the risk exposure when the portfolio wealth moves closer to the barrier and increases the risk exposure when the portfolio moves away from the barrier. In this way, we introduce a feedback from the portfolio results to the current portfolio decisions and adapt the risk aversion to the loss incurring capacity.

The backtesting results will be compared in future research to other well known ALM techniques such as immunization or fixed-mix allocations. These comparisons could further illustrate the strength and weakness of this approach.

ACKNOWLEDGEMENTS

The authors wish to acknowledge the comments of three anonymous referees and thank the Editor-in-Chief for his detailed comments and corrections which materially improved the chapter.

REFERENCES

Alexander, C., *Market Models*, 2001 (Wiley: Chichester).

Artzner, P., Delbaen, F., Eber, J. and Heath, D., Coherent measures of risk. *Math. Finance*, 1999, **9**(3), 203–228.

Batina, I., Stoorvogel, A.A. and Weiland, S., Model predictive control of linear stochastic systems with state and input constraints. In *Conference on Decision and Control* pp. 1564–1569, 2002 (Las Vegas, NV).

Bertsekas, D.P., *Dynamic Programming and Optimal Control*, Vol. I, 1995 (Althena Scientific: Belmont, MA).

Bertsekas, D.P. and Shreve, S.E., *Stochastic Optimal Control: The Discrete-Time Case*, 1978 (Academic Press: New York).

Birge, J.R., Donohue, C., Holmes, D. and Svintsitski, O., A parallel implementation of the nested decomposition algorithm for multistage stochastic linear programs. *Math. Program.*, 1994, **75**, 327–352.

Bratley, P. and Fox, B., Algorithm 659: Implementing Sobol's quasirandom sequence generator *ACM Trans. Math. Software*, 1988, **14**(1), 88–100.

Brennan, M.J., Schwartz, E.S. and Lagnado, R., Strategic asset allocation *J. Econ. Dyn. Control*, 1997, **21**(8–9), 1377–1403.

Carino, D.R. and Ziemba, W.T., Formulation of the Russell–Yasuda Kasai financial planning model *Operat. Res.*, 1998, **46**(4), 433–449.

Casella, G. and Berger, R.L., *Statistical Inference*, 2nd edn, 2002 (Duxbury: Pacific Grove, CA).

Dempster, M.A.H., Sequential importance sampling algorithms for dynamic stochastic programming. *Trans. St. Petersburg Steklov Math. Inst.*, 2004, **312**, 94–129.

Dempster, M.A.H. and Consigli, G., The CALM stochastic programming model for dynamic asset and liability management. In *Worldwide Asset and Liability Modeling*, edited by W. Ziemba and J. Mulvey, pp. 464–500, 1998 (Cambridge University Press: Cambridge).

Dempster, M.A.H., Germano, M., Medova, E., Rietbergen, M., Sandrini, F. and Scrowston, M., Managing guarantees. *J. Portfolio Manag.*, 2006, **32**(2), 51–61.

Dempster, M.A.H., Germano, M., Medova, E., Rietbergen, M.I., Sandrini, F. and Scrowston, M., Designing minimum guaranteed return funds. *Quant. Finance*, 2007, **7**(2), (this issue).

Dempster, M.A.H. and Thompson, R., EVPI-based importance sampling solution procedures for multistage stochastic linear programmes on parallel MIMD architectures. *Ann. Operat. Res.*, 1999, **90**, 161–184.

Dondi, G., Models and dynamic optimisation for the asset and liability management for pension funds. PhD thesis, Dissertation ETH No. 16257, ETH Zurich, 2005.

Dondi, G., Herzog, F., Schumann, L.M. and Geering, H.P., Dynamic asset & liability management for Swiss pension funds. In *Handbook of Asset and Liability Management*, Vol. 2, edited by S. Zenios and W. Ziemba, 2007 (North-Holland: Oxford) in press.

Dupacova, J., Consigli, G. and Wallace, S., Scenario for multistage stochastic programs. *Ann. Operat. Res.*, 2000, **100**, 25–53.

Fabozzi, F.J., (Ed.), *The Handbook of Fixed Income Securities*, 7th edn, 2005 (McGraw-Hill: New York).

Geyer, A., Herold, W., Kontriner, K. and Ziemba, W.T., The Innovest Austrian pension fund financial planning model: InnoALM. In *Conference on Asset and Liability Management: From Institutions to Households*, May 2004 (Nicosia).

Givens, G. and Hoeting, J., *Computational Statistics*, 2005 (Wiley Interscience: Hoboken).

Glasserman, P., *Monte Carlo Methods in Financial Engineering*, 2004 (Springer: New York).

Hoyland, K., Kaut, M. and Wallace, S., A heuristic for moment-matching scenario generation. *Comput. Optim. Appl.*, 2003, **24**, 169–185.

Illien, M., Optimal strategies for currency portfolios. Diploma thesis, IMRT. *Introduction to Stochastic Programming*, 1997 (Springer: New York).

McNeil, A.J., Frey, R. and Embrechts, P., *Quantitative Risk Management: Concepts, Techniques and Tools*, 2005 (Princeton University Press: Princeton, NJ).

Mulvey, J., Gould, G. and Morgan, C., An asset and liability management system for Towers Perrin-Tillinghast. *Interfaces*, 2000, **30**, 96–114.

Mulvey, J.M., Generating scenarios for the Towers Perrin investment system. *Interfaces*, 1995, **26**, 1–15.

Mulvey, J.M. and Shetty, B., Financial planning via multi-stage stochastic optimization. *Comput. Operat. Res.*, 2004, **31**, 1–20.

Pennanen, T., Epi-convergent discretizations of multistage stochastic programs. *Math. Operat. Res.*, 2005, **30**, 245–256.

Pennanen, T. and Koivu, M., Epi-convergent discretizations of stochastic programs via integration quadratures. *Numer. Math.*, 2005, **100**, 141–163.

Riedel, F., Dynamic coherent risk measures. *Stochast. Process. Appl.*, 2004, **112**, 185–200.

Rockafellar, R. and Uryasev, S., Optimization of conditional value-at-risk. *J. Risk*, 2000, **2**(3), 21–41.

Rockafellar, R. and Uryasev, S.P., Conditional value-at-risk for general loss distributions. *J. Bank. Finance*, 2002, **26**, 1443–1471.

Römisch, W., Stability of stochastic programming problems. In *Handbooks in Operations Research and Management Science*, Vol. 10, edited by A. Ruszczynski and A. Shapiro, pp. 483–554, 2003 (Elsevier: Amsterdam).

Zenios, S.A. and Ziemba, W.T., (Eds), *Handbook of Asset and Liability Management*, Vol. 1, 2007 (Oxford: North-Holland), in press.

Ziemba, W.T., *The Stochastic Programming Approach to Asset Liability and Wealth Management*, 2003 (AIMR: Charlottesville, VA).

Ziemba, W.T. and Rockafellar, T., Modified risk measures and acceptance sets. In *Working Paper*, 2000 (Sauder School of Business, University of British Columbia).

APPENDIX 9.A: ADDITIONAL DATA FOR THE CASE STUDY

In Table 9.A1 we give the factors of the case study reported. Factors that were never selected, such as the dividend yield or macroeconomic variables (GNP), are omitted from the table.

Table 9.A1 All Factors for Case Study 3 with Swiss and EU Data

Factor no.	Factor name
1	10-year Swiss Government bond interest rate
2	10-year EU Benchmark Government bond (GB) interest rate
3	log (EP ratio) − log (10-year GB int. rate) Switzerland
4	log (EP ratio) − log (10-year GB int. rate) EU
5	10-year GB rate − 3-month LIBOR Switzerland
6	10-year GB rate − 3-month LIBOR EU
7	FX sopt rate CHF/Euro (DM)
8	3 months momentum. Swiss stock market
9	3 months momentum. EU stock market
10	3 months momentum. Swiss bond market
11	3 months momentum. EU bond market
12	FX 3-monthforward rate CHF/Euro (DM)

APPENDIX 9.B: DYNAMIC PROGRAMMING RECURSION FOR THE SAMPLE APPROXIMATION

Proof of Theorem 9.1: Let $\Pi^s(\tau) := [\pi^s(\tau), \pi^s(\tau + 1), \dots, \pi^s(T - 1)]$ and insert (9.3) into (9.5) which yields

$$\hat{J}^s(\tau, y(\tau)) = \max_{\Pi^s(\tau)} \left\{ \hat{E} \left[\sum_{i=\tau}^{T-1} L(i, y^s(i), \pi^s(i)) + M(T, \pi^s(T)) \right] \right\}$$

$$= \max_{\pi^s(\tau), \Pi^s(\tau+1)} \left\{ \hat{E} \left[L(\tau, y^s(\tau), \pi^s(\tau)) + \sum_{i=\tau+1}^{T-1} L(i, y^s(i), \pi^s(i)) + M(T, \pi^s(T)) \right] \right\}$$

$$\text{s.t. } y^s(\tau + 1) = D(\tau, y^s, \pi^s) + S(\tau, y^s, \pi^s)\epsilon^s(\tau),$$

$$(9.\text{B}1)$$

and nonanticipativity. Since $L(\tau, y^s(\tau), \pi^s(\tau))$ is independent of the future decisions $\Pi^s(t + 1)$ and the scenarios for $\epsilon^s(i)$, $i > \tau$ are independent of $\pi^s(\tau)$, move the maximization operator over $\Pi^s(t + 1)$ inside the bracket to obtain

$$\hat{J}^s(\tau, y(\tau)) = \max_{\pi^s(\tau)} \left\{ \hat{E} \left[L(\tau, y^s(\tau), \pi^s(\tau)) \right. \right.$$

$$\left. \left. + \max_{\Pi^s(t+1)} \left\{ \hat{E} \left[\sum_{i=\tau+1}^{T-1} L(i, y^s(i), \pi^s(i)) + M(T, \pi^s(T)) \right] \right\} \right] \right\}$$

$$= \max_{\pi^s(\tau)} \left\{ \hat{E} \left[L(\tau, y^s(\tau), \pi^s(\tau)) + \hat{J}^s(\tau + 1, y^s(\tau + 1)) \right] \right\},$$

subject to the dynamical constraints and nonanticipativity. From the first of the dynamical constraints $y^s(\tau + 1) = D(\tau, y^s, \pi^s) + S(\tau, y^s, \pi^s)\epsilon^s(\tau)$ and the basic idea of Bellman's principle

$$\hat{J}^s(\tau + 1, y(\tau + 1)) = \max_{\Pi^s(t+1)} \left\{ \hat{E} \left[\sum_{i=\tau+1}^{T-1} L(i, y^s(i), \pi^s(i)) + M(T, \pi^s(T)) \right] \right\},$$

subject to the remaining dynamical constraints and nonanticipativity, to give

$$\hat{J}^s(\tau, y(\tau)) = \max_{\pi^s(\tau)} \left\{ \hat{E} \left[L(\tau, y^s(\tau), \pi^s(\tau)) + \hat{J}^s(\tau + 1, D(\tau, y^s, \pi^s) + S(\tau, y^s, \pi^s)\,\epsilon^s(\tau)) \right] \right\}$$

$$= \max_{u^s(\tau) \in \mathcal{U}} \left\{ \hat{E} \left[L(\tau, y^s(\tau), u^s(\tau)) + \hat{J}^s(\tau + 1, D(\tau, y^s, u^s) + S(\tau, y^s, u^s)\,\epsilon^s(\tau)) \right] \right\},$$

subject to the appropriate dynamical constraints and nonanticipativity, where we convert the maximization over $\pi^s(\tau)$ to a maximization over $u^s(\tau)$, using the fact that for any function f of x and u it is true that

$$\max_{\pi \in \mathcal{Q}}\{f(x, \pi(x))\} = \max_{u \in \mathcal{U}}\{f(x, u)\},$$

where $\mathcal{Q}$ is the set of all functions $\pi(x)$ such that $\pi(x) \in \mathcal{U}$ $\forall x$. This statement can be found in Bertsekas (1995, Chapter 2). □

CHAPTER 10

Designing Minimum Guaranteed Return Funds

M. A. H. DEMPSTER, M. GERMANO, E. A. MEDOVA,
M. I. RIETBERGEN, F. SANDRINI and M. SCROWSTON

CONTENTS

10.1 INTRODUCTION

IN RECENT YEARS THERE HAS BEEN A SIGNIFICANT GROWTH of investment products aimed at attracting investors who are worried about the downside potential of the financial markets for pension investments. The main feature of these products is a minimum guaranteed return together with exposure to the upside movements of the market.

There are several different guarantees available in the market. The most common one is the nominal guarantee which guarantees a fixed percentage of the initial investment. However there also exist funds with a guarantee in real terms which is linked to an inflation index. Another distinction can be made between fixed and flexible guarantees, with the fixed guarantee linked to a particular rate and the flexible to, for instance, a capital market index. Real guarantees are a special case of flexible guarantees. Sometimes the guarantee of a minimum rate of return is even set relative to the performance of other pension funds.

Return guarantees typically involve hedging or insuring. Hedging involves eliminating the risk by sacrificing some or all of the potential for gain, whereas insuring involves paying an insurance premium to eliminate the risk of losing a large amount.

Many government and private pension schemes consist of defined benefit plans. The task of the pension fund is to guarantee benefit payments to retiring clients by investing part of their current wealth in the financial markets. The responsibility of the pension fund is to hedge the client's risk, while meeting the solvency requirements in such a way that all benefit payments are met. However at present there are significant gaps between fund values, contributions made by employees and pension obligations to retirees.

One way in which a guarantee can be achieved is by investing in zero-coupon Treasury bonds with a maturity equal to the time horizon of the investment product in question. However using this option foregoes all upside potential. Even though the aim is protect the investor from the downside, a reasonable expectation of returns higher than guaranteed needs to remain.

In this chapter we will consider long-term nominal minimum guaranteed return plans with a fixed time horizon. They will be closed end guarantee funds; after the initial contribution there is no possibility of making any contributions during the lifetime of the product. The main focus will be on how to optimally hedge the risks involved in order to avoid having to buy costly insurance.

However, this task is not straightforward, as it requires long-term forecasting for all investment classes and dealing with a stochastic liability. *Dynamic stochastic programming* is the technique of choice to solve this kind of problem as such a model will automatically hedge current portfolio allocations against the future uncertainties in asset returns and liabilities over a long horizon (see e.g. Consigli and Dempster 1998; Dempster *et al.* 2000, 2003). This will lead to more robust decisions and previews of possible future benefits and problems contrary to, for instance, static portfolio optimization models such as the Markowitz (1959) mean-variance allocation model.

Consiglio *et al.* (2007) have studied fund guarantees over single investment periods and Hertzog *et al.* (2007) treat dynamic problems with a deterministic risk barrier. However, a practical method should have the flexibility to take into account multiple time periods, portfolio constraints such as prohibition of short selling and varying degrees of risk aversion. In addition, it should be based on a realistic representation of the dynamics of the relevant factors such as asset prices or returns and should model the changing market dynamics of risk management. All these factors have been carefully addressed here and are explained further in the sequel.

The rest of the chapter is organized as follows. In Section 10.2 we describe the stochastic optimization framework, which includes the problem set up, model constraints and possible objective functions. Section 10.3 presents a three-factor term structure model and its application to pricing the bond portfolio and the liability side of the fund on individual scenarios. As our portfolio will mainly consist of bonds, this area has been extensively researched. Section 10.4 presents several historical backtests to show how the framework would have performed had it been implemented in practice, paying particular attention to the effects of using different objective functions and varying tree structures. Section 10.5 repeats the backtest when the stock index is modelled as a jumping diffusion so that the corresponding returns exhibit fat tails and Section 10.6 concludes. Throughout this chapter boldface is used to denote random entities.

10.2 STOCHASTIC OPTIMIZATION FRAMEWORK

In this section we describe the framework for optimizing minimum guaranteed return funds using stochastic optimization. We will focus on risk management as well as strategic asset allocation concerned with allocation across broad asset classes, although we will allow specific maturity bond allocations.

10.2.1 Set Up

This section looks at several methods to optimally allocate assets for a minimum guaranteed return fund using expected average and expected maximum shortfall risk measures relative to the current value of the guarantee. The models will be applied to eight different assets: coupon bonds with maturity equal to 1, 2, 3, 4, 5, 10 and 30 years and an equity index, and the home currency is the euro. Extensions incorporated into these models are the presence of coupon rates directly dependent on the term structure of bond returns and the annual rolling over of the coupon-bearing bonds.

We consider a discrete time and space setting. The time interval considered is given by $\{0, (1/12), (2/12), \ldots, T\}$, where the times indexed by $t = 0, 1, \ldots, T-1$ correspond to decision times at which the fund will trade and T to the planning horizon at which no decision is made (see Figure 10.1). We will be looking at a five-year horizon.

Uncertainty Ω is represented by a *scenario tree*, in which each path through the tree corresponds to a *scenario* ω in Ω and each node in the tree corresponds to a time along one or more scenarios. An example scenario tree is given in Figure 10.2. The probability $p(\omega)$ of scenario ω in Ω is the reciprocal of the total number of scenarios as the scenarios are generated by Monte Carlo simulation and are hence equiprobable.

The stock price process $\mathbf{S}$ is assumed to follow a geometric Brownian motion, i.e.

$$\frac{d\mathbf{S}_t}{\mathbf{S}_t} = \mu_S dt + \sigma_S d\mathbf{W}_t^S, \tag{10.1}$$

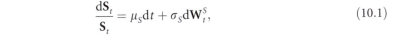

FIGURE 10.1 Time and stage setting.

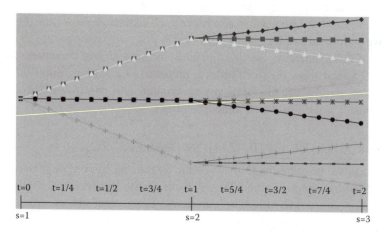

FIGURE 10.2 Graphical representation of scenarios.

where $d\mathbf{W}_t^S$ is correlated with the $d\mathbf{W}_t$ terms driving the three term structure factors discussed in Section 10.3.

10.2.2 Model Constraints

Let (see Table 10.1)

- $h_t(\omega)$ denote the *shortfall* at time t and scenario ω, i.e.

$$h_t(\omega) := \max(0, L_t(\omega) - W_t(\omega)) \qquad \forall \omega \in \Omega \quad t \in T^{\text{total}}. \tag{10.2}$$

- $H(\omega) := \max_{t \in T^{\text{total}}} h_t(\omega)$ denote the *maximum shortfall* over time for scenario ω. The constraints considered for the minimum guaranteed return problem are:
- *Cash balance constraints.* These constraints ensure that the net cash flow at each time and at each scenario is equal to zero

$$\sum_{a \in A} fP_{0,a}^{\text{buy}}(\omega) x_{0,a}^+(\omega) = W_0, \quad \omega \in \Omega, \tag{10.3}$$

$$\sum_{a \in A \setminus \{S\}} \frac{1}{2} \delta_{t-1}^a(\omega) F^a x_{t,a}^-(\omega) + \sum_{a \in A} gP_{t,a}^{\text{sell}}(\omega) x_{t,a}^-(\omega)$$
$$= \sum_{a \in A} fP_{t,a}^{\text{buy}}(\omega) x_{t,a}^+(\omega), \quad \omega \in \Omega \quad t \in T^d \setminus \{0\}. \tag{10.4}$$

In (10.4) the left-hand side represents the cash freed up to be reinvested at time $t \in T^d \setminus \{0\}$ and consists of two distinct components. The first term represents the semi-annual coupons received on the coupon-bearing Treasury bonds held between time $t - 1$ and t, the second term represents the cash obtained from selling part of the portfolio. This must equal the value of the new assets bought given by the right-hand side of (10.4).

TABLE 10.1 Variables and Parameters of the Model

Time sets

$T^{\text{total}} = \{0, \frac{1}{12}, \ldots, \}$	*Set of all times* considered in the stochastic program
$T^d = \{0, 1, \ldots, T - 1\}$	*Set of decision times*
$T^i = T^{\text{total}} \backslash T^d$	*Set of intermediate times*
$T^c = \{\frac{1}{2}, \frac{3}{2}, \ldots, T - \frac{1}{2}\}$	*Times when a coupon is paid out* in-between decision times

Instruments

$S_t(\omega)$	*Dow Jones Eurostoxx 50 index level* at time t in scenario ω
$B_t^T(\omega)$	*EU Treasury bond with maturity T* at time t in scenario ω
$\delta_t^{B^T}(\omega)$	*coupon rate* of EU Treasury bond with maturity T at time t in scenario ω
F^{B^T}	*face value* of EU Treasury bond with maturity T
$Z_t(\omega)$	*EU zero-coupon Treasury bond price* at time t in scenario ω

Risk management barrier

$y_{t,T}(\omega)$	*EU zero-coupon Treasury yield* with maturity T at time t in scenario ω
G	*Annual guaranteed return*
$L_t^N(\omega)$	*Nominal barrier* at time t in scenario ω

Portfolio evolution

A	*Set of all assets*
$P_{t,a}^{buy}(\omega)/P_{t,a}^{sell}(\omega)$	*Buy/sell price* of asset $a \in A$ at time t in scenario ω
$f g$	*Transaction costs* for buying/selling
$x_{t,a}(\omega)$	*Quantity held* of asset $a \in A$ between time t and $t + 1/12$ in scenario ω
$x_{t,a}^+(\omega)/x_{t,a}^-(\omega)$	*Quantity bought/sold* of asset $a \in A$ at time t in scenario ω
W_0	*Initial portfolio wealth*
$W_t(\omega)$	*Portfolio wealth before rebalancing* at time $t \in T$ in scenario ω
$w_t(\omega)$	*Portfolio wealth after rebalancing* at time $t \in T^c \cup T^d \backslash \{T\}$ in scenario ω
$h_t(\omega) := \max(0, L_t(\omega) - W_t(\omega))$	*Shortfall* at time t in scenario ω

- *Short sale constraints.* In our model we assume no short selling of any stocks or bonds

$$x_{t,a}(\omega) \geq 0, \quad a \in A, \quad \omega \in \Omega, \quad t \in T^{\text{total}}, \tag{10.5}$$

$$x_{t,a}^+(\omega) \geq 0, \quad \forall a \in A, \quad \forall \omega \in \Omega, \quad \forall t \in T^{\text{total}} \backslash \{T\}, \tag{10.6}$$

$$x_{t,a}^-(\omega) \geq 0, \quad \forall a \in A, \quad \forall \omega \in \Omega, \quad \forall t \in T^{\text{total}} \backslash \{0\}. \tag{10.7}$$

- *Information constraints.* These constraints ensure that the portfolio allocation cannot be changed during the period from one decision time to the next and hence that no decisions with perfect foresight can be made

$$x_{t,a}^+(\omega) = x_{t,a}^-(\omega) = 0, \quad a \in A, \quad \omega \in \Omega, \quad t \in T^i \backslash T^c. \tag{10.8}$$

- *Wealth constraint.* These constraints determine the portfolio wealth at each point in time prior to and after rebalancing

$$w_t(\omega) = \sum_{a \in A} P_{t,a}^{\text{buy}}(\omega) x_{t,a}(\omega), \qquad \omega \in \Omega \quad t \in T^{\text{total}} \backslash \{T\}, \tag{10.9}$$

$$W_t(\omega) = \sum_{a \in A} P_{t,a}^{\text{sell}}(\omega) x_{t-(1/12),a}(\omega), \qquad \omega \in \Omega \quad t \in T^{\text{total}} \backslash \{0\}, \tag{10.10}$$

$$w_T(\omega) = \sum_{a \in A} g P_{T,a}^{\text{sell}}(\omega) x_{T-(1/12),a}(\omega) + \sum_{a \in A \backslash \{S\}} \frac{1}{2} \delta_{T-1}^a(\omega) F^a x_{T-(1/12),a}(\omega), \qquad \omega \in \Omega. \tag{10.11}$$

- *Accounting balance constraints.* These constraints give the quantity invested in each asset at each time and for each scenario

$$x_{0,a}(\omega) = x_{0,a}^+(\omega), \qquad a \in A, \ \omega \in \Omega, \tag{10.12}$$

$$x_{t,a}(\omega) = x_{t-1/12,a}(\omega) + x_{t,a}^+(\omega) - x_{t,a}^-(\omega), \qquad a \in A, \ \omega \in \Omega, \ t \in T^{\text{total}} \backslash \{0\}. \tag{10.13}$$

The total quantity invested in asset $a \in A$ between time t and $t + (1/12)$ is equal to the total quantity invested in asset $a \in A$ between time $t - (1/12)$ and t plus the quantity of asset $a \in A$ bought at time t minus the quantity of asset $a \in A$ sold at time t.

- *Annual rolling constraint.* This constraint ensures that at each decision time all the coupon-bearing Treasury bond holdings are sold

$$x_{t,a}^-(\omega) = x_{t-(1/12),a}(\omega), \qquad a \in A \backslash \{S\}, \ \omega \in \Omega, \ t \in T^d \backslash \{0\}. \tag{10.14}$$

- *Coupon re-investment constraints.* We assume that the coupon paid every six months will be re-invested in the same coupon-bearing Treasury bond

$$x_{t,a}^+(\omega) = \frac{(1/2)\delta_t^a(\omega) F^a x_{t-(1/12),a}(\omega)}{f P_{t,a}^{\text{buy}}(\omega)}, \qquad x_{t,a}^-(\omega) = 0, \quad x_{t,S}^+(\omega) = x_{t,S}^-(\omega) = 0,$$

$$a \in A \backslash \{S\}, \ \omega \in \Omega, \ t \in T^c. \tag{10.15}$$

- *Barrier constraints.* These constraints determine the shortfall of the portfolio at each time and scenario as defined in Table 10.1

$$h_t(\omega) + W_t(\omega) \geq L_t(\omega), \quad \omega \in \Omega, \ t \in T^{\text{total}}, \tag{10.16}$$

$$h_t(\omega) \geq 0, \quad \omega \in \Omega, \ t \in T^{\text{total}}. \tag{10.17}$$

As the objective of the stochastic program will put a penalty on any shortfall, optimizing will ensure that $h_t(\omega)$ will be zero if possible and as small as possible otherwise, i.e.

$$h_t(\omega) = \max(0, L_t(\omega) - W_t(\omega)), \quad \omega \in \Omega, \ t \in T^{\text{total}}, \tag{10.18}$$

which is exactly how we defined $h_t(\omega)$ in (10.2).

To obtain the maximum shortfall for each scenario, we need to add one of the following two sets of constraints:

$$H(\omega) \geq h_t(\omega), \quad \omega \in \Omega, \ t \in T^{\text{d}} \cup \{T\} \tag{10.19}$$

$$H(\omega) \geq h_t(\omega), \quad \omega \in \Omega, \ t \in T^{\text{total}}. \tag{10.20}$$

Constraint (10.19) needs to be added if the max shortfall is to be taken into account on a yearly basis and constraint (10.20) if max shortfall is calculated on a monthly basis.

10.2.3 Objective Functions: Expected Average Shortfall and Expected Maximum Shortfall

Starting with an initial wealth W_0 and an annual *nominal guarantee* of G, the liability at the planning horizon at time T is given by

$$W_0(1 + G)^T. \tag{10.21}$$

To price the liability at time $t < T$ consider a zero-coupon Treasury bond which pays 1 at time T, i.e. $Z_T(\omega) = 1$, for all scenarios $\omega \in \Omega$. The *zero-coupon Treasury bond price* at time t in scenario ω assuming continuous compounding is given by

$$Z_t(\omega) = e^{-y_{t,T}(\omega)(T-t)}, \tag{10.22}$$

where $y_{t,T}(\omega)$ is the *zero-coupon Treasury yield* with maturity T at time t in scenario ω.

This gives a formula for the value of the *nominal (fixed) guarantee barrier* at time t in scenario ω as

$$L_t^N(\omega) := W_0(1+G)^T Z_t(\omega) = W_0(1+G)^T e^{-y_{t,T}(\omega)(T-t)}. \tag{10.23}$$

In a minimum guaranteed return fund the objective of the fund manager is twofold; firstly to manage the investment strategies of the fund and secondly to take into account the guarantees given to all investors. Investment strategies must ensure that the guarantee for all participants of the fund is met with a high probability.

In practice the guarantor (the parent bank of the fund manager) will ensure the investor guarantee is met by forcing the purchase of the zero coupon bond of (10.22) when the fund is sufficiently near the barrier defined by (10.23). Since all upside potential to investors is thus foregone, the aim of the fund manager is to fall below the barrier with acceptably small if not zero probability.

Ideally we would add a constraint limiting the probability of falling below the barrier in a VaR-type minimum guarantee constraint, i.e.

$$P\left(\max_{t \in T^{\text{total}}} h_t(\omega) > 0 \right) \le \alpha \tag{10.24}$$

for α small. However, such scenario-based probabilistic constraints are extremely difficult to implement, as they may without further assumptions convert the convex large-scale optimization problem into a non-convex one. We therefore use the following two convex approximations in which we trade off the risk of falling below the barrier against the return in the form of the expected sum of wealth.

Firstly, we look at the *expected average shortfall* (EAS) model in which the objective function is given by

$$\max_{\substack{ \{x_{t,a}(\omega), x_{t,a}^+(\omega), x_{t,a}^-(\omega): \\ a \in A, \omega \in \Omega, t \in T^d \cup T\}}} \left\{ \sum_{\omega \in \Omega} \sum_{t \in T^d \cup T} p(\omega) \left((1-\beta) W_t(\omega) - \beta \frac{h_t(\omega)}{|T^d \cup T|} \right) \right\}$$

$$= \max_{\substack{ \{x_{t,a}(\omega), x_{t,a}^+(\omega), x_{t,a}^-(\omega): \\ a \in A, \omega \in \Omega, t \in T^d \cup T\}}} \left\{ (1-\beta) \left(\sum_{\omega \in \Omega} p(\omega) \sum_{t \in T^d \cup T} W_t(\omega) \right) - \beta \left(\sum_{\omega \in \Omega} p(\omega) \sum_{t \in T^d \cup T} \frac{h_t(\omega)}{|T^d \cup T|} \right) \right\}.$$

$$\tag{10.25}$$

In this case we maximize the expected sum of wealth over time while penalizing each time the wealth falls below the barrier. For each scenario $\omega \in \Omega$ we can calculate the average shortfall over time and then take expectations over all scenarios.

In this case only shortfalls at decision times are taken into account and any serious loss in portfolio wealth in-between decision times is ignored. However, from the fund manager's and guarantor's perspective the position of the portfolio wealth relative to the fund's barrier is significant on a continuous basis and serious or repeated drops below this barrier might force the purchase of expensive insurance. To capture this feature specific to minimum guaranteed return funds, we also consider an objective function in which the shortfall of the portfolio is considered on a monthly basis.

For the *expected average shortfall with monthly checking* (EAS MC) model the objective function is given by

$$
\max_{\substack{\{x_{t,a}(\omega),x_{t,a}^+(\omega),x_{t,a}^-(\omega):\\ a\in A,\omega\in\Omega,t\in T^d\cup T\}}}\left\{(1-\beta)\left(\sum_{\omega\in\Omega}p(\omega)\sum_{t\in T^d\cup T}W_t(\omega)\right)-\beta\left(\sum_{\omega\in\Omega}p(\omega)\sum_{t\in T^{total}}\frac{h_t(\omega)}{|T^{total}|}\right)\right\}. \quad (10.26)
$$

Note that although we still only rebalance once a year shortfall is now being measured on a monthly basis in the objective and hence the annual decisions must also take into account the possible effects they will have on the monthly shortfall.

The value of $0\leq\beta\leq1$ can be chosen freely and sets the level of risk aversion. The higher the value of β, the higher the importance given to shortfall and the less to the expected sum of wealth, and hence the more risk-averse the optimal portfolio allocation will be. The two extreme cases are represented by $\beta=0$, corresponding to the 'unconstrained' situation, which is indifferent to the probability of falling below the barrier, and $\beta=1$, corresponding to the situation in which the shortfall is penalized and the expected sum of wealth ignored.

In general short horizon funds are likely to attract more risk-averse participants than long horizon funds, whose participants can afford to tolerate more risk in the short run. This natural division between short and long-horizon funds is automatically incorporated in the problem set up, as the barrier will initially be lower for long-term funds than for short-term funds as exhibited in Figure 10.3. However, the importance of closeness to the barrier can be adjusted by the choice of β in the objective.

The second model we consider is the *expected maximum shortfall* (EMS) model given by

$$
\max_{\substack{\{x_{t,a}(\omega),x_{t,a}^+(\omega),x_{t,a}^-(\omega):\\ a\in A,\omega\in\Omega,t\in T^d\cup\{T\}\}}}\left\{(1-\beta)\left(\sum_{\omega\in\Omega}p(\omega)\sum_{t\in T^d\cup\{T\}}W_t(\omega)\right)-\beta\left(\sum_{\omega\in\Omega}p(\omega)H(\omega)\right)\right\} \quad (10.27)
$$

using the constraints (10.19) to define $H(\omega)$.

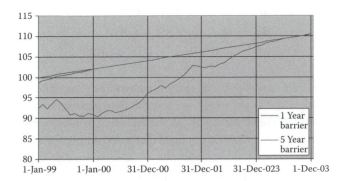

FIGURE 10.3 Barrier for one-year and five-year 2% guaranteed return fund.

For the *expected maximum shortfall with monthly checking* (EMS MC) model the objective function remains the same but $H(\omega)$ is now defined by (10.20).

In both variants of this model we penalize the expected maximum shortfall, which ensures that $H(\omega)$ is as small as possible for each scenario $\omega \in \Omega$. Combining this with constraints (10.19)/(10.20) it follows that $H(\omega)$ is exactly equal to the maximum shortfall.

The constraints given in Section 10.2.2 apply to both the expected average shortfall and expected maximum shortfall models.

The EAS model incurs a penalty every time portfolio wealth falls below the barrier, but it does not differentiate between a substantial shortfall at one point in time and a series of small shortfalls over time. The EMS model on the other hand, focuses on limiting the maximum shortfall and therefore does not penalize portfolio wealth falling just slightly below the barrier several times. So one model limits the *number of times* portfolio wealth falls below the barrier while the other limits *any substantial shortfall.*

10.3   BOND PRICING

In this section we present a three-factor term structure model which we will use to price both our bond portfolio and the fund's liability. Many interest-rate models, like the classic one-factor Vasicek (1977) and Cox *et al.* (1985) class of models and even more recent multi-factor models like Anderson and Lund (1997), concentrate on modelling just the short-term rate.

However, for minimum guaranteed return funds we have to deal with a long-term liability and bonds of varying maturities. We therefore must capture the dynamics of the whole term structure. This has been achieved by using the economic factor model described below in Section 10.3.1. In Section 10.3.2 we describe the pricing of coupon-bearing bonds and Section 10.3.3 investigates the consequences of rolling the bonds on an annual basis.

10.3.1   Yield Curve Model

To capture the dynamics of the whole term structure, we will use a Gaussian *economic factor model* (EFM) (see Campbell 2000 and also Nelson and Siegel 1987) whose evolution under the risk-neutral measure Q is determined by the stochastic differential equations

$$\mathbf{dX}_t = (\mu_X - \lambda_X X_t)dt + \sigma_X \mathbf{dW}_t^X, \tag{10.28}$$

$$\mathbf{dY}_t = (\mu_Y - \lambda_Y Y_t)dt + \sigma_Y \mathbf{dW}_t^Y, \tag{10.29}$$

$$\mathbf{dR}_t = k(X_t + Y_t - R_t)dt + \sigma_R \mathbf{dW}_t^R, \tag{10.30}$$

where the $\mathbf{dW}$ terms are correlated. The three unobservable Gaussian factors $\mathbf{R}$, $\mathbf{X}$ and $\mathbf{Y}$ represent respectively a *short rate*, a *long rate* and the *slope* between an *instantaneous short rate* and the long rate. Solving these equations the following formula for the yield at time t with time to maturity equal to $T - t$ is obtained (for a derivation, see Medova *et al.* 2005)

$$y_{t,T} = \frac{A(t, T)R_t + B(t, T)X_t + C(t, T)Y_t + D(t, T)}{T}, \tag{10.31}$$

where

$$A(t, T) := \frac{1}{k}\left(1 - e^{-k(T-t)}\right), \tag{10.32}$$

$$B(t, T) := \frac{k}{k - \lambda_X}\left\{\frac{1}{\lambda_X}\left(1 - e^{-\lambda_X(T-t)}\right) - \frac{1}{k}\left(1 - e^{-k(T-t)}\right)\right\}, \tag{10.33}$$

$$C(t, T) := \frac{k}{k - \lambda_Y}\left\{\frac{1}{\lambda_Y}\left(1 - e^{-\lambda_Y(T-t)}\right) - \frac{1}{k}\left(1 - e^{-k(T-t)}\right)\right\}, \tag{10.34}$$

$$D(t, T) := \left(T - t - \frac{1}{k}\left(1 - e^{-k(T-t)}\right)\right)\left(\frac{\mu_X}{\lambda_X} + \frac{\mu_Y}{\lambda_Y}\right) - \frac{\mu_X}{\lambda_X}B(t, T) - \frac{\mu_Y}{\lambda_Y}C(t, T)$$

$$-\frac{1}{2}\sum_{i=1}^{3}\left\{\frac{m_{X_i}^2}{2\lambda_X}\left(1 - e^{-2\lambda_X(T-t)}\right) + \frac{m_{Y_i}^2}{2\lambda_Y}\left(1 - e^{-2\lambda_Y(T-t)}\right) + \frac{n_i^2}{2k}\left(1 - e^{-2k(T-t)}\right) + p_i^2(T - t)\right.$$

$$+\frac{2m_{X_i}m_{Y_i}}{\lambda_X + \lambda_Y}\left(1 - e^{-(\lambda_X+\lambda_Y)(T-t)}\right) + \frac{2m_{X_i}n_i}{\lambda_X + k}\left(1 - e^{-(\lambda_X+k)(T-t)}\right) + \frac{2m_{X_i}p_i}{\lambda_X}\left(1 - e^{-\lambda_X(T-t)}\right)$$

$$+\frac{2m_{Y_i}n_i}{\lambda_Y + k}\left(1 - e^{-(\lambda_Y+k)(T-t)}\right) + \frac{2m_{Y_i}p_i}{\lambda_Y}\left(1 - e^{-\lambda_Y(T-t)}\right) + \frac{2n_ip_i}{k}\left(1 - e^{-k(T-t)}\right)\right\} \tag{10.35}$$

and

$$m_{X_i} := -\frac{k\sigma_{X_i}}{\lambda_X(k - \lambda_X)},$$

$$m_{Y_i} := -\frac{k\sigma_{Y_i}}{\lambda_Y(k - \lambda_Y)},$$

$$n_i := \frac{\sigma_{X_i}}{k - \lambda_X} + \frac{\sigma_{Y_i}}{k - \lambda_Y} - \frac{\sigma_{R_i}}{k},$$

$$p_i := -\left(m_{X_i} + m_{Y_i} + n_i\right). \tag{10.36}$$

Bond pricing must be effected under the *risk-neutral* measure Q. However, for the model to be used for forward simulation the set of stochastic differential equations must be adjusted to capture the model dynamics under the real-world or *market* measure P. We therefore have to model the market prices of risk which take us from the risk-neutral measure Q to the real-world measure P.

Under the market measure P we adjust the drift term by adding the *risk premium* given by the *market price of risk γ* in terms of the quantity of risk. The effect of this is a change in the long-term mean, e.g. for the factor $\mathbf{X}$ the long-term mean now equals $(\mu_X + \gamma_X \sigma_X)/\lambda_X$. It is generally assumed in a Gaussian world that the quantity of risk is given by the *volatility* of each factor.

This gives us the following set of processes under the market measure

$$d\mathbf{X}_t = (\mu_X - \lambda_X X_t + \gamma_X \sigma_X)dt + \sigma_X d\mathbf{W}_t^X, \qquad (10.37)$$

$$d\mathbf{Y}_t = (\mu_Y - \lambda_Y Y_t + \gamma_Y \sigma_Y)dt + \sigma_Y d\mathbf{W}_t^Y, \qquad (10.38)$$

$$d\mathbf{R}_t = \{k(X_t + Y_t - R_t) + \gamma_R \sigma_R\}dt + \sigma_R d\mathbf{W}_t^R, \qquad (10.39)$$

where all three factors contain a market price of risk γ in volatility units.

The yields derived in the economic factor model are continuously compounded while most yield data are annually compounded. So for appropriate comparison when estimating the parameters of the model we will have to convert the annually compounded yields into continuously compounded yields using the transformation

$$y^{\text{continuous}} = \ln(1 + y^{\text{annual}}). \qquad (10.40)$$

In the limit as T tends to infinity it can be shown that expression (10.31) derived for the yield does not tend to the 'long rate' factor X, but to a constant. This suggests that the three factors introduced in this term structure model may really be unobservable. To handle the unobservable state variables we formulate the model in state space form, a detailed description of which can be found in Harvey (1993) and use the Kalman filter to estimate the parameters (see e.g. Dempster *et al.* 1999 or Medova *et al.* 2005).

10.3.2 Pricing Coupon-Bearing Bonds

As sufficient historical data on Euro coupon-bearing Treasury bonds is difficult to obtain we use the zero-coupon yield curve to construct the relevant bonds. Coupons on newly-issued bonds are generally closely related to the ccorresponding spot rate at the time, so the current zero-coupon yield with maturity T is used as a proxy for the coupon rate of a coupon-bearing Treasury bond with maturity T. For example, on scenario ω the coupon rate $\delta_2^{B^{10}}(\omega)$ on a newly issued 10-year Treasury bond at time $t=2$ will be set equal to the projected 10-year spot rate $y_{2,10}(\omega)$ at time $t=2$.

Generally

$$\delta_t^{B^{(T)}}(\omega) = y_{t,T}(\omega), \quad \forall t \in T^d, \ \forall \omega \in \Omega, \qquad (10.41)$$

$$\delta_t^{B^{(T)}}(\omega) = \delta_{\lfloor t \rfloor}^{(T)}(\omega), \quad \forall t \in T^i, \ \forall \omega \in \Omega, \qquad (10.42)$$

where $\lfloor \cdot \rfloor$ denotes integral part. This ensures that as the yield curve falls, coupons on newly-issued bonds will go down correspondingly and each coupon cash flow will be discounted at the appropriate zero-coupon yield.

The bonds are assumed to pay coupons semi-annually. Since we roll the bonds on an annual basis, a coupon will be received after six months and again after a year just before the bond is sold. This forces us to distinguish between the price at which the bond is sold at rebalancing times and the price at which the new bond is purchased.

Let $P_{t,B^{(T)}}^{(sell)}$ denote the selling price of the bond $B^{(T)}$ at time t, assuming two coupons have now been paid out and the time to maturity is equal to $T-1$, and let $P_{t,B^{(T)}}^{(buy)}$ denote the price of a newly issued coupon-bearing Treasury bond with a maturity equal to T.

The 'buy' bond price at time t is given by

$$B_t^T(\omega) = F^{B^T} e^{-(T+\lfloor t \rfloor - t) y_{t,T+\lfloor t \rfloor - t}(\omega)}$$

$$+ \sum_{s=\frac{\lfloor 2t \rfloor}{2}+\frac{1}{2}, \frac{\lfloor 2t \rfloor}{2}+1,...,\lfloor t \rfloor + T} \frac{\delta_t^{B^T}(\omega)}{2} F^{B^T} e^{-(s-t)y_{t,(s-t)}(\omega)}, \quad \omega \in \Omega, \ t \in T^{total}, \quad (10.43)$$

where the principal F^{B^T} of the bond is discounted in the first term and the stream of coupon payments in the second.

At rebalancing times t the sell price of the bond is given by

$$B_t^T(\omega) = F^{B^T} e^{-(T-1)y_{t,T-1}(\omega)}$$

$$+ \sum_{s=\frac{1}{2},1,...,T-1} \frac{\delta_{t-1}^{B^T}(\omega)}{2} F^{B^T} e^{-(s-t)y_{t,(s-t)}(\omega)} \quad \omega \in \Omega \quad t \in \{T^d \setminus \{0\}\} \cup \{T\} \quad (10.44)$$

with coupon rate $\delta_{t-1}^{B^T}(\omega)$. The coupon rate is then reset for the newly-issued Treasury bond of the same maturity. We assume that the coupons paid at six months are re-invested in the on-the-run bonds. This gives the following adjustment to the amount held in bond B^T at time t.

$$x_{t,B^T}(\omega) = x_{t-\frac{1}{12}, B^T}(\omega) + \frac{\frac{1}{2}\delta_t^{B^T}(\omega) F^{B^T} x_{t-\frac{1}{12}, B^T}(\omega)}{f P_{t,B^T}^{buy}(\omega)}, \quad t \in T^c, \ \omega \in \Omega. \quad (10.45)$$

10.4 HISTORICAL BACKTESTS

We will look at an *historical backtest* in which statistical models are fitted to data up to a trading time t and scenario trees are generated to some chosen horizon $t + T$. The optimal root node decisions are then implemented at time t and compared to the historical returns at time $t + 1$. Afterwards the whole procedure is rolled forward for T trading times. Our backtest will involve a *telescoping horizon* as depicted in Figure 10.4.

At each decision time t the parameters of the stochastic processes driving the stock return and the three factors of the term structure model are re-calibrated using historical data up to and including time t and the initial values of the simulated scenarios are given by the actual historical values of the variables at these times. Re-calibrating the simulator

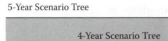

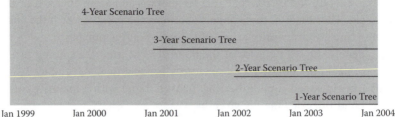

FIGURE 10.4 Telescoping horizon backtest schema.

TABLE 10.2 Tree Structures for Different Horizon Backtests

Jan 1999	6.6.6.6.6 = 7776	32.4.4.4.4 = 8192	512.2.2.2.2 = 8192
Jan 2000	9.9.9.9 = 6561	48.6.6.6 = 10368	512.2.2.2 = 4096
Jan 2001	20.20.20 = 8000	80.10.10 = 8000	768.3.3 = 6912
Jan 2002	88.88 = 7744	256.32 = 8192	1024.8 = 8192
Jan 2003	7776	8192	8192

parameters at each successive initial decision time t captures information in the history of the variables up to that point.

Although the optimal second and later-stage decisions of a given problem may be of 'what-if' interest, manager and decision maker focus is on the implementable first-stage decisions which are hedged against the simulated future uncertainties. The reasons for implementing stochastic optimization programs in this way are twofold. Firstly, after one year has passed the actual values of the variables realized may not coincide with any of the values of the variables in the simulated scenarios. In this case the optimal investment policy would be undefined, as the model only has optimal decisions defined for the nodes on the simulated scenarios. Secondly, as one more year has passed new information has become available to re-calibrate the simulator's parameters. Relying on the original optimal investment strategies will ignore this information. For more on backtesting procedures for stochastic optimization models see Dempster *et al.* (2003).

For our backtests we will use three different tree structures with approximately the same number of scenarios, but with an increasing initial branching factor. We first solve the five-year problem using a 6.6.6.6.6 tree, which gives 7776 scenarios. Then we use 32.4.4.4.4 = 8192 scenarios and finally the extreme case of 512.2.2.2.2 = 8192 scenarios.

For the subsequent stages of the telescoping horizon backtest we adjust the branching factors in such a way that the total number of scenarios stays as close as possible to the original number of scenarios and the same ratio is maintained. This gives us the tree structures set out in Table 10.2.

Historical backtests show how specific models would have performed had they been implemented in practice. The reader is referred to Rietbergen (2005) for the calibrated parameter values employed in these tests. Figures 10.5 to 10.10 show how the various optimal portfolios' wealth would have evolved historically relative to the barrier. It is also

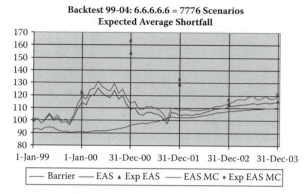

FIGURE 10.5 Backtest 1999–2004 using expected average shortfall for the 6.6.6.6.6 tree.

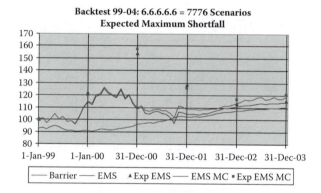

FIGURE 10.6 Backtest 1999–2004 using expected maximum shortfall for the 6.6.6.6.6 tree.

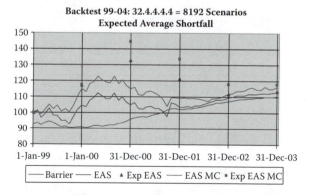

FIGURE 10.7 Backtest 1999–2004 using expected average shortfall for the 32.4.4.4.4 tree.

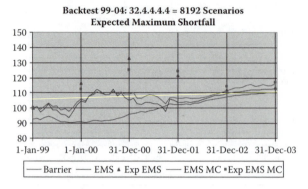

FIGURE 10.8 Backtest 1999–2004 using expected maximum shortfall for the 32.4.4.4.4 tree.

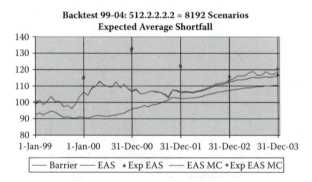

FIGURE 10.9 Backtest 1999–2004 using expected average shortfall for the 512.2.2.2.2 tree.

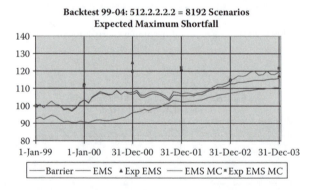

FIGURE 10.10 Backtest 1999–2004 using expected maximum shortfall for the 512.2.2.2.2 tree.

important to determine how the models performed in-sample on the generated scenario trees and whether or not they had realistic forecasts with regard to future historical returns. To this end one-year-ahead in-sample expectations of portfolio wealth are shown as points in the backtest performance graphs. Implementing the first-stage decisions *in-sample*, the portfolio's wealth is calculated one year later for each scenario in the simulated tree after which an expectation is taken over the scenarios.

From these graphs a first observation is that the risk management monitoring incorporated into the model appears to work well. In all cases the only time portfolio wealth dips below the barrier, if at all, is on 11 September 2001. The initial in-sample wealth overestimation of the models is likely to be due mainly to the short time series available for initial parameter estimation which led to hugely inflated stock return expectations during the equity bubble. However as time progresses and more data points to re-calibrate the model are obtained, the models' expectations and real-life realizations very closely approximate each other.

For reference we have included the performance of the Eurostoxx 50 in Figure 10.11 to indicate the performance of the stock market over the backtesting period. Even though this was a difficult period for the optimal portfolios to generate high historical returns, it provides an excellent demonstration that the risk management incorporated into the models operates effectively. It is in periods of economic downturn that one wants the portfolio returns to survive.

Tables 10.3 and 10.4 give the optimal portfolio allocations for the 32.4.4.4.4 tree using the two maximum shortfall objective functions. In both cases we can observe a for the portfolio to move to the safer, shorter-term assets as time progresses. This is naturally built into the model as depicted in Figure 10.3.

For the decisions made in January 2002/2003, the portfolio wealth is significantly closer to the barrier for the EMS model than it is for the EMS MC model. This increased risk for the fund is taken into account by the EMS model and results in an investment in safer short-term bonds and a negligible equity component. Whereas the EMS model stays in the one to three year range the EMS MC model invests mainly in bonds with a maturity in the range of three to five years and for both models the portfolio wealth manages to stay above the barrier.

From Figures 10.5 to 10.10 it can be observed that in all cases the method with monthly checking outperforms the equivalent method with just annual shortfall checks. Similarly as the initial branching factor is increased, the models' out-of-sample performance is generally improved. For the 512.2.2.2.2 = 8192 scenario tree, all four objective functions give optimal portfolio allocations which keep the portfolio wealth above the barrier at all times, but the models with the monthly checking still outperform the others. The more important difference however seems to lie in the deviation of the expected in-sample portfolio's wealth from the actual historical realization of the portfolio value. Table 10.5 displays this annual deviation averaged over the five rebalances and shows a clear reduction in this deviation for three of the four models as the initial branching factor is increased. Again the model that uses the expected maximum shortfall with monthly checking as its objective function outperforms the rest.

Overall the historical backtests have shown that the described stochastic optimization framework carefully considers the risks created by the guarantee. The EMS MC model produces well-diversified portfolios that do not change drastically from one year to the next and results in optimal portfolios which even through a period of economic downturn and uncertainty remained above the barrier.

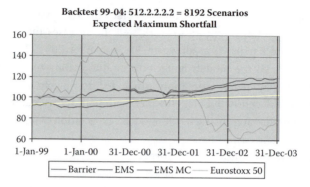

FIGURE 10.11 Comparison of the fund's portfolio performance to the Eurostoxx 50.

TABLE 10.3 Portfolio Allocation Expected Maximum Shortfall Using the 32.4.4.4.4 Tree

	1y	2y	3y	4y	5y	10y	30y	Stock
Jan 99	0	0	0	0	0	0.23	0.45	0.32
Jan 00	0	0	0	0	0	0	0.37	0.63
Jan 01	0.04	0	0	0	0	0.39	0.53	0.40
Jan 02	0.08	0.16	0.74	0	0	0	0	0.01
Jan 03	0.92	0	0	0	0	0.07	0	0.01

TABLE 10.4 Portfolio Allocation Expected Maximum Shortfall with Monthly Checking Using the 32.4.4.4.4 Tree

	1y	2y	3y	4y	5y	10y	30y	Stock
Jan 99	0	0	0	0	0.49	0.27	0	0.24
Jan 00	0	0	0	0	0.25	0.38	0	0.36
Jan 01	0	0	0	0	0.49	0.15	0	0.36
Jan 02	0	0	0	0.47	0.44	0	0	0.10
Jan 03	0	0	0.78	0.22	0	0	0	0.01

TABLE 10.5 Average Annual Deviation

	EAS	EAS MC	EMS	EMS MC
6.6.6.6.6	9.87%	13.21%	9.86%	10.77%
32.4.4.4.4	10.06%	9.41%	9.84%	7.78%
512.2.2.2.2	10.22%	8.78%	7.78%	6.86%

10.5 ROBUSTNESS OF BACKTEST RESULTS

Empirical equity returns are now well known not to be normally distributed but rather to exhibit complex behaviour including fat tails. To investigate how the EMS MC model performs with more realistic asset return distributions we report in this section experiments using a geometric Brownian motion with Poisson jumps to model equity returns. The stock price process **S** is now assumed to follow

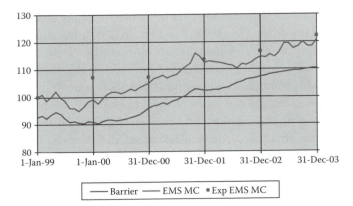

FIGURE 10.12 Expected maximum shortfall with monthly checking using the 512.2.2.2.2 tree for the GBM with jumps equity index process.

TABLE 10.6 Portfolio Allocation Expected Maximum Shortfall with Monthly Checking Using the 512.2.2.2.2 Tree

	1y	2y	3y	4y	5y	10y	30y	Stock
Jan 99	0	0	0	0	0.69	0.13	0	0.18
Jan 00	0	0	0	0	0.63	0	0	0.37
Jan 01	0	0	0	0	0.37	0.44	0	0.19
Jan 02	0	0	0	0	0.90	0	0	0.10
Jan 03	0	0	0.05	0	0.94	0	0	0.01

TABLE 10.7 Portfolio Allocation Expected Maximum Shortfall with Monthly Checking Using the 512.2.2.2.2 Tree for the GBM with Poisson Jumps Equity Index Process

	1y	2y	3y	4y	5y	10y	30y	Stock
Jan 99	0	0	0	0	0.12	0.77	0	0.11
Jan 00	0	0	0	0	0	0.86	0	0.14
Jan 01	0	0	0	0	0.43	0.56	0	0.01
Jan 02	0	0	0	0	0.70	0.11	0	0.19
Jan 03	0	0	0	0	0.04	0.81	0	0.15

$$\frac{dS_t}{S_t} = \tilde{\mu}_S dt + \tilde{\sigma}_S d\tilde{W}_t^S + J_t dN_t, \tag{10.46}$$

where N is an independent Poisson process with intensity λ and the jump saltus J at Poisson epochs is an independent normal random variable.

As the EMS MC model and the 512.2.2.2.2 tree provided the best results with Gaussian returns the backtest is repeated for this model and treesize. Figure 10.12 gives the historical backtest results and Tables 10.6 and 10.7 represent the allocations for the 512.2.2.2.2 tests with the EMS MC model for the original GBM process and the GBM with Poisson jumps process respectively. The main difference in the two tables is that the investment in equity is substantially lower initially when the equity index volatility is high (going down to 0.1% when the volatility is 28% in 2001), but then increases as the volatility comes down to 23% in 2003. This is born out by Figure 10.12 which shows much more realistic in-sample one-year-ahead portfolio wealth predictions (cf. Figure 10.10) and a 140 basis point increase in terminal historical fund return over the Gaussian model. These phenomena are the result of the calibration of the normal jump saltus distributions to have negative means and hence more downward than upwards jumps resulting in downwardly skewed equity index return distributions, but with the same compensated drift as in the GBM case. As a consequence the optimal portfolios are more sensitive to equity variation and take benefit from its lower predicted values in the last two years.

Although much more complex equity return processes are possible, these results show that the historical backtest performance of the EMS MC model is only improved in the presence of downwardly skewed asset equity return distributions possessing fat tails due to jumps.

10.6 CONCLUSIONS

This chapter has focused on the design of funds to support investment products which give a minimum guaranteed return. We have concentrated here on the design of the liability side of the fund, paying particular attention to the pricing of bonds using a three-factor term structure model with reliable results for long-term as well as the short-term yields. Several objective functions for the stochastic optimization of portfolios have been constructed using expected average shortfall and expected maximum shortfall risk measures to combine risk management with strategic asset allocation. We also introduced the concept of monthly shortfall checking which improved the historical backtesting results considerably. In addition to the standard GBM model for equity returns we reported experiments using a GBM model with Poisson jumps to create downwardly skewed fat tailed equity index return distributions. The EMS MC model responded well with more realistic expected portfolio wealth predictions and the historical fund portfolio wealth staying significantly above the barrier at all times.

The models of this chapter have been extended in practice to open ended funds which allow for contributions throughout the lifetime of the corresponding investment products. In total funds of the order of 10 billion euros have been managed with these extended models. In future research we hope to examine open multi-link pension funds constructed using several unit linked funds of varying risk aversion in order to allow the application of individual risk management to each client's portfolio.

ACKNOWLEDGEMENTS

The authors wish to thank Drs. Yee Sook Yong and Ning Zhang for their able assistance with computational matters. We also acknowledge the helpful comments of two anonymous referees which led to material improvements in the chapter.

REFERENCES

Andersen, T.G. and Lund, J., Estimating continuous-time stochastic volatility models of the short-term interest rate. *J. Econometrics*, 1997, **77**(2), 343–377.

Campbell, R., The economic factor model: Theory and applications. *Lehman Brothers Presentation to Europlus Research and Management Dublin*, 31 March 2000.

Consigli, G. and Dempster, M.A.H., The CALM stochastic programming model for dynamic asset-liability management. In *Worldwide Asset and Liability Modeling*, edited by J.M. Mulvey and W.T. Ziemba, pp. 464–500, 1998 (Cambridge University Press: Cambridge).

Consiglio, A., Cocco, F. and Zenios, S.A., The PROMETEIA model for managing insurance policies with guarantees. In *Handbook of Asset and Liability Management*, edited by S. Zenios and W.T. Ziemba, Vol. 2, 2007 (North-Holland: Oxford); in press.

Cox, J.C., Ingersoll, J.E. and Ross, S.A., A theory of the term structure of interest rates. *Econometrica*, 1985, **53**, 385–407.

Dempster, M.A.H., Germano, M., Medova, E.A. and Villaverde, M., Global asset liability management. *Br. Actuarial J.*, 2003, **9**(1), 137–216.

Dempster, M.A.H., Hicks-Pedron, N., Medova, E.A., Scott, J.E. and Sembos, A., Planning logistics operations in the oil industry. *J. Operat. Res. Soc*, 2000, **51**, 1271–1288.

Dempster, M.A.H., Jones, C.M., Khokhar, S.Q. and Hong, G.S-S., Implementation of a model of the stochastic behaviour of commodity prices. Report to Rio Tinto, Centre for Financial Research, Judge Institute of Management, University of Cambridge, 1999.

Harvey, A.C., *Time Series Models*, 1993 (Harvester Wheatsheaf: Hemel Hempstead).

Hertzog, F., Dondi, G., Keel, S., Schumani, L.M. and Geering, H.P., Solving ALM problems via sequential stochastic programming. *Quant. Finance*, 2007, **7**(2), 231–244.

Hull, J.C., *Options, Futures and Other Derivatives*, 1997 (Prentice Hall: Upper Saddle River, NJ).

Markowitz, H.M., *Portfolio Selection*, 1959 (Wiley: New York).

Medova, E.A., Rietbergen, M.I., Villaverde, M. and Yong, Y.S., Modelling the long-term dynamics of the yield curve. *Working Paper*, 2005 (Centre for Financial Research, Judge Business School, University of Cambridge).

Nelson, C.R. and Siegel, A.F., Parsimonious modeling of yield curves. *J. Bus*, 1987, **60**, 473–489.

Rietbergen, M.I., Long term asset liability management for minimum guaranteed return funds. PhD Dissertation, Centre for Financial Research, Judge Business School, University of Cambridge, 2005.

Vasicek, O., An equilibrium characterization of the term structure. *J. Financ. Econ*, 1977, **5**(2), 177–188.

PART 2

Portfolio Construction and Risk Management

CHAPTER 11

DC Pension Fund Benchmarking with Fixed-Mix Portfolio Optimization

M. A. H. DEMPSTER, M. GERMANO, E. A. MEDOVA, M. I. RIETBERGEN, F. SANDRINI, M. SCROWSTON and N. ZHANG

CONTENTS

11.1   INTRODUCTION

C ORPORATE SPONSORED *DEFINED BENEFIT* (DB) pension schemes have recently found themselves in hot water. Accounting practices that led to over-exposure to equity markets, increases in longevity of the scheme participants and low interest rates have all contributed to the majority of schemes in the EU and the U.K. finding themselves underfunded. In essence, a DB scheme promises to pay its participants an annuity at retirement that gives them a pension equal to a proportion of their final salary (the proportion depending on the number of years of service). Therefore the responsibility to meet these promises (liabilities) rests firmly with the scheme's trustees and ultimately with the corporate sponsor.

The management of these corporate schemes was greatly affected in the past by quarterly earnings reports which directly impacted stock prices in the quest for 'shareholder value.'

Consequently DB scheme sponsors resorted to a management style that was able to keep the liabilities, if not off the balance sheet, then at least to a minimum. One sanctioned tactic that achieved these aims was the ability to discount liabilities by the expected return of the constituent asset classes of the fund. In other words, by holding a large part of the fund in equities, the liabilities could be discounted away at over 10% p.a. The recent performance of the equity markets and the perception of equity as a long-horizon asset class assisted in justifying this asset-mix in the eyes of the scheme's trustees. However, with the collapse of the equity-market bubble in 2001, many funds found their schemes grossly underfunded and were forced to crystallize their losses by panicked trustees. Consequent tightening of the regulations has made the situation even worse (e.g. all discounting must be done by the much lower AA credit quality bond yield rates in the UK FRS17 standard).

As a result many DB schemes have closed and are now being replaced with *defined contribution* (DC) schemes.[1] In this world of corporate sponsored DC pension schemes the liability is separated from the sponsor and the market risk is placed on the shoulders of the participants. The scheme is likely to be overseen by an investment consultant and if the scheme invests in funds that perform badly over time a decision may be made by the consultant to move the capital to another fund. However, any losses to the fund will be borne by the participants in the scheme and not by the corporate sponsor.

Since at retirement date scheme participants will wish to either purchase an annuity or invest their fund payout in a self-managed portfolio, an obvious need arises in the market place for real return guaranteed schemes which are similar to those often found in life insurance policies. These guarantees will typically involve inflation protection plus some element of capital growth, for example, inflation rate plus 1% per annum. From the DC fund manager's viewpoint provision of the relevant guarantee requires very tight risk, control, as the recent difficulties at Equitable Life so graphically illustrate.

The question addressed in this chapter is how consultants or DC fund-managers can come to a sensible definition of an easily understandable liability-related benchmark against which the overall fund performance for a DC scheme can be measured. Performance of both fund and benchmark must be expressed to fund participants in easy-to-understand concepts such as probability of achieving some target wealth level above the scheme guarantee—a measure easily derived from the solutions of the models discussed in this chapter.

11.2 CURRENT MARKET PRACTICE

Currently, DC pension funds are market-benchmarked against either a fixed-mix of defined asset classes (total return bond and equity indices) or against some average performance of their peers. The benchmark is not defined in terms of the pension liability and investment is not liability-driven. The standard definitions of investment risk—standard deviation, semi-variance and downside risk—do not convey information regarding the probability of missing the scheme participants' investment goals and obligations.

---

[1] DC pension scheme participants typically make a lump sum initial payment and regular contributions to the pension fund which are employer matched.

For example, the macro-asset benchmark may be defined as 20% of certain equity indices and 80% of particular bond indices, but may not reflect the risk that the scheme participants are willing to take in order to attain a specific *substitution rate* between their final salary and pension income (pension earnings/final salary).

11.3 OPTIMAL BENCHMARK DEFINITION FOR DC FUNDS

In line with current market practice, we wish to find a definition of a fixed asset mix benchmark similar to that given in the example above but with an asset mix that optimizes returns against user-defined risk preferences. Specifically, for participants that are willing to take a certain amount of risk in order to aim for a given substitution rate between final salary and pension, we should be able to 'tune' the asset mix in an optimal way to reflect the participants desire to reach this substitution rate. The risk could then be defined as the probability of not reaching that substitution rate.

In contrast to dynamic multi-stage portfolio optimization, where the asset-mix is changed dynamically over time to reflect changing attitudes to risk as well as market performance (dynamic utility), a *fixed-mix rebalance strategy* benchmark in some sense reflects an average of this dynamic utility over the fund horizon. For such a strategy the realized portfolio at each decision stage is rebalanced back to a fixed set of portfolio weights. In practice for DC pension schemes we want the returns to be in line with salary inflation in the sense that the required substitution rate is reached with a given probability. The solution to this problem will entail solving a fixed-mix dynamic stochastic programme that reflects the long run utility of the scheme participants.

In general multi-period *dynamic stochastic optimization* will be more appropriate for long-term investors. Single-period models construct optimal portfolios that remain unchanged over the planning horizon while fixed mix rebalance strategies fail to consider possible investment opportunities that might arise due to market conditions over the course of the investment horizon. Dynamic stochastic programmes on the other hand capture optimally an investment policy in the face of the uncertainty about the future given by a set of scenarios.

Carino and Turner (1998) compare a multi-period stochastic programming approach to a fixed-mix strategy employing traditional mean-variance efficient portfolios. Taking a portfolio from the mean-variance efficient frontier, it is assumed that the allocations are rebalanced back to that mix at each decision stage. They also highlight the inability of the mean-variance optimization to deal with derivatives such as options due to the skewness of the resulting return distributions not being taken into account. The objective function of the stochastic programme is given by maximizing expected wealth less a measure of risk given by a convex cost function. The stochastic programming approach was found to dominate fixed-mix in the sense that for any given fixed-mix rebalance strategy, there is a strategy that has either the same expected wealth and lower shortfall cost, or the same shortfall cost and higher expected wealth. Similar results were found by Hicks-Pedrón (1998) who also showed the superiority in terms of final Sharpe ratio of both methods to the constant proportion portfolio insurance (CPPI) strategy over long horizons.

Fleten *et al.* (2002) compare the performance of four-stage stochastic models to fixed-mix strategies of in- and out-of-sample, using a set of 200 flat scenarios to obtain the out-of-sample results. They show that the dynamic stochastic programming solutions dominate the fixed-mix solutions both in- and out-of-sample, although to a lesser extent out-of-sample. This is due to the ability of the stochastic programming model to adapt to the information in the scenario tree in-sample, although they do allow the fixed-mix solution to change every year once new information has become available, making this sub-optimal strategy inherently more dynamic.

Mulvey *et al.* (2003) compare buy-and-hold portfolios to fixed-mix portfolios over a ten-year period, showing that in terms of expected return versus return standard deviation, the fixed-mix strategy generates a superior efficient frontier, where the excess returns are due to portfolio rebalancing. Dempster *et al.* (2007a) discuss the theoretical cause of this effect (and the historical development of its understanding) under the very general assumption of stationary ergodic returns. A similar result was found in Mulvey *et al.* (2004) with respect to including alternative investments into the portfolio. In particular they looked at the use of the Mt. Lucas Management index in multi-period fixed-mix strategies. A multi-period optimization will not only identify these gains but also take advantage of volatility by suggesting solutions that are optimal in alternative market scenarios. In Mulvey *et al.* (2007) the positive long term performance effects of new asset classes, leverage and various overlay strategies are demonstrated for both fixed-mix and dynamically optimized strategies.

11.4   FUND MODEL

The dynamic optimal portfolio construction problem for a DC fund with a performance guarantee is modelled here at the strategic level with annual rebalancing. The objective is to maximize the expected sum of accumulated wealth while keeping the expected maximum shortfall of the portfolio relative to the guarantee over the 5 year planning horizon as small as possible. A complete description of the *dynamic stochastic programming model* can be found in Dempster *et al.* (2006). In the *fixed-mix model* the portfolio is rebalanced to fixed proportions at all future decision nodes, but not at the intermediate time stages used for shortfall checking.

This results in annual rebalancing while keeping the risk management function monthly, leading to the objective function for both problems as

$$
\max_{\left\{ \substack{x_{t,a}(\omega),x_{t,a}^+(\omega),x_{t,a}^-(\omega): \\ a\in A,\omega\in\Omega, t\in T^d\cup\{T\}} \right\}} \left\{ (1-\beta)\left(\sum_{\omega\in\Omega} p(\omega) \sum_{t\in T^d\cup\{T\}} W_t(\omega) \right) \right.
$$
$$
\left. -\beta\left(\sum_{\omega\in\Omega} p(\omega)\, \max_{t\in T^{\text{total}}} h_t(\omega) \right) \right\}, \qquad (11.1)
$$

where

- $p(\omega)$ denotes the *probability* of scenario ω in Ω—here $p(\omega):=1/N$ with N scenarios,
- $W_t(\omega)$ denotes the *portfolio wealth* at time $t\in T^{\text{total}}$ in scenario ω,
- $h_t(\omega)$ denotes the *shortfall* relative to the barrier at time t in scenario ω.

For the *nominal* or *fixed guarantee*, the *barrier* at time t in scenario ω, below which the fund will be unable to meet the guarantee, is given by

$$L_t^F(\omega) = W_0(1+G)^T Z_t(\omega) = W_0(1+G)^T e^{-y_{t,T}(\omega)(T-t)}, \qquad (11.2)$$

where

- G denotes the *annual nominal guarantee*
- $Z_t(\omega)$ denotes the *zero-coupon Treasury bond* price at time t in scenario ω.

For simplicity we model closed-end funds here, but see Dempster *et al.* (2006, 2007) and Rietbergen (2005) for the treatment of contributions. We employ a five-period (stage) model with a total of 8192 scenarios to obtain the solutions for the dynamic optimization and fixed mix approaches.[2]

For this chapter, five different experiments were run on a five-year closed-end fund with a minimum nominal guarantee of 2% and an initial wealth of 100 using a 512.2.2.2.2 tree.[3] (See Dempster *et al.* (2006, 2007) for more details on this problem). The parameter of risk aversion β is set to 0.99 and the parameter values used were estimated over the period June 1997–December 2002. The Pioneer CASM simulator was used to generate the problem data at monthly intervals.

The five experiments run were as follows.

- Experiment 1: *No* fixed-mix constraints. Objective function: fund wealth less expected maximum shortfall with monthly checking.
- Experiment 2: Arbitrary fixed-mix: 30% equity and 10% in each of the bonds.
- Experiment 3: The fixed-mix is set equal to the root node decision of Experiment 1.
- Experiment 4: The fixed-mix is set equal to the root node decision of Experiment 1 but only applied after the first stage. The root node decision is optimized.
- Experiment 5: The fixed-mix is determined optimally.

Experiments 2–4 with fixed-mixed constraints are '*fixed* fixed-mix' problems in which the fixed-mix is specified in advance in order to keep the optimization problem convex. Finally Experiment 5 uses fixed-mix constraints without fixing them in advance. This renders the optimization problem non-convex so that a global optimization technique needs to be used. In preliminary experiments we found that although the resulting unconstrained problems are multi-extremal they are 'near-convex' and can be globally optimized by a search routine followed by a local convex optimizer. For this purpose we used Powell's (1964) algorithm followed by the SNOPT solver. Function evaluations involving all fixed-mix rebalances were evaluated by linear programming using CPLEX. This method is described in detail in Scott (2002).

As the fixed-mix policy remains the same at all rebalances, theoretically there is no reason to have a scenario tree which branches more than once at the beginning of the first

---

[2] In practice 10 and 15 year horizons have also been employed.
[3] Assets employed are Eurobonds of 1, 2, 3, 4, 5, 10 and 30 year maturities and equity represented by the Eurostock 50 index.

year. A simple fan tree structure would be perfectly adequate as the fixed-mix approach is unable to exploit the perfect foresight implied after the first stage in this tree. However for comparison reasons we use the same tree for both the dynamic stochastic programme (Experiment 1) and the fixed-mix approach (Experiments 2–5).

11.5   NOMINAL GUARANTEE RESULTS

Table 11.1 shows the expected terminal wealth and expected maximum shortfall for the five experiments.

As expected, Experiment 1 with no fixed-mix constraints results in the highest expected terminal wealth and lowest expected maximum shortfall. Whereas Experiments 2 and 3 underperform, Experiment 3 in which the initial root node solution of Experiment 1 is used as the fixed-mix is a significant improvement on Experiment 2 (arbitrary fixed-mix) and might serve as an appropriate benchmark. Experiment 4 resulted in a comparable expected terminal wealth to Experiment 3, but the expected maximum shortfall is now an order of magnitude smaller. Finally in Experiment 5 global optimization was used which correctly resulted in an improvement relative to Experiment 3 in both the expected terminal wealth and the expected maximum shortfall.

Table 11.2 shows the optimal root node decisions for all five experiments. With the equity market performing badly and declining interest rates over the 1997–2002 period, we see a heavy reliance on bonds in all portfolios.

Figures 11.1 and 11.2 show the efficient frontiers for the dynamic stochastic programme and the fixed-mix solution, where the risk measure is given by expected maximum shortfall. Figure 11.1 shows that the dynamic stochastic programme generates a much bigger range of possible risk return trade-offs and even if we limit the range of risk parameters to that given for the fixed-mix experiments as in Figure 11.2, we see that the DSP problems clearly outperform the fixed-mix problems.

TABLE 11.1 Expected Terminal Wealth and Maximum Shortfall for the Nominal Guarantee

	Expected terminal wealth	Expected maximum shortfall
Experiment 1	126.86	8.47 E-08
Experiment 2	105.58	14.43
Experiment 3	120.69	0.133
Experiment 4	119.11	0.014
Experiment 5	122.38	0.122

TABLE 11.2 Root Node Solutions for the Nominal Guarantee

	1y	2y	3y	4y	5y	10y	30y	Stock
Exp 1	0	0	0.97	0	0	0	0.02	0.01
Exp 2	0.10	0.10	0.10	0.10	0.10	0.10	0.10	0.30
Exp 3	0	0	0.97	0	0	0	0.02	0.01
Exp 4	0	0	0.06	0.94	0	0	0	0
Exp 5	0	0	0.96	0.04	0	0	0	0

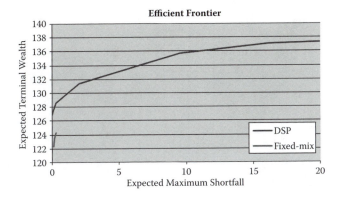

FIGURE 11.1 Efficient frontier for the nominal guarantee.

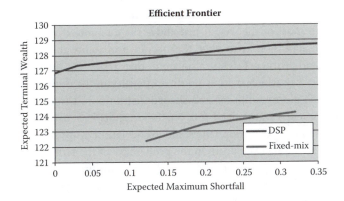

FIGURE 11.2 Efficient frontier for the nominal guarantee.

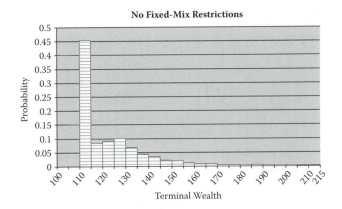

FIGURE 11.3 Terminal wealth distribution for optimal dynamic stochastic policy of Experiment 1.

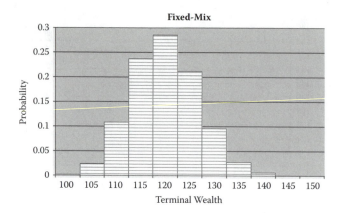

FIGURE 11.4 Terminal wealth distribution for the optimal fixed-mix policy of Experiment 5.

We also considered the distribution of the terminal wealth as shown in Figures 11.3 and 11.4. From Figure 11.3 we observe a highly skewed terminal wealth distribution for Experiment 1 with most of the weight just above the guaranteed wealth of 110.408. The effect of dynamic allocation is to alter the overall probability distribution of the final wealth. Carino and Turner (1998) also note this result in their experiments. For Experiments 2 and 3 in which there is no direct penalty in the optimization problem for shortfall, we see the more traditional bell-shaped distribution. Using the initial root node solution of Experiment 1 as the fixed-mix portfolio results in a distribution with a higher mean and lower standard deviation (the standard deviation drops from 20.51 to 6.43). In Experiment 4 we see an increase again in the probability of the terminal wealth ending up just above the minimum guarantee of 110 as the optimization problem has flexibility at the initial stage. The standard deviation is further reduced in this experiment to 4.04. The mean and standard deviation of Experiment 5 is comparable to that of Experiment 3, which is as expected since the portfolio allocations of the two experiments are closely related.

11.6 INFLATION-LINKED GUARANTEE RESULTS

In the case of *an inflation-indexed guarantee* the final guarantee at time T is given by

$$W_0 \prod_{s=1/12}^{T} \left(1 + i_s^{(m)}(\omega)\right), \tag{11.3}$$

where $i_s^{(m)}(\omega)$ represents the *monthly inflation* rate at time s in scenario ω.

However, unlike the nominal guarantee, at time $t < T$ the final inflation-linked guarantee is still *unknown*. We propose to *approximate* the final guarantee by using the inflation rates which are known at time t, combined with the expected inflation at time t for the period $[t + (1/12), T]$.

The *inflation-indexed barrier* at time t is then given by

$$
\begin{aligned}
L_t^I(\omega) &= W_0 \left(\prod_{s=\frac{1}{12}}^{t} \left(1 + i_s^{(m)}(\omega)\right) \right) \left(\prod_{s=t+\frac{1}{12}}^{T} \left(1 + i_s^{(m)}(\omega)\right) \right) Z_t(\omega) \\
&= W_0 \left(\prod_{s=\frac{1}{12}}^{t} \left(1 + i_s^{(m)}(\omega)\right) \right) \left(\prod_{s=t+\frac{1}{12}}^{T} \left(1 + i_s^{(m)}(\omega)\right) \right) e^{-y_{t,T}(\omega)(T-t)}.
\end{aligned}
\tag{11.4}
$$

In general the expected terminal wealth is higher for the inflation-linked barrier, but we also see an increase in the expected maximum shortfall (see Table 11.3). This reflects the increased uncertainty related to the inflation-linked guarantee which also forces us to increase the exposure to more risky assets. With an inflation-linked guarantee the final guarantee is only known for certain at the end of the investment horizon. Relative to the nominal guarantee results of Table 11.2, Table 11.4 shows that the initial portfolio allocations for the inflation-linked guarantee are more focused on long-term bonds.

TABLE 11.3 Expected Terminal Wealth and Maximum Shortfall for the Inflation-Linked Guarantee

	Expected terminal wealth	Expected maximum shortfall
Experiment 1	129.88	0.780
Experiment 2	122.81	13.60
Experiment 3	129.34	1.580
Experiment 4	129.54	1.563
Experiment 5	128.23	1.456

TABLE 11.4 Root Node Solution for the Inflation-Linked Guarantee

	1y	2y	3y	4y	5y	10y	30y	Stock
Exp 1	0	0	0	0	0.77	0.23	0	0
Exp 2	0.10	0.10	0.10	0.10	0.10	0.10	0.10	0.30
Exp 3	0	0	0	0	0.77	0.23	0	0
Exp 4	0	0	0	0	0.88	0.12	0	0
Exp 5	0	0	0	0	0.94	0.06	0	0

As in Figure 11.3, Figure 11.5 shows that there is a noticeable pattern of asymmetry in the final wealth outcomes for the dynamic stochastic programme of Experiment 1. However the skewness is not so marked. This is due to the fact that for the inflation-linked guarantee problems inflation rates differ on each scenario and the final guarantee is scenario dependent which results in a different value of the barrier being pursued along each scenario. This symmetrizing effect is even more marked for the inflation-linked guarantee as shown in Figure 11.6 (*cf.* Figure 11.4).

Figure 11.7 shows for the inflation-linked guarantee problem similar out-performance of DSP relative to the optimal fixed-mix policy as in Figure 11.2.

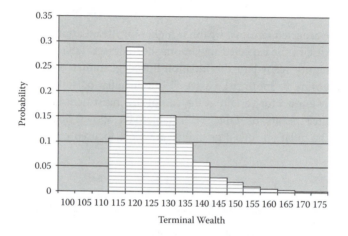

FIGURE 11.5 Terminal wealth distribution for Experiment 1 for the inflation-linked guarantee.

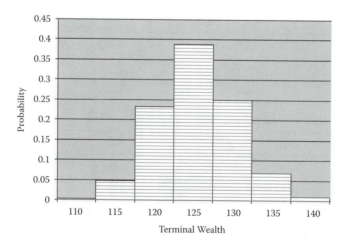

FIGURE 11.6 Terminal wealth distribution for Experiment 5 for the inflation-linked guarantee.

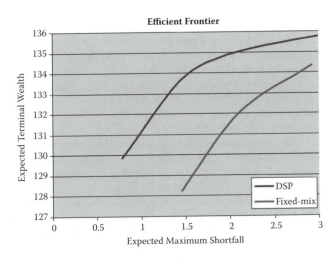

FIGURE 11.7 Efficient frontier for the inflation-linked guarantee.

11.7   CONCLUSION

In this chapter we have compared the performance of two alternative versions of a dynamic portfolio management model for a DC pension scheme which accounts for the liabilities arising from a guaranteed fund return. The results show that a fixed-mixed rebalance policy can be used as a benchmark for the dynamic stochastic programming optimal solution with less complexity and lower computational cost. Whereas the risk-return trade-off for a fixed-mix portfolio rebalancing strategy is constant over the planning horizon, for the dynamic stochastic programming solutions portfolio allocations shift to less volatile assets as the excess over the liability barrier is reduced. The resulting guarantee shortfall risk for the easy-to-explain fixed-mix portfolio rebalancing strategy is therefore higher and its portfolio returns are lower than those of the dynamic optimal policy. On a percentage basis however these differences are sufficiently small to be able to use the easier-to-compute fixed-mix results as a conservative performance benchmark for both in-sample (model) and actual out-of-sample fund performance. For out-of-sample historical backtests of *optimal* dynamic stochastic programming solutions for these and related problems the reader is referred to Dempster *et al.* (2006, 2007b). Perhaps the easiest way to explain both benchmark and actual fund performances to DC pension scheme participants is to give probabilities of achieving (expected) guaranteed payouts and more. These are easily estimated *a priori* by scenario counts in both models considered in this paper for fund design and risk management. It is of course also possible to link final fund payouts to annuity costs and substitution rates with the corresponding probability estimates.

REFERENCES

Carino, D.R. and Turner, A.L., Multiperiod asset allocation with derivative assets. In *Worldwide Asset and Liability Modeling*, edited by W.T. Ziemba and J.M. Mulvey, pp. 182–204, 1998 (Cambridge University Press: Cambridge).

Dempster, M.A.H., Germano, M., Medova, E.A., Rietbergen, M.I., Sandrini, F. and Scrowston, M., Managing guarantees. *J. Portfolio Manag.*, 2006, **32**(2), 51–61.

Dempster, M.A.H., Evstigneev, I.V. and Schenk-Hoppé, K.R., Volatility-induced financial growth. *Quant. Finance*, 2007a, **7**(2), 151–160.

Dempster, M.A.H., Germano, M., Medova, E.A., Rietbergen, M.I., Sandrini, F. and Scrowston, M., Designing minimum guaranteed return funds. *Quant. Finance*, 2007b, **7**(2), 245–256.

Fleten, S.E., Hoyland, K. and Wallace, S.W., The performance of stochastic dynamic and fixed mix portfolio models. *Eur. J. Oper. Res.*, 2002, **140**(1), 37–49.

Hicks-Pedrón, N., Model-based asset management: A comparative study. PhD dissertation, Centre for Financial Research, Judge Institute of Management, University of Cambridge, 1998.

Mulvey, J.M., Pauling, W.R. and Madey, R.E., Advantages of multiperiod portfolio models. *J. Portfolio Manag.*, 2003, **29**(2), 35–45.

Mulvey, J.M., Kaul, S.S.N. and Simsek, K.D., Evaluating a trend-following commodity index for multi-period asset allocation *J. Alternat. Invest.*, 2004, **7**(1), 54–69.

Mulvey, J.M., Ural, C. and Zhang, Z., Improving performance for long-term investors: Wide diversification, leverage and overlay strategies. *Quant. Finance*, 2007, **7**(2), 175–187.

Powell, M.J.D., An efficient method for finding the minimum of a function of several variables without calculating derivatives. *Comput. J.*, 1964, **7**, 303–307.

Rietbergen, M.I., Long term asset liability management for minimum guaranteed return funds. PhD dissertation, Centre for Financial Research, Judge Business School, University of Cambridge, 2005.

Scott, J.E., Modelling and solution of large-scale stochastic programmes. PhD dissertation, Centre for Financial Research, Judge Institute of Management, University of Cambridge, 2002.

CHAPTER 12

Coherent Measures of Risk in Everyday Market Practice

CARLO ACERBI

CONTENTS

12.1   MOTIVATION

T HIS CHAPTER IS A GUIDED TOUR of the recent (sometimes very technical) literature on Coherent Risk Measures (CRMs). Our purpose is to review the theory of CRMs from the perspective of practical risk management applications. We have tried to single out those results from the theory which help to understand which CRMs can today be considered as realistic candidate alternatives to Value at Risk (*VaR*) in financial risk management practice. This has also been the spirit of the author's research line in recent years Acerbi 2002, Acerbi and Simonetti 2002, Acerbi and Tasche 2002a, b (see Acerbi 2003 for a review).

12.2   COHERENCY AXIOMS AND THE SHORTCOMINGS OF *VAR*

In 1997, a seminal paper by Artzner *et al.* (1997, 1999) introduced the concept of a Coherent Measure of Risk, by imposing, via an axiomatic framework, specific mathematical conditions which enforce some basic principles that a sensible risk measure should always satisfy. This cornerstone of financial mathematics was welcomed by many as the first serious attempt to give a precise definition of financial risk itself, via a deductive approach. Among the four celebrated axioms of coherency, a special role has always been played by the so-called *subadditivity axiom*

$$\rho(X + Y) \le \rho(X) + \rho(Y), \tag{12.1}$$

where $\rho(\cdot)$ represents a measure of risk acting on portfolios' profit–loss r.v.'s X, Y over a chosen time horizon. The reason why this condition has been long debated is probably due to the fact that VaR—the most popular risk measure for capital adequacy purposes—-turned out to be not subadditive and consequently not coherent. As a matter of fact, since inception, the development of the theory of CRMs has run in parallel with the debate on whether and how VaR should be abandoned by the risk management community.

The subadditivity axiom encodes the risk diversification principle. The quantity

$$H(X, Y; \rho) = \rho(X) + \rho(Y) - \rho(X + Y) \tag{12.2}$$

is the *hedging benefit* or, in capital adequacy terms, the *capital relief* associated with the merging of portfolios X and Y. This quantity will be larger when the two portfolios contain many bets on the same risk driver, but of opposite direction, which therefore hedge each other in the merged portfolio. It will be zero in the limiting case when the two portfolios bet on the same directional move of every common risk factor. But the problem with non-subadditive risk measures such as VaR is that there happen to be cases in which the hedging benefit turns out to be negative, which is simply nonsensical from a risk–theoretical perspective.

Specific examples of subadditivity violations of VaR are available in the literature (Artzner *et al.* 1999; Acerbi and Tasche 2002a), which typically involve discrete distributions. It may be surprising to know however that examples of subadditivity violations of VaR can also be built with very inoffensive distributions. For instance, a subadditivity violation of VaR with confidence level $p = 10^{-2}$ occurs with two portfolios X and Y which are identically distributed as standard univariate Gaussian $\mathcal{N}(0, 1)$, but are dependent through the copula function

$$C(x, y) = \delta(x - y) \, 1_{\{x > q, y > q\}} + \delta(q - x - y) \, 1_{\{x < q, y < q\}} \tag{12.3}$$

with $q = p + \epsilon$. For instance, choosing $\epsilon = 10^{-4}$ one finds

$$H(X, Y; VaR_p) \simeq 2.4 + 2.4 - 5.2 = -0.4. \tag{12.4}$$

This may be surprising for the many people who have always been convinced that VaR is subadditive in the presence of 'normal distributions.' This is in fact true, provided that the full *joint* distribution is a multivariate Gaussian, but it is false, as in this case, when only the two *marginal* distributions are Gaussian. This leads to the conclusion that it is never sufficient to study the marginals to ward off a VaR violation of subadditivity, because the trigger of such events is a subtler copula property.

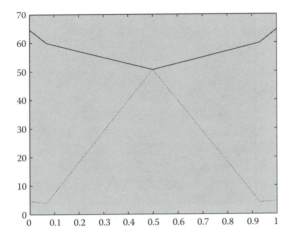

FIGURE 12.1 The one-dimensional non-convex $VaR_{5\%}(\Pi_\lambda)$ risk surface (dotted line) and the convex $ES_{5\%}(\Pi_\lambda)$ risk surface (solid line) for a specific portfolio Π_λ $\lambda \in [0,1]$. From Acerbi (2003), Example 2.15.

Other examples of subadditivity violation by *VaR* (see Acerbi 2003, Examples 2.15 and 4.4) allow us to display the connection between the coherence of a risk measure and the *convexity* of risk surfaces. By *risk surface*, we mean the function $\vec{w} \mapsto \rho(\Pi(\vec{w}))$ which maps the vector of weights $\vec{w}$ of the portfolio $\Pi(\vec{w}) = \sum_i w_i X_i$ onto the risk $\rho(\Pi(\vec{w}))$ of the portfolio. The problem of ρ-portfolio optimization amounts to a global search for minima of the surface. An elementary consequence of coherency is the convexity of risk surfaces (see Figure 12.1)

$$\rho \text{ coherent} \Rightarrow \rho(\Pi(\vec{w})) \text{ convex.} \qquad (12.5)$$

This immediate result tells us that risk-optimization—if we just define carefully our variables—is an intrinsically convex problem. This bears enormous practical consequences, because the border between convex and non-convex optimization in fact delimits solvable and unsolvable problems when things are complex enough, whatever supercomputer you may have. In examples, *VaR* exhibits non-convex risk surfaces, infested with local minima, which can easily be recognized to be just artefacts of the chosen (non-coherent) risk measure (see Figure 12.2).

In the same examples, thanks to convexity, any CRM displays conversely a single global minimum, which corresponds undoubtedly to the correct optimal portfolio, as can be shown by symmetry arguments.

The lesson we learn is that adopting a non-coherent measure as a decision-making tool for asset allocation means choosing to face formidable (and often unsolvable) computational problems related to the minimization of risk surfaces plagued by a plethora of risk–nonsensical local minima. As a matter of fact we are persuaded that no bank in the world has actually ever performed a *true VaR* minimization in its portfolios, if we exclude multivariate Gaussian frameworks á la Riskmetrics where *VaR* is actually just a disguised version of standard deviation and hence convex.

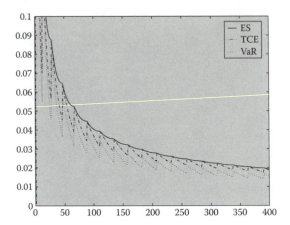

FIGURE 12.2 The risk of an equally weighted portfolio of i.i.d. defaultable bonds, as a function of the number of bonds N. The global minimum is expected at $N = \infty$ by diversification principle and symmetry arguments. Notice however that *VaR* and *TCE* (Tail Conditional Expectation (Artzner *et al.* 1997)), being non-coherent, display also an infinite number of nonsensical local minima. Do not expect convexity for the *ES* plot because it is not a 'risk surface' (which at $N = 400$ would be a 400-dimensional hypersurface in $\mathbb{R}^{401}$). From Acerbi (2003) Example 4.4.

Nowadays, sacrificing the huge computational advantage of convex optimization only for the sake of *VaR* fanaticism is pure masochism.

12.3   THE OBJECTIVIST PARADIGM

The general representation of CRMs is well known (Artzner *et al.* 1999; Delbaen 2000). Any CRM $\rho_{\mathcal{F}}$ is in one-to-one correspondence with a family $\mathcal{F}$ of probability measures $\mathbb{P}$. The formula is strikingly simple

$$\rho_{\mathcal{F}}(X) = \sup_{\mathbb{P} \in \mathcal{F}} \mathbb{E}^{\mathbb{P}}[-X]. \tag{12.6}$$

But this representation is of little help for a risk manager. It gives him too much freedom. And more importantly, it generates a sort of philosophical impasse as it assumes an intrinsically *subjectivist* point of view that is opposite to the typical risk manager's philosophy, which is *objectivist*. The formula defines the CRM $\rho_{\mathcal{F}}$ as the worst case expected loss of the portfolio in a family $\mathcal{F}$ of 'parallel universes' $\mathbb{P}$.

Objectivists are, in a word, those statisticians who believe that a unique real probability measure of future events must necessarily exist somewhere, and whose principal aim is to try to estimate it empirically. Subjectivists on the contrary, are intransigent statisticians who, starting from the observation that even if this real probability measure existed it would be unknowable, simply prefer to give up considering this concept at all and think of probability measures as mere mathematical instruments. Representation (12.6) is manifestly subjectivist as it is based on families of probability measures.

Risk Managers are objectivists because the algorithm they use to assess capital adequacy via *VaR* is intrinsically objectivist. We can in fact split this process into two clearly distinct steps:

1. model *the* probability distribution of your portfolio;
2. compute *VaR* on this distribution.

An overwhelmingly larger part of the computational effort (data mining, multivariate risk-factors distribution modelling, asset pricing, . . .) is done in step 1, which has no relation with *VaR* and is just an objectivist paradigm. The computation of *VaR*, given the distribution, is typically a single last code line. Hence, in this scheme, replacing *VaR* with any other CRM is immediate, but it is clear that for this purpose it is necessary to identify those CRMs that fit the objectivist paradigm.

If we look for something better than *VaR* we cannot forget that despite its shortcomings, this risk measure brought a real revolution into risk management practice thanks to some features, which were innovative at the time of its advent, and that nobody would be willing to give up today.

- *universality* (*VaR* applies to risks of any nature);
- *globality* (*VaR* condenses multiple risks into a single figure);
- *probability* (*VaR* contains probabilistic information on the measured risks);
- *right units of measure* (*VaR* is simply expressed in 'lost money').

The last two features explain why *VaR* is worshipped by any firm's boss, whose daily refrain is: 'how much money do we risk and with what probability?' Remember that risk sensitivities (aka 'greeks,' namely partial derivatives of the portfolio value with respect to a specific risk factor) do not share any of the above features and you will immediately understand why *VaR* became so popular. As a matter of fact a bank's greeks-based risk report is immensely more cumbersome and less communicative than a *VaR*-based one.

But if we look more closely at the features that made the success of *VaR*, we notice that in fact they have nothing to do with *VaR* in particular, but rather with the objectivist paradigm above. In other words, if in step 2 above, we replace *VaR* with any sensible risk measure defined in terms of some monetary statistics of the portfolio distribution, we automatically preserve these features. That is why looking for CRMs that fit the objectivist paradigm is so crucial.

In our opinion, the real lasting heritage of *VaR* in the development of the theory and practice of risk management is precisely the very fact that it served to introduce for the first time the objectivist paradigm into the market practice. Risk managers started to plot the distribution of their portfolio's values and learned to fear its left tail thanks to the lesson of *VaR*.

12.4   ESTIMABILITY

The property that characterizes the subset of those CRMs that fit the objectivist paradigm is *law invariance*, first studied in this context by Kusuoka (2001). A measure of risk ρ is said to be law-invariant (LI) if it is a functional of *the* portfolio's distribution function

$F_X(\cdot)$ only. The concept of law invariance therefore can be defined only with reference to a single chosen probability space,

$$\rho \; law \; invariant \; \Leftrightarrow \; \rho(X) = \rho[F_X(\cdot)] \tag{12.7}$$

or equivalently

$$\rho \; law \; invariant \; \Leftrightarrow \; [F_X(\cdot) = F_Y(\cdot) \Rightarrow \rho(X) = \rho(Y)]. \tag{12.8}$$

It is easy to see that law invariance means *estimability* from empirical data.

Theorem 12.1:

$$\rho \; law \; invariant \; \Leftrightarrow \; \rho \; estimable \tag{12.9}$$

Proof ($\Leftarrow$): suppose ρ estimable and let X and Y be r.v.'s with identical probability distribution function. Consider N i.i.d. realizations $\{x_i\}_{i=1,\ldots,N}$ and $\{y_i\}_{i=1,\ldots,N}$ and an estimator $\hat{\rho}$. We will have

$$\hat{\rho}(\{x_i\}) \xrightarrow{N \to \infty} \rho(X),$$

$$\hat{\rho}(\{y_i\}) \xrightarrow{N \to \infty} \rho(Y).$$

But for large N, the samples $\{x_i\}$ and $\{y_i\}$ are indistinguishable, hence $\rho(X) = \rho(Y)$.
Proof ($\Rightarrow$): suppose ρ LI. Then a (canonical) estimator is defined by

$$\hat{\rho}(\{x_i\}) := \rho(\hat{F}_X(\{x_i\})) \tag{12.10}$$

where $\hat{F}_X(\{x_i\})$ represents the empirical distribution estimated from the data $\{x_i\}$.   □
     It is then clear that for CRMs to be measurable with respect to a single given probability distribution it must also be LI. That is why, unless an unlikely subjectivistic revolution takes place in the market, risk managers will always focus their attention on just the subset of LI CRMs for any practical application. Law invariance, in other words, is a sort of unavoidable 'fifth axiom' for practitioners.
     Popular examples of LI CRMs are for instance α-*expected shortfall* (ES_α) (aka CVaR, aka AV@R,...) (Rockafellar and Uryasev 2000; Acerbi and Tasche 2002a)

$$ES_\alpha(X) = -\frac{1}{\alpha} \int_0^\alpha F_X^{-1}(p)dp, \quad \alpha \in (0, 1), \tag{12.11}$$

namely the '*average loss of the portfolio in the worst α cases*' or the family of CRMs based on one-sided moments (Fischer 2001)

$$\rho_{p,a}(X) = -\mathbb{E}[X] + a\|(X - \mathbb{X})^-\|_p, \quad a \in [0, 1], \quad p \geq 1 \qquad (12.12)$$

among which we recognize *semivariance* (when $a = 1$, $p = 2$).

12.5 THE DIVERSIFICATION PRINCIPLE REVISITED

There is one aspect of the diversification principle which subadditivity does not capture. It is related to the limiting case when we sum two portfolios X and Y which are *comonotonic*. This means that we can write $X = f(Z)$ and $Y = g(Z)$ where f and g are non-decreasing functions driven by the same random risk factor Z. Such portfolios go up and down together in all cases, and hence provide no mutual hedge at all, i.e. no diversification. For comonotonic random variables people speak also of 'perfect dependence' because it turns out that the dependence structure of such variables is in fact the same (*copula maxima*) that links any random variable X to itself.

The diversification principle tells us that for a measure of risk ρ the hedging benefit $H(X, Y; \rho)$ should be exactly zero when X and Y are comonotonic. This property of ρ is termed *comonotonic additivity* (CA)

$$\rho \; comonotonic \; additive \Leftrightarrow [X, Y \; comonotonic \Rightarrow \rho(X + Y) = \rho(X) + \rho(Y)]. \quad (12.13)$$

Subadditivity does not imply CA. There are in fact CRMs that are not comonotonic additive, such as (12.12), for instance.

We think that the diversification principle is well embodied only by requiring both subadditivity and CA in a risk measure. Each property separately is not enough. To understand this fact the clearest explanation we know is showing that in the absence of each of these conditions, there exists a specific *regulatory arbitrage* (RA) allowing a risk manager to reduce the capital requirement for a portfolio without reducing at all the risks he runs.

RA$_1$ Lack of Subadditivity: split your portfolio into suitable subportfolios and compute capital adequacy for each one.

RA$_2$ Lack of Comonotonic Additivity: merge your portfolio with the one of some 'comonotone' partner and compute capital adequacy on the global portfolio.

CA is therefore a natural further condition to add to the list of properties of a good risk measure. It becomes a sort of 'sixth axiom,' because it is a condition distinct from LI when imposed on a CRM. There exist CRMs which satisfy LI and not CA (for instance (12.12)) and vice versa (for instance Worst Conditional Expectation as defined in Artzner *et al.* (1997)).

The above arguments support the description of the class of CRMs that satisfy also LI and CA (LI CA CRMs).

12.6   SPECTRAL MEASURES OF RISK

The class of LI CA CRMs was first described exhaustively by Kusuoka (2001). They have the general representation

$$\rho_\mu(X) = \int_0^1 d\mu(p) \, ES_p(X) \quad d\mu \; any \; measure \; on \; [0,1]. \tag{12.14}$$

The same class were termed *spectral measures of risk* independently in Acerbi (2002) with an equivalent representation

$$\rho_\phi(X) = -\int_0^1 \phi(p) F_X^{-1}(p) dp, \tag{12.15}$$

where the function $\phi : [0,1] \mapsto \mathbb{R}$, named the *risk spectrum*, satisfies the coherence conditions

(1) $\phi(p) \geq 0$,
(2) $\int_0^1 \phi(p) dp = 1$,
(3) $\phi(p_1) \geq \phi(p_2)$ if $p_1 \leq p_2$.

Despite the complicated formula, a spectral measure ρ_ϕ is nothing but *the ϕ-weighted average of all outcomes of the portfolio, from the worst ($p = 0$) to the best ($p = 1$)*. This is the most general form that a LI CA CRM can assume. The only residual freedom is in the choice of the weighting function ϕ within the above conditions.

Condition 3 is related to subadditivity. It just says that in general worse cases must be given a larger weight when we measure risk, and this seems actually very reasonable. This is also where *VaR* fails, as it measures the severity of the loss associated with the quantile threshold, forgetting to give a weight to the losses in the tail beyond it. *VaR* can also in fact be expressed via formula (12.15), but with a Dirac-delta risk spectrum peaked on the confidence level which therefore does not fulfil condition 3 (see Figure 12.3). Expected

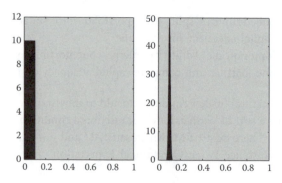

FIGURE 12.3 The risk spectrum ϕ_{ES} of ES_q (left plot) and the risk spectrum ϕ_{VaR} of VaR_q (right plot), for $q = 0.1$. The latter, being a (here stylized) Dirac-delta peaked on q, violates condition 3.

shortfall ES_α is a special case of a spectral measure of risk whose risk spectrum is a constant function with domain $[0, \alpha]$.

Spectral measures of risk turned out to be strictly related to the class of *distortion risk measures* introduced in actuarial mathematics in 1996 by Wang (1996) in a different setting.

12.7   ESTIMATORS OF SPECTRAL MEASURES

It is easy to provide estimators of spectral measures. Given N i.i.d. scenario outcomes $\{x_i^{(k)}\}_{i=1,\ldots,N}$ for the vector of market variables (possibly assets) $\vec{X} = X^{(k)}$ and given any portfolio function of them $Y = Y(\vec{X})$, we can just exploit law-invariance and use the $\Rightarrow$ proof of Theorem 12.1 to obtain canonical estimators as

$$\hat{\rho}_\phi^{(N)}(Y) = \rho_\phi\left[\hat{F}_Y^{(N)}(\{\vec{x}_i\})\right], \tag{12.16}$$

where we have defined the empirical marginal distribution function of Y

$$\hat{F}_Y^{(N)}(\{\vec{x}_i\})(t) := \frac{1}{N}\sum_{i=1}^{N}\theta(t - y_i), \tag{12.17}$$

which is nothing but the cumulative empirical histogram of the outcomes $y_i := Y(\vec{x}_i)$. Equation (12.16) can in fact be solved to give simply

$$\hat{\rho}_\phi^{(N)}(Y) = -\sum_{i=1}^{N} y_{i:N} \bar{\phi}_i, \tag{12.18}$$

where we have adopted the notation $y_{i:N}$ of *order statistics* (i.e. the sorted version of the vector y_i) and defined the weights

$$\bar{\phi}_i := \int_{(i-1)/N}^{i/N} \phi(p)\mathrm{d}p. \tag{12.19}$$

In (12.18) we see in a very transparent language that a spectral measure is nothing but a weighted average of all the outcomes of a portfolio sorted from the worst to the best.

In the case of ES_α the estimator (12.18) specializes to

$$\widehat{ES}_\alpha^{(N)}(Y) = -\frac{1}{N\alpha}\left(\sum_{i=1}^{\lfloor N\alpha \rfloor} y_{i:N} + (N\alpha - \lfloor N\alpha \rfloor)\, y_{\lfloor N\alpha \rfloor+1:N}\right). \tag{12.20}$$

All these estimators can be easily implemented in a simple spreadsheet or in any programming language.

We remark that these estimators not only converge for large N to the estimated measure, but also preserve coherency for every finite N by construction.

12.8 OPTIMIZATION OF CRMs: EXPLOITING CONVEXITY

As we have already stressed, CRMs' surfaces are convex. Nonetheless, setting up an optimization program using, say, the estimator (12.20) of *ES*, requires some clever trick. In fact, suppose you want to minimize the $ES_\alpha(Y_{\vec{w}})$ of a portfolio $Y_{\vec{w}} = \sum_k w_k X^{(k)}$ choosing the optimal weights w_k under some given constraints. A naive minimization procedure using (12.20) will simply fail, because the order statistics $y_{i:N} = \{\text{sort}_{j=1,\dots,N}[\sum_k w_k \, x_j^{(k)}]\}_i$ reshuffle discretely when the parameters $\vec{w}$ change continuously. In other words, $\widehat{ES}_\alpha^{(N)}(Y_{\vec{w}})$ is not analytical in the weights $\vec{w}$ and this creates big troubles for any standard minimization routine. Moreover, the `sort` algorithm is a very slow and memory consuming one for large samples.

The problem of finding efficient optimization routines for ES_α was however elegantly solved by Pflug (2000), and Rockafellar and Uryasev (2000, 2002) who mapped it onto the equivalent problem of finding the minima of the functional

$$\Gamma_\alpha^{(N)}(Y_{\vec{w}}, \psi) := -\psi + \frac{1}{N\alpha} \sum_{i=1}^{N} (\psi - y_i)^+ \tag{12.21}$$

with $y_i = \sum_k w_k \, x_i^{(k)}$. Despite the presence of an additional parameter ψ, Γ is a much simpler objective function to minimize, thanks to the manifest analyticity and convexity w.r.t the weights $\vec{w}$. Notice, in particular, that expression (12.21) is free from order statistics.

The main result of Rockafellar and Uryasev (2000) is that $\widehat{ES}_\alpha(Y_{\vec{w}})$ (as a function of $\vec{w}$) and $\Gamma_\alpha^{(N)}(Y_{\vec{w}}, \psi)$ (as a function of both $\vec{w}$ and ψ) attain their minima on the same argument weights $\vec{w}$. So, we can find ES_α optimal portfolios minimizing Γ_α instead, which is dramatically easier. Furthermore, it can be shown (Rockafellar and Uryasev 2000) that this convex non-linear program can be mapped onto an equivalent *linear* program, at the price of introducing further additional parameters $\vec{z}$.

$$\min_{\vec{w}, \psi, \vec{z}} \left\{ -\psi + \frac{1}{N\alpha} \sum_{i=1}^{N} z_i \right\}$$

$$\begin{aligned}
\text{s.t.} \quad & z_i \geq \psi - y_i & i = 1, \dots, N \\
& \vec{w} \in \mathcal{W} & \text{weights domain} \\
& \psi \in \mathbb{R} \\
& z_i \geq 0 \\
& y_i = \sum_k w_k \, x_i^{(k)}
\end{aligned} \tag{12.22}$$

It is for this linearized version that the most efficient routines are obtained, making it possible to set up an optimization procedure for portfolios of essentially any size and complexity.

It is difficult to overestimate the importance of this result. It allows the full exploitation of the advantages of convex optimization with ES_α and opens the way to efficient routines for large and complex portfolios, under any distributional assumptions. In practice, using this methodology with ES_α, risk managers have at last the possibility of solving problems that using VaR they had just dreamed of.

This methodology was extended from ES_α to any spectral measure ρ_ϕ (i.e. any LI CA CRM) in Acerbi and Simonetti (2002). Also in the general case, the main problem to tackle is the presence of sorting routines induced in (12.18) by the order statistics. In parallel to the above result, one introduces the functional

$$\Gamma_\phi^{(N)}(Y_{\vec{w}}, \vec{\psi}) := \sum_{j=1}^{N-1}(\bar{\phi}_{j+1} - \bar{\phi}_j)\left\{ j\,\psi_j - \sum_{i=1}^{N}(\psi_j - y_i)^+ \right\} - \bar{\phi}_N \sum_{i=1}^{N} y_i, \qquad (12.23)$$

whose minimum (as a function of both $\vec{w}$ and $\vec{\psi}$) can be shown to occur at the same argument $\vec{w}$ that minimizes $\hat{\rho}_\phi^{(N)}(Y_{\vec{w}})$. Therefore we see that generalizing from ES_α to any spectral measure ρ_ϕ the only difference is that the additional parameter ψ has become a vector of additional parameters $\vec{\psi} = (\psi_1, \ldots, \psi_J)$, where J is essentially the number of 'steps' of the chosen risk spectrum ϕ. Also for this extended methodology, it is possible to map the non-linear convex problem onto an equivalent linear one:

$$\min_{\vec{w}, \vec{\psi}, \vec{z}}\left\{ \sum_{j=1}^{J}\Delta\bar{\phi}_j\left\{ j\,\psi_j - \sum_{i=1}^{N} z_{ij} \right\} - \bar{\phi}_N \sum_{i=1}^{N} y_i \right\}$$

$$\text{s.t.} \quad z_{ij} \geq \psi_j - y_i \qquad i = 1, \ldots, N \qquad j = 1, \ldots, J$$
$$\vec{w} \in \mathcal{W} \qquad (12.24)$$
$$\vec{\psi} \in \mathbb{R}^J$$
$$z_{ij} \geq 0$$
$$y_i = \sum_k w_k\, x_i^{(k)}$$

This extends the efficiency of the ES_α optimization routines available also for any other LI CA CRMs. See Acerbi and Simonetti (2002), and Acerbi (2003) for more details.

12.9 CONCLUSIONS

We have discussed why, in our opinion, the class of CRMs is too large from the perspective of practical risk management applications. If the practice of risk management remains intrinsically objectivistic, the additional constraint of law-invariance will always be implicitly assumed by the market. A further restriction is provided by a closer look at the risk diversification principle, which naturally introduces the condition of comonotonic additivity.

The subset of CRMs which possess also LI and CA coincide with the class of spectral measures. This class lends itself to immediate transparent representation, straightforward estimation and—adopting nontrivial tricks—to powerful optimization techniques which exploit the convexity of the risk minimization problems and allow, probably for the first time, risk managers to face the problem of finding optimal portfolios with virtually no restrictions of size, complexity and distributional assumptions.

REFERENCES

Acerbi, C., Spectral measures of risk: A coherent representation of subjective risk aversion *J. Bank. Finance*, 2002, **26**, 1505–1518.

Acerbi, C., Coherent representations of subjective risk aversion. In *Risk Measures for the XXI Century*, edited by G. Szego, 2003 (Wiley: New York).

Acerbi, C. and Simonetti, P., 2002, Portfolio optimization with spectral measures of risk. Abaxbank preprint. Available online at: www.gloriamundi.org.

Acerbi, C. and Tasche, D., On the coherence of expected shortfall. *J. Bank. Finance*, 2002a, **26**, 1487–1503.

Acerbi, C. and Tasche, D., Expected shortfall: A natural coherent alternative to value at risk. *Econ. Notes*, 2002b, **31**(2), 379–388.

Artzner, P., Delbaen, F., Eber, J.-M. and Heath, D., Thinking coherently. *RISK*, 1997, **10**(11), 68–71.

Artzner, P., Delbaen, F., Eber, J.-M. and Heath, D., Coherent measures of risk. *Math. Fin.*, 1999, **9**(3), 203–228.

Delbaen, F., Coherent risk measures on general probability spaces. Preprint, 2000 (ETH Zürich)

Fischer, T., Examples of coherent risk measures depending on one-sided moments, Working Paper, 2001 (Darmstadt University of Technology).

Kusuoka, S., On law invariant coherent risk measures. *Adv. Math. Econ.*, 2001, **3**, 83–95.

Pflug, G., Some remarks on the value-at-risk and the conditional value-at-risk. In *Probabilistic Constrained Optimization: Methodology and Applications*, edited by S. Uryasev, 2000 (Kluwer Academic Publishers: Dordrecht).

Rockafellar, R.T. and Uryasev, S., Optimization of conditional value-at-risk. *J. Risk*, 2000, **2**(3), 21–41.

Rockafellar, R.T. and Uryasev, S., Conditional value-at-risk for general loss distributions. *J. Bank. Finance*, 2002, **26**(7), 1443–1471.

Wang, S., Premium calculation by transforming the layer premium density *Astin Bull.*, 1996, **26**, 71–92.

CHAPTER 13

Higher Moment Coherent Risk Measures

PAVLO A. KROKHMAL

CONTENTS

13.1   INTRODUCTION

RESEARCH AND PRACTICE OF PORTFOLIO management and optimization is driven to a large extent by tailoring the measures of reward (satisfaction) and risk (unsatisfaction/regret) of the investment venture to the specific preferences of an investor. While there exists a common consensus that an investment's reward may be adequately associated with its expected return, the methods for proper modelling and measurement of an investment's risk are subject to much more pondering and debate. In fact, the *risk-reward* or *mean-risk* models constitute an important part of the investment science subject and, more generally, the field of decision making under uncertainty.

The cornerstone of modern portfolio analysis was set up by Markowitz (1952, 1959), who advocated identification of the portfolio's risk with the volatility (variance) of its returns. On the other hand, Markowitz's work led to formalization of the fundamental view that any decision under uncertainties may be evaluated in terms of its *risk* and *reward*. Markowitz's seminal ideas are still widely used today in many areas of decision making, and the entire paradigm of bi-criteria '*risk-reward*' optimization has received extensive development in both directions of increasing the computational efficiency and enhancing the models for risk measurement and estimation.

At the same time, it has been recognized that the symmetric attitude of the classical Mean-Variance (MV) approach, where both the 'positive' and 'negative' deviations from the expected level are penalized equally, does not always yield an adequate estimation of risks induced by the uncertainties. Hence, significant effort has been devoted to the development of *downside* risk measures and models. Replacing the variance by the *lower standard semi-deviation* as a measure of investment risk so as to take into account only 'negative' deviations from the expected level has been proposed as early as by Markowitz (1959); see also more recent works by Ogryczak and Ruszczyński (1999, 2001, 2002).

Among the popular downside risk models we mention the *Lower Partial Moment* and its special case, the *Expected Regret*, which is also known as *Integrated Chance Constraint* in stochastic programming (Bawa 1975; Fishburn 1977; Dembo and Rosen 1999; Testuri and Uryasev 2003; van der Vlerk 2003). Widely known in the finance and banking industry is the *Value-at-Risk* measure (JP Morgan 1994; Jorion 1997; Duffie and Pan 1997). Being simply a quantile of loss distribution, the Value-at-Risk (VaR) concept has its counterparts in stochastic optimization (*probabilistic*, or *chance programming*, see Prékopa 1995), reliability theory, etc. Yet, minimization or control of risk using the VaR measure proved to be technically and methodologically difficult, mainly due to VaR's notorious non-convexity as a function of the decision variables. A downside risk measure that circumvents the shortcomings of VaR while offering a similar quantile approach to estimation of risk is the *Conditional Value-at-Risk* measure (Rockafellar and Uryasev 2000, 2002; Krokhmal *et al.* 2002a). Risk measures that are similar to CVaR and/or may coincide with it, are *Expected Shortfall* and *Tail VaR* (Acerbi and Tasche 2002), see also *Conditional Drawdown-at-Risk* (Krokhmal *et al.* 2002b; Chekhlov *et al.* 2005). A simple yet effective risk measure closely related to CVaR is the so-called *Maximum Loss*, or *Worst-Case Risk* (Young 1998; Krokhmal *et al.* 2002b), whose use in problems with uncertainties is also known as the *robust optimization* approach (see, e.g. Kouvelis and Yu 1997).

In the last few years, the formal theory of risk measures received a major impetus from the works of Artzner *et al.* (1999) and Delbaen (2002), who introduced an *axiomatic* approach to definition and construction of risk measures by developing the concept of *coherent risk measures*. Among the risk measures that satisfy the coherency properties, there are Conditional Value-at-Risk, Maximum Loss (Pflug 2000; Acerbi and Tasche 2002), coherent risk measures based on one-sided moments (Fischer 2003), etc. Recently, Rockafellar *et al.* (2006) have extended the theory of risk measures to the case of *deviation measures*, and demonstrated a close relationship between the coherent risk measures and deviation measures; *spectral measures of risk* have been proposed by Acerbi (2002).

An approach to decision making under uncertainty, different from the risk-reward paradigm, is embodied by the von Neumann and Morgenstern (vNM) utility theory, which exercises a mathematically sound axiomatic description of preferences and construction of the corresponding decision strategies. Along with its numerous modifications and extensions, the vNM utility theory is widely adopted as a basic model of rational choice, especially in economics and social sciences (see, e.g. Fishburn 1970, 1988; Karni and Schmeidler 1991, etc). Thus, substantial attention has been paid in the literature to the development of risk-reward optimization models and risk measures that are consistent with expected utility maximization. In particular, it has been shown that under certain conditions the Markovitz MV framework is consistent with the vNM theory (Kroll *et al.* 1984). Ogryczak and Ruszczyński (1999, 2001, 2002) developed mean-semideviation models that are consistent with stochastic dominance concepts (Fishburn 1964; Rothschild and Stiglitz 1970; Levy 1998); a class of risk-reward models with SSD-consistent coherent risk measures was discussed in De Giorgi (2005). Optimization with stochastic dominance constraints was recently considered by Dentcheva and Ruszcsyński (2003); stochastic dominance-based portfolio construction was discussed in Roman *et al.* (2006).

In this chapter we aim to offer an additional insight into the properties of axiomatically defined measures of risk by developing a number of representations that express risk measures via solutions of stochastic programming problems (Section 13.2.1); using the developed representations, we construct a new family of higher-moment coherent risk (HMCR) measures. In Section 13.2.2 it is demonstrated that the suggested representations are amenable to seamless incorporation into stochastic programming problems. In particular, implementation of the HMCR measures reduces to p-order conic programming, and can be approximated via linear programming. Section 13.2.3 shows that the developed results are applicable to deviation measures, while Section 13.2.4 illustrates that the HMCR measures are compatible with the second-order stochastic dominance and utility theory. The conducted case study (Section 13.3) indicates that the family of HMCR measures has a strong potential for practical application in portfolio selection problems. Finally, the appendix contains the proofs of the theorems introduced in the chapter.

13.2 MODELLING OF RISK MEASURES AS STOCHASTIC PROGRAMS

The discussion in Section 13.1 has illustrated the variety of approaches to definition and estimation of risk. Arguably, the recent advances in risk theory are associated with the axiomatic approach to construction of risk measures pioneered by Artzner *et al.* (1999). The present endeavor essentially exploits this axiomatic approach in order to devise simple computational recipes for dealing with several types of risk measures by representing them in the form of stochastic programming problems. These representations can be used to create new risk measures to be tailored to specific risk preferences, as well as to incorporate these preferences into stochastic programming problems. In particular, we present a new family of Higher Moment Coherent Risk (HMCR) measures. It will be shown that the HMCR measures are well behaved in terms of theoretical properties, and demonstrate very promising performance in test applications.

Within the axiomatic framework of risk analysis, risk measure $\mathscr{R}(X)$ of a random outcome X from some probability space $(\Omega, \mathscr{F}, \mu)$ may be defined as a mapping $\mathscr{R}: \mathcal{X} \mapsto \overline{\mathbb{R}}$, where $\mathcal{X}$ is a linear space of $\mathscr{F}$-measurable functions $X: \Omega \mapsto \mathbb{R}$. In a more general setting one may assume $\mathcal{X}$ to be a separated locally convex space; for our purposes it suffices to consider $\mathcal{X} = \mathscr{L}_p(\Omega, \mathscr{F}, P)$, $1 \leq p \leq \infty$, where the particular value of p shall be clear from the context. Traditionally to convex analysis, we call function $f : \mathcal{X} \mapsto \overline{\mathbb{R}}$ *proper* if $f(X) > -\infty$ for all $X \in \mathcal{X}$ and dom $f \neq \emptyset$, i.e. there exists $X \in \mathcal{X}$ such that $f(X) < +\infty$ (see, e.g. Rockafellar 1970, Zălinescu 2002). In the remainder of the chapter, we confine ourselves to risk measures that are proper and not identically equal to $+\infty$. Also, throughout the chapter it is assumed that X represents a *loss function*, i.e. small values of X are 'good', and large values are 'bad.'

13.2.1 Convolution-Type Representations for Coherent Measures of Risk

A *coherent risk measure*, according to Artzner *et al.* (1999) and Delbaen (2002), is defined as a mapping $\mathscr{R}: \mathcal{X} \mapsto \overline{\mathbb{R}}$ that further satisfies the next four properties (axioms):

(A1) *monotonicity*: $X \leq 0 \Rightarrow \mathscr{R}(X) \leq 0$ for all $X \in \mathcal{X}$,
(A2) *sub-additivity*: $\mathscr{R}(X + Y) \leq \mathscr{R}(X) + \mathscr{R}(Y)$ for all $X, Y \in \mathcal{X}$,
(A3) *positive homogeneity*: $\mathscr{R}(\lambda X) = \lambda \mathscr{R}(X)$ for all $X \in \mathcal{X}$, $\lambda > 0$,
(A4) *translation invariance*: $\mathscr{R}(X + a) = \mathscr{R}(X) + a$ for all $X \in \mathcal{X}$, $a \in \mathbb{R}$.

Observe that given the positive homogeneity (A3), the requirement of sub-additivity (A2) in the above definition can be equivalently replaced with the requirement of convexity (see also Schied and Follmer 2002):

(A2') *convexity*: $\mathscr{R}(\lambda X + (1 - \lambda)Y) \leq \lambda \mathscr{R}(X) + (1 - \lambda)\mathscr{R}(Y), X, Y \in \mathcal{X}, 0 \leq \lambda \leq 1$.

From the axioms (A1)–(A4) one can easily derive the following useful properties of coherent risk measures (see, e.g. Delbaen 2002, Ruszczyński and Shapiro 2006):

(C1) $\mathscr{R}(0) = 0$ and, in general, $\mathscr{R}(a) = a$ for all $a \in \mathbb{R}$,
(C2) $X \leq Y \Rightarrow \mathscr{R}(X) \leq \mathscr{R}(Y)$, and, in particular, $X \leq a \Rightarrow \mathscr{R}(X) \leq a$, $a \in \mathbb{R}$,
(C3) $\mathscr{R}(X - \mathscr{R}(X)) = 0$,
(C4) if $\mathcal{X}$ is a Banach lattice, $\mathscr{R}(X)$ is continuous in the interior of its effective domain, (where the inequalities $X \geq a$, $X \leq Y$, etc., are assumed to hold almost surely). From the definition of coherent risk measures it is easy to see that, for example, EX and ess.sup X, where

$$\text{ess.sup } X = \begin{cases} \min\{\eta \in \mathbb{R} \mid X \leq \eta\}, & \text{if } \{\eta \in \mathbb{R} \mid X \leq \eta\} \neq \emptyset, \\ \infty, & \text{otherwise,} \end{cases}$$

are coherent risk measures; more examples can be found in Rockafellar *et al.* (2006). Below we present simple computational formulas that aid in construction of coherent risk measures and their incorporation into stochastic programs. Namely, we execute the idea that one of the axioms (A3) or (A4) can be relaxed and then 'reinstated' by solving an appropriately defined mathematical programming problem. In other words, one can construct a coherent risk measure via solution of a stochastic programming problem that involves a function $\phi: \mathcal{X} \mapsto \mathbb{R}$ satisfying only three of the four axioms (A1)–(A4).

First we present a representation for coherent risk measures that is based on the relaxation of the translation invariance axiom (A4). The next theorem shows that if one selects a function $\phi: \mathcal{X} \mapsto \mathbb{R}$ satisfying axioms (A1)–(A3) along with additional technical conditions, then there exists a simple stochastic optimization problem involving ϕ whose optimal value would satisfy (A1)–(A4).

Theorem 13.1 *Let function $\phi: \mathcal{X} \mapsto \mathbb{R}$ satisfy axioms (A1)–(A3), and be a lower semicontinuous (lsc) function such that $\phi(\eta) > \eta$ for all real $\eta \neq 0$. Then the optimal value of the stochastic programming problem*

$$\rho(X) = \inf_{\eta} \; \eta + \phi(X - \eta) \tag{13.1}$$

is a proper coherent risk measure, and the infimum is attained for all X, so $\inf_{\eta}$ in (13.1) may be replaced by $\min_{\eta \in \mathbb{R}}$.

For proof of Theorem 13.1, as well as other theorems introduced in the chapter, see the appendix.

Remark 13.1.1 It is all-important that the stochastic programming problem (13.1) is *convex*, due to the convexity of the function ϕ. Also, it is worth mentioning that one cannot substitute a coherent risk measure itself for function ϕ in (1), as it will violate the condition $\phi(\eta) > \eta$ of the Theorem.

Corollary 13.1.1 *The set $\arg\min_{\eta \in \mathbb{R}}\{\eta + \phi(X - \eta)\} \subset \mathbb{R}$ of optimal solutions of (13.1) is closed.*

Example 13.1.1 (Conditional Value-at-Risk): A famous special case of (13.1) is the optimization formula for Conditional Value-at-Risk (Rockafellar and Uryasev 2000, 2002):

$$\mathrm{CVaR}_\alpha(X) = \min_{\eta \in \mathbb{R}} \; \eta + (1 - \alpha)^{-1} E(X - \eta)^+, \quad 0 < \alpha < 1, \tag{13.2}$$

where $(X)^\pm = \max\{\pm X, 0\}$, and function $\phi(X) = (1 - \alpha)^{-1} E(X)^+$ evidently satisfies the conditions of Theorem 1. The space in this case can be selected as $\mathcal{L}_2(\Omega, \mathcal{F}, P)$. One of the many appealing features of (2) is that it has a simple intuitive interpretation: if X represents loss/unsatisfaction, then $\mathrm{CVaR}_\alpha(X)$ is, roughly speaking, the conditional expectation of losses that may occur in $(1 - \alpha) \cdot 100\%$ of the worst cases. In the case of a continuously distributed X, this rendition is exact: $\mathrm{CVaR}_\alpha(X) = E[X \mid X \geq \mathrm{VaR}_\alpha(X)]$, where $\mathrm{VaR}_\alpha(X)$ is defined as the α-quantile of X: $\mathrm{VaR}_\alpha(X) = \inf\{\zeta \mid P[X \leq \zeta] \geq \alpha\}$. In the general case the formal definition of $\mathrm{CVaR}_\alpha(X)$ becomes more intricate (Rockafellar and Uryasev 2002) but representation (13.2) still applies.

Example 13.1.2 A generalization of (13.2) can be constructed as (see also Ben-Tal and Teboulle 1986)

$$\mathscr{R}_{\alpha,\beta}(X) = \min_{\eta \in \mathbb{R}} \eta + \alpha\, E(X - \eta)^+ - \beta\, E(X - \eta)^- \qquad (13.3)$$

where, in accordance with the conditions of Theorem 13.1, one has to put $\alpha > 1$ and $0 \le \beta < 1$.

Example 13.1.3 (Maximum loss): If the requirement of finiteness of ϕ in (1) is relaxed, i.e. the image of ϕ is $(-\infty, +\infty]$, then the optimal value of (1) still defines a coherent risk measure, but the infimum may not be achievable. An example is served by the so-called MaxLoss measure,

$$\text{MaxLoss}(X) = \text{ess.sup}\, X = \inf_{\eta} \eta + \phi^*(X - \eta),$$
$$\text{where } \phi^*(X) = \begin{cases} 0, & X \le 0, \\ \infty, & X > 0. \end{cases}$$

It is easy to see that ϕ^* is positive homogeneous convex, non-decreasing, lsc, and satisfies $\phi^*(\eta) > \eta$ for all $\eta \neq 0$, but is not finite.

Example 13.1.4 (Higher moment coherent risk measures): Let $\mathcal{X} = \mathscr{L}_p(\Omega, \mathscr{F}, P)$, and for some $0 < \alpha < 1$ consider $\phi(X) = (1 - \alpha)^{-1}\|(X)^+\|_p$, where $\|X\|_p = \left(E|X|^p\right)^{1/p}$. Clearly, ϕ satisfies the conditions of Theorem (13.1), thus we can define a family of *higher-moment coherent risk measures* (HMCR) as

$$\text{HMCR}_{p,\alpha}(X) = \min_{\eta \in \mathbb{R}} \eta + (1 - \alpha)^{-1}\|(X - \eta)^+\|_p, \quad p \ge 1,\ \alpha \in (0, 1). \qquad (13.4)$$

From the fact that $\|X\|_p \le \|X\|_q$ for $1 \le p < q$, it immediately follows that the HMCR measures are monotonic with respect to the order p:

$$\text{HMCR}_{p,\alpha}(X) \le \text{HMCR}_{q,\alpha}(X) \quad \text{for} \quad p < q \quad \text{and} \quad X \in \mathscr{L}_q. \qquad (13.5)$$

Of special interest is the case $p = 2$ that defines the *second-moment coherent risk measure* (SMCR):

$$\text{SMCR}_{\alpha}(X) = \min_{\eta \in \mathbb{R}} \eta + (1 - \alpha)^{-1}\|(X - \eta)^+\|_2, \quad 0 < \alpha < 1. \qquad (13.6)$$

We will see below that $\text{SMCR}_{\alpha}(X)$ is quite similar in properties to $\text{CVaR}_{\alpha}(X)$ while measuring the risk in terms of the second moments of loss distributions. Implementation-wise, the SMCR measure can be incorporated into a mathematical programming problem via the second-order cone constraints (see Section 13.3). The second-order cone programming (SOCP) is a well-developed topic in the field of convex optimization (see, for example, a review by Alizadeh and Goldfarb 2003), and a number of commercial off-the-shelf software packages are available for solving convex problems with second-order cone constraints.

Now we comment briefly on the relation between the above-introduced HMCR family and other known in the literature risk measures that involve higher moments of distributions. The Lower Partial Moment (see, e.g. Bawa 1975, Fishburn 1977)[1]

$$\mathrm{LPM}_p(X; t) = \mathrm{E}\big((X - t)^+\big)^p, \quad p \geq 1, \ t \in \mathbb{R}, \tag{13.7}$$

is convex in X, but not positive homogeneous or translation invariant. In the context of axiomatically defined risk measures, an interesting example of spectral risk measure that corresponds to 'pessimistic manipulation' of X and is sensitive to higher moments was considered by Tasche (2002). Closely related to the proposed here HMCR measures are the so-called coherent measures based on one-sided moments, or coherent measures of semi-$\mathscr{L}_p$ type (Fischer 2003; Rockafellar *et al.* 2006):[2]

$$\mathscr{R}(X) = \mathrm{E}X + \beta \big\| (X - \mathrm{E}X)^+ \big\|_p, \quad p \geq 1, \ \beta \geq 0. \tag{13.8}$$

A key difference between (13.8) and the HMCR measures (13.4) is that the HMCR family are *tail* risk measures, while the measures of type (13.8) are based on *central* semi-moments (see Example 13.2.3).

Example 13.1.5 (Composition of risk measures): Formula (13.1) readily extends to the case of multiple functions $\phi_i, i = 1, \ldots, n,$ that are cumulatively used in measuring the risk of element $X \in \mathcal{X}$ and conform to the conditions of Theorem 13.1. Namely, one has that

$$\rho_n(X) = \min_{\eta_i \in \mathbb{R}, \ i=1,\ldots,n} \sum_{i=1}^{n} (\eta_i + \phi_i(X - \eta_i)), \tag{13.9}$$

is a proper coherent risk measure.

The value of η that delivers minimum in (13.1) does also possess some noteworthy properties as a function of X. In establishing these properties the following notation is convenient. Assuming that the set $\mathrm{argmin}_{x \in \mathbb{R}} f(x)$ is closed for some function $f: \mathbb{R} \mapsto \mathbb{R}$, we denote its left endpoint as

$$\mathop{\mathrm{Arg\,min}}_{x \in \mathbb{R}} f(x) = \min\{y \mid y \in \mathrm{arg\,min}_{x \in \mathbb{R}} f(x)\}.$$

---

[1] Here, the traditional terminology is preserved: according to the convention adopted in this chapter, X denotes losses and therefore the proper term for (7) would be the *upper* partial moment.

[2] Interestingly, Fischer (2003) restricted the range of values for the constant β in (8) to $\beta \in [0,1]$, whereas Rockafellar *et al.* (2006) allowed β to take values in $(0, \infty)$.

Theorem 13.2 *Let function* $\phi \colon \mathcal{X} \mapsto \mathbb{R}$ *satisfy the conditions of Theorem (13.1). Then function*

$$\eta(X) = \underset{\eta \in \mathbb{R}}{\operatorname{Arg\,min}} \ \eta + \phi\,(X - \eta) \tag{13.10}$$

exists and satisfies properties (A3) and (A4). If, additionally, $\phi\,(X) = 0$ *for every* $X \le 0$, *then* $\eta(X)$ *satisfies* (A1), *along with the inequality* $\eta\,(X) \le \rho\,(X)$, *where* $\rho(X)$ *is the optimal value of* (13.1).

Remark 13.2.1 If ϕ satisfies all the conditions of Theorem 13.2, the optimal solution $\eta(X)$ of the stochastic optimization problem (13.1) complies with all axioms for coherent risk measures, except (A2), thereby failing to be convex.

Example 13.2.1 (Value-at-Risk): A well-known example of two risk measures obtained by solving a stochastic programming problem of type (1) is again provided by formula (2) due to Rockafellar and Uryasev (2000, 2002), and its counterpart

$$\mathrm{VaR}_\alpha(X) = \underset{\eta \in \mathbb{R}}{\operatorname{Arg\,min}} \ \eta + (1 - \alpha)^{-1} \mathrm{E}(X - \eta)^+. \tag{13.11}$$

The Value-at-Risk measure $\mathrm{VaR}_\alpha(X)$, despite being adopted as a *de facto* standard for measurement of risk in finance and banking industries, is notorious for the difficulties it presents in risk estimation and control.

Example 13.2.2 (Higher moment coherent risk measures): For higher-moment coherent risk measures, the function ϕ in (13.10) is taken as $\phi(X) = (1 - \alpha)^{-1} \|(X)^+\|_p$, and the corresponding optimal $\eta_{p,\alpha}(X)$ satisfies the equation

$$(1 - \alpha)^{-1/(p-1)} = \frac{\left\| \left(X - \eta_{p,\alpha}(X) \right)^+ \right\|_p}{\left\| \left(X - \eta_{p,\alpha}(X) \right)^+ \right\|_{p-1}}, \quad p > 1. \tag{13.12}$$

A formal derivation of equality (13.12) can be carried out using the techniques employed in Rockafellar and Uryasev (2002) to establish formula (13.11). Although the optimal $\eta_{p,\alpha}(X)$ in (12) is determined implicitly, Theorem 13.2 ensures that it has properties similar to those of VaR_α (monotonicity, positive homogeneity, etc). Moreover, by plugging relation (13.12) with $p = 2$ into (13.6), the second-moment coherent risk measure (SMCR) can be presented in the form that involves only *the first moment of losses* in the tail of the distribution:

$$\begin{aligned}
\mathrm{SMCR}_\alpha(X) &= \eta_{2,\alpha}(X) + (1 - \alpha)^{-2} \left\| \left(X - \eta_{2,\alpha}(X) \right)^+ \right\|_1 \\
&= \eta_{2,\alpha}(X) + (1 - \alpha)^{-2} \mathrm{E}\left(X - \eta_{2,\alpha}(X) \right)^+.
\end{aligned} \tag{13.13}$$

Note that in (13.13) the second-moment information is concealed in the $\eta_{2,\alpha}(X)$. Further, by taking a CVaR measure with the confidence level $\alpha^* = 2\alpha - \alpha^2$, one can write

$$
\begin{aligned}
\mathrm{SMCR}_\alpha(X) &= \eta_{\mathrm{smcr}} + \frac{1}{1-\alpha^*}\,\mathrm{E}\big(X - \eta_{\mathrm{smcr}}\big)^+ \\
&\geq \eta_{\mathrm{cvar}} + \frac{1}{1-\alpha^*}\,\mathrm{E}\big(X - \eta_{\mathrm{cvar}}\big)^+ \\
&= \mathrm{CVaR}_{\alpha^*}(X),
\end{aligned}
\tag{13.14}
$$

where $\eta_{\mathrm{smcr}} = \eta_{p,\alpha}(X)$ as in (13.12) with $p = 2$, and $\eta_{\mathrm{cvar}} = \mathrm{VaR}_{\alpha^*}(X)$; note that the inequality in (13.14) holds due to the fact that $\eta_{\mathrm{cvar}} = \mathrm{VaR}_{\alpha^*}(X)$ minimizes the expression $\eta + (1-\alpha^*)^{-1}\mathrm{E}(X-\eta)^+$. In other words, with the above selection of α and α^*, the expressions for $\mathrm{SMCR}_\alpha(X)$ and $\mathrm{CVaR}_{\alpha^*}(X)$ differ only in the choice of η that delivers the minimum to the corresponding expressions (13.6) and (13.2). For Conditional Value-at-Risk, it is the α-quantile of the distribution of X, whereas the optimal $\eta_{2,\alpha}(X)$ for the SMCR measure incorporates the information on the second moment of losses X.

Example 13.2.3 (HMCR as tail risk measures): It is easy to see that the HMCR family are tail risk measures, namely that $0 < \alpha_1 < \alpha_2 < 1$ implies $\eta_{p,\alpha_1}(X) \leq \eta_{p,\alpha_2}(X)$, where $\eta_{p,\alpha_i} = \operatorname{Arg\,min}_\eta\{\eta + (1-\alpha_i)^{-1}\|(X-\eta)^+\|_p\}$; in addition, one has $\lim_{\alpha \to 1} \eta_{p,\alpha}(X) = \mathrm{ess.sup}\,X$, at least when $\mathrm{ess.sup}\,X$ is finite (see the appendix).

These properties puts the HMCR family in a favorable position comparing to the coherent measures of type (13.8) (Fischer 2003; Rockafellar et al. 2006), where the 'tail cutoff' point, about which the partial moments are computed, is always fixed at $\mathrm{E}X$. In contrast to (13.8), the location of tail cutoff in the HMCR measures is determined by the optimal $\eta_{p,\alpha}(X)$ and is adjustable by means of the parameter α. In a special case, for example, the HMCR measures (13.4) can be reduced to form (13.8) with $\beta = (1-\alpha_p)^{-1} > 1$ by selecting α_p according to (13.12) as $\alpha_p = 1 - \big(\|(X - \mathrm{E}X)^+\|_{p-1}\big/\|(X - \mathrm{E}X)^+\|_p\big)^{p-1}$, $p > 1$.

13.2.1.1 Representations based on relaxation of (A3) Observe that formula (1) in Theorem 13.1 is analogous to the operation of *infimal convolution*, well known in convex analysis:

$$
(f \,\square\, g)(x) = \inf_y \, f(x - y) + g(y).
$$

Continuing this analogy, consider the operation of right scalar multiplication

$$
(\phi\eta)(X) = \eta\phi(\eta^{-1}X), \quad \eta \geq 0,
$$

where for $\eta = 0$ we set $(\phi 0)(X) = (\phi 0^+)(X)$. If ϕ is proper and convex, then it is known that $(\phi\eta)(X)$ is a convex proper function in $\eta \geq 0$ for any $X \in \mathrm{dom}\,\phi$ (see, e.g. Rockafellar 1970). Interestingly enough, this fact can be pressed into service to formally define a coherent risk measure as

$$\underline{\rho}(X) = \inf_{\eta \geq 0} \eta \phi (\eta^{-1} X), \qquad (13.15)$$

if function ϕ, along with some technical conditions similar to those of Theorem 13.1, satisfies axioms (A1), (A2′) and (A4). Note that excluding the positive homogeneity (A3) from the list of properties of ϕ denies also its convexity, thus one must replace (A2) with (A2′) to ensure convexity in (13.15). In the terminology of convex analysis the function $\underline{\rho}(X)$ defined by (13.15) is known as the *positively homogeneous convex function generated by* ϕ. Likewise, by direct verification of conditions (A1)–(4) it can be demonstrated that

$$\overline{\rho}(X) = \sup_{\eta > 0} \eta \phi(\eta^{-1} X), \qquad (13.16)$$

is a proper coherent risk measure, provided that $\phi(X)$ satisfies (A1), (A2) and (A4). By (C1), axioms (A1) and (A2) imply that $\phi(0) = 0$, which allows one to rewrite (13.16) as

$$\overline{\rho}(X) = \sup_{\eta > 0} \frac{\phi(\eta X + 0) - \phi(0)}{\eta} = \phi 0^{+}(X), \qquad (13.17)$$

where the last equality in (13.17) comes from the definition of the recession function (Rockafellar 1970; Zălinescu 2002).

13.2.2   Implementation in Stochastic Programming Problems

The developed results can be efficiently applied in the context of stochastic optimization, where the random outcome $X = X(x, \omega)$ can be considered as a function of the decision vector $\mathbf{x} \in \mathbb{R}^m$, convex in $\mathbf{x}$ over some closed convex set $\mathcal{C} \subset \mathbb{R}^m$. Firstly, representation (1) allows for efficient minimization of risk in stochastic programs. For a function ϕ that complies with the requirements of Theorem 13.1, denote

$$\Phi(\mathbf{x}, \eta) = \eta + \phi\big(X(\mathbf{x}, \omega) - \eta\big) \quad \text{and} \quad R(\mathbf{x}) = \rho\big(X(\mathbf{x}, \omega)\big) = \min_{\eta \in \mathbb{R}} \Phi(\mathbf{x}, \eta). \qquad (13.18)$$

Then, clearly,

$$\min_{\mathbf{x} \in \mathcal{C}} \rho\big(X(\mathbf{x}, \omega)\big) \quad \Longleftrightarrow \quad \min_{(\mathbf{x}, \eta) \in \mathcal{C} \times \mathbb{R}} \Phi(\mathbf{x}, \eta), \qquad (13.19)$$

in the sense that both problems have the same optimal objective values and optimal vector $\mathbf{x}^*$. The last observation enables seamless incorporation of risk measures into stochastic programming problems, thereby facilitating the modelling of risk-averse preferences. For example, a generalization of the classical 2-stage stochastic linear programming (SLP) problem (see, e.g. Prékopa 1995; Birge and Louveaux 1997) where the outcome of the second-stage (recourse) action is evaluated by its *risk* rather than the expected value, can be formulated by replacing the expectation operator in the second-stage problem with a coherent risk measure $\mathcal{R}$:

$$\min_{x \geq 0} \ c^\top x + \mathscr{R}\left[\min_{y \geq 0} \ q(\omega)^\top y(\omega)\right]$$

$$\text{s.t.} \quad Ax = b, \quad T(\omega)\, x + W(\omega)\, y(\omega) = h(\omega). \qquad (13.20)$$

Note that the expectation operator is a member of the class of coherent risk measures defined by (A1)–(A4), whereby the classical 2-stage SLP problem is a special case of (13.20). Assuming that the risk measure $\mathscr{R}$ above is amenable to representation (1) via some function ϕ, problem (13.20) can be implemented by virtue of Theorem 13.1 as

$$\min \ c^\top x + \eta + \phi\big(q(\omega)^\top y(\omega) - \eta\big)$$

$$\text{s.t.} \quad Ax = b, \quad T(\omega)\, x + W(\omega)\, y(\omega) = h(\omega),$$

$$x \geq 0, \quad y(\omega) \geq 0, \quad \eta \in \mathbb{R},$$

with all the distinctive properties of the standard SLP problems (e.g. the convexity of the recourse function, etc.) being preserved.

Secondly, representation (1) also admits implementation of *risk constraints* in stochastic programs. Namely, let $g(x)$ be a function that is convex then, the following two problems are equivalent, as demonstrated by Theorem 13.3 below:

$$\min_{x \in C} \ \{ g(x) \mid R(x) \leq c \}, \qquad (13.21a)$$

$$\min_{(x,\eta) \in C \times \mathbb{R}} \ \{ g(x) \mid \Phi(x, \eta) \leq c \}. \qquad (13.21b)$$

Theorem 13.3 *Optimization problems* (13.21a) *and* (13.21b) *are equivalent in the sense that they achieve minima at the same values of the decision variable* x *and their optimal objective values coincide. Further, if the risk constraint in* (13.21a) *is binding at optimality,* (x^*, η^*) *achieves the minimum of* (13.21b) *if and only if* x^* *is an optimal solution of* (13.21a) *and* $\eta^* \in \arg\min_\eta \Phi(x^*, \eta)$.

In other words, one can implement the risk constraint $\rho(X(x, \omega)) \leq c$ by using representation (1) for the risk measure ρ with the infimum operator omitted.

13.2.2.1 HMCR measures in stochastic programming problems The introduced higher moment coherent risk measures can be incorporated in stochastic programming problems via conic constraints of order $p > 1$. Namely, let $\{\omega_1, \ldots, \omega_J\} \subseteq \Omega$ where $P\{\omega_j\} = \pi_j \in (0, 1)$ be the scenario set of a stochastic programming model. Observe that a HMCR-based objective or constraint can be implemented via the constraint $\text{HMCR}_{p,\alpha}(X(x, \omega))$, with u being either a variable or a constant, correspondingly. By virtue of Theorem 13.3, the latter constraint admits a representation by the set of inequalities

$$u \geq \eta + (1 - \alpha)^{-1} t, \qquad (13.22a)$$

$$t \geq \left(w_1^p + \cdots + w_J^p\right)^{1/p}, \tag{13.22b}$$

$$w_j \geq \pi_j^{1/p}\left(X(\mathbf{x}, \omega_j) - \eta\right), \quad j = 1, \ldots, J, \tag{13.22c}$$

$$w_j \geq 0, \quad j = 1, \ldots, J. \tag{13.22d}$$

Note that the convexity of X as a function of the decision variables $\mathbf{x}$ implies convexity of (13.22c), and, consequently, convexity of the set (13.22). Constraint (13.22b) defines a $J + 1$-dimensional cone of order p, and is central to practical implementation of constraints (13.22) in mathematical programming models. In the special case of $p = 2$, it represents a second-order (quadratic) cone in $\mathbb{R}^{J+1}$, and well-developed methods of second-order cone programming (SOCP) can be invoked to handle the constructions of type (13.22). In the general case of $p \in (1, \infty)$, the p-order cone within the positive orthant

$$\left(w_1^p + \cdots + w_J^p\right)^{1/p} \leq t, \quad t, w_j \geq 0, \ j = 1, \ldots, J, \tag{13.23}$$

can be approximated by linear inequalities when $J = 2^d$. Following Ben-Tal and Nemirovski (1999), the $2^d + 1$-dimensional p-order conic constraint (13.23) can be represented by a set of 3-dimensional p-order conic inequalities

$$\left[\left(w_{2j-1}^{(k-1)}\right)^p + \left(w_{2j}^{(k-1)}\right)^p\right]^{1/p} \leq w_j^{(k)}, \quad j = 1, \ldots, 2^{d-k}, \quad k = 1, \ldots, d, \tag{13.24}$$

where $w_1^{(d)} \equiv t$ and $w_j^{(0)} \equiv w_j$ $(j = 1, \ldots, 2^d)$. Each of the 3-dimensional p-order cones (13.24) can then be approximated by a set of linear equalities. For any partition $0 \equiv \alpha_0 < \alpha_1 < \cdots < \alpha_m \equiv \pi/2$ of the segment $[0, \pi/2]$, an internal approximation of the p-order cone in the positive orthant of $\mathbb{R}^3$

$$\xi_3 \geq \left(\xi_1^p + \xi_2^p\right)^{1/p}, \quad \xi_1, \xi_2, \xi_3 \geq 0, \tag{13.25}$$

can be written in the form

$$\xi_3 \left(\sin^{2/p}\alpha_{i+1}\cos^{2/p}\alpha_i - \cos^{2/p}\alpha_{i+1}\sin^{2/p}\alpha_i\right)$$
$$\geq \xi_1 \left(\sin^{2/p}\alpha_{i+1} - \sin^{2/p}\alpha_i\right) + \xi_2\left(\cos^{2/p}\alpha_i - \cos^{2/p}\alpha_{i+1}\right),$$
$$i = 0, \ldots, m-1, \tag{13.26a}$$

and an external approximation can be constructed as

$$\xi_3\left(\cos^p\alpha_i + \sin^p\alpha_i\right)^{(p-1)/p} \geq \xi_1\cos^{p-1}\alpha_i + \xi_2\sin^{p-1}\alpha_i, \quad i = 0, \ldots, m. \tag{13.26b}$$

For example, the uniform partition $\alpha_i = \pi i/2m$ $(i = 0, \ldots, m)$ generates the following approximations of a 3-dimensional second-order cone:

$$\xi_3 \cos\frac{\pi}{4m} \geq \xi_1 \cos\frac{\pi(2i+1)}{4m} + \xi_2 \sin\frac{\pi(2i+1)}{4m}, \quad i = 0, \ldots, m-1,$$

$$\xi_3 \geq \xi_1 \cos\frac{\pi i}{2m} + \xi_2 \sin\frac{\pi i}{2m}, \quad i = 0, \ldots, m.$$

13.2.3 Application to Deviation Measures

Since being introduced in Artzner *et al.* (1999), the axiomatic approach to construction of risk measures has been repeatedly employed by many authors for the development of other types of risk measures tailored to specific preferences and applications (see Acerbi 2002; Rockafellar *et al.* 2006; Ruszczyński and Shapiro 2006). In this subsection we consider *deviation measures* as introduced by Rockafellar *et al.* (2006). Namely, deviation measure is a mapping $\mathscr{D}: \mathcal{X} \mapsto [0, +\infty]$ that satisfies

(D1) $\mathscr{D}(X) > 0$ for any non-constant $X \in \mathcal{X}$, whereas $\mathscr{D}(X) = 0$ for constant X,

(D2) $\mathscr{D}(X + Y) \leq \mathscr{D}(X) + \mathscr{D}(Y)$ for all $X, Y \in \mathcal{X}$, $0 \leq \lambda \leq 1$,

(D3) $\mathscr{D}(\lambda X) = \lambda\mathscr{D}(X)$ for all $X \in \mathcal{X}$, $\lambda > 0$,

(D4) $\mathscr{D}(X + a) + \mathscr{D}(X)$ for all $X \in \mathcal{X}$, $a \in \mathbb{R}$.

Again, from axioms (D1) and (D2) follows convexity of $\mathscr{D}(X)$. In Rockafellar *et al.* (2006) it was shown that deviation measures that further satisfy

(D5) $\mathscr{D}(X) \leq$ ess.sup $X - EX$ for all $X \in \mathcal{X}$,

are characterized by the one-to-one correspondence

$$\mathscr{D}(X) = \mathscr{R}(X - EX) \tag{13.27}$$

with *expectation-bounded* coherent risk measures, i.e. risk measures that satisfy (A1)–(A4) and an additional requirement (A5) $\mathscr{R}(X) > EX$, for all non-constant $X \in \mathcal{X}$, whereas $\mathscr{R}(X) = EX$ for all constant X.

Using this result, it is easy to provide an analogue of formula (1) for deviation measures.

Theorem 13.4 *Let function* $\phi: \mathcal{X} \mapsto \mathbb{R}$ *satisfy axioms* (A1)–(A3), *and be a lsc function such that* $\phi(X) > EX$ *for all* $X \neq 0$. *Then the optimal value of the stochastic programming problem*

$$\mathscr{D}(X) = -EX + \inf_{\eta} \{\eta + \phi(X - \eta)\} \tag{13.28}$$

is a deviation measure, and the infimum is attained for all X, so that $\inf_{\eta}$ *in* (13.28) *may be replaced by* $\min_{\eta \in \mathbb{R}}$.

Given the close relationship between deviation measures and coherent risk measures, it is straightforward to apply the above results to deviation measures.

13.2.4 Connection with Utility Theory and Second-Order Stochastic Dominance

As has been mentioned in Section 13.1, considerable attention has been devoted in the literature to the development of risk models and measures compatible with the utility theory of von Neumann and Morgenstern (1944), which represents one of the cornerstones of the decision-making science.

The vNM theory argues that when the preference relation '$\succeq$' of the decision-maker satisfies certain axioms (completeness, transitivity, continuity and independence), there exists a function $u: \mathbb{R} \mapsto \mathbb{R}$, such that an outcome X is preferred to outcome Y ('$X \succeq Y$') if and only if $E[u(X)] \geq E[u(Y)]$. If the function u is non-decreasing and concave, the corresponding preference is said to be risk averse. Rothschild and Stiglitz (1970) have bridged the vNM utility theory with the concept of the second-order stochastic dominance by showing that X dominating Y by the second-order stochastic dominance, $X \succeq_{SSD} Y$, is equivalent to the relation $E[u(X)] \geq E[u(Y)]$ holding true for all concave non-decreasing functions u, where the inequality is strict for at least one such u. Recall that a random outcome X dominates outcome Y by the second-order stochastic dominance if

$$\int_{-\infty}^{z} P[X \leq t]\, dt \leq \int_{-\infty}^{z} P[Y \leq t]\, dt \quad \text{for all } z \in \mathbb{R}.$$

Since coherent risk measures are generally inconsistent with the second-order stochastic dominance (see an example in De Giorgi (2005)), it is of interest to introduce risk measures that comply with this property. To this end, we replace the monotonicity axiom (A1) in the definition of coherent risk measures by the requirement of *SSD isotonicity* (Pflug 2000; De Giorgi 2005):

$$(-X) \succeq_{SSD} (-Y) \implies \mathcal{R}(X) \leq \mathcal{R}(Y).$$

Namely, we consider risk measures $\mathcal{R}: \mathcal{X} \mapsto \overline{\mathbb{R}}$ that satisfy the following set of axioms:[3]

(A1′) *SSD isotonicity*: $(-X) \succeq_{SSD} (-Y) \implies \mathcal{R}(X) \leq \mathcal{R}(Y)$ for $X, Y \in \mathcal{X}$,
(A2′) *convexity*: $\mathcal{R}(\lambda X + (1 - \lambda)Y) \leq \lambda \mathcal{R}(X) + (1 - \lambda)\mathcal{R}(Y)$, $X, Y \in \mathcal{X}$, $0 \leq \lambda \leq 1$,
(A3) *positive homogeneity*: $\mathcal{R}(\lambda X) = \lambda \mathcal{R}(X)$, $X \in \mathcal{X}$, $\lambda > 0$,
(A4) *translation invariance*: $\mathcal{R}(X + a) = \mathcal{R}(X) + a$, $X \in \mathcal{X}$, $a \in \mathbb{R}$.

Note that unlike the system of axioms (A1)–(A4), the above axioms, and in particular (A1′), require X and Y to be integrable, i.e. one can take the space in (A1′)–(A4) to be $\mathcal{L}_1$ (for a discussion of topological properties of sets defined by stochastic dominance relations, see, e.g. Dentcheva and Ruszczyński (2004).

---

[3] See Mansini *et al.* (2003) and Ogryczak and Opolska-Rutkowska (2006) for conditions under which SSD-isotonic measures also satisfy the coherence properties.

Again, it is possible to develop an analogue of formula (1), which would allow for construction of risk measures with the above properties using functions that comply with (A1′), (A2′) and (A3).

Theorem 13.5 *Let function $\phi: \mathcal{X} \mapsto \mathbb{R}$ satisfy axioms (A1′), (A2′), and (A3), and be a lsc function such that $\phi(\eta) > \eta$ for all real $\eta \neq 0$. Then the optimal value of the stochastic programming problem*

$$\rho(X) = \min_{\eta \in \mathbb{R}} \eta + \phi(X - \eta) \tag{13.29}$$

exists and is a proper function that satisfies (A1′), (A2′), (A3) and (A4).

Obviously, by solving the risk-minimization problem

$$\min_{\mathbf{x} \in C} \rho(X(\mathbf{x}, \omega)),$$

where ρ is a risk measure that is both coherent and SSD-compatible in the sense of (A1′), one obtains a solution that is SSD-efficient, i.e. acceptable to *any* risk-averse rational utility maximizer, and also bears the lowest risk in terms of coherence preference metrics. Below we illustrate that functions ϕ satisfying the conditions of Theorem 13.5 can be easily constructed in the scope of the presented approach.

Example 13.5.1 Let $\phi(X) = E[u(X)]$, where $u: \mathbb{R} \mapsto \mathbb{R}$ is a convex, positively homogeneous, non-decreasing function such that $u(\eta) > \eta$ for all $\eta \neq 0$. Obviously, function $\phi(X)$ defined in this way satisfies the conditions of Theorem 13.5. Since $-u(-\eta)$ is concave and non-decreasing, one has that $-E[u(X)] \geq -E[u(Y)]$, and, consequently, $\phi(X) \leq \phi(Y)$, whenever $(-X) \succeq_{SSD} (-Y)$. It is easy to see that, for example, function ϕ of the form

$$\phi(X) = \alpha E(X)^+ - \beta E(X)^-, \quad \alpha \in (1, +\infty), \ \beta \in [0, 1),$$

satisfies the conditions of Theorem 13.5. Thus, in accordance with Theorems 13.1 and 13.5, the coherent risk measure $\mathcal{R}_{\alpha,\beta}$ (3) is also consistent with the second-order stochastic dominance. A special case of (3) is Conditional Value-at-Risk, which is known to be compatible with the second-order stochastic dominance (Pflug 2000).

Example 13.5.2 (Higher moment coherent risk measures) SMCR and, in general, the family of higher-moment coherent risk measures constitute another example of risk measures that are both coherent and compatible with the second-order stochastic dominance. Indeed, function $u(\eta) = ((\eta)^+)^p$ is convex and non-decreasing, whence $(E[u(X)])^{1/p} \leq (E[u(Y)])^{1/p}$ for any $(-X) \succeq_{SSD} (-Y)$. Thus, the HMCR family, defined by (29) with $\phi(X) = (1 - \alpha)^{-1} \|(X)^+\|_p$,

$$\text{HMCR}_{p,\alpha}(X) = \min_{\eta \in \mathbb{R}} \eta + (1-\alpha)^{-1} \left\| (X-\eta)^+ \right\|_p, \quad p \geq 1,$$

is both coherent and SSD-compatible, by virtue of Theorems 13.1 and 13.5. Implementation of such a risk measure in stochastic programming problems enables one to introduce risk preferences that are consistent with both concepts of coherence and second-order stochastic dominance.

The developed family of higher-moment risk measures (4) possesses all the outstanding properties that are sought after in the realm of risk management and decision making under uncertainty: compliance with the coherence principles, amenability for an efficient implementation in stochastic programming problems (e.g. via second-order cone programming), and compatibility with the second-order stochastic dominance and utility theory. The question that remains to be answered is whether these superior properties translate into an equally superior performance in practical risk management applications.

The next section reports a pilot study intended to investigate the performance of the HMCR measures in real-life risk management applications. It shows that the family of HMCR measures is a promising tool for tailoring risk preferences to the specific needs of decision-makers, and can be compared favourably with some of the most widely used risk management frameworks.

13.3 PORTFOLIO CHOICE USING HMCR MEASURES: AN ILLUSTRATION

In this section we illustrate the practical utility of the developed HMCR measures on the example of portfolio optimization, a typical testing ground for many risk management and stochastic programming techniques. To this end, we compare portfolio optimization models that use the HMCR measures with $p = 2$ (SMCR) and $p = 3$ against portfolio allocation models based on two well-established, and theoretically as well as practically proven methodologies, the Conditional Value-at-Risk measure and the Markowitz Mean-Variance framework.

This choice of benchmark models is further supported by the fact that the HMCR family is similar in the construction and properties to CVaR (more precisely, CVaR is a HMCR measure with $p = 1$), but, while CVaR measures the risk in terms of the first moment of losses residing in the tail of the distribution, the SMCR measure quantifies risk using the second moments, in this way relating to the MV paradigm. The HMCR measure with $p = 3$ demonstrates the potential benefits of using higher-order tail moments of loss distributions for risk estimation.

13.3.1 Portfolio Optimization Models and Implementation

The portfolio selection models employed in this case study have the general form

$$\min_{\mathbf{x}} \quad \mathcal{R}(-\mathbf{r}^\top \mathbf{x}) \tag{13.30a}$$

$$\text{s.t.} \quad \mathbf{e}^\top \mathbf{x} = 1, \tag{13.30b}$$

$$\mathbf{E}\mathbf{r}^\top \mathbf{x} \geq r_0, \tag{13.30c}$$

$$\mathbf{x} \geq 0, \tag{13.30d}$$

where $\mathbf{x} = (x_1, \ldots, x_n)^\top$ is the vector of portfolio weights, $\mathbf{r} = (r_1, \ldots, r_n)^\top$ is the random vector of assets' returns, and $\mathbf{e} = (1, \ldots, 1)^\top$. The risk measure $\mathscr{R}$ in (13.30a) is taken to be either SMCR (6), HMCR with $p = 3$ (4), CVaR (2), or variance σ^2 of the negative portfolio return, $-\mathbf{r}^\top \mathbf{x} = X$. In the above portfolio optimization problem, (13.30b) represents the budget constraint, which, together with the no-short-selling constraint (13.30d) ensures that all the available funds are invested, and (13.30c) imposes the minimal required level r_0 for the expected return of the portfolio.

We have deliberately chosen not to include any additional trading or institutional constraints (transaction costs, liquidity constraints, etc.) in the portfolio allocation problem (13.30) so as to make the effect of risk measure selection in (13.30a) onto the resulting portfolio rebalancing strategy more marked and visible.

Traditionally to stochastic programming, the distribution of random return r_i of asset i is modelled using a set of J discrete equiprobable scenarios $\{r_{i1}, \ldots, r_{iJ}\}$. Then, optimization problem (13.30) reduces to a linear programming problem if CVaR is selected as the risk measure $\mathscr{R}$ in (13.30a) (see, for instance, Rockafellar and Uryasev 2000; Krokhmal et al. 2002a). Within the Mean-Variance framework, (13.30) becomes a convex quadratic optimization problem with the objective

$$\mathscr{R}(-\mathbf{r}^\top \mathbf{x}) = \sum_{i,k=1}^n \sigma_{ik} x_i x_k, \quad \text{where}$$
$$\sigma_{ik} = \frac{1}{J-1} \sum_{j=1}^J (r_{ij} - \bar{r}_i)(r_{kj} - \bar{r}_k), \quad \bar{r}_i = \frac{1}{J} \sum_{j=1}^J r_{ij}. \tag{13.31}$$

In the case of $\mathscr{R}(X) = \mathrm{HMCR}_{p,\alpha}(X)$, problem (13.30) transforms into a linear programming problem with a single p-order cone constraint (13.32e):

$$\min \quad \eta + \frac{1}{1-\alpha} \frac{1}{J^{1/p}} t \tag{13.32a}$$

$$\text{s.t.} \quad \sum_{i=1}^n x_i = 1, \tag{13.32b}$$

$$\frac{1}{J} \sum_{j=1}^J \sum_{i=1}^n r_{ij} x_i \geq r_0, \tag{13.32c}$$

$$w_j \geq -\sum_{i=1}^n r_{ij} x_i - \eta, \quad j = 1, \ldots, J, \tag{13.32d}$$

$$t \geq \left(w_1^p + \cdots + w_J^p \right)^{1/p}, \tag{13.32e}$$

$$x_i \geq 0, \quad i = 1, \ldots, n, \tag{13.32f}$$

$$w_j \geq 0, \quad j = 1, \ldots, J. \tag{13.32g}$$

When $p = 2$, i.e. $\mathscr{R}$ is equal to SMCR, (32) reduces to a SOCP problem. In the case when $\mathscr{R}$ is selected as HMCR with $p = 3$, the 3rd-order cone constraint (13.32e) has been approximated via linear inequalities (13.26a) with $m = 500$, thereby transforming problem (13.32) to a LP. The resulting mathematical programming problems have been implemented in C++, and we used CPLEX 10.0 for solving the LP and QP problems, and MOSEK 4.0 for solving the SOCP problem.

13.3.2 Set of Instruments and Scenario Data

Since the introduced family of HMCR risk measures quantifies risk in terms of higher tail moments of loss distributions, the portfolio optimization case studies were conducted using a data set that contained return distributions of fifty S&P 500 stocks with the so-called 'heavy tails.' Namely, for scenario generation we used 10-day historical returns over $J = 1024$ overlapping periods, calculated using daily closing prices from 30 October 1998 to 18 January 2006. From the set of S&P 500 stocks (as of January 2006) we selected $n = 50$ instruments by picking the ones with the highest values of kurtosis of biweekly returns, calculated over the specified period. In such a way, the investment pool had the average kurtosis of 51.93, with 429.80 and 17.07 being the maximum and minimum kurtosis, correspondingly. The particular size of scenario model, $J = 1024 = 2^{10}$, has been chosen so that the linear approximation techniques (13.26) can be employed for the HMCR measure with $p = 3$.

13.3.3 Out-of-Sample Simulations

The primary goal of our case study is to shed light on the potential 'real-life' performance of the HMCR measures in risk management applications, and to this end we conducted the so-called *out-of-sample* experiments. As the name suggests, the out-of-sample tests determine the merits of a constructed solution using *out-of-sample* data that have *not* been included in the scenario model used to generate the solution. In other words, the out-of-sample setup simulates a common situation when the true realization of uncertainties happens to be outside of the set of the 'expected,' or 'predicted' scenarios (as is the case for most portfolio optimization models). Here, we employ the out-of-sample method to compare simulated historic performances of four self-financing portfolio rebalancing strategies that are based on (13.30) with $\mathscr{R}$ chosen either as a member of the HMCR family with $\alpha = 0.90$, namely, $\mathrm{CVaR}_{0.90}(\cdot)$, $\mathrm{SMCR}_{0.90}(\cdot)$ $\mathrm{HMCR}_{3,0.90}(\cdot)$, or as variance $\sigma^2(\cdot)$.

It may be argued that in practice of portfolio management, instead of solving (13.30), it is of more interest to construct investment portfolios that maximize the expected return subject to risk constraint(s), e.g.

$$\max_{\mathbf{x} \geq 0} \left\{ \mathrm{E}\mathbf{r}^{\mathsf{T}}\mathbf{x} \mid \mathscr{R}(-\mathbf{r}^{\mathsf{T}}\mathbf{x}) \leq c_0, \mathbf{e}^{\mathsf{T}}\mathbf{x} = 1 \right\}. \tag{13.33}$$

Indeed, many investment institutions are required to keep their investment portfolios in compliance with numerous constraints, including constraints on risk. However, our main point is to gauge the effectiveness of the HMCR risk measures in portfolio optimization by comparing them against other well-established risk management methodologies, such as the CVaR and MV frameworks. And since these risk measures yield risk estimates on different scales, it is not obvious which risk tolerance levels c_0 should be selected in (13.33) to make the resulting portfolios comparable.

Thus, to facilitate a fair 'apple-to-apple' comparison, we construct self-financing portfolio rebalancing strategies by solving the risk-minimization problem (13.30), so that the resulting portfolios all have the same level r_0 of expected return, and the success of a particular portfolio rebalancing strategy will depend on the actual amount of risk borne by the portfolio due to the utilization of the corresponding risk measure.

The out-of-sample experiments have been set up as follows. The initial portfolios were constructed on 11 December 2002 by solving the corresponding variant of problem (13.30), where the scenario set consisted of 1024 overlapping bi-weekly returns covering the period from 30 October 1998 to 11 December 2002. The duration of the rebalancing period for all strategies was set at two weeks (ten business days). On the next rebalancing date of 26 December 2002,[4] the 10-day *out-of-sample* portfolio returns, $\hat{\mathbf{r}}^{\mathsf{T}}\mathbf{x}^*$, were observed for each of the three portfolios, where $\hat{\mathbf{r}}$ is the vector of out-of-sample (11 December 2002–26 December 2002) returns and $\mathbf{x}^*$ is the corresponding optimal portfolio configuration obtained on 11 December 2002. Then, all portfolios were rebalanced by solving (13.30) with an updated scenario set. Namely, we included in the scenario set the ten vectors of overlapping biweekly returns that realized during the ten business days from 11 December 2002 to 26 December 2002, and discarded the oldest ten vectors from October–November of 1998. The process was repeated on 26 December 2002, and so on. In such a way, the out-of-sample experiment consisted of 78 biweekly rebalancing periods covering more than three years. We ran the out-of-sample tests for different values of the minimal required expected return r_0 and typical results are presented in Figures 13.1 to 13.3.

Figure 13.1 reports the portfolio values (in percent of the initial investment) for the four portfolio rebalancing strategies based on (13.30) with $\mathrm{SMCR}_{0.90}(\cdot)$ $\mathrm{CVaR}_{0.90}(\cdot)$, variance $\sigma^2(\cdot)$, and $\mathrm{HMCR}_{3,0.90}(\cdot)$ as $\mathscr{R}(\cdot)$, and the required level r_0 of the expected return being set at 0.5%. One can observe that in this case the clear winner is the portfolio based on the HMCR measure with $p = 3$, the SMCR-based portfolio is runner-up, with the CVaR- and MV-based portfolios falling behind these two. This situation is typical for smaller values of r_0; as r_0 increases and the rebalancing strategies become more aggressive, the CVaR- and MV-portfolios become more competitive, while the HMCR ($p = 3$) portfolio remains dominant most of the time (Figure 13.2).

---

[4] Holidays were omitted from the data.

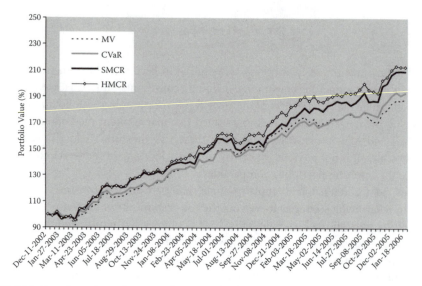

FIGURE 13.1 Out-of-sample performance of conservative ($r_0 = 0.5\%$) self-financing portfolio rebalancing strategies based on the MV model, and CVaR, SMCR (p = 2) and HMCR (p = 3) measures of risk with $\alpha = 0.90$.

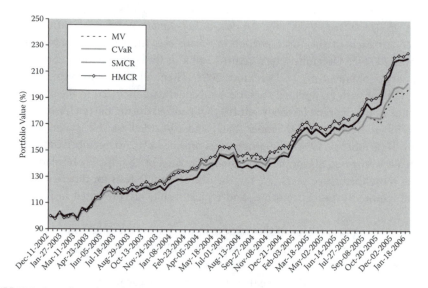

FIGURE 13.2 Out-of-sample performance of self-financing portfolio rebalancing strategies that have expected return $r_0 = 1.0\%$ and are based on the MV model, and CVaR, SMCR (p = 2) and HMCR (p = 3) risk measures with $\alpha = 0.90$.

An illustration of typical behaviour of more aggressive rebalancing strategies is presented in Figure 13.3, where r_0 is set at 1.3% (in the dataset used in this case study, infeasibilities in (13.30) started to occur for values of $r_0 > 0.013$). As a general trend, the HMCR (p = 2, 3) and CVaR portfolios exhibit similar performance (which can be

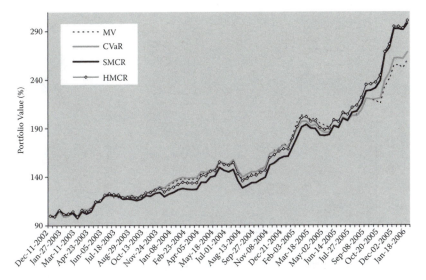

FIGURE 13.3 Out-of-sample performance of aggressive ($r_0 = 1.3\%$) self-financing portfolio rebalancing strategies based on SMCR ($p = 2$), CVaR, MV and HMCR ($p = 3$) measures of risk.

explained by the fact that at high values of r_0 the set of instruments capable of providing such a level of expected return is rather limited). Still, it may be argued that the HMCR ($p = 3$) strategy can be preferred to the other three, based on its overall stable performance throughout the duration of the run.

Finally, to illuminate the effects of taking into account higher-moment information in estimation of risks using the HMCR family of risk measures, we compared the simulated historical performances of the SMCR measure with parameter $\alpha = 0.90$ and CVaR measure with confidence level $\alpha^* = 0.99$, so that the relation $\alpha^* = 2\alpha - \alpha^2$ holds (see the discussion in Example 13.2.2). Recall that in such a case the expressions for SMCR and CVaR differ only in the location of the optimal η^* (see (13.14)). For CVaR, the optimal η equals to VaR, and for SMCR the corresponding value of η is determined by (13.12) and depends on the second tail moment of the distribution.

Figure 13.4 presents a typical outcome for mid-range values of the expected return level r_0: most of the time, the SMCR portfolio dominates the corresponding CVaR portfolio. However, for smaller values of expected return (e.g. $r_0 \leq 0.005$), as well as for values approaching $r_0 = 0.013$, the SMCR- and CVaR-based rebalancing strategies demonstrated very close performance. This can be explained by the fact that lower values of r_0 lead to rather conservative portfolios, while the values of r_0 close to the infeasibility barrier of 0.013 lead to very aggressive and poorly diversified portfolios that are comprised of a limited set of assets capable of achieving this high level of the expected return.

Although the obtained results are data-specific, the presented preliminary case studies indicate that the developed HMCR measures demonstrate very promising performance, and can be successfully employed in the practice of risk management and portfolio optimization.

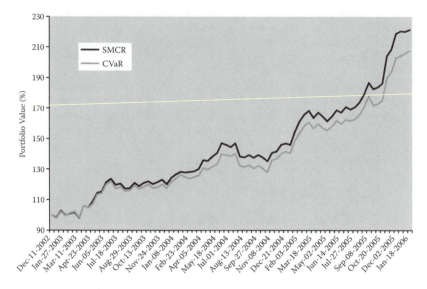

FIGURE 13.4 Out-of-sample performance of self-financing portfolio rebalancing strategies based on $SMCR_{0.90}$ and $CVaR_{0.99}$ measures of risk ($r_0 = 1.0\%$).

13.4   CONCLUSIONS

In this chapter we have considered modelling of risk-averse preferences in stochastic programming problems using risk measures. We utilized the axiomatic approach to construction of coherent risk measures and deviation measures in order to develop simple representations for these risk measures via solutions of specially designed stochastic programming problems. Using the developed general representations, we introduced a new family of higher-moment coherent risk measures (HMCR). In particular, we considered the second-moment coherent risk measure (SMCR), which is implementable in stochastic optimization problems using the second-order cone programming, and the third-order HMCR ($p = 3$). The conducted numerical studies indicate that the HMCR measures can be effectively used in the practice of portfolio optimization, and compare well with the well-established benchmark models such as Mean-Variance framework or CVaR.

ACKNOWLEDGEMENT

This work was supported in part by NSF grant DMI 0457473.

REFERENCES

Acerbi, C., Spectral measures of risk: A coherent representation of subjective risk aversion. *J. Bank. Finan.*, 2002, **26**(7), 1487–1503.

Acerbi, C. and Tasche, D., On the coherence of expected shortfall. *J. Bank. Finan.*, 2002, **26**(7), 1487–1503.

Alizadeh, F. and Goldfarb, D., Second-order cone programming. *Math. Program.*, 2003, **95**(1), 3–51.

Artzner, P., Delbaen, F., Eber, J.M. and Heath, D., Coherent measures of risk. *Math. Finan.*, 1999, **9**(3), 203–228.

Bawa, V.S., Optimal rules for ordering uncertain prospects. *Rev. Financ. Stud.*, 1975, **2**(1), 95–121.

Ben-Tal, A. and Nemirovski, A., On polyhedral approximations of the second-order cone. *Math. Operat. Res.*, 1999, **26**(2), 193–205.

Ben-Tal, A. and Teboulle, M., Expected utility, penalty functions, and duality in stochastic nonlinear programming. *Manag. Sci.*, 1986, **32**(11), 1145–1466.

Birge, J.R. and Louveaux, F., *Introduction to Stochastic Programming*, 1997 (Springer: New York).

Chekhlov, A., Uryasev, S. and Zabarankin, M., Drawdown measure in portfolio optimization. *Int. J. Theoret. Appl. Finan.*, 2005, **8**(1), 13–58.

De Giorgi, E., Reward-risk portfolio selection and stochastic dominance. *J. Bank. Finan.*, 2005, **29**(4), 895–926.

Delbaen, F., Coherent risk measures on general probability spaces. In *Advances in Finance Stochast.: Essays in Honour of Dieter Sondermann*, edited by K. Sandmann and P.J. Schönbucher, pp. 1–37, 2002 (Springer: Berlin).

Dembo, R.S. and Rosen, D., The practice of portfolio replication: A practical overview of forward and inverse problems. *Ann. Oper. Res.*, 1999, **85**, 267–284.

Dentcheva, D. and Ruszczyński, A., Optimization with stochastic dominance constraints. *SIAM J. Optim.*, 2003, **14**(2), 548–566.

Dentcheva, D. and Ruszczyński, A., Semi-infinite probabilistic optimization: First order stochastic dominance constraints. *Optimization*, 2004, **53**(5–6), 583–601.

Duffie, D. and Pan, J., An overview of value-at-risk. *J. Deri.*, 1997, **4**, 7–49.

Fischer, T., Risk capital allocation by coherent risk measures based on one-sided moments. *Insur.: Math. Econ.*, 2003, **32**(1), 135–146.

Fishburn, P.C., Stochastic dominance and moments of distributions. *Math. Oper. Res.*, 1964, **5**, 94–100.

Fishburn, P.C., *Utility Theory for Decision-Making*, 1970 (Wiley: New York).

Fishburn, P.C., Mean-risk analysis with risk associated with below-target returns. *The Amer. Eco. Rev.*, 1977, **67**(2), 116–126.

Fishburn, P.C., *Non-Linear Preference and Utility Theory*, 1988 (Johns Hopkins University Press: Baltimore).

Jorion, P., *Value at Risk: The New Benchmark for Controlling Market Risk*, 1997 (McGraw-Hill: New York).

Karni, E. and Schmeidler, D., Utility theory with uncertainty. In *Handbook of Mathematical Economics*, Vol. IV, edited by W. Hildenbrand and H. Sonnenschein, 1991 (North-Holland: Amsterdam).

Kouvelis, P. and Yu, G., *Robust Discrete Optimization and Its Applications*, 1997 (Kluwer Academic Publishers: Dordrecht).

Krokhmal, P., Palmquist, J. and Uryasev, S., Portfolio optimization with conditional value-at-risk objective and constraints. *J. Risk*, 2002a, **4**(2), 43–68.

Krokhmal, P., Uryasev, S. and Zrazhevsky, G., Risk management for hedge fund portfolios: A comparative analysis of linear rebalancing strategies. *J. Alter. Invest.*, 2002b, **5**(1), 10–29.

Kroll, Y., Levy, H. and Markowitz, H.M., Mean-variance versus direct utility maximization. *J. Finan.*, 1984, **39**(1), 47–61.

Levy, H., *Stochastic Dominance*, 1998 (Kluwer Academic Publishers: Dordrecht).

Mansini, R., Ogryczak, W. and Speranza, M.G., LP solvable models for portfolio optimization: A LP solvable models for portfolio optimization: A classification and computational comparison. *IMA J. Manag. Math.*, 2003, **14**(3), 187–220.

Markowitz, H.M., Portfolio selection. *J. Finan.*, 1952, **7**(1), 77–91.

Markowitz, H.M., *Portfolio Selection*, 1st, 1959 (Wiley and Sons: New York).

Ogryczak, W. and Opolska-Rutkowska, M., SSD consistent criteria and coherent risk measures, In *System Modeling and Optimization. Proceedings of the 22nd IFIP TC7 Conference*, Vol. 199, in the series IFIP International Federation for Information Processing, edited by F. Ceragioli, A. Dontchev, H. Furuta, M. Marti and L. Pandolfi, 2006, pp. 227–237 (IFIP).

Ogryczak, W. and Ruszczyński, A., From stochastic dominance to mean-risk models: Semideviations as risk measures. *Eur. J. Operat. Res.*, 1999, **116**, 33–50.

Ogryczak, W. and Ruszczyński, A., On consistency of stochastic dominance and mean-semidevia-tion models. *Math. Program.*, 2001, **89**, 217–232.

Ogryczak, W. and Ruszczyński, A., Dual stochastic dominance and related mean-risk models. *SIAM J. Optim.*, 2002, **13**(1), 60–78.

Pflug, G., Some remarks on the value-at-risk and the conditional value-at-risk. In *Probabilistic Constrained Optimization: Methodology and Applications,* edited by S. Uryasev, 2000 (Kluwer Academic Publishers: Dordrecht).

Prékopa, A., *Stochastic Programming*, 1995 (Kluwer Academic Publishers: Dordrecht).

Riskmetrics, 1994 (JP Morgan: New York).

Rockafellar, R.T., Convex Analysis. In the series, *Princeton Mathematics*, Vol. 28, 1970 (Princeton University Press: Princeton, NJ).

Rockafellar, R.T. and Uryasev, S., Optimization of conditional value-at-risk. *J. Risk*, 2000, **2**, 21–41.

Rockafellar, R.T. and Uryasev, S., Conditional value-at-risk for general loss distributions. *J. Bank. Finan.*, 2002, **26**(7), 1443–1471.

Rockafellar, R.T., Uryasev, S. and Zabarankin, M., Generalized deviations in risk analysis. *Finan. Stochast.*, 2006, **10**(1), 51–74.

Roman, D., Darby-Dowman, K. and Mitra, G., Portfolio construction based on stochastic dominance and target return distributions. *Math. Program.*, 2006, **108**(1), 541–569.

Rothschild, M. and Stiglitz, J., Increasing risk I: A definition. *J. Econ. Theory*, 1970, **2**(3), 225–243.

Ruszczyński, A. and Shapiro, A., Optimization of convex risk functions. *Math. Operat. Res.*, 2006, **31**(3), 433–452.

Schied, A. and Follmer, H., Robust preferences and convex measures of risk. In *Advances in Finance Stochast.: Essays in Honour of Dieter Sondermann*, edited by K. Sandmann and P.J. Schönbucher, pp. 39–56, 2002 (Springer: Berlin).

Tasche, D., *Expected shortfall and beyond*, Working Paper. http://arxiv.org/abs/cond-mat/0203558, 2002.

Testuri, C. and Uryasev, S., On relation between expected regret and conditional value-at-risk. In *Handbook of Numerical Methods in Finance*, edited by Z. Rachev, 2003 (Birkhauser: Boston).

van der Vlerk, M.H., *Integrated chance constraints in an ALM model for pension funds*, Working Paper, 2003. Available online at: http://irs.ub.rug.nl/ppn/252290909.

von Neumann, J. and Morgenstern, O., *Theory of Games and Economic Behavior*, 1953rd, 1944 (Princeton University Press: Princeton, NJ).

Young, M.R., A minimax portfolio selection rule with linear programming solution. *Manag. Sci.*, 1998, **44**(5), 673–683.

Zălinescu, C., *Convex Analysis in General Vector Spaces*, 2002 (World Scientific: Singapore).

APPENDIX 13.A

Proof of Theorem 13.1 Convexity, lower semicontinuity, and sublinearity of ϕ in $\mathcal{X}$ imply that the function $\phi_X(\eta) = \eta + \phi(X - \eta)$ is also convex, lsc, and proper in $\eta \in \mathbb{R}$ for each fixed $X \in \mathcal{X}$. For the infimum of $\phi_X(\eta)$ to be achievable at finite η, its recession function has to be positive: $\phi_X 0^+(\pm 1) > 0$, which is equivalent to $\phi_X 0^+(\xi) > 0$, $\xi \neq 0$, due to the positive homogeneity of ϕ. By definition of the recession function (Rockafellar 1970; Zălinescu 2002) and positive homogeneity of ϕ, we have that the last condition holds if $\phi(\xi) > \xi$ for all $\xi \neq 0$:

$$\phi_X 0^+(\xi) = \lim_{\tau \to \infty} \frac{\eta + \tau\xi + \phi(X - \eta - \tau\xi) - \eta - \phi(X - \eta)}{\tau} = \xi + \phi(-\xi).$$

Hence, $\rho(X)$ defined by (1) is a proper lsc function, and minimum in (1) is attained at finite η for all $X \in \mathcal{X}$. Below we verify that $\rho(X)$ satisfies axioms (A1)–(A4).

(A1) Let $X \leq 0$. Then $\phi(X) \leq 0$ as ϕ satisfies (A1) which implies

$$\min_{\eta \in \mathbb{R}} \eta + \phi(X - \eta) \leq 0 + \phi(X - 0) \leq 0.$$

(A2) For any $Z \in \mathcal{X}$ let $\eta_Z \in \arg\min_{\eta \in \mathbb{R}}\{\eta + \phi(Z - \eta)\} \subset \mathbb{R}$ then

$$\begin{aligned}
\rho(X) + \rho(Y) &= \eta_X + \phi(X - \eta_X) + \eta_Y + \phi(Y - \eta_Y)\\
&\geq \eta_X + \eta_Y + \phi(X + Y - \eta_X - \eta_Y)\\
&\geq \eta_{X+Y} + \phi(X + Y - \eta_{X+Y}) = \rho(X + Y).
\end{aligned}$$

(A3) For any fixed $\lambda > 0$ we have

$$\rho(\lambda X) = \min_{\eta \in \mathbb{R}} \{\eta + \phi(\lambda X - \eta)\} = \lambda \min_{\eta \in \mathbb{R}} \{\eta/\lambda + \phi(X - \eta/\lambda)\} = \lambda\rho(X) \qquad (13.A1)$$

(A4) Similarly, for any fixed $a \in \mathbb{R}$,

$$\begin{aligned}
\rho(X + a) &= \min_{\eta \in \mathbb{R}} \{\eta + \phi(X + a - \eta)\}\\
&= a + \min_{\eta \in \mathbb{R}} \{(\eta - a) + \phi(X - (\eta - a))\} = a + \rho(X). \qquad (13.A2)
\end{aligned}$$

Thus, $\rho(X)$ defined by (1) is a proper coherent risk measure. □

Proof of Theorem 13.2 Conditions on function ϕ ensure that the set of optimal solutions of problem (1) is closed and finite, whence follows the existence of $\eta(X)$ in (13.10). Property (A3) is established by noting that for any $\lambda > 0$ equality (13.A1) implies

$$\eta(\lambda X) = \operatorname*{Arg\,min}_{\eta \in \mathbb{R}}\{\eta + \phi(\lambda X - \eta)\} = \operatorname*{Arg\,min}_{\eta \in \mathbb{R}}\{\eta/\lambda + \phi(\lambda X - \eta/\lambda)\},$$

from which follows that $\eta(\lambda X) = \lambda\eta(X)$. Similarly, by virtue of (13.A2), we have

$$\eta(X + a) = \operatorname*{Arg\,min}_{\eta \in \mathbb{R}}\{\eta + \phi(X + a - \eta)\} = \operatorname*{Arg\,min}_{\eta \in \mathbb{R}}\{(\eta - a) + \phi(\lambda X - (\eta - a))\},$$

which leads to the sought relation (A4): $\eta(X + a) = \eta(X) + a$. To validate the remaining statements of the Theorem, consider ϕ to be such that $\phi(X) = 0$ for every $X \leq 0$. Then, (C2) immediately yields $\phi(X) \geq 0$ for all $X \in \mathcal{X}$, which proves

$$\eta(X) \leq \eta(X) + \phi(X - \eta(X)) = \rho(X).$$

By the definition of $\eta(X)$, we have for all $X \leq 0$

$$\eta(X) + \phi(X - \eta(X)) \leq 0 + \phi(X - 0) = 0, \quad \text{or} \quad \eta(X) \leq -\phi(X - \eta(X)). \quad (13.\text{A}3)$$

Assume that $\eta(X) > 0$, which implies $\phi(-\eta(X)) = 0$. From (A2) it follows that $\phi(X - \eta(X)) \leq \phi(X) + \phi(-\eta(X)) = 0$, leading to $\phi(X - \eta(X)) = 0$, and, consequently, to $\eta(X) \leq 0$ by (13.A3). The contradiction furnishes the statement of the theorem. $\square$

Proof of Theorem 13.3 Denote the feasible sets of (13.21a) and (13.21b), respectively, as

$$S_a = \{\mathbf{x} \in C \mid R(\mathbf{x}) \leq c\} \quad \text{and} \quad S_b = \{(\mathbf{x}, \eta) \in C \times \mathbb{R} \mid \Phi(\mathbf{x}, \eta) \leq c\}.$$

Now observe that projection $\Pi_C(S_b)$ of the feasible set of (21b) onto ,

$$\Pi_C(S_b) = \{\mathbf{x} \in C \mid (\mathbf{x}, \eta) \in S_b \text{ for some } \eta \in \mathbb{R}\}$$
$$= \{\mathbf{x} \in C \mid \Phi(\mathbf{x}, \eta) \leq c \text{ for some } \eta \in \mathbb{R}\},$$

coincides with the feasible set of (13.21a):

$$S_a = \Pi_C(S_b).$$

Indeed, $\mathbf{x}' \in S_a$ means that $\mathbf{x}' \in C$ and $R(\mathbf{x}') = \min_\eta \Phi(\mathbf{x}', \eta) \leq c$. By virtue of Theorem 13.1 there exists $\eta' \in \mathbb{R}$ such that $\Phi(\mathbf{x}', \eta') = \min_\eta \Phi(\mathbf{x}', \eta)$, whence $(\mathbf{x}', \eta') \in S_b$, and, consequently, $\mathbf{x}' \in \Pi_C(S_b)$. If, on the other hand, $\mathbf{x}'' \in \Pi_C(S_b)$, then there exists $\eta'' \in \mathbb{R}$ such that $(\mathbf{x}'', \eta'') \in S_b$ and therefore $\phi(\mathbf{x}'', \eta'') \leq c$. By definition of $R(\cdot)$, $R(\mathbf{x}'') \leq \phi(\mathbf{x}'', \eta'') \leq c$, thus $\mathbf{x}'' \in S_a$.

Given (13.37), it is easy to see that (13.21a) and (13.21b) achieve minima at the same values of $\mathbf{x} \in C$ and their optimal objective values coincide. Indeed, if $\mathbf{x}^*$ is an optimal solution of (13.21a) then $\mathbf{x}^* \in S_a$ and $g(\mathbf{x}^*) \leq g(\mathbf{x})$ holds for all $\mathbf{x} \in S_a$. By (13.37), if $\mathbf{x} \in S_a$ then there exists some $\eta \in \mathbb{R}$ such $(\mathbf{x}, \eta) \in S_b$. Thus, for all $(\mathbf{x}, \eta) \in S_b$ one has $g(\mathbf{x}^*) \leq g(\mathbf{x})$, meaning that $(\mathbf{x}^*, \eta^*)$ is an optimal solution of (13.21b), where $\eta^* \in \mathbb{R}$ is such that $(\mathbf{x}^*, \eta^*) \in S_b$. Inversely, if $(\mathbf{x}^*, \eta^*)$ solves (13.21b), then $(\mathbf{x}^*, \eta^*) \in S_b$ and $g(\mathbf{x}^*) \leq g(\mathbf{x})$ for all $(\mathbf{x}, \eta) \in S_b$. According to (13.A4), $(\mathbf{x}, \eta) \in S_b$ also yields $\mathbf{x} \in S_a$, hence for all $\mathbf{x} \in S_a$ one has $g(\mathbf{x}^*) \leq g(\mathbf{x})$, i.e. $\mathbf{x}^*$ is an optimal solution of (13.21a).

Finally, assume that the risk constraint in (13.21a) is binding at optimality. If $(\mathbf{x}^*, \eta^*)$ achieves the minimum of (13.21b), then $\Phi(\mathbf{x}^*, \eta^*) \leq c$ and, according to the above, $\mathbf{x}^*$ is an optimal solution of (13.21a), whence $c = R(\mathbf{x}^*) \leq \Phi(\mathbf{x}^*, \eta^*) \leq c$. From the last relation we have $\Phi(\mathbf{x}^*, \eta^*) = R(\mathbf{x}^*)$ and thus $\eta^* \in \text{argmin}_\eta \Phi(\mathbf{x}^*, \eta)$. Now consider $\mathbf{x}^*$ that solves (13.21a) and η^* such that $\eta^* \in \arg\min_\eta \Phi(x^*, \eta)$. This implies that $\Phi(x^*, \eta^*) = R(x^*) = c$, or $(x^*, \eta^*) \in S_b$. Taking into account that $g(x^*) \leq g(x)$ for all $x \in S_a$ and consequently for all $(x, \eta) \in S_b$, one has that (x^*, η^*) is an optimal solution of (13.21b).

Proof of Theorem 13.4 Since formula (13.28) differs from (13.1) by the constant summand $(-EX)$, we only have to verify that $\mathcal{R}(X) = \inf_\eta \{\eta + \phi(X - \eta)\}$ satisfies (A5). As $\phi(X) > EX$ for all $X \neq 0$, we have that $\phi(X - \eta_X) > E(X - \eta_X)$ for all non-constant $X \in \mathcal{X}$, where $\eta_X \in \arg\min_\eta \eta + \phi(X - \eta)$. From the last inequality it follows that $\eta_X + \phi(X - \eta_X) > EX$, or $\mathcal{R}(X) > EX$ for all non-constant $X \in \mathcal{X}$. Thus, $\mathcal{D}(X) > 0$ for all non-constant X. For $a \in \mathbb{R}$, $\inf_\eta \{\eta + \phi(a - \eta)\} = a$, whence $\mathcal{D}(a) = 0$. $\square$

Proof of Theorem 13.5 The proof of existence and all properties except (A1′) is identical to that of Theorem 13.1. Property (A1′) follows elementarily: if $(-X) \succeq_{SSD} (-Y)$, then $(-X + c) \succeq_{SSD} (-Y + c)$, and consequently, $\phi(X - c) \leq \phi(Y - c)$ for $c \in \mathbb{R}$, whence

$$\rho(X) = \eta_X + \phi(X - \eta_X) \leq \eta_Y + \phi(X - \eta_Y) \leq \eta_Y + \phi(Y - \eta_Y) = \rho(Y), \quad (13.A4)$$

where, as usual, $\eta_Z \in \arg\min_\eta \{\eta + \phi(Z - \eta)\} \subset \mathbb{R}$, for any $Z \in \mathcal{X}$. $\square$

Example 13.2.3 (Additional details): To demonstrate the monotonicity of $\eta_{p,\alpha}(X)$ with respect to $\alpha \in (0, 1)$, observe that by definition of $\eta_{p,\alpha_1}(X)$

$$\eta_{p,\alpha_1}(X) + (1 - \alpha_1)^{-1} \left\| \left(X - \eta_{p,\alpha_1}\right)^+ \right\|_p \leq \eta_{p,\alpha_2}(X) + (1 - \alpha_1)^{-1} \left\| \left(X - \eta_{p,\alpha_2}\right)^+ \right\|_p. \quad (13.A5)$$

Now, assume that $\eta_{p,\alpha_1}(X) > \eta_{p,\alpha_2}(X)$ for some $\alpha_1 < \alpha_2$, then (13.A5) yields

$$0 < \eta_{p,\alpha_1}(X) - \eta_{p,\alpha_2}(X) \leq (1 - \alpha_1)^{-1} \times \left(\left\| \left(X - \eta_{p,\alpha_2}\right)^+ \right\|_p - \left\| \left(X - \eta_{p,\alpha_1}\right)^+ \right\|_p \right)$$

$$< (1 - \alpha_2)^{-1} \left(\left\| \left(X - \eta_{p,\alpha_2}\right)^+ \right\|_p - \left\| \left(X - \eta_{p,\alpha_1}\right)^+ \right\|_p \right).$$

From the last inequality it follows directly that

$$\eta_{p,\alpha_1}(X) + (1 - \alpha_2)^{-1} \left\| \left(X - \eta_{p,\alpha_1}\right)^+ \right\|_p < \eta_{p,\alpha_2}(X) + (1 - \alpha_2)^{-1} \left\| \left(X - \eta_{p,\alpha_2}\right)^+ \right\|_p,$$

which contradicts the definition of $\eta_{p,\alpha_2}(X)$.

The limiting behaviour of $\eta_{p,\alpha}(X)$ can be verified by noting first that for $1 \leq p < \infty$

$$\lim_{\alpha \to 1} \text{HMCR}_{p,\alpha}(X) = \text{ess.sup } X. \quad (13.A6)$$

Indeed, using the notation of Example 13.1.3 one has

$$\lim_{\alpha \to 1} \inf_{\eta} \left\{ \eta + (1 - \alpha)^{-1} \left\| (X - \eta)^+ \right\|_p \right\} \leq \inf_{\eta} \lim_{\alpha \to 1} \left\{ \eta + (1 - \alpha)^{-1} \left\| (X - \eta)^+ \right\|_p \right\}$$
$$= \inf_{\eta} \ \eta + \phi^*(X - \eta) = \text{ess.sup } X.$$

On the other hand, from the inequality (see Example 13.1.4)

$$\text{HMCR}_{p,\alpha}(X) \leq \text{HMCR}_{q,\alpha}(X) \quad \text{for} \quad 1 \leq p < q,$$

and the fact that $\lim_{\alpha \to 1} \text{CVaR}_\alpha(X) = \text{ess.sup } X$ (see, e.g. Rockafellar *et al.* 2006) we obtain

$$\text{ess.sup } X = \lim_{\alpha \to 1} \text{CVaR}_\alpha(X) \leq \lim_{\alpha \to 1} \text{HMCR}_{p,\alpha}(X),$$

which verifies (13.A6). The existence of $\lim_{\alpha \to 1} \eta_{p,\alpha}(X) \in \overline{\mathbb{R}}$ follows from the monotonicity of $\eta_{p,\alpha}(X)$ with respect to α. Theorem 13.2 maintains that $\eta_{p,\alpha}(X) \leq \text{HMCR}_{p,\alpha}(X)$, whence

$$\lim_{\alpha \to 1} \eta_{p,\alpha}(X) \leq \text{ess.sup } X.$$

In the case of finite ess.sup X, by rewriting (13.A6) in the form

$$\lim_{\alpha \to 1} \left\{ \eta_{p,\alpha}(X) + (1 - \alpha)^{-1} \left\| (X - \eta_{p,\alpha}(X))^+ \right\|_p \right\} = \text{ess.sup } X,$$

and assuming that $\lim_{\alpha \to 1} \eta_{p,\alpha}(X) = \text{ess.sup} X - \varepsilon$ for some $\varepsilon \geq 0$, it is easy to see that the above equality holds only in the case of $\varepsilon = 0$.

CHAPTER 14

On the Feasibility of Portfolio Optimization under Expected Shortfall

STEFANO CILIBERTI, IMRE KONDOR and MARC MÉZARD

CONTENTS

14.1   INTRODUCTION

A MONG THE SEVERAL EXISTING RISK MEASURES in the context of portfolio optimization, expected shortfall (ES) has certainly gained increasing popularity in recent years. In several practical applications, ES is starting to replace the classical Value-at-Risk (VaR). There are a number of reasons for this. For a given threshold probability β, the VaR is defined so that with probability β the loss will be smaller than VaR. This definition only gives the minimum loss one can reasonably expect but does not tell anything about the typical value of that loss which can be measured by the *conditional* value-at-risk (CVaR, which is the same as ES for the continuous distributions that we consider here).[1] We will be more precise with these definitions below. The point we want to stress here is that the VaR measure, lacking the mandatory properties of subadditivity and convexity, is not

---

[1] See Acerbi and Tasche (2002) for the subtleties related to a discrete distribution.

coherent (Artzner *et al.* 1999). This means that summing the VaRs of individual portfolios will not necessarily produce an upper bound for the VaR of the combined portfolio, thus contradicting the holy principle of diversification in finance. A nice practical example of the inconsistency of VaR in credit portfolio management is reported by Frey and McNeil (2002). On the other hand, it has been shown (Acerbi and Tasche 2002) that ES is a coherent measure with interesting properties (Pflug 2000). Moreover, the optimization of ES can be reduced to linear programming (Rockafellar and Uryasev 2000) (which allows for a fast implementation) and leads to a good estimate for the VaR as a byproduct of the minimization process. To summarize, the intuitive and simple character, together with the mathematical properties (coherence) and the fast algorithmic implementation (linear programming), are the main reasons behind the growing importance of ES as a risk measure.

In this chapter we focus on the feasibility of the portfolio optimization problem under the ES measure of risk. The control parameters of this problem are (i) the imposed threshold in probability, β and (ii) the ratio N/T between the number N of financial assets making up the portfolio and the time series length T used to sample the probability distribution of returns. It is curious, albeit trivial, that the scaling in N/T had not been explicitly pointed out before (Pafka and Kondor 2002). It was reported by Kondor *et al.* (2007) that, for certain values of these parameters, the optimization problem does not have a finite solution because, even if convex, it is not bounded from below. Extended numerical simulations allowed these authors to determine the feasibility map of the problem. Here, in order to better understand the root of the problem and to study the transition from a feasible to an unfeasible regime (corresponding to an ill-posed minimization problem) we address the same problem from an analytical point of view.

The chapter is organized as follows. In Section 14.2 we briefly recall the basic definitions of β-VaR and β-CVaR and we show how the portfolio optimization problem can be reduced to linear programming. We introduce a 'cost function' to be minimized under linear constraints and we discuss the rationale for a statistical mechanics approach. In Section 14.3 we solve the problem of optimizing large portfolios under ES using the replica approach. Our results and a comparison with numerics are reported in Section 14.4, and our conclusions are summarized in Section 14.5.

14.2   THE OPTIMIZATION PROBLEM

We consider a portfolio of N financial instruments $\mathbf{w} = \{w_1, \ldots, w_N\}$ where w_i is the position of asset i. The global budget constraint fixes the sum of these numbers: we impose, for example,

$$\sum_{i=1}^{N} w_i = N. \tag{14.1}$$

We do not stipulate any constraint on short selling, so that w_i can be any negative or positive number. This is, of course, unrealistic for liquidity reasons, but considering this case allows us to demonstrate the essence of the phenomenon. If we imposed a constraint that would render the domain of w_i bounded (such as a ban on short selling, for example), this would evidently prevent the weights from diverging, but a vestige of the transition would still remain in the form of large, although finite, fluctuations of the weights, and in a large number of them sticking to the 'walls' of the domain.

We denote the returns on the assets by $\mathbf{x} = \{x_1, x_2, \ldots, x_N\}$ and we assume that there exists an underlying probability distribution function $p(\mathbf{x})$ of the returns. The loss of portfolio $\mathbf{w}$ given the returns $\mathbf{x}$ is $\ell(\mathbf{w} \mid \mathbf{x}) = -\sum_{i=1}^{N} w_i x_i$, and the probability of that loss being smaller than a given threshold α is

$$\mathcal{P}_<(\mathbf{w}, \alpha) = \int d\mathbf{x}\, p(\mathbf{x})\theta(\alpha - \ell(\mathbf{w} \mid \mathbf{x})), \tag{14.2}$$

where $\theta(\cdot)$ is the Heaviside step function, equal to 1 if its argument is positive and 0 otherwise. The β-VaR of this portfolio is formally defined by

$$\beta\text{-VaR}(\mathbf{w}) = \min\{\alpha : \mathcal{P}_<(\mathbf{w}, \alpha) \geq \beta\} \tag{14.3}$$

(see Figure 14.1), While the CVaR (or ES, in this case) associated with the same portfolio is the average loss on the tail of the distribution,

$$
\begin{aligned}
\beta\text{-CVaR}(\mathbf{w}) &= \frac{\int d\mathbf{x}\, p(\mathbf{x})\ell(\mathbf{w}|\mathbf{x})\theta(\ell(\mathbf{w}|\mathbf{x}) - \beta\text{-VaR}(\mathbf{w}))}{\int d\mathbf{x}\, p(\mathbf{x})\theta(\ell(\mathbf{w}|\mathbf{x}) - \beta\text{-VaR}(\mathbf{w}))} \\
&= \frac{1}{1 - \beta} \int d\mathbf{x}\, p(\mathbf{x})\ell(\mathbf{w}|\mathbf{x})\theta(\ell(\mathbf{w}|\mathbf{x}) - \beta\text{-VaR}(\mathbf{w})). \tag{14.4}
\end{aligned}
$$

The threshold β then represents a confidence level. In practice, the typical values of β which one considers are $\beta = 0.90, 0.95$, and 0.99, but we will address the problem for any

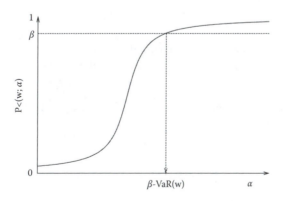

FIGURE 14.1 Schematic representation of the VaR measure of risk. $\mathcal{P}_<(\mathbf{w})$ is the probability of a loss associated with the portfolio $\mathbf{w}$ being smaller than α. The conditional VaR β-CVaR (or ES) is the average loss when this is constrained to be greater than the β-VaR.

$\beta \in [0, 1]$. What is usually called 'exceedance probability' in previous literature would correspond here to $(1 - \beta)$.

As mentioned in the Introduction, the ES measure can be obtained from a variational principle (Rockafellar and Uryasev 2000). The minimization of a properly chosen objective function leads directly to (14.4)

$$\beta\text{-CVaR}(\mathbf{w}) = \min_v F_\beta(\mathbf{w}, v), \qquad (14.5)$$

$$F_\beta(\mathbf{w}, v) \equiv v + (1 - \beta)^{-1} \int d\mathbf{x} \, p(\mathbf{x})[\ell(\mathbf{w}|\mathbf{x}) - v]^+. \qquad (14.6)$$

Here, $[a]^+ \equiv (a + |a|)/2$. The external parameter v over which one has to minimize is claimed to be relevant in itself (Rockafellar and Uryasev 2000), since its optimal value may represent a good estimate for the actual value-at-risk of the portfolio. We will come back to this point when we discuss our results. We stress here that minimizing (14.6) over $\mathbf{w}$ *and* v is equivalent to minimizing (14.4) over the portfolio vectors $\mathbf{w}$.

Of course, in practical cases the probability distribution of the loss is not known and must be inferred from past data. In other words, we need an 'in-sample' estimate of the integral in (14.6) which would turn a well-posed (but useless) optimization problem into a practical approach. We thus approximate the integral by sampling the probability distributions of returns. If we have a time series $\mathbf{x}^{(1)}, \ldots, \mathbf{x}^{(T)}$, our objective function simply becomes

$$\hat{F}_\beta(\mathbf{w}, v) = v + \frac{1}{(1 - \beta)T} \sum_{\tau=1}^{T} [\ell(\mathbf{w} \mid \mathbf{x}^{(\tau)}) - v]^+$$

$$= v + \frac{1}{(1 - \beta)T} \sum_{\tau=1}^{T} \left[-v - \sum_{i=1}^{N} w_i x_{i\tau} \right]^+, \qquad (14.7)$$

where we denote by $x_{i\tau}$ the return of asset i at time τ.

Minimizing this risk measure is the same as the following linear programming problem:

- given one data sample, i.e. a matrix $x_{i\tau}$, $i = 1, \ldots, N, \tau = 1, \ldots, T$,
- minimize the *cost function*

$$E_\beta[\mathbf{Y}; \{x_{i\tau}\}] = E_\beta[v, \{w_i\}, \{u_\tau\}; \{x_{i\tau}\}] = (1 - \beta)Tv + \sum_{t=\tau}^{T} u_\tau, \qquad (14.8)$$

- over the $(N + T + 1)$ variables $\mathbf{Y} \equiv \{w_1, \ldots, w_N, u_1, \ldots, u_T, v\}$,

- under the $(2T + 1)$ constraints

$$u_\tau \geq 0, \quad u_\tau + v + \sum_{i=1}^{N} x_{i\tau} w_i \geq 0, \quad \forall \tau \quad \text{and} \quad \sum_{i=1}^{N} w_i = N. \qquad (14.9)$$

Since we allow short positions, not all the w_i are positive, which makes this problem different from standard linear programming. To keep the problem tractable, we impose the condition that $w_i \geq -W$, where W is a very large cutoff, and the optimization problem will be said to be ill-defined if its solution does not converge to a finite limit when $W \to \infty$. It is now clear why constraining all the w_i to be non-negative would eliminate the feasibility problem: a finite solution will always exist because the weights are by definition bounded, the worst case being an optimal portfolio with only one non-zero weight taking care of the total budget. The control parameters that govern the problem are the threshold β and the ratio N/T of assets to data points. The resulting 'phase diagram' is then a line in the $\beta - N/T$ plane separating a region in which, with high probability, the minimization problem is not bounded and thus does not admit a finite solution, and another region in which a finite solution exists with high probability. These statements are non-deterministic because of the intrinsic probabilistic nature of the returns. We will address this minimization problem in the non-trivial limit where $T \to \infty$, $N \to \infty$, while N/T stays finite. In this 'thermodynamic' limit, we shall assume that extensive quantities (like the average loss of the optimal portfolio, i.e. the minimum cost function) do not fluctuate, namely that their probability distribution is concentrated around the mean value. This 'self-averaging' property has been proved for a wide range of similar statistical mechanics models (Guerra and Toninelli 2004). Then, we will be interested in the average value of the minimum of the cost function (14.8) over the distribution of returns. Given the similarity of portfolio optimization to the statistical physics of disordered systems, this problem can be addressed analytically by means of a replica approach (Mézard et al. 1987).

14.3   THE REPLICA APPROACH

For a given history of returns x_{it}, one can compute the minimum of the cost function, $\min_Y E_\beta[Y; \{x_{it}\}]$. In this section we show how to compute analytically the expectation value of this quantity over the histories of returns. For simplicity we shall keep to the case in which the x_{it} are independent identically distributed (iid) normal variables, so that a history of returns x_{it} is drawn from the distribution

$$p(\{x_{it}\}) \sim \prod_{it} e^{-Nx_{it}^2/2}. \qquad (14.10)$$

This assumption of an iid normal distribution of returns is very restrictive, but we would like to emphasize that the method that we use can be generalized easily to iid variables with other distributions, and also in some cases to correlated variables. Certainly, the precise location of the critical value of N/T separating an infeasible from a feasible phase depends on the distribution of returns. But we expect that broad features like the existence of this critical value, or the way the fluctuations in the portfolio diverge when approaching the transition, should not depend on this distribution. This property, called universality, has been one of the major discoveries of statistical mechanics in the last 50 years.

Instead of focusing only on the minimal cost, the statistical mechanics approach makes a detour: it considers, for a given history of returns x_{it}, a probability distribution in the space of variables $\mathbf{Y}$, defined by $P_\gamma(\mathbf{Y}) = 1/Z_\gamma[\{x_{it}\}]\exp[-\gamma E_\beta[\mathbf{Y};\{x_{it}\}]]$. The parameter γ is an auxiliary parameter. In physics it is the inverse of the temperature, however in the present case it is just a parameter that we introduce in order to have a probability distribution on $\mathbf{Y}$ that interpolates between the uniform probability ($\gamma = 0$) and a probability that is peaked on the value of $\mathbf{Y}$ which minimizes the cost $E_\beta[\mathbf{Y};\{x_{it}\}]$ (the case where $\gamma = \infty$).

The normalization constant $Z_\gamma[\{x_{it}\}]$ is called the partition function at inverse temperature γ it is defined as

$$Z_\gamma[\{x_{it}\}] = \int_V d\mathbf{Y}\exp[-\gamma E_\beta[\mathbf{Y};\{x_{it}\}]],\qquad(14.11)$$

where V is the convex polytope defined by (14.9).

The partition function contains a lot of information on the problem. For instance, the minimal cost can be expressed as $\lim_{\gamma\to\infty}(-1)/(N\gamma)\log Z_\gamma[\{x_{it}\}]$. We shall be interested in computing the large N limit of the minimal cost per variable:

$$\varepsilon[\{x_{it}\}] = \lim_{N\to\infty}\frac{\min E[\{x_{it}\}]}{N} = \lim_{N\to\infty}\lim_{\gamma\to\infty}\frac{-1}{N\gamma}\log Z_\gamma[\{x_{it}\}].\qquad(14.12)$$

In the following we will compute the average value of this quantity over the choice of the sample x_{it}. Using Equation (14.12) we can compute this average minimum cost if we are able to compute the average of the *logarithm* of Z. This is a difficult problem that is usually circumvented by means of the so-called 'replica trick:' one computes the average of Z^n, where n is an integer, and then the average of the logarithm is obtained from

$$\overline{\log Z} = \lim_{n\to\infty}\frac{\partial\overline{Z^n}}{\partial n},\qquad(14.13)$$

thus assuming that Z^n can be analytically continued to real values of n. The overline indicates an average over different samples, i.e. over the probability distribution (14.10). This technique has a long history in the physics of spin glasses (Mézard *et al.* 1987): the proof that it leads to the correct solution has recently been reported (Talagrand 2002).

The partiton function (14.11) can be written more explicity as

$$Z_\gamma[\{x_{it}\}] = \int_{-\infty}^{+\infty} dv \int_{0}^{+\infty} \prod_{t=1}^{T} du_t \int_{-\infty}^{+\infty} \prod_{i=1}^{N} dw_i$$

$$\times \int_{-i\infty}^{+i\infty} d\lambda \exp\left[\lambda\left(\sum_{i=1}^{N} w_i - N\right)\right] \int_{0}^{+\infty} \prod_{t=1}^{T} d\mu_t$$

$$\times \int_{-i\infty}^{+i\infty} \prod_{t=1}^{T} d\hat{\mu}_t \exp\left[\sum_{t=1}^{T} \hat{\mu}_t\left(u_t + v + \sum_{i=1}^{N} x_{it} w_i - \mu_t\right)\right]$$

$$\times \exp\left[-\gamma(1-\beta)Tv - \gamma\sum_{t=1}^{T} u_t\right], \tag{14.14}$$

where the constraints are imposed by means of the Lagrange multipliers λ, μ, $\hat{\mu}$. The replica trick is based on the idea that the nth power of the partition function appearing in (14.13) can be written as the partition function for n independent replicas $\mathbf{Y}^1, \dots, \mathbf{Y}^n$ of the system: all the replicas correspond to the *same* history of returns $\{x_{it}\}$, and their joint probability distribution function is $P_\gamma(\mathbf{Y}^1, \dots, \mathbf{Y}^n) = 1/Z_\gamma^n[\{x_{it}\}]$ $\exp[-\gamma \sum_{a=1}^{n} E_\beta[\mathbf{Y}^a; \{x_{it}\}]]$. It is not difficult to write down the expression for Z^n and average it over the distribution of samples x_{it}. One introduces the *overlap* matrix

$$Q^{ab} = \frac{1}{N} \sum_{i=1}^{N} w_i^a w_i^b, \quad a, b = 1, \dots, n, \tag{14.15}$$

as well as its conjugate $\hat{Q}^{ab}$ (the Lagrange multiplier imposing (14.15)), where a and b are replica indexes. This matrix characterizes how the portfolios in different replicas differ: they provide some indication of how the measure $P\gamma(\mathbf{Y})$ is spread. After (several) Gaussian integrations, one obtains

$$Z_\gamma^n[\{x_{it}\}] \sim \int_{-\infty}^{+\infty} \prod_{a=1}^{n} dv^a \int_{-\infty}^{+\infty} \prod_{a,b} dQ^{ab} \int_{-i\infty}^{+i\infty} \prod_{a,b} d\hat{Q}^{ab}$$

$$\times \exp\left\{ N \sum_{a,b} Q^{ab} \hat{Q}^{ab} - N \sum_{a,b} \hat{Q}^{ab} - \gamma(1-\beta)T \sum_a v^a \right.$$

$$- Tn \log \gamma + T \log \hat{Z}_\gamma(\{v^a\}, \{Q^{ab}\})$$

$$\left. - \frac{T}{2} \mathrm{Tr} \log Q - \frac{N}{2} \mathrm{Tr} \log \hat{Q} - \frac{nN}{2} \log 2 \right\}, \tag{14.16}$$

where

$$\hat{Z}_\gamma(\{v^a\}, \{Q^{ab}\}) \equiv \int_{-\infty}^{+\infty} \prod_{a=1}^{n} dy^a \exp\left[-\frac{1}{2} \sum_{a,b=1}^{n} (Q^{-1})^{ab}(y^a - v^a) \right.$$

$$\left. \times (y^b - v^b) + \gamma \sum_{a=1}^{n} y^a \theta(-y^a) \right]. \tag{14.17}$$

We now write $T = tN$ and work with fixed t while $N \to \infty$.

The most natural solution is obtained by realizing that all the replicas are identical. Given the linear character of the problem, the symmetric solution should be the correct one. The replica-symmetric solution corresponds to the *ansatz*

$$Q^{ab} = \begin{cases} q_1, & \text{if } a = b, \\ q_0, & \text{if } a \neq b, \end{cases} \qquad \hat{Q}^{ab} = \begin{cases} \hat{q}_1, & \text{if } a = b, \\ \hat{q}_0, & \text{if } a \neq b, \end{cases} \tag{14.18}$$

and $v^a = v$ for any a. As we discuss in detail in appendix 14.A, one can show that the optimal cost function, computed from Equation (14.12), is the minimum of

$$\varepsilon(v, q_0, \Delta) = \frac{1}{2\Delta} + \Delta\left[t(1 - \beta)v - \frac{q_0}{2} + \frac{t}{2\sqrt{\pi}} \int_{-\infty}^{+\infty} ds\, e^{-s^2} g\left(v + s\sqrt{2q_0}\right) \right], \tag{14.19}$$

where $\Delta \equiv \lim_{\gamma \to \infty} \gamma\Delta q$ and the function $g(\cdot)$ is defined as

$$g(x) = \begin{cases} 0, & x \geq 0, \\ x^2, & -1 \leq x < 0, \\ -2x - 1, & x < -1. \end{cases} \tag{14.20}$$

Note that this function and its derivative are continuous. Moreover, v and q_0 in (14.19) are solutions of the saddle point equations

$$1 - \beta + \frac{1}{2\sqrt{\pi}} \int ds\, e^{-s^2} g'(v + s\sqrt{2q_0}) = 0, \tag{14.21}$$

$$-1 + \frac{t}{\sqrt{2\pi q_0}} \int ds\, e^{-s^2} sg'(v + s\sqrt{2q_0}) = 0. \tag{14.22}$$

We require that the minimum of (14.19) occurs at a finite value of Δ. In order to understand this point, we recall the meaning of Δ (see also (14.18)):

$$\Delta/\gamma \sim \Delta q = (q_1 - q_0) = \frac{1}{N} \sum_{i=1}^{N} \left(w_i^{(1)}\right)^2 - \frac{1}{N} \sum_{i=1}^{N} w_i^{(1)} w_i^{(2)} \sim \overline{w^2} - \overline{w}^2, \tag{14.23}$$

where the superscripts (1) and (2) represent two generic replicas of the system. We then find that Δ is proportional to the fluctuations in the distribution of the w's. An infinite value of Δ would then correspond to a portfolio which is infinitely short on some particular positions and, because of the global budget constraint (1), infinitely long on others.

Given (14.19), the existence of a solution at finite Δ translates into the following condition:

$$t(1 - \beta)v - \frac{q_0}{2} + \frac{t}{2\sqrt{\pi}} \int_{-\infty}^{+\infty} ds\, e^{-s^2} g(v + s\sqrt{2q_0}) \geq 0, \tag{14.24}$$

which defines, along with Equations (14.21) and (14.22), our phase diagram.

14.4 THE FEASIBLE AND INFEASIBLE REGIONS

We can now chart the feasibility map of the expected shortfall problem. Following the notation of Kondor et al. (2007), we will use as control parameters $N/T \equiv 1/t$ and β. The limiting case $\beta \to 1$ can be determined analytically and one can show that the critical value t^* is given by

$$\frac{1}{t^*} = \frac{1}{2} - \mathcal{O}\left[(1 - \beta)^3 e^{-(4\pi(1-\beta)^2)^{-1}}\right]. \tag{14.25}$$

This limit corresponds to the over-pessimistic case of maximal loss, in which the single worst loss contributes to the risk measure. The optimization problem is the following:

$$\min_{\mathbf{w}}\left[\max_{t}\left(-\sum_i w_i x_{it}\right)\right]. \tag{14.26}$$

A simple 'geometric' argument of Kondor et al. (2007) leads to the critical value $1/t^* = 0.5$ in this extreme case. The idea is the following. According to Equation (14.26), one has to look for the minimum of a polytope made by a large number of planes, whose normal vectors (the x_{it}) are drawn from a symmetric distribution. The simplex is convex, but with some probability it can be unbounded from below and then the optimization problem is ill-defined. Increasing T means that the probability of this event decreases, because there are more planes and thus it is more likely that for large values of w_i the max over t has a positive slope in the ith direction. The exact law for this probability can be obtained by induction on N and T (Kondor et al. 2007) and, as we said, it jumps in the thermodynamic limit from 1 to 0 at $N/T = 0.5$. The example of the max-loss risk measure is also helpful because it allows us to stress two aspects of the problem: (1) even for finite N and T there is a finite chance that the risk measure is unbounded from below in some samples, and (2) the phase transition occurs in the thermodynamic limit when N/T is strictly smaller than 1, i.e. much before the covariance matrix develops zero modes. The very nature of the problem is that the risk measure there is simply not bounded from below. As for the ES risk measure, the threshold value $N/T = 0.5$ can be thought of as a good approximation of the actual value for many cases of practical interest (i.e. $\beta \gtrsim 0.9$), since the corrections to this limit case are exponentially small (Equation (14.25)).

For finite values of β we solve numerically Equations (14.21), (14.22) and (14.24) using the following procedure. We first solve the two Equations (14.21) and (14.22), which always admit a solution for (v, q_0). We then plot the l.h.s. of Equation (14.24) as a function of $1/t$ for a fixed value of β. This function is positive at small $1/t$ and becomes negative beyond a threshold $1/t^*$. By keeping track of $1/t^*$ (obtained numerically via linear interpolations) for each value of β we build up the phase diagram (Figure 14.2, left). This diagram is in agreement with the numerical results obtained by Kondor et al. (2007). We show in the right panel of Figure 14.2 the divergence of the order parameter Δ versus $1/t - 1/t^*$. The critical exponent is found to be $1/2$:

$$
\Delta \sim \left(\frac{1}{t} - \frac{1}{t^*(\beta)} \right)^{-1/2},
\tag{14.27}
$$

again in agreement with the scaling found by Kondor et al. (2007). We performed extensive numerical simulations in order to check the validity of our analytical findings. For a given realization of the time series, we solve the optimization problem (14.8) by standard linear programming (Press et al. 1992). We impose a large negative cutoff for the w's, that is $w_i > -W$, and we say that a feasible solution exists if it stays finite for $W \to \infty$. We then repeat the procedure for a certain number of samples, and then average our final results (optimal cost, optimal v, and the variance of the w's in the optimal portfolio) over those that produced a finite solution. In Figure 14.3 we show how the probability of finding a finite solution depends on the size of the problem. Here, the probability is simply defined in terms of the frequency. We see that the convergence towards the expected $1-0$ law is fairly slow, and a finite size scaling analysis is shown in

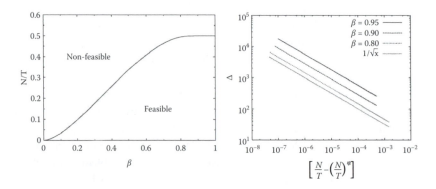

FIGURE 14.2 Left: phase diagram of the feasibility problem for the expected shortfall. Right: the order parameter Δ diverges with exponent $1/2$ as the transition line is approached. A curve of slope $-1/2$ is also shown for comparison.

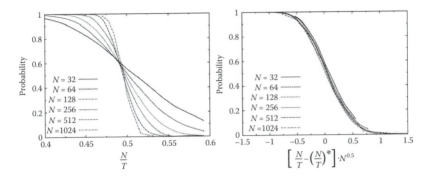

FIGURE 14.3 Left: the probability of finding a finite solution as obtained from linear programming at increasing values of N and with $\beta = 0.8$. Right: scaling plot of the same data. The critical value is set equal to the analytical value, $N/T = 0.4945$, and the critical exponent is 1/2, i.e. that obtained by Kondor *et al.* (2007) for the limit case $\beta \to 1$. The data do not collapse perfectly, and better results can be obtained by slightly changing either the critical value or the exponent.

the right panel. Without loss of generality, we can summarize the finite-N numerical results by writing the probability of finding a finite solution as

$$p(N, T, \beta) = f\left[\left(\frac{1}{t} - \frac{1}{t^*(\beta)}\right) \cdot N^{\alpha(\beta)}\right], \tag{14.28}$$

where $f(x) \to 1$ if $x \gg 1$ and $f(x) \to 0$ if $x \ll 1$, and where $\alpha(1) = 1/2$. It is interesting to note that these results do not depend on the initial conditions of the algorithm used to solve the problem: for a given sample, the algorithm finds, in linear time, the minimum of the polytope by looking at all its vertexes exhaustively. The statistics are taken by repeating such a deterministic procedure on a large number of samples chosen at random.

In Figure 14.4 (left panel) we plot, for a given value of β, the optimal cost found numerically for several values of the size N compared with the analytical prediction at infinite N. One can show that the cost vanishes as $\Delta^{-1} \sim (1/t - 1/t^*)^{1/2}$. The right panel

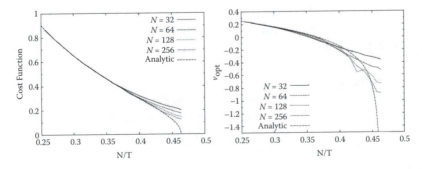

FIGURE 14.4 Numerical results from linear programming and comparison with analytical predictions at large N. Left: the minimum cost of the optimization problem versus N/T at increasing values of N. The thick line is the analytical solution (14.19). Here, $\beta = 0.7$, $(N/T)^* \simeq 0.463$. Right: the optimal value of v as found numerically for several values of N compared with the analytical solution.

of the same figure shows the behaviour of the value of v which leads to the optimal cost versus N/T for the same fixed value of β. Also in this case, the analytical $(N \to \infty)$ limit is plotted for comparison. We note that this quantity has been suggested (Rockafellar and Uryasev 2000) to be a good approximation of the VaR of the optimal portfolio: we find here that v_{opt} diverges at the critical threshold and becomes negative at an even smaller value of N/T.

14.5   CONCLUSIONS

We have shown that the problem of optimizing a portfolio under the expected shortfall measure of risk using empirical distributions of returns is not well-defined when the ratio N/T of assets to data points is larger than a certain critical value. This value depends on the threshold β of the risk measure in a continuous way and this defines a phase diagram. The smaller the value of β the longer the length of the time series needed for portfolio optimization. The analytical approach we have discussed in this chapter allows us to have a clear understanding of this phase transition. The mathematical reason for the non-feasibility of the optimization problem is that, with a certain probability $p(N, T, \beta)$, the linear constraints in (14.9) define a simplex which is not bounded from below, thus leading to a solution which is not finite ($\Delta q \to \infty$ in our language), in the same way as occurs in the extreme case $\beta \to 1$ discussed by Kondor *et al.* (2007). From a more physical point of view, it is reasonable that the feasibility of the problem depends on the number of data points we take from the time series with respect to the number of financial instruments of our portfolio. The probabilistic character of the time series is reflected in the probability $p(N, T, \beta)$. Interestingly, this probability becomes a threshold function at large N if $N/T \equiv 1/t$ is finite, and its general form is given by (14.28).

These results have practical relevance in portfolio optimization. The order parameter discussed in this chapter is tightly related to the relative estimation error Kondor *et al.* (2007). The fact that this order parameter has been found to diverge means that, in some regions of parameter space, the estimation error blows up, which makes the task of portfolio optimization completely meaningless. The divergence of the estimation error is not limited to the case of expected shortfall: as shown by Kondor *et al.* (2007), it also occurs in the case of variance and absolute deviation, but the noise sensitivity of the expected shortfall turns out to be even greater than that of these more conventional risk measures.

There is nothing surprising about the fact that if there are insufficient data, the estimation error is large and we cannot make a good decision. What is surprising is the fact that there is a sharply defined threshold where the estimation error actually diverges.

For a given portfolio size, it is important to know that a minimum number of data points is required in order to perform an optimization based on empirical distributions. We also note that the divergence of the parameter Δ at the phase transition, which is directly related to the fluctuations of the optimal portfolio, may play a dramatic role in practical cases. To stress this point, we can define a sort of 'susceptibility' with respect to the data,

$$\chi_{ij}^t = \frac{\partial \langle w_j \rangle}{\partial x_{it}}, \tag{14.29}$$

and one can show that this quantity diverges at the critical point, since $\chi_{ij} \sim \Delta$. A small change (or uncertainty) in x_{it} becomes increasingly relevant as the transition is approached, and the portfolio optimization could then be very unstable even in the feasible region of the phase diagram. We stress that the susceptibility we have introduced might be considered as a measure of the effect of noise on portfolio selection and is very reminiscent of the measure proposed by Pafka and Kondor (2002).

In order to present a clean, analytic picture, we have made several simplifying assumptions in this work. We have omitted the constraint on the returns, liquidity constraints, correlations between the assets, non-stationary effects, etc. Some of these can be systematically taken into account and we plan to return to these finer details in subsequent work.

ACKNOWLEDGEMENTS

We thank O.C. Martin and M. Potters for useful discussions, and particularly J.P. Bouchaud for critical reading of the manuscript. S.C. is supported by the EC through the network MTR 2002-00319, STIPCO and I.K. by the National Office of Research and Technology under grant No. KCKHA005.

REFERENCES

Acerbi, C. and Tasche, D., On the coherence of expected shortfall. *J. Bank. Finan.*, 2002, **26**, 1487–1503.

Artzner, P., Delbaen, F., Eber, J.M. and Heath, D., Coherent measures of risk. *Math. Finan.*, 1999, **9**, 203–228.

Frey, R. and McNeil, A.J., VaR and expected shortfall in portfolios of dependent credit risks: Conceptual and practical insights. *J. Bank. Finan.*, 2002, **26**(7), 1317–1334.

Guerra, F. and Toninelli, F.L., The high temperature region of the Viana-Bray diluted spin glass model: In honor of Gianni Jana-Lasinio's 70th birthday. *J. Statist. Phys.*, 2004, **115**, 531–555.

Kondor, I., Pafka, S. and Nagy, G., Noise sensitivity of portfolio selection under various risk measures. *J. Bank. Finan.*, 2007, **31**(5), 1545–1573.

Mézard, M., Parisi, G. and Virasoro, M.A., *Spin Glass Theory and Beyond*, Lecture Notes in Physics, Vol. 9, 1987 (World Scientific: Singapore).

Pafka, S. and Kondor, I., Noisy covariance matrices and portfolio optimization. *Eur. Phys. J.*, 2002, **B27**, 277.

Pflug, G.C., Some remarks on the value-at-risk and the conditional value-at-risk. In *Probabilistic Constrained Optimization: Methodology and Applications*, edited by S. Uryasev, 2000 (Kluwer Academic: Dordrecht).

Press, W.H., Teukolsky, S.H., Wetterling, W.T. and Flannery, B.P., *Numerical Recipes in C*, 1992 (Cambridge University Press: Cambridge).

Rockafellar, R. and Uryasev, S., Optimisation of conditional value-at-risk. *J. Risk*, 2000, **2**, 21–41.

Talagrand, M., *Spin Glasses: A Challenge for Mathematicians*, 2002 (Springer: Berlin).

APPENDIX 14.A: THE REPLICA SYMMETRIC SOLUTION

We show in this appendix how the minimum cost function corresponding to the replica-symmetric ansatz is obtained.

The 'Tr log Q' term in (14.16) is computed by realizing that the eigenvalues of such a symmetric matrix are $(q_1 + (n-1)q_0)$, with multiplicity 1, and $(q_1 - q_0)$, with multiplicity $n - 1$. Then,

$$\text{Tr} \log Q = \log \det Q = \log(q_1 + (n-1)q_0)$$
$$+ (n-1) \log(q_1 - q_0) = n\left(\log \Delta q + \frac{q_1}{\Delta q}\right) + \mathcal{O}(n^2), \qquad (14.A1)$$

where $\Delta q \equiv q_1 - q_0$. The effective partition function in (14.17) depends on Q^{-1}, the elements of which are

$$(Q^{-1})^{ab} = \begin{cases} (\Delta q - q_0)/(\Delta q)^2 + \mathcal{O}(n), & \text{if } a = b, \\ -q_0/(\Delta q)^2 + \mathcal{O}(n), & \text{if } a \neq b. \end{cases} \qquad (14.A2)$$

By introducing a Gaussian measure $dP_{q_0}(s) \equiv (ds/\sqrt{2\pi q_0})e^{-s^2/2q_0}$, one can show that

$$\frac{1}{n} \log \hat{Z}(v, q_1, q_0) = \frac{1}{n} \log \left\{ \int \prod_a dx_a \, e^{-(1/2\Delta q) \sum_a (x^a)^2 + \gamma \sum_a (x^a + v)\theta(-x^a - v)} \right.$$
$$\left. \times \int dP_{q_0}(s) \, e^{(s/\Delta q) \sum_a x^a} \right\}$$
$$= \frac{q_0}{2\Delta q} + \int dP_{q_0}(s) \log B_\gamma(s, v, \Delta q) + \mathcal{O}(n), \qquad (14.A3)$$

where we have defined

$$B_\gamma(s, v, \Delta q) \equiv \int dx \exp\left(-\frac{(x-s)^2}{2\Delta q} + \gamma(x+v)\theta(-x-v)\right). \qquad (14.A4)$$

The exponential in (14.16) now reads $\exp Nn[S(q_0, \Delta q, \hat{q}_0, \Delta\hat{q}) + \mathcal{O}(n)]$, where

$$S(q_0, \Delta q, \hat{q}_0, \Delta\hat{q}) = q_0\Delta\hat{q} + \hat{q}_0\Delta q + \Delta q\Delta\hat{q} - \Delta\hat{q} - \gamma t(1-\beta)v$$
$$- t\log\gamma + t\int dP_{q_0}(s)\log B_\gamma(s, v, \Delta q)$$
$$- \frac{t}{2}\log\Delta q - \frac{1}{2}\left(\log\Delta\hat{q} + \frac{\hat{q}_0}{\Delta\hat{q}}\right) - \frac{\log 2}{2}. \qquad (14.A5)$$

The saddle point equations for $\hat{q}_0$ and $\Delta\hat{q}$ then allow us to simplify this expression. The free energy $(-\gamma)f_\gamma = \lim_{n\to 0} \partial Z^n / \partial n$ is given by

$$-\gamma f_\gamma(v, q_0, \Delta q) = \frac{1}{2} - t \log \gamma + \frac{1-t}{2} \log \Delta q + \frac{q_0 - 1}{2\Delta q}$$

$$- \gamma t(1 - \beta)v + t \int dP_{q_0}(s) \log B_\gamma(s, v, \Delta q), \qquad (14.A6)$$

where the actual values of v, q_0 and Δq are fixed by the saddle point equations

$$\frac{\partial f_\gamma}{\partial v} = \frac{\partial f_\gamma}{\partial q_0} = \frac{\partial f_\gamma}{\partial \Delta q} = 0. \qquad (14.A7)$$

Close inspection of these saddle point equations allows one to perform the low temperature $\gamma \to \infty$ limit by assuming that $\Delta q = \Delta/\gamma$ while v and q_0 do not depend on the temperature. In this limit, one can show that

$$\lim_{\gamma \to \infty} \frac{1}{\gamma} \log B_\gamma(s, v, \Delta/\gamma) = \begin{cases} s + v + \Delta/2, & s < -v - \Delta, \\ -(v+s)^2/2\Delta, & -v - \Delta \le s < -v, \\ 0, & s \ge -v. \end{cases} \qquad (14.A8)$$

If we plug this expression into Equation (14.A6) and perform the large-γ limit we obtain the minimum cost:

$$E = \lim_{\gamma \to \infty} f_\gamma = -\frac{q_0 - 1}{2\Delta} + t(1 - \beta)v - t \times \int_{-\infty}^{-\Delta} \frac{dx}{\sqrt{2\pi q_0}} e^{-(x-v)^2/2q_0} \left(x + \frac{\Delta}{2} \right)$$

$$+ \frac{t}{2\Delta} \int_{-\Delta}^{0} \frac{dx}{\sqrt{2\pi q_0}} e^{-(x-v)^2/2q_0} x^2. \qquad (14.A9)$$

We rescale $x \to x\Delta$, $v \to v\Delta$, and $q_0 \to q_0\Delta^2$, and after some algebra we obtain Equation (14.19).

CHAPTER 15

Stability Analysis of Portfolio Management with Conditional Value-at-Risk

MICHAL KAUT, HERCULES VLADIMIROU, STEIN W. WALLACE and STAVROS A. ZENIOS

CONTENTS

15.1 INTRODUCTION

T HE CONVENTIONAL MEAN-VARIANCE APPROACH, which constitutes the primary basis for portfolio selection, assumes that asset returns follow normal distributions and/or that the investor has a quadratic utility function. Despite the long and widespread use of the mean-variance method in portfolio management, its fundamental assumptions often do not hold in practice. The returns of many financial securities exhibit skewed and leptokurtic distributions. Derivatives, or securities with embedded options, have, by construction, highly skewed return distributions. Many other investments are exposed to multiple risk factors whose joint effect on portfolio returns often cannot be modelled by a normal distribution.

Substantial research effort has been directed toward the development of models that properly capture asymmetries and dynamic effects in the observed behaviour of asset returns. At the same time, alternative risk metrics have been sought. Such measures are concerned with other, or additional, characteristics of the return distribution (e.g. the tails) besides the variance, and can accommodate a wide range of investor priorities and regulatory requirements for risk management. Value-at-risk (VaR) has essentially attained the status of a de-facto standard in financial practice (see, e.g. Jorion 2001). VaR is defined as the maximal loss (or minimal return) of a portfolio over a specific time horizon at a specified confidence level; VaR corresponds to a percentile of the portfolio's loss (or return) distribution at a specified confidence level.

Despite its widespread popularity in recent years, VaR suffers from a number of theoretical and practical limitations. Although its calculation for a certain portfolio indicates that shortfall returns, below VaR, will occur only with a prespecified likelihood, it provides no information on the extent of the distribution's tail which may be quite long; in such cases, the portfolio return may take substantially lower values than VaR and result in severe losses. More importantly, VaR is not a coherent risk measure in the sense defined by Artzner *et al.* (1999). It fails to reward diversification, as it is not subadditive; hence, the VaR of a diversified portfolio can be larger than the sum of the VaRs of its constituent asset components. Moreover, when the returns of assets are expressed in terms of discrete distributions (i.e. scenarios) VaR is a non-smooth and non-convex function of the portfolio positions and exhibits multiple local extrema (see, e.g. Rockafellar and Uryasev 2002). Incorporating such functions in mathematical programs is very difficult, thus making impractical the use of VaR in portfolio optimization models.

To overcome the deficiencies of VaR, suitable alternative risk metrics have been sought. Artzner *et al.* (1999) discuss such metrics and specify the properties that sound risk measures should satisfy, which they characterize as *coherent risk measures*. A family of closely related risk metrics—termed as *expected shortfall, mean excess loss, tail VaR, conditional VaR*—have been suggested that quantify the mass in the tail of the distribution beyond VaR. Tasche (2002) examines the properties of this family of measures; he shows that it characterizes the smallest coherent risk measures to dominate VaR and that it can incorporate higher moment effects. Acerbi and Tasche (2002) show that the alternative definitions of these measures lead to the same results when applied to continuous loss distributions. They note that differences appear when the underlying distribution has discontinuities and they demonstrate that, in such cases, care must be exercised in the details of the definition to maintain the desired properties of coherence. The theoretical underpinnings of coherent risk measures and their properties are thoroughly discussed in Foellmer and Schied (2004).

Rockafellar and Uryasev (2002) introduced a definition of the *conditional value-at-risk* (CVaR) measure for general distributions, including discrete distributions that exhibit discontinuities, and showed that CVaR is a continuous and convex function of the portfolio positions. They also showed that a CVaR optimization model can be formulated as a linear program in the case of discrete distributions of the stochastic input parameters. CVaR is defined as the conditional expectation of losses exceeding VaR; it is a coherent risk measure that quantifies the worst (lowest) portfolio returns below the respective VaR.[1] As CVaR is concerned with the tail of the distribution it is a suitable risk measure when the distribution is asymmetric and/or heavy-tailed.

As a result, CVaR models are seeing increasing use in various financial management applications. For example, CVaR models have been suggested by Bogentoft *et al.* (2001) for asset-liability management of pension funds, by Krokhmal *et al.* (2002, 2003) for hedge fund portfolios and by Anderson *et al.* (2001) for credit risk optimization. Jobst and Zenios (2001) showed that CVaR models are effective for modelling credit risk and accounting for default events in the tails. Topaloglou *et al.* (2002, 2008) applied CVaR models to international portfolio management problems to account for asymmetric and leptokurtic distributions of exchange rates and asset returns.

A specification of the distribution of stochastic parameters (asset returns) is a critical input for all portfolio management models. In parametric models, the multivariate distribution is specified by the values of key statistics that are usually estimated using historical data, analytical methods, analysts' forecasts and other methods. In nonparametric models, the distribution is usually represented in terms of a discrete set of plausible outcomes (scenarios) that are generated by simulation, bootstrapping historical data or even subjective estimates in some cases. In all cases, the reliability of the model's results depends on the accuracy with which the postulated distribution approximates the true distribution of the random variables—which is never actually observable. Hence, the models are inevitably exposed to estimation errors. Consequently, it is important to understand the sensitivity of a model to mis-specifications of distributional characteristics, and to assess the relative effects that mis-specifications of various statistical properties have on the results. This can guide analysts in their choice among alternative estimation methods, as well as in the relative effort they invest to obtain robust estimates of the various model inputs.

As CVaR is concerned with the tail of the portfolio's return distribution, models that employ this measure are expected to be sensitive to higher moments of the constituent assets' random returns. This chapter aims specifically to study the stability of a CVaR portfolio management model with respect to changes in input specifications. In this respect, we follow previous studies on stability of mean-variance models.

Most notably, Kallberg and Ziemba (1984), Broadie (1993) and Chopra and Ziemba (1993) examined the relative effects of estimation errors in the mean, variance and covariance of asset returns on mean-variance efficient portfolios. They found that the model results are most sensitive to mis-specifications in the means of asset returns. They reported that the impact of errors in the variance of asset returns was about an order of magnitude lower than that of errors in the means, while errors in covariance values had about half the impact of errors in the variance. Chopra and Ziemba found that the sensitivity of the model's results to estimation errors of statistical properties of asset returns varies with the level of risk aversion. Broadie pointed out that the impact of estimation errors on the mean-variance model increases with the number of securities included in the portfolio.

---

[1] When defined in terms of portfolio return this risk measure is commonly referred to as *Return-at-Risk* (RaR).

Here, we extend these studies as we similarly investigate the effects of mis-specifications of statistical properties—including higher moments—of asset returns on the results of a model based on the CVaR measure. As a test case we use a portfolio optimization model for international investments. The portfolio is exposed to market risk in multiple countries and to currency risk. We use discrete scenarios to model the uncertainty in asset returns and spot exchange rates. The scenarios are generated by the moment matching method of Høyland et al. (2003) so that in the set of generated scenarios the random variables have statistical properties that match specific target values as determined from historical market data.

First, we define in-sample and out-of-sample stability and we demonstrate that the scenario generation procedure does not bias the results of the optimization model. That is, for sufficiently large scenario sets, the portfolio model produces stable solutions that are not dependent on the specific scenario sets (i.e. the results are stable with respect to sample). We then conduct extensive computational experiments to assess the effects on the model's results due to variations in the target statistics: mean, standard deviation, skewness, kurtosis and correlations of the random variables. We demonstrate that the CVaR model is indeed sensitive to the higher moments of the stochastic inputs. Moreover, we quantify the relative impact of mis-specifications in the various statistical properties of the inputs on the model's results.

The chapter is organized as follows. Section 15.2 presents the CVaR model for international portfolio management that we use as a test case in this study. Section 15.3 describes the scenario generation method, the input data, and the tests to verify the stability of the optimization model with respect to the scenario generation procedure. In Section 15.4 we describe the computational experiments involving mis-specifications of the statistical properties of stochastic input parameters and we present the effects of these errors on the model's results. Finally, Section 15.5 concludes.

15.2   CVaR MODEL FOR INTERNATIONAL PORTFOLIO MANAGEMENT

We test a CVaR model for international portfolio management. We view the problem from the perspective of a US investor who may construct a portfolio composed of domestic and foreign securities starting with an initial endowment in US dollars. Thus, we have a simple portfolio construction problem with a holding period of one month. To purchase foreign securities, the investor must first convert funds to the respective currency; the current spot exchange rates apply in the currency exchange transactions.

The asset set includes a stock index (Stk), a short-term (Bnd1) and a long-term (Bnd7) government bond index in each of four countries: United States (USA), United Kingdom (UK), Germany (Ger) and Japan (Jap). The values of the assets and the exchange rates at the end of the holding period are uncertain; their joint distribution is modelled in terms of a scenario set (i.e. a set of discrete outcomes with associated probabilities).[2] At the end of the holding period we compute the scenario-dependent value of each investment using

---

[2]In this model instance we have 15 random variables: the returns of the 12 indices (3 for each of the 4 countries) during the holding period, and the exchange rates of the 3 foreign currencies to USD at the end of the holding period.

its projected price under the respective scenario. The USD-equivalent value is determined by applying the estimate of the appropriate spot exchange rate to USD at the end of the period under the same scenario.

The portfolio is exposed to market risk in the various countries, as well as to currency risk. To (partly) hedge the currency risk, the investor may enter into forward currency exchange contracts. The monetary amounts (in USD) of forward contracts are decided at the time of portfolio selection, but the currency exchanges are executed at the end of the holding period.

We define the following notation.

User-specified parameters:

α confidence (percentile) level for VaR and CVaR
ϑ minimal allowable CVaR of portfolio returns

Sets and indices:

M set of markets (synonymously, countries, currencies)
$\ell \in M$ index of investor's base (reference) currency in the set of currencies
M_f set of foreign markets; $M_f = M\backslash\{\ell\}$
I_j set of available asset classes (stock and bond indices) in market $j \in M$
S set of scenarios: $S = \{1, \ldots, S\}$

Deterministic input data:

c_ℓ amount of initially available cash in base currency ℓ, $(c_\ell = 100)$
π_{ij}^0 current market price of asset $i \in I_j$, $j \in M$; in units of local currency j
γ_{ij} transaction cost rate for purchases of asset $i \in I_j$, $j \in M$, $(\gamma_{\text{Stk},j} = 0.001, \; \forall j \in M;$ $\gamma_{\text{Bnd1},j} = \gamma_{\text{Bnd7},j} = 0.0005, \; \forall j \in M)$
λ transaction cost rate for spot currency exchanges, $(\lambda = 0.0001)$
e_j^0 current spot exchange rate of currency $j \in M$
ϕ_j current one-moth forward exchange rate of currency $j \in M$ (i.e. the market-quoted rate for a currency exchange to be executed at the end of the holding period)

Scenario dependent data:

S number of scenarios, $S = |S|$
p_s probability of scenario $s \in S$—in our tests, scenarios are equiprobable (i.e. $p_s = 1/S$)
π_{ij}^s price of asset $i \in I_j$, $j \in M$ at the end of the holding period under scenario $s \in S$; in units of local currency j
e_j^s spot exchange rate of currency $j \in M$ at the end of the holding period under scenario $s \in S$

Decision variables:

x_{ij} number of assets $i \in I_j$, $j \in M$ in the portfolio, in units of face value
f_j amount of base currency collected from sale of currency $j \in M_f$ in the forward market (i.e. amount of forward contract, in units of the base currency)

Auxiliary variables:

v^s total value of the portfolio at the end of the holding period under scenario $s \in S$, in units of the base currency

r^s portfolio return under scenario $s \in S$

z variable in definition of CVaR—equals VaR at the optimal solution

y_s return shortfall below VaR under scenario $s \in S$, $y_s = [0, z - r_s]^+$

All exchange rates are expressed as the equivalent amount of the base currency for one unit of the foreign currency. Obviously, the exchange rate of the base currency to itself is trivially equal to one $e_\ell^0 = e_\ell^s = 1$, $\forall s \in S$.

We formulate the international portfolio selection model as follows:

$$\text{maximize} \quad \sum_{s \in S} p_s r^s \tag{15.1a}$$

$$\text{s.t.} \quad c_\ell = \sum_{i \in I_\ell} x_{i\ell}\pi_{i\ell}^0(1 + \gamma_{i\ell}) + \sum_{j \in M_f} e_j^0 (1 + \lambda)\left(\sum_{i \in I_j} x_{ij}\pi_{ij}^0 (1 + \gamma_{ij})\right) \tag{15.1b}$$

$$v^s = \sum_{i \in I_\ell} x_{i\ell}\pi_{i\ell}^s + \sum_{j \in M_f}\left\{f_j + e_j^s\left(\sum_{i \in I_j} x_{ij}\pi_{ij}^s - \frac{f_j}{\varphi_j}\right)\right\} \quad \forall s \in S \tag{15.1c}$$

$$r^s = (v^s - c_\ell)/c_\ell, \quad \forall s \in S \tag{15.1d}$$

$$y_s \geq z - r^s, \quad \forall s \in S \tag{15.1e}$$

$$z - \frac{1}{1 - \alpha}\sum_{s \in S} p_s y_s \geq \vartheta \tag{15.1f}$$

$$y_s \geq 0, \quad \forall s \in S \tag{15.1g}$$

$$f_j \geq 0, \; x_{ij} \geq 0, \quad \forall j \in M, \forall i \in I_j \tag{15.1h}$$

This is a simplified version of the international portfolio management model in Topaloglou *et al.* (2002). The model in that chapter accounts for an initial portfolio—including cash and asset positions in any currency—and determines transactions—asset sales and purchases, as well as spot currency exchanges—so as to obtain a revised portfolio. A multi-stage extension of that model to address dynamic international portfolio management problems is developed in Topaloglou *et al.* (2008). Here, we consider a simple portfolio construction model that selects a portfolio starting with an initial cash endowment in the base currency only.

The objective function (15.1a) maximizes the expected portfolio return over the holding period. Equation (15.1b) is the budget constraint; it indicates that the cost for the purchase of domestic and foreign assets is covered by the available cash (c_l). Linear transaction costs (γ_{ij}) are charged for asset purchases (x_{ij}); a linear transaction cost (λ) is also charged for spot currency exchanges that are effected in order to purchase foreign assets. Note that the entire budget is allocated to the available assets; a simple extension of the model can allow investments in money market accounts in the various currencies.

Equation (15.1c) determines the scenario-dependent values of the portfolio (v^s), in units of the base currency, at the end of the holding period. These valuation equations account for the revenues from the liquidation of all portfolio positions at the end of the holding period at the projected asset prices (π_{ij}^s) for the corresponding scenario. The contribution of foreign investments to the total value of the portfolio accounts for the settlement of any outstanding forward contracts (f_j). The residual amount in a foreign currency is valued in terms of the base currency by using the projected spot exchange rates (e_j^s) at the end of the holding period. Equation (15.1d) determines the return of the portfolio under each scenario.

Constraints (15.1e) and (15.1g) determine the excess shortfall returns, beyond VaR, under each scenario. Constraint (1f) imposes a minimal allowable value (ϑ) on the CVaR of portfolio returns over the holding period at the $(1 - \alpha)100$th percentile. At the optimal solution, the variable z is equal to the VaR at the same percentile—when constraint (15.1f) is active, which is always the case in this model. The constraints in (15.1h) disallow short positions.

The linear programming formulation of CVaR models when the stochastic inputs follow a discrete distribution is due to Rockafellar and Uryasev (2002). The model here maximizes the expected portfolio return while constraining the CVaR value of portfolio returns; with constraint (15.1f) the expected excess loss in the tail of the distribution, beyond VaR, is bounded by the parameter ϑ. Financial optimization models with CVaR constraints are reported, for example, in Anderson *et al.* (2001), Bogentoft *et al.* (2001), and Krokhmal *et al.* (2002, 2003). Alternatively, we could have opted to maximize the CVaR of portfolio returns and impose a minimal target on expected return, as is done in Topaloglou *et al.* (2002). We chose this formulation as it is more natural to interpret the impact of estimation errors in stochastic inputs on the expected portfolio return, rather than on the value of a risk measure.

As was shown in Topaloglou *et al.* (2002, 2008) the monthly variations of exchange rates—also the returns of several stock indices—exhibit skewed and fat-tailed distributions. The use of the CVaR metric is appropriate in the context of the international portfolio management model, as it can accommodate the skewed and leptokurtic distributions of the stochastic inputs (see Table 15.A1 in Appendix 15.A). As we noted earlier, many other portfolio management models involve securities with asymmetric and leptokurtic return distributions, for which a CVaR model would be suitable.

15.3   SCENARIO GENERATION

We used the method of Høyland *et al.* (2003) to generate scenarios of asset returns and spot currency exchange rates. The method generates a set of discrete scenarios for the random variables so that the first four moments of the marginal distributions (mean, standard deviation, skewness and kurtosis), as well as the correlation coefficients match specified targets. We estimate the target values for these statistics from historical data. However, this is not a prerequisite for the scenario generation procedure. We could, as easily, use subjective estimates for the target statistics, as well as target values determined with alternative estimation procedures.

The moment-matching method allows full control of the moments when generating scenarios. This capability is essential for the purposes of this study. To investigate the impact on the model of variations in the values of some moment of the random variables, we need a procedure that can generate scenarios effected differing only in terms of the moment studied, while keeping all other statistical properties of the stochastic inputs unchanged. The moment-matching method provides this capability.

15.3.1   Data

The data for the stock indices were obtained from the Morgan Stanley Capital International, Inc. database (www.mscidata.com). The data for the bond indices and the currency exchange rates were collected from *DataStream*. All time series have a monthly time-step and cover the period from January 1990 to April 2001 (i.e. a total of 136 monthly observations). The statistical properties of these data series are reported in Tables 15.A1 and 15.A2 in Appendix 15.A.

15.3.2   Assessment of the Scenario Generation Method

In Section 15.4, we investigate the behaviour of the CVaR model with respect to the number of scenarios and with respect to mis-specifications in the statistical properties of stochastic inputs. To ensure the reliability of the results, however, we must first show that the scenario generation method used does not influence the results by causing instability of the solutions. That is, if the solutions changed for different scenario sets then the results of Section 15.4 would be suspect.

Ideally, we would like to determine that the scenario generation procedure can effectively produce robust solutions with respect to the true distribution of the random variables. This is not an attainable goal as the true distribution is not observable. Hence, we assess the scenario generation method in terms of its ability to closely approximate a benchmark distribution, and the stability of the results with respect to the benchmark. It is important that the benchmark distribution is provided exogenously, that is, it is not generated by the same method which we are testing.

We use as benchmark a discrete distribution (scenario set) generated by a method based on principal component analysis as described in Topaloglou *et al.* (2002). The benchmark distribution has 15 000 scenarios that jointly depict the co-variation of the 15 random variables in the international portfolio management problem. We note that the scenarios of the benchmark are not equiprobable. From the benchmark scenario set, we compute the moments and correlations of the random variables. We use these values as the target statistics to match with the scenario generation procedure.

First, we verify that moments of the random variables in the scenario sets that we generate match the target values. We also check that the generated scenario sets reproduce other distributional characteristics (e.g. the entire marginal distributions).

15.3.2.1 Matching marginal distributions The easiest to check is a match of the marginal distributions. We generated scenario sets ranging in size from 250 to 5000 scenarios. For each set we determined the marginal distributions of the random variables from generated scenario sets and compared them to the corresponding distributions from the benchmark. The comparison in the case of the US stock index (Stk.USA) is depicted in Figure 15.1. The reproduction of the marginal distributions of the remaining random variables is quite similar.

We observe that even with moderate-size scenario sets (> 250 scenarios) we can closely reproduce the marginal distributions from the 1st to the 99th percentile. At the extreme tails the distribution is not as accurately matched unless a sufficiently large number of scenarios is generated. This is understandable, as we should expect more samples in the tails as the number of scenarios increases.

The desired degree of matching the distributions depends on the decision model in which the scenarios will be used. For example, if we are to apply a mean-variance model then the accuracy of match at the tails will not make any difference, as long as the first two marginal moments and the correlations are matched. A close match of the tails becomes relevant for the CVaR model which is concerned with the tail of the portfolio's return distribution.

The match of the marginal distributions of the random variables is illustrative. Yet, it is not sufficient, even though the generated scenario sets also match the desired correlations as well. We need to establish that the portfolio optimization model produces stable results regardless of the specific scenario set generated in any given run—i.e. that it is stable with respect to sample. Evidently, scenario sets of sufficiently large size are needed to ensure

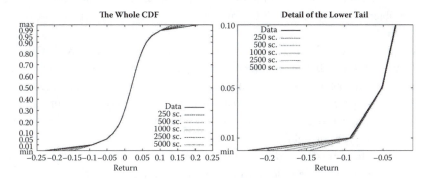

FIGURE 15.1 Match of the distribution function for the US stock index. Comparison of the distribution function for the monthly returns of the US stock index in the benchmark (data) and in generated scenarios sets of different size. *The scale of the vertical axis is not linear, the two outer intervals are prolonged.*

such stability. We need to test jointly the scenario generation method and the optimization model in order to verify that the scenario generation method does not cause instability of the solutions.

15.3.2.2 Joint stability test of scenario generation and the CVaR model We generate 25 different scenario sets of a given size, each matching the moments and correlations of the benchmark distribution. We solve the optimization model with each scenario set and record the optimal portfolio composition and the expected portfolio return. The confidence level in all tests is $\alpha = 0.95$; thus, CVaR is the expected return for the 5% worst scenarios. The bound on CVaR is $\vartheta = -1\%$. As the constraint (15.1f) is always active, the CVaR of portfolio returns is always at its minimal value (-1%) at the optimal solution; thus, the expected excess loss over the 5% worst scenarios is 1%.

As stochastic programs tend to have multiple optimal or near-optimal solutions, we study the stability in terms of the optimal value; we do not compare the optimal portfolio compositions. We then simulate all the solutions on the benchmark distribution and record the out-of-sample values of both the expected return and CVaR.

As in Kaut and Wallace (2007) we examine two types of stability:

> **In-sample stability:** The solutions should not vary across scenario sets of the same size. We examine the in-sample variation of the optimal values (expected return) across the 25 scenario sets of a given size; ideally these should be equal.
>
> **Out-of-sample stability:** We examine the variation of the expected portfolio returns and CVaR values obtained when the solutions are simulated on the benchmark distribution. These out-of-sample values should ideally be equal for all scenario sets. They should also be equal to the in-sample values.

The two notions of stability are not equivalent. We can have in-sample stability without out-of-sample stability. Consider, for example, a case in which all the scenario sets are identical but incorrect in comparison to the benchmark. On the other hand, we can have alternative scenario sets for which the model yields the same optimal solution. Then the in-sample objective values could differ for different scenario sets, but the out-of-sample values would be equal.

Verifying out-of-sample stability in terms of a benchmark indicates that the model yields robust solutions that do not vary with respect to sample. This is essential in this study as we need to ensure that the variations in the solutions that are observed in the tests of the next section are caused by the variations in the statistics of the inputs, and not by instability with respect to sample. Hence, the purpose of our tests is to assess whether we can achieve both types of stability.

Results of the tests are depicted in Figure 15.2. We see that as we increase the number of scenarios to around 5000 we indeed achieve both in-sample and out-of-sample stability. Thus, the scenario generation method is effective, in the sense that it does not cause instability in the solutions of the CVaR model.

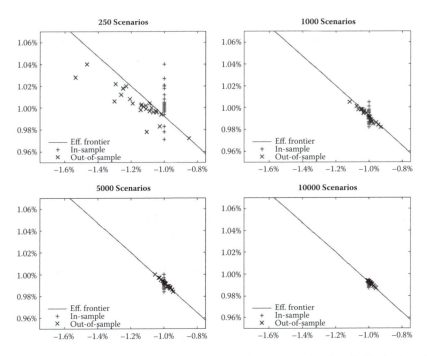

FIGURE 15.2 Stability of the CVaR model with respect to the exogenous benchmark. The horizontal axis shows the CVaR value, the vertical axis the expected portfolio return (monthly).

15.4 SENSITIVITY TESTS OF THE CVAR MODEL

This section studies the sensitivity of the CVaR model with respect to mis-specifications in the statistics of stochastic parameters. Again we need a benchmark as a reference. Here we calibrate the scenario generation method using the statistical properties estimated from historical market data, as reported in Tables 15.A1 and 15.A2 of Appendix 15.A; all scenario sets in the following tests are generated so as to match these target statistics. First, we generate a benchmark with 20,000 scenarios. This scenario set is sufficiently large (according to the findings of the previous section), to ensure both in-sample and out-of-sample stability of the solutions, while it is still easily solvable so as to trace the reference efficient frontier. The efficient frontier, depicting the tradeoff between expected portfolio return and the CVaR risk metric, is obtained by repeatedly solving the parametric optimization model for different allowable limits ϑ on CVaR.

To interpret the results of the tests, we must understand the source of differences between the in-sample and out-of-sample expected return of a given portfolio. If the portfolio is invested solely in domestic assets, the expected return would depend only on the portfolio composition and the means (expected values) of asset returns. The contribution of a foreign asset, however, on the portfolio's return depends on the product of the asset return (in its local currency) and the change of the exchange rate to the reference currency. When foreign investments are present in a portfolio, the return depends on products of random variables; hence, the expected portfolio return depends not only on the means of the random variables, but also on their covariances. Thus, for a given portfolio, the in-sample and the out-of-sample expected portfolio returns would be equal only if the random variables have the same means, standard deviations and correlations in the respective scenario sets (i.e. the test set and the benchmark). This condition is satisfied by construction in our scenario generation method as the random variables have matching moments and correlations in the benchmark and in the test sets. Hence, a portfolio has the same in-sample and out-of-sample expected return, but its CVaR value is different when it is simulated on the benchmark scenario set in comparison to its value on a test set.

15.4.1   Determining a Sufficient Number of Scenarios

Before proceeding to the sensitivity tests, we verify that we employ sufficiently large scenario sets in our tests to ensure stability with respect to sample. That is, we must ensure that variations observed in the model's results stem from changes in the statistical properties of the stochastic inputs and not from insufficiency of the scenario test sets. In the tests of Section 15.3.2 we found that at least 5000 scenarios were needed to attain both in-sample and out-of-sample stability to an acceptable level. As the benchmark is now different we repeat the same tests here; the statistics of both the benchmark as well as the test sets in this section were estimated from time series of market data. The results of the tests are summarized in Figure 15.3. Again, we observe that we need at least 5000 scenarios to ensure adequate stability of the CVaR model.

Some comments on the figures follow.

- The in-sample results always lie on a vertical line, as the CVaR value is always equal to its minimal allowable limit ϑ at the optimal solution. The range of this line indicates the in-sample variation of expected return with respect to scenario sets of a given size.
- As the in-sample values are computed on the respective scenario sets—and not the benchmark—they can cross the reference efficient frontier that is generated using the benchmark.

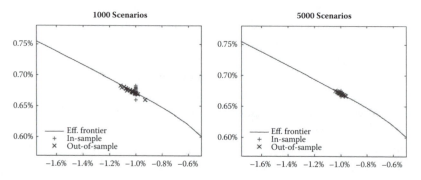

FIGURE 15.3 Stability of CVaR model with respect to the number of scenarios. The horizontal axis shows the CVaR value, the vertical axis the expected portfolio return (monthly).

- Because the random variables have the same moments and correlations in the test sets and in the benchmark, the in-sample and the out-of-sample expected portfolio returns are the same for a given portfolio, as we explained at the start of this section. Only the CVaR values of a portfolio change when it is simulated on the benchmark scenario set.

Table 15.1 presents measures of variation of expected portfolio returns for tests using scenario sets of increasing size. Variations in expected return decrease monotonically with increasing number of scenarios; this is a necessary consistency check.

15.4.2   Effects of Mis-Specifications in Statistical Properties

In this section we assess the sensitivity of the CVaR model with respect to mis-specifications in the statistical properties of the stochastic inputs. Controlled errors are systematically introduced to the target statistics (first four moments and correlations) of the random variables at the scenario generation phase. Multiple scenario test sets are then generated to match the perturbed statistical properties. Similar tests for the mean-variance model are presented in Kallberg and Ziemba (1984) and Chopra and Ziemba (1993).

We quantify the induced errors by means of the following approach. We compute the moments and correlations of the random variables based on subsets of our data set, using a moving time window of half the size of the available time series. Thus, we obtain a series of plausible estimates for the moments and correlations of the random variables. For each statistic, we take the interval from the minimal to the maximal estimated value which we call the *variation interval* for the corresponding statistic. These variation intervals for moments and correlations are reported in Tables 15.A3 and 15.A4 in Appendix 15.A. We term the value of the respective statistic, calculated on the basis of the entire data set, the *true value*. We define a δ-percent error in a statistical property as

$$\text{true value} + \varepsilon \frac{\delta}{100} \text{ length } (\textit{variation interval}), \tag{15.2}$$

where ε is a random number from the uniform distribution on the interval $[-1, 1]$. With this definition, the average absolute error is

$$\frac{1}{2} \frac{\delta}{100} \text{length } (\textit{variation interval}).$$

Note that this is different from the corresponding definition in Chopra and Ziemba (1993). There, the δ-percent error was defined as $\textit{true value}(1 + \varepsilon(\delta/100)), \varepsilon \in \mathcal{N}(0, 1)$. If we have a statistic (e.g. skewness or correlation) with a *true value* equal to zero, then this

TABLE 15.1 Standard Deviation and Range of Out-of-Sample Expected Portfolio Returns

# of scenarios	250	500	1000	2500	5000	10000
Standard deviation	0.010%	0.007%	0.005%	0.002%	0.002%	0.001%
Range (max-min)	0.034%	0.026%	0.023%	0.009%	0.006%	0.003%

statistic would never be changed if we introduced errors using the approach of Chopra and Ziemba; variations of statistics with very small values would also be very small. For this reason, we chose to calculate controlled errors by means of (15.2).

A potential problem when introducing random errors to statistical properties is that we may specify a property, or a combination of properties, that is not feasible. For example, we may end up with specifications that may violate the condition, $kurt > 1 + skew^2$, or we may specify a correlation matrix that is not positive definite. When this happens, we simply discard these particular specifications.

To test the impact of mis-specifications in each statistic we generate 100 scenario sets (with 5000 scenarios each) by randomly varying the value of the statistic. Every test proceeds as follows.

i. For each random variable, perturb the selected statistic using (15.2).
ii. Generate a scenario set, matching the perturbed statistical properties.
iii. Solve the portfolio optimization model and record the expected portfolio return and the value of CVaR at the optimal solution.
iv. Simulate the solution on the benchmark—which was generated with unperturbed statistics—estimating the expected portfolio return and the value of CVaR.

15.4.2.1 Results of the sensitivity tests We ran tests for 10% and 25% errors (i.e. for $\delta = 0.10$ and 0.25). In the case of errors in marginal moments, we never obtained an infeasible specification. In the case of 25% errors in correlations, however, many of the generated correlation matrices were not positive definite, and were discarded and replaced. The discarded cases resulted from samples that introduced the larger levels of errors, i.e. $\varepsilon \approx \pm 1$. As the large error instances were discarded, the effective errors in correlations in this case are somewhat smaller.

The results of the tests with parameter settings $\alpha = 0.95$ and $\vartheta = -1.0\%$ are shown in Figures 15.4 and 15.5 for 10% and 25% errors, respectively. We observe that the larger levels of error, in any statistic, have a discernibly higher impact on the expected portfolio return. Estimation errors in the means clearly exhibit the highest impact on the solutions, followed by errors in standard deviations.

To examine the effects of estimation errors at different levels of risk aversion, we repeated the tests at different levels of the parameters α and ϑ; the combination of these parameters relates to the level of risk aversion. Increasing values of the percentile level α refer to more extreme tails of the return distribution. The parameter ϑ controls the allowable mass in the tail of the distribution; thus, lower values of this parameter (in absolute terms) constrain more tightly the size of the tail (beyond the percentile level specified by α), and reflect higher risk aversion. Results of the tests are summarized in Table 15.2.

Table 15.2 summarizes the impacts of estimation errors in statistics of the stochastic inputs to the CVaR model. It reports the sample ranges of out-of-sample expected (monthly) return estimates when errors were introduced to the respective statistics. For all statistics, the impact of errors increases with the level of the error; the larger rate of increase results from errors in the means, followed by errors in the standard deviations.

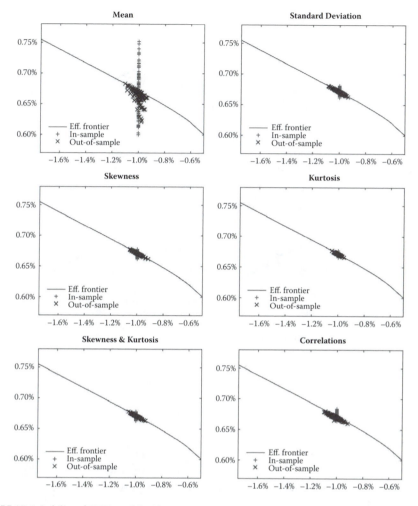

FIGURE 15.4 Stability of CVaR model with parameters $\alpha = 95\%$; $\vartheta = -1\%$; 10% estimation errors in the respective statistics. The horizontal axis shows the CVaR value, the vertical axis the expected monthly return.

We assess the impacts of estimation errors in statistics of the stochastic inputs in terms of the variation they induce to the out-of-sample estimates of expected portfolio return.

We summarize our observations on the results as follows.

- In all tests, the CVaR model is most sensitive to errors in the means of the stochastic inputs. The mean is by far the most important statistical property to estimate accurately, as errors in the means of the random variables have the most significant impact on the results—about 2 to 10 times higher than the effect of errors in any other statistical property.

- After errors in the means, mis-specifications of standard deviations have the next most important impact on the out-of-sample estimates of expected portfolio return —about 2 to 5 times lower than the impacts of the means. Next is the impact of estimation

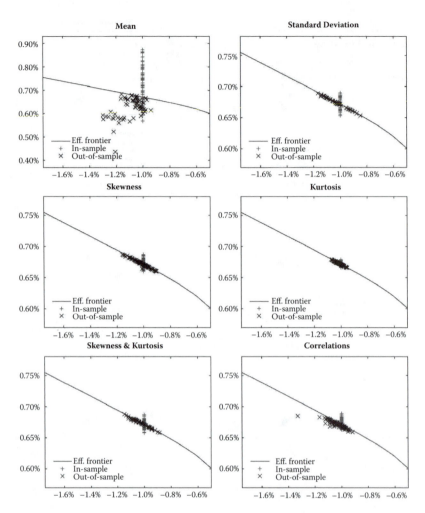

FIGURE 15.5 Stability of CVaR model with parameters $\alpha = 95\%$; $\vartheta = -1\%$; 25% estimation errors in the respective statistics. The horizontal axis shows the CVaR value, the vertical axis the expected monthly return. *The graph for errors in means has a bigger scale on the vertical axis.*

TABLE 15.2 Sample Ranges (max–min) of Out-of-Sample Expected (Monthly) Portfolio Return Estimates Caused by Estimation Errors in Statistics of the Stochastic Inputs

				Range of out-of-sample expected return estimates caused by estimation errors in respective statistics				
	α	ϑ	Benchmark expected return	Mean	Std dev.	Skew	Kurt	Corr
10% error	95%	−0.5%	0.601%	0.065%	0.035%	0.031%	0.024%	0.022%
	95%	−2.0%	0.782%	0.045%	0.025%	0.022%	0.020%	0.027%
	99%	−2.0%	0.707%	0.060%	0.034%	0.030%	0.022%	0.031%
25% error	95%	−0.5%	0.601%	0.255%	0.073%	0.058%	0.039%	0.031%
	95%	−2.0%	0.782%	0.247%	0.058%	0.028%	0.016%	0.039%
	99%	−2.0%	0.707%	0.247%	0.057%	0.034%	0.036%	0.043%

errors in values of skewness—about 2 to 9 times lower than the impacts of the means—followed by the impact of errors in the estimates of kurtosis—about 3 to 12 times lower compared to the impacts of the means. The effects of estimation errors in correlations are between those of the standard deviations and skewness.[3]

- Errors in the values of kurtosis have a detectable and non-negligible influence on the results of the CVaR model, though their impact is lower in comparison to that of the other statistics.
- For all marginal moments, the effect of estimation errors seems to increase for more risk averse settings of the model's parameters, i.e. for smaller allowable tails (as controlled by the parameter ϑ).
- The mean and the correlations are the only statistics whose mis-specifications result in portfolios that deviate significantly from the efficient frontier. Errors—especially when they are relatively small—in the other statistics seem to result in portfolios with different CVaR values, which are, however, still close to the efficient frontier.

The results demonstrate that estimation errors in higher-order moments of stochastic inputs do indeed affect in measurable ways portfolio management models that use the CVaR risk measure. The results imply that it pays to devote care and effort so as to accurately estimate the values of higher-order moments when employing risk measures concerned with the tails of the return distribution in portfolio management models.

15.5 CONCLUSIONS

We tested a risk management model for international portfolios based on the CVaR risk metric. We employed a scenario generation procedure based on principles of moment matching. We showed that this scenario generation method is effective and 'unbiased,' in the sense that it can closely reproduce the characteristics of a desired distribution and it leads to stable solutions of the portfolio optimization model.

We investigated the sensitivity of the CVaR model with respect to errors introduced to the statistical properties of stochastic inputs, as represented by discrete scenario sets. The statistical properties investigated include the first four marginal moments and the correlations of the random variables (assets returns and spot currency exchange rates). The tests quantify the relative effects of errors in these statistics on the model's results. The results confirm that the mean value of the random variables is the most important statistic to accurately estimate; the CVaR model exhibits high sensitivity to mis-specifications of the means. But, unlike the mean-variance model, the CVaR model shows sensitivity to errors in the estimates of higher-order moments as well. Errors in the standard deviation, correlations and skewness of the random variables have considerable impact on the model's results, in this order of importance. Estimation errors in the values of kurtosis have lesser, yet non-negligible, effects.

---

[3] We note that the effects of mis-specifications in correlation values at 25% errors are underestimated. In these tests, samples with the larger levels of errors in correlations yielded non-positive definite correlation matrices and were discarded. Hence, the effective estimation errors in this case are somewhat lower, and the corresponding effects are underestimated.

When assessing the potential effects of estimation errors in statistical properties, we should have a sense of the magnitude of such errors in practice. Much more care is exercised in generating reliable estimates of the means, variances and covariances of random financial variables, and more effective tools are available for their estimation, in comparison to higher-order moments. This is because the mean-variance model continues to be the primary paradigm for portfolio management, and because the importance of the first two moments is well understood. The need for accurate estimates of higher-order moments is often overlooked as their potential impact in portfolio management models is not as well understood and appreciated. This study sheds some light in this respect, by indicating the relative importance of accurate estimates of higher-order moments for random variables in risk management models that employ risk measures tailored to control the tails of the portfolio's return distribution.

REFERENCES

Acerbi, C. and Tasche, D., On the coherence of expected shortfall. *J. Bank. Finance*, 2002, **27**(6), 1487–1503.

Anderson, F., Mausser, H., Rosen, D. and Uryasev, S., Credit risk optimization with conditional value-at-risk criterion. *Math. Program.*, 2001, **89**(2), 273–291.

Artzner, P., Delbaen, F., Eber, J.M. and Heath, D., Coherent measures of risk. *Math. Finance*, 1999, **9**(3), 203–228.

Bogentoft, E., Romeijn, H.E. and Uryasev, S., Asset/liability management for pension funds using CVaR constraints. *J. Risk Finance*, 2001, **3**(1), 57–71.

Broadie, M., Computing efficient frontiers using estimated parameters. *Ann. Oper. Res.*, 1993, **45**, 21–58.

Chopra, V. and Ziemba, W.T., The effects of errors in means, variances, and covariances on optimal portfolio choice. *J. Portfolio Manag.*, 1993, **Winter**, 6–11.

Foellmer, H. and Schied, A., *Stochastic Finance: An Introduction in Discrete Time*, 2004 (W. de Gruyter, Berlin).

Høyland, K., Kaut, M. and Wallace, S.W., A heuristic for moment matching scenario generation. *Comput. Optim. Appl.*, 2003, **24**(2–3), 169–185.

Jarque, C.M. and Berra, A.K., Efficient tests for normality, homoscedasticity and serial independence of regression residuals. *Econ. Lett.*, 1980, **6**(5), 255–259.

Jobst, N. and Zenios, S.A., The tail that wags the dog: Integrating credit risk in asset portfolios. *J. Risk Finance*, 2001, **3**(1), 31–43.

Jorion, P., *Value at Risk: The New Benchmark for Managing Financial Risk*, 2001 (McGraw-Hill: New York).

Kallberg, J.G. and Ziemba, W.T., Mis-specifications in portfolio selection problems. In *Risk and Capital*, edited by G. Bamberg and K. Spremann, pp. 74–87, 1984 (Springer Verlag: Berlin).

Kaut, M. and Wallace, S.W., Evaluation of scenario-generation methods for stochastic programming. *Pacific J. Optim.*, 2007, **3**(2), 257–271.

Krokhmal, P., Uryasev, S. and Zrazhevsky, G., Risk management for hedge fund portfolios: A comparative analysis of linear portfolio rebalancing strategies. *J. Alternative Invest.*, 2002, **5**(1), 10–29.

Krokhmal, P., Uryasev, S. and Zrazhevsky, G., Numerical comparison of CVaR and CDaR approaches: Application to hedge funds. In *The Stochastic Programming Approach to Asset-Liability and Wealth Management*, edited by W.T. Ziemba, 2003 (AIMR/Blackwell: Charlottesville, VA).

Rockafellar, R.T. and Uryasev, S., Conditional value-at-risk for general distributions. *J. Bank. Finance*, 2002, **26**(7), 1443–1471.

Tasche, D., Expected shortfall and beyond. *J. Bank. Finance*, 2002, **26**(7), 1253–1272.

Topaloglou, N., Vladimirou, H. and Zenios, S.A., CVaR models with selective hedging for international asset allocation. *J. Bank. Finance*, 2002, **26**(7), 1535–1561.

Topaloglou, N., Vladimirou, H. and Zenios, S.A., A dynamic stochastic programming model for international portfolio management. *Eur. J. Oper. Res.*, DOI: 10.1016/j.ejor.2005.07.035, 2008, **185**(3), 1501–1524.

APPENDIX 15A: PROPERTIES OF THE DATA

Tables 15.A1 and 15.A2 present the first four marginal moments and the correlation matrix of the monthly differentials in the historical market data of the random variables (returns of the stock and bond indices, as well as of the spot currency exchange rates). These statistics constitute the targets matched in the scenario sets used in the empirical tests of Section 15.4. Note that the random variables have skewness ranging from −1.00 to 1.36 and kurtosis ranging from 2.78 to 7.39. The historical observations indicate that the random variables in the international portfolio management problem are not normally distributed—Jarque–Berra (1980) tests reject the normality hypothesis for these data (see Topaloglou *et al.* 2008). This was a primary factor behind modelling choices in this study. That is,

(1) we adopted the CVaR risk measure as it is suitable to accommodate higher-order moments and measures risk in the tail of the portfolio's return distribution;

(2) we employed a scenario generation method based on principles of moment-matching as it provides full control of the moments in the generation of scenarios.

Tables 15.A3 and 15.A4 present the lengths of the *variation intervals* for the moments and correlations. The *variation intervals* are defined and used in the model sensitivity tests in Section 15.4.2.

TABLE 15.A1 Moments of Monthly Differentials of the Historical Market Data

	Stk.USA	Stk.UK	Stk.Ger	Stk.Jap	Bnd1.USA	Bnd7.USA	Bnd1.UK	Bnd7.UK
Mean	0.01296	0.01047	0.01057	−0.00189	0.00553	0.00702	0.00718	0.00894
SD	0.04101	0.04150	0.05796	0.06184	0.00467	0.01620	0.00688	0.01884
Skewness	−0.47903	−0.19051	−0.47281	0.04768	−0.18341	−0.07482	1.36036	0.12127
Kurtosis	3.76519	3.11399	4.11970	3.62119	2.77801	3.23974	7.38764	3.52858

Bnd1.Ger	Bnd7.Ger	Bnd1.Jap	Bnd7.Jap	ExR.UK	ExR.Ger	ExR.Jap
0.00535	0.00671	0.00318	0.00622	−0.00077	−0.00152	0.00179
0.00455	0.01368	0.00506	0.01681	0.02801	0.03021	0.03607
0.55214	−0.87820	0.54803	−0.53562	−0.99772	−0.25505	1.09286
5.13927	4.42483	4.28775	5.23964	6.51592	3.80887	6.75996

TABLE 15.A2 Correlations of Monthly Differentials of the Historical Market Data

	Stk.USA	Stk.UK	Stk.Ger	Stk.Jap	Bnd1.USA	Bnd7.USA	Bnd1.UK
Stk.UK	0.6651						
Stk.Ger	0.5573	0.5911					
Stk.Jap	0.3568	0.3601	0.3429				
Bnd1.USA	0.1965	0.0844	−0.0578	−0.0105			
Bnd7.USA	0.2656	0.1150	0.0027	0.0205	0.8768		
Bnd1.UK	0.0853	0.4014	0.0276	0.0018	0.3600	0.3176	
Bnd7.UK	0.2258	0.5075	0.1714	0.0392	0.4314	0.4815	0.8175
Bnd1.Ger	0.0556	0.2642	0.0536	0.0081	0.3466	0.3574	0.6121
Bnd7.Ger	0.1687	0.3066	0.2326	0.0408	0.4385	0.5453	0.4639
Bnd1.Jap	0.0557	0.0814	−0.0005	0.0226	0.2513	0.2186	0.3274
Bnd7.Jap	0.0463	0.0493	0.0140	−0.0029	0.2831	0.3235	0.1815
ExR.UK	0.0247	−0.2177	−0.1062	0.1162	0.2422	0.1911	−0.2811
ExR.Ger	−0.0643	−0.2263	−0.2651	−0.0828	0.2716	0.2129	−0.1429
ExR.Jap	0.1126	0.0945	−0.1414	0.0475	0.1319	0.0975	0.0927

	Bnd7.UK	Bnd1.Ger	Bnd7.Ger	Bnd1.Jap	Bnd7.Jap	ExR.UK	ExR.Ger
Bnd1.Ger	0.5688						
Bnd7.Ger	0.6627	0.7779					
Bnd1.Jap	0.2645	0.4008	0.2853				
Bnd7.Jap	0.2100	0.3025	0.3093	0.7827			
ExR.UK	−0.1588	−0.2227	−0.0948	0.0145	0.0262		
ExR.Ger	−0.1332	−0.0638	−0.0598	0.1021	0.1375	0.6949	
ExR.Jap	0.0680	0.0825	−0.0072	0.0334	−0.0122	0.2680	0.4236

TABLE 15.A3 Lengths of the Variation Intervals of Moments

	Stk.USA	Stk.UK	Stk.Ger	Stk.Jap	Bnd1.USA	Bnd7.USA	Bnd1.UK	Bnd7.UK
Mean	0.00946	0.00764	0.01836	0.01055	0.00166	0.00435	0.00351	0.00591
SD	0.02035	0.01217	0.02153	0.02321	0.00171	0.00293	0.00485	0.00952
Skewness	1.62256	0.8916	1.0513	0.54337	0.79795	0.62041	2.32511	0.78578
Kurtosis	3.87492	1.82066	2.56178	0.99143	1.2739	0.891	7.2035	1.04044

	Bnd1.Ger	Bnd7.Ger	Bnd1.Jap	Bnd7.Jap	ExR.UK	ExR.Ger	ExR.Jap
	0.00366	0.00284	0.00433	0.00487	0.0057	0.0107	0.01108
	0.00174	0.00262	0.00249	0.00351	0.01533	0.00954	0.0132
	1.18387	0.55199	1.21049	1.48868	1.62235	1.22593	1.55678
	3.12805	2.33646	2.18066	5.13353	6.35287	2.086	4.0713

TABLE 15.A4 Lengths of the Variation Intervals of Correlations

	Stk.USA	Stk.UK	Stk.Ger	Stk.Jap	Bnd1.USA	Bnd7.USA	Bnd1.UK
Stk.UK	0.28245						
Stk.Ger	0.41613	0.1814					
Stk.Jap	0.33692	0.27907	0.38272				
Bnd1.USA	0.48044	0.43639	0.3613	0.21907			
Bnd7.USA	0.43832	0.44669	0.37637	0.28547	0.07029		
Bnd1.UK	0.25762	0.63827	0.47922	0.29948	0.32842	0.3272	
Bnd7.UK	0.18449	0.53845	0.3226	0.18788	0.30047	0.29125	0.10638
Bnd1.Ger	0.32837	0.53526	0.44861	0.20951	0.23945	0.21754	0.23587
Bnd7.Ger	0.28289	0.4632	0.52714	0.18048	0.25918	0.25712	0.1254
Bnd1.Jap	0.39658	0.30524	0.38022	0.47467	0.35529	0.34925	0.42718
Bnd7.Jap	0.53725	0.42657	0.43082	0.53627	0.1201	0.19085	0.34086
ExR.UK	0.32612	0.2294	0.23331	0.34925	0.37893	0.34466	0.48376
ExR.Ger	0.34274	0.23943	0.30395	0.41129	0.4128	0.42665	0.49304
ExR.Jap	0.35474	0.3389	0.22861	0.29166	0.23326	0.21832	0.19624

	Bnd7.UK	Bnd1.Ger	Bnd7.Ger	Bnd1.Jap	Bnd7.Jap	ExR.UK	ExR.Ger
Bnd1.Ger	0.18615						
Bnd7.Ger	0.19871	0.08346					
Bnd1.Jap	0.39379	0.41627	0.31011				
Bnd7.Jap	0.39649	0.35563	0.26651	0.11535			
ExR.UK	0.41516	0.57385	0.39195	0.26009	0.36195		
ExR.Ger	0.43402	0.33271	0.3127	0.22496	0.23136	0.34187	
ExR.Jap	0.18262	0.24788	0.25119	0.44653	0.5184	0.32304	0.27371

CHAPTER 16

Stress Testing for VaR and CVaR

JITKA DUPAČOVÁ and JAN POLÍVKA

CONTENTS

16.1   STRESS TESTING AND CONTAMINATION

S TRESS TESTING IS A TERM used in financial practice without any generally accepted definition. It appears in the context of quantification of losses or risks that may appear under special, mostly extremal circumstances (Kupiec 2002). Such circumstances are described by certain scenarios which may come from historical experience (a crisis observed in the past)—*historical stress test*, or may be judged to be possible in the future given changes of macroeconomic, socioeconomic or political factors—*prospective stress test*, etc. The performance of the obtained optimal decision is then evaluated along these, possibly dynamic, scenarios or the model is solved with an alternative input. Stress testing approaches differ among institutions and also due to the nature of the tested problem and the way in which the stress scenarios have been selected. In this chapter, we focus on the stress testing of two risk measures, VaR and CVaR, giving the 'test' a more precise meaning. This is made possible by the exploitation of parametric sensitivity results and the contamination technique.

The contamination approach was initiated in mathematical statistics as one of the tools for the analysis of the robustness of estimators with respect to deviations from the assumed probability distribution and/or its parameters. It goes back to von Mises and the concepts are briefly described, for example, in Serfling (1980). In stochastic programming, it was developed in a series of papers; see, for example, Dupačová (1986, 1996) for results applicable to two-stage stochastic linear programs. For application of contamination bounds, it is important that the stochastic program is reformulated as

$$\min_{x \in \mathcal{X}} F(x, P) := \int_{\Omega} f(x, \omega) P(\mathrm{d}\omega), \tag{16.1}$$

with $\mathcal{X}$ independent of P.

Via contamination, robustness analysis with respect to changes in the probability distribution P is reduced to a much simpler analysis with respect to scalar parameter λ. Assume that (16.1) is solved for probability distribution P. Denote $\varphi(P)$ the optimal value and $\mathcal{X}^*(P)$ the set of optimal solutions. The possible changes in the probability distribution P are modeled using contaminated distributions P_λ,

$$P_\lambda := (1 - \lambda)P + \lambda Q, \quad \lambda \in [0, 1], \tag{16.2}$$

with Q another *fixed* probability distribution. Limiting the analysis to a selected direction $Q - P$ only, the results are directly applicable, but they are less general than quantitative stability results with respect to arbitrary (but small) changes in P, summarized, for example, in Römisch (2003).

The objective function in (16.1) is linear in P, hence

$$F(x, \lambda) := \int_{\Omega} f(x, \omega) P_\lambda(\mathrm{d}\omega) = (1 - \lambda)F(x, P) + \lambda F(x, Q)$$

is linear in λ. Suppose that stochastic program (16.1) has an optimal solution for all considered distributions $P_\lambda, 0 \le \lambda \le 1$, of the form (16.2). Then the optimal value function

$$\varphi(\lambda) := \min_{x \in \mathcal{X}} F(x, \lambda)$$

is concave on $[0, 1]$, which implies its continuity and the existence of directional derivatives on $(0, 1)$. Continuity at the point $\lambda = 0$ is a property related to the stability results for the stochastic program in question. In general, one needs a non-empty, bounded set of optimal solutions $\mathcal{X}^*(P)$ of the initial stochastic program (16.1). This assumption, together with the stationarity of derivatives $dF(x, \lambda)/d\lambda = F(x, Q) - F(x, P)$, is used to derive the form of the directional derivative,

$$\varphi'(0^+) = \min_{x \in \mathcal{X}^*(P)} F(x, Q) - \varphi(0), \qquad (16.3)$$

which enters the upper bound for the optimal value function $\varphi(\lambda)$:

$$\varphi(0) + \lambda\varphi'(0^+) \geq \varphi(\lambda) \geq (1 - \lambda)\varphi(0) + \lambda\varphi(1), \quad \lambda \in [0, 1]; \qquad (16.4)$$

for details, see Dupačová (1986, 1996) and references therein.

If $x^*(P)$ is the *unique* optimal solution of (16.1), $\varphi'(0^+) = F(x^*(P), Q) - \varphi(0)$, i.e. the *local change in the optimal value function caused by a small change in P in direction $Q - P$ is the same as that of the objective function at $x^*(P)$.* If there are multiple optimal solutions, each of them leads to an upper bound $\varphi'(0^+) \leq F(x(P), Q) - \varphi(0)$, $x(P) \in \mathcal{X}^*(P)$. Contamination bounds can then be written as

$$(1 - \lambda)\varphi(P) + \lambda F(x(P), Q) \geq \varphi(P_\lambda) \geq (1 - \lambda)\varphi(P) + \lambda\varphi(Q), \qquad (16.5)$$

valid for an arbitrary optimal solution $x(P) \in \mathcal{X}^*(P)$ and for all $\lambda \in [0,1]$.

Contamination bounds (16.4) and (16.5) help to quantify the change in the optimal value due to the considered perturbations of (16.1). They exploit the optimal value $\varphi(Q)$ of the problem solved under the alternative probability distribution Q and the expected performance $F(x(P), Q)$ of the optimal solution $x(P)$ obtained for the original probability distribution P in situations where Q applies. Note that both of these values appear under the heading of stress testing methods.

The contaminated probability distribution P_λ may also be understood as a result of contaminating Q by P. Provided that the set of optimal solutions $x(Q)$ of the problem $\min_{x \in \mathcal{X}} F(x, Q)$ is non-empty and bounded, an alternative upper bound may be constructed in a similar way. Together with the original upper bound from (16.5), one may use a tighter upper bound

$$\min \left\{ (1 - \lambda)\varphi(P) + \lambda F(x(P), Q), \lambda\varphi(Q) + (1 - \lambda)F(x(Q), P) \right\}, \qquad (16.6)$$

for $\varphi(\lambda)$.

The contamination bounds are global, valid for all $\lambda \in [0,1]$. They are suitable for post-optimality analysis, out-of-sample analysis and stress testing in various disparate situations. For example, the choice of a degenerated distribution $Q = \delta_{\{\omega^*\}}$ may correspond to an additional stress or out-of-sample scenario ω^* or to increasing probability of an already considered scenario ω^*. Contamination bounds (16.4), (16.5) and (16.6) then provide information concerning the influence of including an additional scenario on the optimal results, etc. For stability studies with respect to small changes in the underlying probability distribution P, small values of the contamination parameter λ are typical. The choice of λ may reflect the degree of confidence in the expert's opinion, represented as the contaminating probability distribution Q, or the wish to obtain equiprobable scenarios, atoms of the contaminated distribution P_λ, and so on.

Contamination bounds were applied, *inter alia*, in Dupačová *et al.* (1998) for post-optimality analysis for multi-period two-stage bond portfolio management problems with respect to additional scenarios. In the present chapter they will be exploited for the stress testing of various optimization problems related to risk measures CVaR and VaR. There are results on the stability of optimal solutions of contaminated stochastic programs and also results for the case where the set $\mathcal{X}$ depends on P. They are not ready for direct application, but possibilities will be explained in the context of VaR.

Section 16.2 includes definitions of CVaR and VaR and the basic formulae from Rockafellar and Uryasev (2001), which open up the possibility of applying the contamination technique to the stress testing of these risk measures with respect to changes in the probability distribution. Section 16.3 is devoted to the stress testing of CVaR and of its optimal value. The results are illustrated numerically. Finally, the problems encountered in the exploitation of the contamination technique to CVaR-mean return efficient solutions are explained.

Stress testing for VaR is substantially more complicated. This can be attributed to the fact that VaR is one of the optimal solutions of an auxiliary optimization problem and that its definition involves a probability constraint. Applicable contamination results can then be obtained only under additional assumptions concerning the probability distribution P. In Section 16.4 we present stress testing for parametric VaR with respect to changes in the covariance matrix and with respect to an additional scenario. The section is concluded by an illustrative result dealing with contamination of the non-parametric VaR.

16.2   BASIC FORMULAE

Let $\mathcal{X} \subset R^n$ be a non-empty, closed set of feasible decisions x, and $\omega \in \Omega \subset R^m$ be a random vector with probability measure P on Ω which does not depend on x. Denote further

- $g(x, \omega)$ the random loss defined on $\mathcal{X} \times \Omega$,
- $G(x, P; v) := P\{\omega : g(x, \omega) \leq v\}$ the distribution function of the loss associated with a *fixed* decision $x \in \mathcal{X}$, and
- $\alpha \in (0,1)$ the selected confidence level.

Value at Risk (VaR) was introduced and recommended as a generally applicable risk measure to quantify, monitor and limit financial risks, to identify losses that occur with an acceptably small probability. There exist several slightly different formal definitions of VaR that coincide for continuous probability distributions. Here, we shall also deal with VaR for discrete distributions and we shall use the definition from Rockafellar and Uryasev (2001).

The *Value at Risk* at confidence level α is defined as

$$\text{VaR}_\alpha(x, P) = \min\{v \in R: G(x, P; v) \geq \alpha\}, \qquad (16.7)$$

and the 'upper' Value at Risk is

$$\mathrm{VaR}_\alpha^+(x, P) = \inf\{v \in R: G(x, P; v) > \alpha\}.$$

Hence, a random loss greater than VaR_α occurs with probability equal to (or less than) $1 - \alpha$. This interpretation is well understood in financial practice.

However, VaR_α does not quantify the loss, it is a qualitative risk measure, and, in general, it lacks the subadditivity property. (An exception is the elliptic distributions G (Embrechts *et al.* 2002), of which the normal distribution is a special case.) Various specific features and weak points of the recommended VaR methodology are summarized and discussed, for example, in Dempster (2002) and in chapter 10 of Rachev and Mittnik (2000). To solve these problems, new risk measures have been introduced; see, for example, Acerbi and Tasche (2002). We shall exploit the results of Rockafellar and Uryasev (2001) to discuss one of them, the Conditional Value at Risk, which may be linked to integrated chance constraints (Klein Haneweld 1986), to constraints involving conditional expectations (Prékopa 1973) and to the absolute Lorenz curve at point α (Ogryczak and Ruszczyński 2002).

According to Rockafellar and Uryasev (2001), CVaR_α, the *Conditional Value at Risk* at confidence level α, is defined as the mean of the α-tail distribution of $g(x, \omega)$, which, in turn, is defined as

$$G_\alpha(x, P; v) = 0, \quad \text{for } v < \mathrm{VaR}_\alpha(x, P),$$

$$G_\alpha(x, P; v) = \frac{G(x, P; v) - \alpha}{1 - \alpha}, \quad \text{for } v \ge \mathrm{VaR}_\alpha(x, P). \tag{16.8}$$

We shall assume below that $g(x, \omega)$ is a continuous function of x for all $\omega \in \Omega$ and $E_P|g(x, \omega)| < \infty$, $\forall x \in \mathcal{X}$. For $v \in R$, define

$$\Phi_\alpha(x, v, P) := v + \frac{1}{1 - \alpha} E_P(g(x, \omega) - v)^+. \tag{16.9}$$

The fundamental minimization formula of Rockafellar and Uryasev (2001) helps to evaluate CVaR for general loss distributions and to analyse its stability, including stress testing.

Theorem 16.2.1 (Rockafellar and Uryasev 2001): *As a function of v, $\Phi_\alpha(x, v, P)$ is finite and convex (hence continuous) with*

$$\min_v \Phi_\alpha(x, v, P) = CVaR_\alpha(x, P), \tag{16.10}$$

and

$$\arg\min{}_v \Phi_\alpha(x, v, P) = [\text{VaR}_\alpha(x, P), \text{VaR}_\alpha^+(x, P)], \tag{16.11}$$

a non-empty compact interval (possibly one point only).

The auxiliary function $\Phi_\alpha(x, v, P)$ is evidently linear in P and convex in v. Moreover, if $g(x, \omega)$ is a *convex* function of x, $\Phi_\alpha(x, v, P)$ is convex jointly in (v, x). In addition, $\text{CVaR}_\alpha(x, P)$ is continuous with respect to α (Rockafellar and Uryasev 2001).

If P is *a discrete probability distribution* concentrated on $\omega^1, \ldots, \omega^S$, with probabilities $p_s > 0$, $s = 1, \ldots, S$, and x a *fixed* element of $\mathcal{X}$, then the optimization problem (16.10) has the form

$$\min_v \left\{ v + \frac{1}{1-\alpha} \sum_s p_s(g(x, \omega^s) - v)^+ \right\}, \tag{16.12}$$

and can be further rewritten as

$$\min_{v, y_1, \ldots, y_S} \left\{ v + \frac{1}{1-\alpha} \sum_s p_s y_s : y_s \geq 0, \quad y_s + v \geq g(x, \omega^s), \; \forall s \right\}.$$

There are various papers that discuss the properties of VaR and CVaR and the relations between them; see, for example, Dempster (2002) and Pflug (2001). We shall focus on contamination-based stress testing for these two risk measures.

16.3 STRESS TESTING FOR CVaR

For a *fixed* vector x we now consider a stress test of $\text{CVaR}_\alpha(x, P)$, i.e. of the optimal value of (16.10). Let Q be the stress probability distribution. We apply the contamination technique and proceed as explained in Section 16.1. According to theorem 16.2.1, $\Phi_\alpha(x, v, P)$ is the corresponding objective function whose minimum equals $\text{CVaR}_\alpha(x, P)$. Evidently, the contaminated objective function

$$\Phi_\alpha(x, v, \lambda) := \Phi_\alpha(x, v, P_\lambda)$$

is linear in λ and convex in v. Its optimal value $\text{CVaR}_\alpha(x, \lambda) := \text{CVaR}_\alpha(x, P_\lambda)$ is concave in λ on $[0, 1]$ and the set of optimal solutions (16.11) of the initial problem (16.10) is bounded. Hence, the derivative of $\text{CVaR}_\alpha(x, \lambda)$, i.e. of the optimal value of the contaminated problem (16.10), at $\lambda = 0^+$ is

$$\frac{d}{d\lambda}\text{CVaR}_\alpha(x, 0^+) = \min_v \Phi_\alpha(x, v, Q) - \text{CVaR}_\alpha(x, P), \tag{16.13}$$

with minimization carried over the set (16.11) of *optimal solutions* of (16.10) formulated and solved for the probability distribution P. An upper bound for the derivative is obtained when minimization over (16.11) is replaced by the evaluation of $\Phi_\alpha(x, v, Q)$ at an arbitrary *optimal solution* $v^*(x, P)$ of (16.10), for example at $v^*(x, P) = \mathrm{VaR}_\alpha(x, P)$.

The *contamination bounds* for $\mathrm{CVaR}_\alpha(x, \lambda)$ for a fixed x follow from the concavity of $\mathrm{CVaR}_\alpha(x, \lambda)$ with respect to λ:

$$(1 - \lambda)\mathrm{CVaR}_\alpha(x, 0) + \lambda\mathrm{CVaR}_\alpha(x, 1)$$
$$\leq \mathrm{CVaR}_\alpha(x, \lambda) \leq \mathrm{CVaR}_\alpha(x, 0) + \lambda\frac{\mathrm{d}}{\mathrm{d}\lambda}\mathrm{CVaR}_\alpha(x, 0^+)$$
$$= (1 - \lambda)\mathrm{CVaR}_\alpha(x, 0) + \lambda \min_v \Phi_\alpha(x, v, Q), \qquad (16.14)$$

for all $0 \leq \lambda \leq 1$. The combined upper bound (16.6) can be constructed in a similar way.

16.3.1 Stress Testing of the Scenario-Based Form of CVaR

Consider first an application of the contamination bounds to the stress testing of the scenario-based form (16.12) of CVaR. Let P be a discrete probability distribution concentrated on $\omega^1, \ldots, \omega^S$ with probabilities p_s, $s = 1, \ldots, S$, x a *fixed* element of $\mathcal{X}$ and Q a discrete probability distribution carried by S' stress or out-of-sample scenarios ω^s, $s = S+1, \ldots, S+S'$, with probabilities p_s, $s = S+1, \ldots, S+S'$. Both $\mathrm{CVaR}_\alpha(x, P)$ and $\mathrm{CVaR}_\alpha(x, Q)$ can be obtained by solving the corresponding linear programs (16.12). Denote by $v^*(x, P)$ an optimal solution of (16.12) for fixed $x \in \mathcal{X}$ and for distribution P.

Bounds for CVaR_α for the contaminated probability distribution P_λ carried by the initial scenarios ω^s, $s = 1, \ldots, S$, with probabilities $(1 - \lambda)p_s$, $s = 1, \ldots, S$, and by the stress scenarios ω^s, $s = S+1, \ldots, S+S'$, with probabilities λp_s, $s = S+1, \ldots, S+S'$, have the form

$$(1 - \lambda)\mathrm{CVaR}_\alpha(x, P) + \lambda\mathrm{CVaR}_\alpha(x, Q) \leq \mathrm{CVaR}_\alpha(x, P_\lambda)$$
$$\leq (1 - \lambda)\mathrm{CVaR}_\alpha(x, P) + \lambda\Phi_\alpha(x, v^*(x, P), Q)$$
$$= \Phi_\alpha(x, v^*(x, P), P_\lambda), \qquad (16.15)$$

and are valid for all $\lambda \in [0,1]$; compare with (16.13) and (16.14).

In the special case of a *degenerate probability distribution* Q carried only by one scenario ω^*, $\mathrm{CVaR}_\alpha(x, Q) = g(x, \omega^*)$ and

$$\Phi_\alpha(x, v^*(x, P), Q) = v^*(x, P) + \frac{1}{1 - \alpha}(g(x, \omega^*) - v^*(x, P))^+.$$

The difference between the upper and lower bounds in (16.14) is

$$\lambda[\Phi_\alpha(x, v^*(x, P), Q) - \mathrm{CVaR}_\alpha(x, Q)]$$
$$= \lambda\left[v^*(x, P) + \frac{1}{1 - \alpha}(g(x, \omega^*) - v^*(x, P))^+ - g(x, \omega^*)\right].$$

In typical applications, the 'stress test' is reduced to evaluating the performance of the already obtained optimal solution along the new scenarios, i.e. the evaluation of $\Phi_\alpha(x, v^*(x, P), Q)$, or obtaining the optimal value such as $\text{CVaR}_\alpha(x, Q)$ for Q carried by the stress scenarios. Contamination bounds (16.15) exploit these criteria simultaneously to *quantify* the influence of the stress scenarios, also taking into account the probability of their occurrence. As a result, they provide a genuine stress test.

16.3.2 Sensitivity Properties of Optimal Solutions

To derive the sensitivity properties of the optimal solutions of (16.10) for *fixed x*, assume that the optimal solution of (16.10) is *unique*, $v^*(x, P)$; hence, it equals $\text{VaR}_\alpha(x, P)$. This also simplifies the form of the derivative of $\text{CVaR}_\alpha(x, \lambda)$ in (16.13) to $\Phi_\alpha(x, \text{VaR}_\alpha(x, P), Q) - \text{CVaR}_\alpha(x, P)$.

The general results concerning the properties of optimal solutions for contaminated distributions (see, for example, Dupačová 1986, 1987 and Shapiro 1990) require additional properties concerning the smoothness of the objective function (16.9) in (16.10). To this end we assume that the probability distribution function $G(x, P; v)$ is continuous, with a positive, continuous density $p(x, P; v)$ on a neighborhood of the unique optimal solution $v^*(x, P) = \text{VaR}_\alpha(x, P)$ of (16.10).

For fixed $x \in \mathcal{X}$ we denote $\eta := g(x, \omega)$, $v := v^*(x, P)$ and use definition (16.9) of $\Phi_\alpha(x, v, P)$. Except for $v = \eta$, the derivative $(d/dv)(\eta - v)^+$ exists and

$$\frac{d}{dv}(\eta - v)^+ = -\frac{1}{2}\left(1 + \frac{\eta - v}{|\eta - v|}\right).$$

Thanks to the assumed properties of the distribution function $G(x, P; v)$, the expected value

$$E_P \frac{d}{dv}(\eta - v)^+ = -P(\eta > v) = -1 + G(x, P; v),$$

and

$$\frac{d}{dv}\Phi_\alpha(x, v, P) = 1 + \frac{G(x, P; v) - 1}{1 - \alpha}.$$

The optimality condition $(d/dv)\Phi_\alpha(x, v, P) = 0$ provides, as expected,

$$\text{VaR}_\alpha(x, P) = v^*(x, P) = G(x, P)^{-1}(\alpha).$$

The second-order derivative $(d^2/dv^2)\Phi_\alpha(x, v, P) = [p(x, P; v)/(1 - \alpha)]$ is positive on a neighborhood of $v^*(x, P)$. Direct application of the implicit function theorem to the system

$$\frac{d}{dv}\Phi_\alpha(x, v, P_\lambda) = 0$$

implies the existence and uniqueness of optimal solution $v^*(x, \lambda) := v^*(x, P_\lambda)$ of the contaminated problem (16.10) for $\lambda > 0$ sufficiently small, and the form of its derivative

$$\frac{d}{d\lambda}v^*(x, P_\lambda) = \frac{d}{d\lambda}\text{VaR}_\alpha(x, P_\lambda) = \frac{\alpha - G(x, Q; v^*(x, P))}{p(x, P; v^*(x, P))}, \qquad (16.16)$$

for $\lambda = 0^+$. Here, $G(x, Q; v)$ denotes the loss distribution function under probability distribution Q. Note that, except for the existence of the expected values, no further assumptions are required concerning Q. Related results for absolutely continuous probability distributions P and Q can be found, for example, in Rau-Bredow (2004).

16.3.3 Optimization Problems with the CVaR$_\alpha(x, P)$ Objective Function

For the next step, let us briefly discuss optimization problems with the CVaR$_\alpha(x, P)$ objective function, which provide the optimal (with respect to the CVaR$_\alpha(x, P)$ criterion) solutions

minimize CVaR$_\alpha(x, P)$ on a closed set $\emptyset \neq \mathcal{X} \subset R^n$.

Using (16.10), the problem is

$$\min_{x,v} \Phi_\alpha(x, v, P), \quad x \in \mathcal{X}. \qquad (16.17)$$

For $\mathcal{X}$ convex, independent of P, and for loss functions $g(\bullet, \omega)$ convex for all ω, $\Phi_\alpha(x, v, P)$ is convex in (x, v) and standard stability results apply. Moreover, if P is the discrete probability distribution considered in Section 16.3.1, $g(\bullet, \omega)$ a linear function of x, say $g(\bullet, \omega) = x^\top \omega$, and $\mathcal{X}$ convex polyhedral, we obtain the linear program

$$\min_{v, y_1,\dots,y_S, x} \left\{ v + \frac{1}{1-\alpha}\sum_s p_s y_s : y_s \geq 0, x^\top \omega^s - v - y_s \leq 0, \forall s, x \in \mathcal{X} \right\}. \qquad (16.18)$$

Let $(v_C^*(P), x_C^*(P))$ be an optimal solution of (16.17) and denote by $\varphi_C(P)$ the optimal value. To obtain contamination bounds for the optimal value of (16.17) with P contaminated by stress probability distribution Q, it is sufficient to assume a compact set $\mathcal{X}$, e.g. $\mathcal{X} = \{x \in R^n : \sum_i x_i = 1, x_i \geq 0, \forall i\}$. The bounds follow the usual pattern (compare with (16.15)):

$$(1-\lambda)\varphi_C(P) + \lambda\varphi_C(Q) \leq \varphi_C(P_\lambda) \leq (1-\lambda)\varphi_C(P) + \lambda\Phi_\alpha(x_C^*(P), v_C^*(P), Q). \qquad (16.19)$$

To apply them, one has to evaluate $\Phi_\alpha(x_C^*(P), v_C^*(P), Q)$ and solve (16.17) with P replaced by the stress distribution Q.

16.3.4 An Illustrative Example

The instruments used in the portfolio management problem (16.18) are the total return indices given in Table 16.1.

The portfolio limits were set in all cases to $x_i \leq 0.3$, hence,

$$\mathcal{X} = \left\{ x \in R^n : \sum_i x_i = 1, 0 \leq x_i \leq 0.3, \forall i \right\}.$$

Assume that the probability distribution P is the distribution of losses under 'normal' conditions, whereas probability distribution Q refers to the situation when adverse conditions prevail on the world market. Both P and Q are distributions of monthly percentage losses to assets $i = 1, \ldots, 12$, which were converted into the home currency (EUR) using the exchange rate mid. We do not consider transaction costs.

The following approach, resembling an historical simulation, was taken to construct discrete distributions P and Q. For asset $i = 1$ (US asset market returns) the percentage returns (not losses) in the home currency were computed. We took the empirical 25% quantile to be the cut-off value for all returns of asset 1. The returns below the cut-off value (and all corresponding returns of other assets on the same date) are attributed to a period of adverse conditions prevailing on the market and hence this data set serves as the input for the approximation of the distribution Q. The rest of the data sample was used for fitting the distribution P.

The two discrete probability distributions P, Q approximating the true continuous distribution of assets' percentage losses in the home currency were constructed using the method of Høyland et al. (2003). We prescribed that both discrete approximations P, Q were carried by 5184 equiprobable scenarios. The empirical means, variances, covariances, skewnesses and kurtoses computed separately from the two data samples enter the scenario fitting procedure for P and Q.

After solving the two CVaR minimization problems with $\alpha = 0.99$, contamination bounds (16.19) sharpened according to (16.6), were constructed. The results of contamination are

TABLE 16.1 Portfolio Assets (MSCI and JP Morgan Indexes)

Asset	Acronym	Description
MSCI Gross Return index US, USD	1	Stock index
MSCI Gross Return index UK, USD	2	Stock index
MSCI Gross Return index Germany, USD	3	Stock index
MSCI Gross Return index Japan, USD	4	Stock index
US Government Bond index (1–3 y mat), USD	5	
US Government Bond index (7–10 y mat), USD	6	
UK Government Bond index (1–3 y mat), GPB	7	
UK Government Bond index (7–10 y mat), GPB	8	
Germany Government Bond index (1–5 y mat), EUR	9	
Germany Government Bond index (7+ y mat), EUR	10	
Japan Government Bond index (1–3 y mat), JPY	11	
Japan Government Bond index (7–10 y mat), JPY	12	

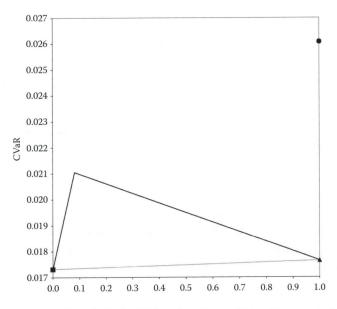

FIGURE 16.1 Contamination bounds for the CVaR optimization problem without constraint on returns.

presented in Figure 16.1 and Table 16.2. The VaR values $v_C^*(P), v_C^*(Q)$ for distributions P, Q calculated for the optimal portfolios $x_C^*(P), x_C^*(Q)$ are obtained as a by-product.

$$(1 - \lambda)\varphi_C(P) + \lambda\varphi_C(Q) \leq \varphi_C(P_\lambda) \leq \min\{(1 - \lambda)\varphi_C(P)$$
$$+ \lambda\Phi_\alpha(x_C^*(P), v_C^*(P), Q), \lambda\varphi_C(Q) + (1 - \lambda)\Phi_\alpha(x_C^*(Q), v_C^*(Q), P)\}, \qquad (16.20)$$

TABLE 16.2 Quantities used in Contamination Bounds (16.20) and Nonzero Components of Optimal Solutions $x_C^*(P)$ and $x_C^*(Q), \alpha = 0.99$

Quantity	Value
$\varphi_C(P)$	0.01731
$\varphi_C(Q)$	0.01765
$\Phi_\alpha(x_C^*(P), v^*(P), Q)$	0.06309
$\Phi_\alpha(x_C^*(Q), v^*(Q), P)$	0.02135
$x_1^*(P)$	0.12880
$x_7^*(P)$	0.20030
$x_9^*(P)$	0.30000
$x_{10}^*(P)$	0.26470
$x_{11}^*(P)$	0.10620
$v_C^*(P)$	0.01365
$x_5^*(Q)$	0.10000
$x_7^*(Q)$	0.30000
$x_9^*(Q)$	0.30000
$x_{10}^*(Q)$	0.30000
$v_C^*(Q)$	0.01588
$CVaR(x_C^*(P), Q)$	0.02607

Some observations are given below.

- The two minimal CVaR values $\varphi C(P)$, $\varphi_C(Q)$ (indicated in the figure by a square and a triangle, respectively) are not very different. This is the result of optimally restructuring the portfolio in the adverse market situation; see the changed composition of the optimal portfolios. The CVaR value for probability distribution Q and for the original optimal portfolio $x_C^*(P)$, i.e. without restructuring the portfolio (indicated by the isolated point in the right upper corner of the figure), is much higher.
- The value $\Phi_\alpha(x_C^*(P), v_C^*(P), Q)$ is relatively large and this determines the steep slope of the left upper bound.
- The contamination bounds in this example are not very tight (see Figure 16.1). The maximal difference between the upper and lower bounds occurs approximately at $\lambda = 0.1$. For $\lambda = 0.5$, i.e. for the distribution carried by the pooled sample of 10 368 equiprobable scenarios, the minimal CVaR value lies in $[0.0175, 0.0195]$. If this precision is sufficient, one does not need to solve the problem with twice the number of scenarios—atoms of the contaminated probability distribution.

16.3.5 Stress Testing for CVaR-Mean Return Problems

Finally, consider stress testing for CVaR-mean return problems, i.e. for bi-criteria problems in which one aims simultaneously for a minimization of $\mathrm{CVaR}_\alpha(x, P)$ and a maximization of the expected return criterion $E_{Pr}(x, \omega)$ on $\mathcal{X}$; see, for example, Rockafellar and Uryasev (2000), Andersson et al. (2001), Pflug (2001), Topaloglou et al. (2002), and Kaut et al. (2007).

To obtain an efficient solution, one minimizes on $\mathcal{X}$ the parametrized objective function

$$\mathrm{CVaR}_\alpha(x, P) - \rho E_{Pr}(x, \omega), \tag{16.21}$$

with parameter $\rho > 0$, or assigns a parametric bound on one of the criteria and solves, for example,

$$\min \mathrm{CVaR}_\alpha(x, P) \text{ on the set } \mathcal{X} \cap \{x : E_{Pr}(x, \omega) \geq r\}. \tag{16.22}$$

The optimal solution and the corresponding values of the two criteria, CVaR_α and the expected return, depend on the chosen parameter values. To obtain the efficient frontier, (16.21) and (16.22) may be solved by parametric programming techniques with scalar parameters ρ or r, respectively. For $g(x, \omega) = x^\top \omega = -r(x, \omega)$, for polyhedral set $\mathcal{X}$ and a discrete probability distribution P, both (16.21) and (16.22) are then parametric linear programs with one scalar parameter; see, for example, Ruszcyński and Vanderbei (2003). By solving (16.22), the efficient frontier is obtained directly. To obtain the efficient frontier in the case of (16.21), values of $E_{Pr}(x, \omega)$ and $\mathrm{CVaR}_\alpha(x, P)$ have to be computed at the optimal solution of (16.21) obtained for a specific value of ρ. Hence, (16.22) is favored for the straightforward possibility of interpreting the trade-off between the two criteria, whereas (16.21) is suitable for developing sensitivity and stability results, including stress testing.

Contamination of the probability distribution P introduces an additional parameter λ into (16.21) and (16.22) and the two problems, in general, lose the readily solvable form of parametric linear programs: nonlinearity with respect to ρ and λ appears in the objective function of (16.21) and both the objective function and the set of feasible solutions of (16.22) depend on the parameters. It is still possible to obtain directional derivatives of the optimal value function for the corresponding contaminated problem. However, the optimal value function is no longer concave, hence the crucial property for the construction of contamination bounds is lost. The same applies also to problem formulations with several CVaR constraints, each with a different confidence level α, called 'risk-shaping' (Rockafellar and Uryasev 2001).

Nevertheless, contamination bounds may be obtained for the special form of the return function $r(x, \omega) = -x^\top \omega$ and for a certain class of probability distributions. Rewrite problem (16.22) as

$$\text{minimize } \mathrm{CVaR}_\alpha(x, P)$$

on the set

$$\mathcal{X}(P, r) = \{x \in \mathcal{X} : -x^\top E_P \omega \geq r\}. \tag{16.23}$$

Let $\varphi_r(P)$ denote the optimal value and $\mathcal{X}_r^*(P)$ the set of optimal solutions and assume that $\mathcal{X}_r^*(P)$ is non-empty and bounded.

Assume, in addition, that the *expected values are equal*, $E_P \omega = E_Q \omega = \bar{\omega}$. (Such an assumption is not typical for stress testing, but it is in agreement with scenario generation methods based on moment fitting (e.g. Høyland *et al.* (2003) and Høyland and Wallace (2001)), and has also been used in the stability studies of Kaut *et al.* (2003).) Then the expected return constraint is $-x^\top \bar{\omega}$, both for the initial probability distribution P and the contaminating distribution Q, as well as for all P_λ, $\lambda \in [0,1]$, and it does not depend on λ. The optimal value function $\varphi_r(P_\lambda) = \varphi_r(\lambda)$ is concave and the contamination bounds have a form similar to (16.19) and (16.20). They are obtained for (16.17) with the set of feasible decisions $\mathcal{X}$ replaced by $\mathcal{X}(P, r) = \{x \in \mathcal{X} : -x^\top \bar{\omega} \geq r\}$. Moreover, there are parametric programming techniques (e.g. Guddat *et al.* 1985) applicable to the contaminated problem (16.23), i.e. the minimization of $\mathrm{CVaR}_\alpha(x, P_\lambda)$ on the set $\mathcal{X}(P, r) = \{x \in \mathcal{X} : -x^\top \bar{\omega} \geq r\}$. They are discussed in Dupačová (2006) and a qualitative conclusion can be summarized as follows.

Under modest non-degeneracy assumptions, a small contamination of P does not influence the composition of CVaR-mean return efficient portfolios.

Note that, for $E_P \omega = E_Q \omega = \bar{\omega}$, problem (16.21) is also simplified, and the objective function is *linear* in the two parameters ρ and λ.

When the *expected loss differs* under P and Q, the optimal value $\varphi_C(P_\lambda)$ is a natural lower bound for $\varphi_r(P_\lambda)$, hence by (16.19),

$$\varphi_r(P_\lambda) \geq (1 - \lambda)\varphi_C(P) + \lambda\varphi_C(Q). \tag{16.24}$$

To construct an upper bound for $\varphi_r(P_\lambda)$ we add the additional constraint $-x^\top E_Q \omega \geq \mathcal{X}(P, r)$. The set of feasible solutions $\mathcal{X}(P, r) \cap \mathcal{X}(Q, r) \subset \mathcal{X}(P_\lambda, r)$ is polyhedral and does not depend on λ. If $\mathcal{X}(P, r) \cap \mathcal{X}(Q, r) \neq \emptyset$ we obtain a *concave* upper bound

$$U_r(\lambda) := \min_{x \in \mathcal{X}(P,r) \cap \mathcal{X}(Q,r)} \text{CVaR}_\alpha(x, P_\lambda) \geq \varphi_r(P_\lambda),$$

which may be bounded from above by the corresponding upper contamination bound. The derivative at the point $\lambda = 0^+$ is of a familiar form—$\min \Phi_\alpha(x, v, Q) - U_r(0)$ with minimization carried over the set of optimal solutions of (16.17) for $\mathcal{X}$ replaced by $\mathcal{X}(P, r) \cap \mathcal{X}(Q, r)$; denote one of them by $\hat{x}_r(P), \hat{v}_r(P)$:

$$(1 - \lambda)U_r(0) + \lambda\Phi_\alpha(\hat{x}_r(P), \hat{v}_r(P), Q) \geq U_r(\lambda) \geq \varphi_r(P_\lambda). \tag{16.25}$$

We have not tested bounds (16.24) and (16.25) numerically, but we expect that they may be quite loose.

16.4   STRESS TESTING FOR VAR

Up to the non-uniqueness of the definitions, $\text{VaR}_\alpha(x,P)$ is the same as the α-*quantile of the loss distribution* $G(x, P, v)$. One can also treat $\text{VaR}_\alpha(x, P)$ as the optimal value of the stochastic program (16.7) with one probabilistic constraint. Such an approach enables us to exploit the existing stability results for stochastic programs of that form (Römisch 2003), which are valid under special distributional and regularity assumptions.

A normal distribution of losses is one of the manageable cases and, initially, *parametric* VaR was developed to quantify the risks associated with normally distributed losses $g(x, \omega)$, the distribution of which at a fixed point x is fully determined by its expectation $\mu(x)$ and variance $\sigma^2(x)$:

$$\text{absolute VaR}_\alpha(x) = \mu(x) + \sigma(x) \cdot u_\alpha,$$
$$\text{and relative VaR}_\alpha(x) = \sigma(x) \cdot u_\alpha,$$

where u_α is the α-quantile of the standard normal $\mathcal{N}(0, 1)$ distribution.

For an arbitrary $\alpha > 0.5$, minimization of the relative VaR_α reduces to minimization of the standard deviation (volatility) of the portfolio losses, and minimization of the absolute VaR_α is minimization of the weighted sum of the standard deviation and the expectation.

16.4.1   Optimization Problem with the Relative VaR$_\alpha(x, P)$ Objective Function

Choose $\alpha > 0.5$ and assume that losses are of the form

$$g(x, \omega) = x^\top \omega,$$

$\mathcal{X}$ is a non-empty, convex polyhedral set, $0 \notin \mathcal{X}$, ω is normally distributed with mean vector μ and a positive definite variance matrix Σ.

The problem is to select portfolio composition $x \in \mathcal{X}$ such that VaR_α is minimal, i.e. to minimize the convex quadratic function $x^\top \Sigma x$ on the set $\mathcal{X}$. In this case, for all values of $\alpha > 0.5$ there is *the same, unique* optimal solution $x^*(\Sigma)$, the composition of the portfolio, which depends on the input variance matrix Σ that was obtained by an estimation procedure and is subject to an estimation error. The same optimal solution is arrived at by minimization of $\text{CVaR}_\alpha(x, P)$ (Rockafellar and Uryasev 2000).

Asymptotic statistics and a detailed analysis of optimal solutions of parametric quadratic programs may help to derive asymptotic results concerning the 'estimated' optimal portfolio composition obtained for an asymptotically normal estimate $\hat{\Sigma}$ of Σ.

Here we follow a suggestion of Kupiec (2002) and rewrite the variance matrix as $\Sigma = DCD$ with the diagonal matrix D of 'volatilities' (standard deviations of the marginal distributions) and the correlation matrix C. Changes in the covariances may then be modeled by 'stressing' the correlation matrix C by a positive semi-definite *stress correlation matrix* $\hat{C}$

$$C(\gamma) = (1 - \gamma)C + \gamma\hat{C}, \tag{16.26}$$

with $\gamma \in [0,1]$ a parameter. This type of perturbation of the initial quadratic program allows us to apply the related stability results of Bank *et al.* (1982) to the perturbed problem,

$$\min_{x \in \mathcal{X}} x^\top DC(\gamma)Dx, \quad \gamma \in [0, 1]: \tag{16.27}$$

- the optimal value $\varphi_V(\gamma)$ of (16.27) is concave and continuous in $\gamma \in [0,1]$;
- the optimal solution $x^*(\gamma)$ is a continuous vector in the range of γ where $C(\gamma)$ is positive definite;
- the directional derivative of $\varphi_V(\gamma)$

$$\varphi_V'(0^+) = x^{*\top}(0)D\hat{C}Dx^*(0) - \varphi_V(0).$$

Contamination bounds constructed as suggested in Section 16.1,

$$(1 - \gamma)x^{*\top}(0)DCDx^*(0) + \gamma x^{*\top}(1)D\hat{C}Dx^*(1) \leq \min_{x \in \mathcal{X}} x^\top DCDx$$
$$\leq (1 - \gamma)x^{*\top}(0)DCDx^*(0) + \gamma x^{*\top}(0)D\hat{C}Dx^*(0),$$

quantify the effect of the considered change in the input data.

16.4.2 Stress Testing of the Relative VaR with Respect to an Additional Scenario ω^*

In this case, the contaminating distribution Q is degenerate, $Q = \delta_{\{\omega^*\}}$. Rewriting (16.16) for the case of a normally distributed loss, we obtain

$$\frac{d}{d\lambda}\mathrm{VaR}_\alpha(x, P_\lambda)|_{\lambda=0^+} = \frac{\alpha - I\{g(x,\omega^*) \le \mathrm{VaR}_\alpha(x, P)\}}{\phi(\mathrm{VaR}_\alpha(x, P))}. \tag{16.28}$$

In the above formula, x is fixed, ϕ denotes the density of the normal distribution $\mathcal{N}(\mu(x), \Sigma(x))$ of $g(x, \omega)$ and I is the indicator function.

Assume, in addition, that $g(x,\omega) = x^\top \omega$. Using the results of Sections 16.4.1 and 16.3.2 for the normal distribution $P \sim \mathcal{N}(\mu, \Sigma)$ and degenerate distribution $Q = \delta_{\{\omega^*\}}$, we have the unique optimal portfolio $x^*(\Sigma)$ for P and both $\mathrm{VaR}_\alpha(x, P_\lambda)$ and its derivative with respect to λ are continuous for $\lambda \ge 0$ sufficiently small. This can be used to derive *sensitivity properties of the minimal relative VaR value,*

$$\varphi(\lambda) := \min_{x \in \mathcal{X}} \mathrm{VaR}_\alpha(x, P_\lambda),$$

in the case of $\mathcal{X} \ne \emptyset$, compact and for small $\lambda > 0$, i.e. when testing the influence of a rare stress scenario. Here, $\mathrm{VaR}_\alpha(x, P_\lambda)$ is not linear in λ. Still, using (16.28) and the general formula for the derivative of the optimal value of nonlinear objective functions from Danskin (1967), we obtain

$$\varphi'(0^+) = \frac{d}{d\lambda}\mathrm{VaR}_\alpha(x^*(\Sigma), P_\lambda)|_{\lambda=0^+}$$
$$= \frac{\alpha - I\{g(x^*(\Sigma), \omega^*) \le \mathrm{VaR}_\alpha(x^*(\Sigma), P)\}}{\phi(\mathrm{VaR}_\alpha(x^*(\Sigma), P))}.$$

Then, the minimal VaR_α value for the stressed distribution P_λ is approximated by

$$\min_{x \in \mathcal{X}} \mathrm{VaR}_\alpha(x, P_\lambda) \cong \mathrm{VaR}_\alpha(x^*(\Sigma), P) + \lambda\varphi'(0^+)$$

for $\lambda > 0$ sufficiently small.

This approach may easily be extended to sensitivity analysis and stress testing of VaR with respect to an additional scenario for a broad class of probability measures P for which the probability distribution of loss $G(x,P;v)$ fulfils the assumptions of Section 16.3.2.

16.4.3 Nonparametric VaR

For general probability distributions the evaluation of VaR_α for a fixed portfolio x is mostly based on a non-parametric approach that is distribution free and also applicable for complicated financial instruments. One exploits a finite number, S, of scenarios so that, for each fixed $x \in \mathcal{X}$, the underlying probability distribution P is replaced by a discrete distribution P_S carried by these scenarios and the probability distribution of the loss $g(x, \omega)$ is discrete with jumps at $g(x, \omega^s)$ $\forall s$.

For a fixed x, let us order $g(x, \omega^s)$ as

$$g^{[1]} < \cdots < g^{[S]}, \tag{16.29}$$

with the probability of $g^{[s]}$ equal to $p^{[s]} > 0$, $\forall s$. Let s_α, P_S be the unique index such that

$$\sum_{s=1}^{s_{\alpha,P_S}} p^{[s]} \geq \alpha > \sum_{s=1}^{s_{\alpha,P_S}-1} p^{[s]}. \tag{16.30}$$

then $\mathrm{VaR}_\alpha(x, P_S) = g^{[s_{\alpha,P_S}]}$.

The consistency of sample quantiles is valid under mild assumptions regarding the smoothness of the distribution function G, and one may even prove their asymptotic normality (Serfling 1980). For example, if there is a positive continuous density $p(x, P; v)$ of $G(x, P; v)$ on a neighborhood of $\mathrm{VaR}_\alpha(x, P)$ and P_S denotes an associated empirical distribution, then $\mathrm{VaR}_\alpha(x, P_S)$ is asymptotically normal,

$$\mathrm{VaR}_\alpha(x, P_S) \sim \mathcal{N}\left(\mathrm{VaR}_\alpha(x, P), \frac{\alpha(1-\alpha)}{Sp^2(x, P; \mathrm{VaR}_\alpha(x, P))}\right).$$

Estimating $\mathrm{VaR}_\alpha(x, P)$ by the *non-parametric* $\mathrm{VaR}_\alpha(x, P_S)$ calls for a large number of scenarios, especially for α close to 1; see Rachev and Mittnik (2000) for extensive numerical results. Moreover, it is evident from (16.30) that, even for fixed x, the inclusion of an additional scenario may cause an abrupt change in VaR_α.

Sensitivity results for VaR_α similar to (16.38) are obtained if the (unique) optimal solution of the CVaR_α problem (16.10) is differentiable (recall Section 16.3.2). Another possibility is to derive them by a direct sensitivity analysis of the simple chance-constrained stochastic program (16.7). In both cases, additional assumptions concerning the probability distribution P are required, such as its continuity properties listed in Section 16.3.2. There is more freedom as to the choice of the contaminating distribution Q. We refer to Dobiáš (2003) and Römisch (2003) for details.

16.4.4   Stress Testing of Nonparametric VaR

The stress testing of non-parametric VaR computed for a discrete probability distribution P carried by a finite number of scenarios ω^s, $s = 1, \ldots, S$, is more involved. To obtain an upper bound for $\mathrm{VaR}_\alpha(x, P_\lambda)$ for a fixed portfolio x, one may use the contamination-based upper bound for $\mathrm{CVaR}_\alpha(x, P_\lambda)$ in (16.15). Formula (16.30) in the definition of the empirical VaR_α implies that, for $\alpha < \sum_{s=1}^{s_{\alpha,P}} p^{[s]}$, the value of VaR_α is robust with respect to small changes in probabilities $p^{[s]}$. This indicates the possibility of covering the interval $[0, 1]$ by a finite number of non-overlapping intervals $[0, \lambda_1], (\lambda_1, \lambda_2], \ldots, (\lambda_{\bar\imath}, 1]$ and constructing bounds for $\mathrm{VaR}_\alpha(x, P_\lambda)$ separately for each of them.

We shall illustrate the approach for the case of one additional 'stress' scenario ω^* with

$$g^{[1]} < \cdots < g^{[s_{\omega^*}-1]} < g(x, \omega^*) < g^{[s_{\omega^*}]} < \cdots < g^{[S]}, \tag{16.31}$$

and with probabilities

$$(1-\lambda)p^{[1]},\ldots,(1-\lambda)p^{[s_{\omega^*}-1]},\lambda,(1-\lambda)p^{[s_{\omega^*}]},\ldots,(1-\lambda)p^{[S]},$$

i.e. for degenerate probability distribution $Q=\delta_{\{\omega^*\}}$.

Suppose that the stress scenario satisfies $g^{[s_{\alpha,P}]}<g^{[s_{\omega^*}-1]}$. It is easy to see that, in the case of $\sum_{s=1}^{s_{\alpha,P}}p^{[s]}>\alpha$, we obtain $\mathrm{VaR}_{\alpha}(x,P_{\lambda})=g^{[s_{\alpha,P}]}=\mathrm{VaR}_{\alpha}(x,P)$ for sufficiently small $\lambda\geq0$. On the other hand, if $\sum_{s=1}^{s_{\alpha,P}}p^{[s]}=\alpha$, then $\mathrm{VaR}_{\alpha}(x,P_{\lambda})=g^{[s_{\alpha,P}+1]}$ for sufficiently small $\lambda>0$. The α-quantile $g^{[s_{\alpha,P_{\lambda}}]}$ of the contaminated distribution fulfills

$$\sum_{s=1}^{s_{\alpha,P_{\lambda}}}(1-\lambda)p^{[s]}\geq\alpha\quad\text{and}\quad\sum_{s=1}^{s_{\alpha,P_{\lambda}}-1}(1-\lambda)p^{[s]}<\alpha.\qquad(16.32)$$

for $\lambda=0$, these inequalities are identical to (16.30). They remain valid with $s_{\alpha,P_{\lambda}}$ replaced by the original $s_{\alpha,P}$ for

$$\lambda\leq1-\frac{\alpha}{\sum_{s=1}^{s_{\alpha,P}}p^{[s]}}\quad\text{and}\quad1-\frac{\alpha}{\sum_{s=1}^{s_{\alpha,P}-1}p^{[s]}}<\lambda.$$

The first inequality provides an upper bound λ_1 and the second is fulfilled for all $\lambda\geq0$.

For $\lambda>\lambda_1$, $\mathrm{VaR}_{\alpha}(x,P_{\lambda})=g^{[s_{\alpha,P}+1]}$, and by solving (16.32) for $s_{\alpha,P_{\lambda}}=s_{\alpha,P}+1$ with respect to λ, we obtain an upper bound λ_2 of the interval on which $\mathrm{VaR}_{\alpha}(x,P_{\lambda})=g^{[s_{\alpha,P}+1]}$ holds true. Note that $\lambda_1=0$ if $\sum_{s=1}^{s_{\alpha,P_S}}p^{[s]}=\alpha$ and, in this case, $\lambda_2>0$.

Similarly for $\lambda>\lambda_i$ with $i<s_{\omega^*}-s_{\alpha,P}$, we obtain an upper bound λ_{i+1} of the interval for which $\mathrm{VaR}_{\alpha}(x,P_{\lambda})=g^{[s_{\alpha,P}+i]}$. This procedure stops when $i=\bar{i}:=s_{\omega^*}-s_{\alpha,P}$. In this case, (16.32) is modified to

$$\sum_{s=1}^{s_{\alpha,P}+\bar{i}-1}(1-\lambda)p^{[s]}+\lambda\cdot1\geq\alpha,$$

valid for all $\lambda\geq0$; hence, $\mathrm{VaR}_{\alpha}(x,P_{\lambda})=g(x,\omega^*)$ for $\lambda_{\bar{i}}<\lambda\leq1$.

To summarize: for contamination by one scenario as in (16.31), setting

$$\lambda_0=0,$$
$$\lambda_i=1-\frac{\alpha}{\sum_{s=1}^{s_{\alpha,P}+i-1}p^{[s]}},\quad\text{for }i=1,\ldots,s_{\omega^*}-s_{\alpha,P},$$
$$\lambda_i=1,\quad\text{for }i>s_{\omega^*}-s_{\alpha,P},$$

we obtain the following theorem.

Theorem 16.4.1: For $g^{[s_{\alpha,P}]}<g^{[s_{\omega^*}-1]}$, $\lambda\in(\lambda_i,\lambda_{i+1}]$, $i=0,1,\ldots,s_{\omega^*}-s_{\alpha,P}-1$,

(a) $\mathrm{VaR}_{\alpha}(x,P_{\lambda})=g^{[s_{\alpha,P}+i]}\leq\mathrm{VaR}_{\alpha}(x,Q)$;

(b) *if* $\sum_{s=1}^{s_{\alpha,P}} p^{[s]} > \alpha$ *or*   *if* $i \geq 2$ *and* $\sum_{s=1}^{s_{\alpha,P}} p^{[s]} = \alpha,$ *then*

$\text{VaR}_\alpha(x, P_{\lambda_i}) = \text{VaR}_\alpha(x, P_\lambda) < \text{VaR}_\alpha(x, P_{\lambda_{i+1}});$ *if* $\sum_{s=1}^{s_{\alpha,P}} p^{[s]} = \alpha,$ *then*

$\text{VaR}_\alpha(x, P_{\lambda_1}) = g^{[s_{\alpha,P}]}$ *and* $\text{VaR}_\alpha(x, P_\lambda) = g^{[s_{\alpha,P}+1]}$ *for* $\lambda \in (\lambda_1, \lambda_2];$

(c) $\text{VaR}_\alpha(x, P_{\bar{\lambda}}) = g(x, \omega^*) = \text{VaR}_\alpha(x, Q),$ *for* $\lambda > \lambda_{\bar{i}},\ \bar{i} = s_{\omega^*} - s_{\alpha,P}.$

This procedure can be extended to stress testing with respect to another discrete probability distribution Q, carried by scenarios $\omega_1^*, \ldots, \omega_{S'}^*$ with probabilities $q^{[1]}, \ldots, q^{[S']}$ and associated losses $g(x, \omega_1^*) < \cdots < g(x, \omega_{S'}^*)$. Now, we have to determine how the support of P is related to the support of Q, e.g. that the following ordering holds:

$$g^{[1]} < \cdots < g^{[s_{\alpha,P}]} < \cdots < g^{[s_{\omega_1^*}-1]} < g(x, \omega_1^*) < g^{[s_{\omega_1^*}]} < \cdots < g^{[s_{\omega_2^*}-1]}$$

$$< g(x, \omega_2^*) < g^{[s_{\omega_2^*}]} < \cdots < g^{[s_{\omega_{S'}^*}-1]} < g(x, \omega_{S'}^*) < g^{[s_{\omega_{S'}^*}]} < \cdots < g^{[S]}$$

The covering of the interval $[0, 1]$ depends on probabilities $q^{[s]}$, namely on the difference in their partial cumulative sums and α. For the obtained λ_i values, statements parallel to (a) and (b) of theorem 16.4.1 can be derived (Polívka 2005).

16.4.5   Minimization of $\text{VaR}_\alpha(x, P)$ with Respect to x

Except for the case of the normal distribution considered in Sections 16.4.1 and 16.4.2, the *minimization* of $\text{VaR}_\alpha(x, P)$ with respect to x is, in general, a non-convex, even discontinuous problem, which may have several local minima. It can be written as

$$\min\{v : P\{\omega : g(x, \omega) \leq v\} \geq \alpha,\ x \in \mathcal{X},\ v \in R\}. \tag{16.33}$$

Stability of the minimal $\text{VaR}_\alpha(P)$ value $v_V^*(P)$ and of the optimal solutions $x_V^*(P)$ with respect to P holds true only under additional, restrictive assumptions (Römisch 2003). For $g(x, \omega)$ jointly continuous in x, ω and $\mathcal{H}(x, v) := \{\omega: g(x, \omega) \leq v\}$, a verifiable sufficient condition is $P(\mathcal{H}(x_V^*(P), v_V^*(P))) > \alpha$, which is fulfilled, for instance, for (non-degenerate) normal distributions, or $\alpha < \sum_{s=1}^{s_{\alpha,P}} p^{[s]}$ in (16.30) for the ordered sample of $g(x_V^*(P), \omega^s)$ with discrete distribution P_S (Dobiáš 2003).

To approximate VaR minimization problems, one may apply the corresponding problems with CVaR criteria, as suggested and tested numerically in Rockfellar and Uryasev (2000): the $v_C^*(P)$ part of the optimal solution of (16.18) is then the value of $\text{VaR}_\alpha(x^*(P),P)$ for the optimal (or efficient) $\text{CVaR}_\alpha(x, P)$ portfolio. Further suggestions are to approximate VaR minimization problems by a sequence of CVaR minimizations (Pflug 2001), to use a smoothed VaR objective (Gaivoronski and Pflug 2004), or to apply the worst-case VaR criterion for the family of probability distributions with given first- and second-order moments (El Ghaoui *et al.* 2003).

16.5   CONCLUSIONS

The application of the contamination technique to CVaR evaluation and optimization is straightforward, and the obtained results provide a genuine stress *quantification*. Stress testing via contamination for CVaR-mean return problems turns out to be more delicate.

The presence of the simple chance constraint in the definition of VaR requires that, for VaR stress testing via contamination, various distributional and structural properties are fulfilled for the unperturbed problem. These requirements rule out direct applications of the contamination technique in the case of discrete distributions, which includes the empirical VaR. Nevertheless, even in this case, it is possible to construct bounds for VaR of the contaminated distribution. In the case of a normal distribution and parametric VaR, one may exploit stability results valid for quadratic programs to stress testing of VaR minimization problems.

Using the contamination technique, we have derived computable bounds which can be extended to stress testing of other risk criteria and risk optimization problems. The presented approaches provide a deeper insight into the stress behaviour of VaR and CVaR than the common numerical evaluations based solely on backtesting and out-of-sample analysis.

ACKNOWLEDGEMENTS

This research was supported by the project 'Methods of modern mathematics and their applications' (MSM 0021620839) and by the Grant Agency of the Czech Republic (grants 201/05/2340 and 402/05/0115).

REFERENCES

Acerbi, C. and Tasche, D., On the coherence of expected shortfall. *J. Bank. Finan.*, 2002, **26**, 1487–1503.

Andersson, F., Mausser, H., Rosen, D. and Uryasev, S., Credit risk optimization with Conditional Value-at-Risk criterion. *Math. Program.*, 2001, **B89**, 273–291.

Bank, B., Guddat, J., Klatte, D., Kummer, B. and Tammer, K., *Non-Linear Parametric Optimization*, 1982 (Akademie: Berlin).

Danskin, J.M., *Theory of Max–Min, Econometrics and Operations Research*, Vol. 5, 1967 (Springer: Berlin).

Dempster, M.A.H., editor, *Risk Management: Value at Risk and Beyond*, 2002 (Cambridge University Press: Cambridge).

Dobiáš, P., Stability in stochastic programming with applications to risk management, PhD thesis, Faculty of Mathematics and Physics, Charles University, Prague, 2003.

Dupačová, J., Stability in stochastic programming with recourse—contaminated distributions. *Math. Prog. Study*, 1986, **27**, 133–144.

Dupačová, J., Stochastic programming with incomplete information: A survey of results on postoptimization and sensitivity analysis. *Optimization*, 1987, **18**, 507–532.

Dupačová, J., Scenario based stochastic programs: Resistance with respect to sample. *Ann. Oper. Res*, 1996, **64**, 21–38.

Dupačová, J., Stress testing via contamination, in *Proceedings of the Workshop on Coping with Uncertainty*, Modeling and Policy Issues, edited by K. Marti *et al.*, 2006 (Springer: Berlin), pp. 29–46.

Dupačová, J., Bertocchi, M. and Moriggia, V., Postoptimality for scenario based financial models with an application to bond portfolio management. In *World Wide Asset and Liability Modeling*, edited by W.T. Ziemba and J. Mulvey, pp. 263–285, 1998 (Cambridge University Press: Cambridge).

El Ghaoui, L., Oks, M. and Oustry, F., Worst-case value-at-risk and robust portfolio optimization: A conic programming approach. *Oper. Res.*, 2003, **51**, 543–556.

Embrechts, P., McNeil, A.J. and Straumann, D., Correlation and dependence in risk management: Properties and pitfalls. In *Risk Management: Value at Risk and Beyond*, edited by M.A.H. Dempster, pp. 176–223, 2002 (Cambridge University Press: Cambridge).

Gaivoronski, A.A. and Pflug, G., Value at Risk in portfolio optimization: Properties and computational approach. *J. Risk*, 2004, **7**(2), 1–31.

Guddat, J., Guerra Vasquez, F., Tammer, K. and Wendler, K., *Multiobjective and Stochastic Optimization Based on Parametric Optimization*, 1985 (Akademie: Berlin).

Høyland, K., Kaut, M. and Wallace, S.W., A heuristic for moment-matching scenario generation *Comput. Optim. Applic.*, 2003, **24**, 169–185.

Høyland, K. and Wallace, S.W., Generating scenario trees for multistage decision problems. *Mgmt Sci.*, 2001, **47**, 295–307.

Kaut, M., Vladimirou, H., Wallace, S.W. and Zenios, S.A., Stability analysis of portfolio management with conditional value-at-risk. *Quant. Fin.*, 2007, **7**(4), 397–409.

Klein Haneweld, W.K., *Duality in Stochastic Linear and Dynamic Programming*, LNEMS 274, 1986 (Springer: Berlin).

Kupiec, P., Stress testing in a Value at Risk framework. In *Risk Management: Value at Risk and Beyond*, edited by M.A.H. Dempster, pp. 76–99, 2002 (Cambridge University Press: Cambridge).

Ogryczak, W. and Ruszczyński, A., Dual stochastic dominance and related mean-risk models. *SIAM J. Optim.*, 2002, **13**, 60–78.

Pflug, G.Ch., Some remarks on the value-at-risk and the conditional value-at-risk. In *Probabilistic Constrained Optimization, Methodology and Applications*, edited by S. Uryasev, pp. 272–281, 2001 (Kluwer Academic: Dordrecht).

Polívka, J., ALM via stochastic programming, PhD thesis, Faculty of Mathematics and Physics, Charles University, Prague, 2005.

Prékopa, A., Contributions to the theory of stochastic programming. *Math. Program*, 1973, **4**, 202–221.

Rachev, S. and Mittnik, S., *Stable Paretian Models in Finance*, 2000 (Wiley: Chichester).

Rau-Bredow, H., Value-at-risk, expected shortfall and marginal risk contribution. In *Risk Measures for the 21st Century*, edited by G. Szegö, pp. 61–68, 2004 (Wiley: Chichester).

Rockafellar, R.T. and Uryasev, S., Optimization of conditional value-at-risk. *J. Risk*, 2000, **2**, 21–41.

Rockafellar, R.T. and Uryasev, S., Conditional value-at-risk for general loss distributions. *J. Bank. Finan.*, 2001, **26**, 1443–1471.

Römisch, W., Stability of stochastic programming problems. In *Handbook on Stochastic Programming*, edited by A. Ruszczynski and A. Shapiro, pp. 483–554, 2003 (Elsevier: Amsterdam).

Ruszczyński, A. and Vanderbei, R.J., Frontiers of stochastically nondominated portfolios. *Econometrica*, 2003, **71**, 1287–1297.

Serfling, R.J., *Approximation Theorems of Mathematical Statistics*, 1980 (Wiley: New York).

Shapiro, A., On differential stability in stochastic programming. *Math. Program.*, 1990, **47**, 107–116.

Topaloglou, N., Vladimirou, H. and Zenios, S.A., CVaR models with selective hedging for international asset allocation. *J. Bank. Finan.*, 2002, **26**, 1535–1561.

Stable Distributions in the Black–Litterman Approach to Asset Allocation

ROSELLA GIACOMETTI, MARIDA BERTOCCHI, SVETLOZAR T. RACHEV and FRANK J. FABOZZI

CONTENTS

17.1   INTRODUCTION

T HE MEAN-VARIANCE MODEL FOR PORTFOLIO management as formulated by Markowitz (1952) is probably one of the most known and cited financial models. Despite its introduction in 1952, there are several reasons cited by academics and practitioners as to why its use is not more widespread. Some of the major reasons are the scarcity of diversification, see Green and Hollifield 1992, or highly concentrated portfolios and the sensitivity of the solution to inputs (especially to estimation errors of the mean, see Kallberg and Ziemba 1981, 1984, Michaud 1989, and Best and Grauer 1991) and the approximation errors in the solution of the maximization problem.

The integration of quantitative asset allocation models and the judgement of portfolio managers and analysts (i.e. qualitative view) has been motivated by various discussions on increasing the usefulness of quantitative models for global portfolio management. The framework dates back to papers by Black and Litterman (1991a, b, 1992) that led to the development of extensions of the framework proposed by members of both the academic and practitioner communities. Subsequent research has explained the advantages of this framework, what is now popularly referred to as the Black–Litterman model (BL *model* hereafter), as well as the model's main characteristics.[1]

The BL model contributes to the asset management literature in two distinct directions. The first direction is the idea that there should exist an equilibrium portfolio, with which one can associate an equilibrium distribution of the market. The equilibrium distribution summarizes neutral information and is not as sensitive to estimation risk as estimations that are purely based on time-series analysis. The second contribution is the process that twists the equilibrium distribution according to the practitioner's views. In particular, the BL model uses a Bayesian argument to perform this step, giving rise to a posterior distribution for the market. The computation of both the equilibrium portfolio and the posterior distribution rely on the assumption of a normal market. As far as the computation of the posterior distribution (under the non-normal assumption) is concerned, results have been obtained in Meucci (2006a, b).

In most of the papers mentioned above, there are explicit or implicit assumptions that returns on N asset classes are multivariate Gaussian distributed, an assumption consistent with other mainstream theories in finance such as the standard Black–Scholes (1973) model. However, there are numerous empirical studies[2] that show that in many cases returns are quite far from being normally distributed, especially for high frequency data. Many recent papers (see Ortobelli *et al.* 2002a, b; Bertocchi *et al.* 2005) show that stable Paretian distributions are suitable for the autoregressive portfolio return process in the framework of the asset allocation problem over a fixed horizon.

As we pointed out above, there are two distinct directions in which the BL model contributes to the field of portfolio management. In this chapter, we investigate further the first direction by exploring more generic distributional assumptions, namely in the computation of the equilibrium portfolio. We investigate whether the BL model can be enhanced by using the stable Paretian distributions as a statistical tool for asset returns. We use as a portfolio of assets a subset, duly constructed, of the S&P500 benchmark. We generalize the procedure of the BL model allowing the introduction of dispersion matrices obtained from the multivariate Gaussian, symmetric t-Student and α-stable distributions for computing the equilibrium returns. Moreover, three different measures of risk (variance, value at risk and conditional value at risk) are considered. Results are reported for monthly data and goodness of the models are tested through a rolling window of fixed size along a fixed horizon. Results for weekly data are also available, however, the BL approach, which is a strategic asset allocation model, is usually adopted for at least monthly data. Finally, our analysis shows that the incorporation of the views of investors

---

[1] See the papers by Fusai and Meucci (2003), Satchell and Scowcroft (2000), He and Litterman (1999) and the books by Litterman (2003) and Meucci (2005).
[2] See Eberlein and Keller (1995), Panorska *et al.* (1995), Mittnik and Paolella (2000), Rachev and Mittnik (2000) and the references therein, Tokat and Schwartz (2002), Embrechts *et al.* (2003), and Tokat *et al.* (2003).

into the model provides information as to how the different distributional hypotheses can impact the optimal composition of the portfolio.

17.2   THE α-STABLE DISTRIBUTION

The α-stable distributions describe a general class of distribution functions.

The α-stable distribution is identified by four parameters: the index of stability $\alpha \in (0,2]$ which is the parameter of the kurtosis, the skewness parameter $\beta \in [-1,1]$; $\mu \in \Re$ and $\gamma \in \Re^+$ which are, respectively, the location and the dispersion parameter. If X is a random variable whose distribution is α-stable, we use the following notation to underline the parameter dependence

$$X \overset{d}{=} S_\alpha(\gamma, \beta, \mu). \tag{17.1}$$

The stable distribution is normal when $\alpha = 2$, and it is leptokurtotic when $\alpha < 2$. A positive skewness $(\beta > 0)$ identifies distributions with right fat tails, while a negative skewness $(\beta < 0)$ typically characterizes distributions with left fat tails. Therefore, the stable density functions synthesize the distributional forms empirically observed in the real financial data. The Maximum Likelihood Estimation (MLE) procedure used to approximate stable parameters is described by Rachev and Mittnik (2000). Unfortunately the density of stable distributions cannot be express in closed form. Thus, in order to value the density function, it is necessary to invert the characteristic function.

In the case where the vector $\mathbf{r} = [r_1, r_2, \ldots, r_n]$ of returns is sub-Gaussian α-stable distributed with $1 < \alpha < 2$, then the characteristic function of $\mathbf{r}$. assumes the following form:

$$\Phi_\mathbf{r}(\mathbf{t}) = E(\exp(i\mathbf{t}'\mathbf{r})) = \exp(-(\mathbf{t}'\mathbf{V}\mathbf{t})^{\alpha/2} + i\mathbf{t}'E(\mathbf{r})). \tag{17.2}$$

For the dispersion matrix $\mathbf{V} = [v_{ij}^2]$ we use the following estimation $\tilde{\mathbf{V}} = [\tilde{v}_{ij}^2]$ (see Ortobelli *et al.* 2004, Lamantia *et al.* 2005):

$$\tilde{v}_{ij}^2 = (\tilde{v}_{jj})^{2-q} A(q) \frac{1}{T} \sum_{k=1}^{T} \tilde{r}_{ik} |\tilde{r}_{ik}|^{q-1} \operatorname{sgn}(\tilde{r}_{jk}), \tag{17.3}$$

where $\tilde{r}_{jk} = r_{jk} - E(r_j)$ is the kth centred observation of the jth asset,

$$A(q) = \frac{\Gamma\left(1 - \dfrac{q}{2}\right)\sqrt{\pi}}{2^q \Gamma\left(1 - \dfrac{q}{\alpha}\right)\Gamma\left(\dfrac{q+1}{2}\right)}$$

and $1 < q < \alpha$

$$\tilde{v}_{jj} = \left(A(p) \frac{1}{n} \sum_{k=1}^{n} |\tilde{r}_{jk}|^p \right)^{2/p}, \quad 1 < p < 2, \tag{17.4}$$

where $\tilde{r}_{jk} = r_{jk} - E(r_j)$ is the kth centred observation of the jth asset.

17.3 THE BLACK–LITTERMAN MODEL FOR ASSET ALLOCATION AND OUR EXTENSION

As previously mentioned, the BL model overcomes the critical step of expected return estimation, using the equilibrium returns defined as the returns implicit in the benchmark. If the Capital Asset Pricing Model holds and if the market is in equilibrium, the weights based on market capitalizations are also the weights of the optimal portfolio. If the benchmark is a good proxy for the market portfolio, its composition is the solution of an optimization problem for a vector of unknown equilibrium returns. Moreover, the equilibrium returns provide a neutral reference point for asset allocation. Black and Litterman argue that the only sensible definition of neutral returns is the set of expected returns that would clear the market if all investors had identical views. In fact, an investor with neutral views should select a passive strategy, tracking the benchmark portfolio. The equilibrium returns Π of the stocks comprising the benchmark are obtained by solving the unconstrained maximization problem faced by an investor with quadratic utility function or assuming normally distributed returns

$$\text{Max } \Pi'\mathbf{x} - \frac{\lambda}{2}\mathbf{x}'\Sigma\mathbf{x}, \tag{17.5}$$

where Σ is the covariance matrix of our stocks' returns.

From Kuhn–Tucker conditions on (17.5), and solving the reverse optimization problem we get

$$\Pi = \lambda\Sigma\mathbf{x}. \tag{17.6}$$

The expected return $E(\mathbf{r})$ is assumed to be normally distributed $E(\mathbf{r}) \sim N(\Pi, \tau\Sigma)$ with the covariance matrix proportional to the historical one, rescaled by a shrinkage factor; since uncertainty of the mean is lower than the uncertainty of the returns themselves, the value of τ should be close to zero.

An active asset manager can deviate from the benchmark tracking strategy, according to his or her economic reasoning in the tactical asset allocation. The major contribution of the BL model is to combine the equilibrium returns with uncertain views about expected returns. In particular, the optimal portfolio weights are moved in the direction of assets favoured by the investor. The investor's views have the effect of modifying $E(\mathbf{r})$ according to the degree of uncertainty. The larger the uncertainty the lesser the deviation from the neutral views. To this aim the new vector of expected returns is computed minimizing the Mahalanobis distance between the expected returns $E(\mathbf{r})$ and the equilibrium returns which are additionally constrained by the investor's view on the expected return. This brings us to the following model:

$$\min(E(\mathbf{r}) - \Pi)'\tau\Sigma(E(\mathbf{r}) - \Pi) \tag{17.7}$$

s.t. constraint

$$\mathbf{P}E(\mathbf{r}) = \mathbf{q} + \mathbf{e}, \tag{17.8}$$

where $\mathbf{P}$ is a matrix with each row corresponding to one view, $\mathbf{q}$ is the vector containing the specific investor views, and $\mathbf{e}$ is a random vector of errors in the view. If all views are independent, the covariance matrix is diagonal. Its diagonal elements are collected in the vector $\mathbf{e}$. This formulation leads to the interpretation of using one view at a time with a certain degree of uncertainty, i.e. scenario by scenario. The idea of seeing the market expected return distribution conditioned on the investor's views as the solution to (17.7) and (17.8) is intuitive and quite general since it does not depend on the type of distribution, see also Zimmermann et al. (2002, Chapter 10). Using Bayes rule, we know that it is possible to compute the distribution of the market conditioned on the investor's views, see Meucci (2005).

In our analysis we consider different problems of optimal allocation among n risky assets with returns $[r_1, r_2, \ldots, r_n]$ using different risk measures—variance, value at risk (*VaR*), and conditional value at risk (*CVaR*). Assume that all portfolios $\mathbf{r}'\mathbf{x}$ are uniquely determined by the neutral mean $\Pi'\mathbf{x}$ and by the risk measure $\rho()$ that is defined alternatively as the dispersion $\mathbf{x}'\mathbf{Vx}$, the $VaR_\delta(\mathbf{r}'\mathbf{x})$ and the $CVaR_\delta(\mathbf{r}'\mathbf{x})$. This means that instead of (17.5) we have

$$\text{Max } \Pi'\mathbf{x} - \frac{\lambda}{2}\rho(\mathbf{r}'\mathbf{x}). \tag{17.9}$$

We recall here that $VaR_\delta(X)$ is implicitly defined by $P(X \leq -VaR_\delta(X)) = \delta$, i.e. the δ percentile of the probability density function of the random variable X such that the probability that the random variable assumes a value less than x is greater than δ, where δ represents, in this framework, the maximum probability of loss that the investor would accept. We also recall that $CVaR_\delta(X)$ for continuous random variables X is defined as $-E(X|X \leq -VaR_\delta(X))$, i.e it measures the expected value of the tail of the distribution for values less than VaR_δ. Note also that $CVaR_\delta(X)$ is a coherent risk measure in the sense of Artzner et al. (1998) while $VaR_\delta(X)$ is not.[3]

Notice that for elliptical distributions, following Embrechts et al. (2003) and Stoyanov et al. (2006), the *CVaR* of portfolio returns is expressed as

$$CVaR_\delta(\mathbf{r}'\mathbf{x}) = \sqrt{\mathbf{x}'\mathbf{V}_t\mathbf{x}}CVaR_\delta(Y) - \mathbf{x}'\mathbf{u}, \tag{17.10}$$

---

[3]For a detailed description of CVaR see for example Rockafeller and Uryasev (2000) and for α-stable see Stoyanov et al. (2006). For comparisons between *CVaR* and *VaR* see Gaivoronski and Pflug (2005).

where $CVaR_\delta(Y)$ for the univariate t-distribution takes the following form:

$$CVaR_\delta(Y) = \frac{\Gamma(v+1)/2}{\Gamma(v/2)} \frac{\sqrt{v}}{(v-1)\delta\sqrt{\pi}} \left(1 + \frac{VaR_\delta(Y)^2}{v}\right)^{(1-v)/2}, \qquad (17.11)$$

where Y is distributed according to a t-student with $v > 1$ degree of freedom.

Using again Stoyanov et al. (2006), we can represent the $CVaR_\delta(\mathbf{X})$ for the multivariate standardized α-stable distribution, $\mathbf{X} \in S_\alpha(\gamma, \beta, \mu)$ as

$$CVaR_\delta(\mathbf{X}) = \frac{\alpha}{1-\alpha} \frac{|VaR_\delta(\mathbf{X})|}{\pi\delta} \times \int_{-\theta_0}^{\pi/2} g(\theta) \exp(-|VaR_\delta(\mathbf{X})|^{\alpha/(\alpha-1)} v(\theta)) d\theta, \qquad (17.12)$$

where

$$g(\theta) = \frac{\sin(\alpha(\theta_0 + \theta) - 2\theta)}{\sin\alpha(\theta_0 + \theta)} - \frac{\alpha \cos^2\theta}{\sin^2\alpha(\theta_0 + \theta)}, \qquad (17.13)$$

$$v(\theta) = (\cos\alpha\theta_0)^{1/(\alpha-1)} \left(\frac{\cos\theta}{\sin\alpha(\theta_0 + \theta)}\right)^{\alpha/(1-\alpha)} \times \frac{\cos(\alpha\theta_0 + (\alpha-1)\theta)}{\cos\theta}, \qquad (17.14)$$

and

$$\theta_0 = \frac{1}{\alpha} \arctan\left(\beta \tan\frac{\pi\alpha}{2}\right), \qquad \beta = -\mathrm{sgn}(VaR_\delta(\mathbf{X}))\beta. \qquad (17.15)$$

In the case where we have a non-standardized α-stable, we need to use the following transformation

$$CVaR_\delta(\gamma\mathbf{X} + \mu) = \gamma CVaR_\delta(\mathbf{X}) - \mu, \qquad (17.16)$$

where

$$\gamma\mathbf{X} + \mu \in S_\alpha(\gamma, \beta, \mu). \qquad (17.17)$$

Properties similar to (17.10) and (17.16) hold for VaR too (see Lamantia et al. 2005).

Notice that the optimization problem for $CVaR$ risk measures, using properties (17.10) and (17.16), can be written as

$$\mathrm{Max}\ \Pi'\mathbf{x} - \frac{\lambda}{2}\left(CVaR_\delta\sqrt{\mathbf{x}'\mathbf{V}\mathbf{x}} - E(\mathbf{r})'\mathbf{x}\right). \qquad (17.18)$$

Similar considerations apply to VaR_δ.

Applying first-order Kuhn–Tucker conditions to (17.18), the reverse optimization model, and using the three different measures of risks and the three different return distributions (Gaussian, t-student, stable) we obtain the following equilibrium returns for the three different dispersion measures **V** characterizing the three distributions:

Risk Measure: variance

$$\Pi = \lambda \mathbf{V}\mathbf{x}. \tag{17.19}$$

Risk Measure: CVaR

$$\Pi = \frac{\lambda}{2}\left(CVaR_\delta \frac{\mathbf{V}\mathbf{x}}{\sqrt{\mathbf{x'Vx}}} - E(\mathbf{r}) \right). \tag{17.20}$$

Risk Measure: VaR

$$\Pi = \frac{\lambda}{2}\left(VaR_\delta \frac{\mathbf{V}\mathbf{x}}{\sqrt{\mathbf{x'Vx}}} - E(\mathbf{r}) \right). \tag{17.21}$$

In formulas (17.19), (17.20), and (17.21), we will substitute the convenient estimate for the dispersion matrix, $CVaR$ and VaR depending on the corresponding distribution.

Notice that the coefficient λ can be interpreted as a coefficient of risk aversion: if λ is zero the investor is risk neutral, if $\lambda > 0$, the investor is risk averse because investments with large dispersion are penalized, if $\lambda < 0$, the investor is a risk seeker because investments with large dispersion are favoured. Once we found the neutral returns implied in the benchmark, we wanted to test the goodness of these equilibrium returns over a 20 month horizon. We thought that a reasonable way was to compute the sum of squared errors between the neutral view return suggested by our model and the day after realization of return for 20 consecutive months, using a rolling window of 110 months for the parameters' estimation. We compare the equilibrium returns obtained under different distributional hypotheses and different risk measures with a naive forecast: the unconditional mean. We recall that for a stationary return process the best forecast of future realizations is the unconditional mean.

But what is the optimal value of λ to be used? Black and Litterman suggest, under the normal distributional hypothesis, using the market risk premium, which is 0.32 in our case.[4]

If we try to determine the value of λ which minimizes the distance between the optimal solution of the portfolio and the weight of the benchmark we get $\lambda = 36.29$, i.e. the risk aversion parameter becomes very large. This may be considered reasonable when we look at it from the equity premium puzzle side (see Fama and French 2002, Mehra and Prescott 1985, 2003). However since we consider three different risk measures we must consider different values of λ. Indeed, in (17.19)–(17.21), the coefficient λ acts as a scaling factor,

---

[4] This is computed as the excess mean return divided by the variance.

i.e. a larger λ increases the equilibrium returns. Large values of λ will eventually scale all the equilbrium returns to very large values that would not be realistic.

We set the value of λ equal to the solution of an optimization problem. For each risk measure, we fix λ equal to the value that minimizes the sum of the squared error under all distributional hypotheses described before, computed for the first day of the out-of-sample analysis. We maintain the same values for all the out-of-sample analysis. Therefore, we choose $\lambda = 0.5$ for the case where risk was measured by dispersion and $\lambda/2 = 0.15$ for the other cases. Finally, we tested how the forcing of a special investor view (both under certainty and under uncertainty) may influence the benchmark composition.

17.4   ANALYSIS OF THE DATA

In this section we analyse the time series data for the benchmark used in the computational part and we estimate the parameters of the different distributions (normal, symmetric t-student and α-stable) and the related dispersion matrices. We selected as the benchmark the S&P 500 and we obtained daily, weekly and monthly data from July 1995 to July 2005 from DataStream. The analysis was done for all the frequencies mentioned above. However, only results for the monthly data are reported here.

We divided the data into two samples: the first 110 data for the parameter estimation and the remaining 20 data for out-of-sample analysis. The out-of-sample analysis is repeated for 20 consecutive months using a rolling window of length 110 to estimate the parameters for each month. In order to better analyse the results, we chose to reduce the dimension of the benchmark, considering the most capitalized shares which account for about 50% of the index. We collected the data for the 50 most capitalized shares and rescaled the weights to sum up to 1. We used the new weights to construct a synthetic index that we will refer to as the S&P 50 in the following. We also tested that the S&P 50 returns are almost perfectly correlated ($\rho = 0.98$) to the S&P 500 along the considered horizon.

In Table 17.1 we report for each of the 50 stocks included in the synthetic index, the ticker, the name of the company, the weight in the S&P 500, the new weight in the S&P 50, the mean, the volatility, the skewness and the kurtosis of each stock in the sample. Recall that we selected monthly data because in an asset allocation problem the reasonable time horizon should not be too short. Because of the frequency selected, we tested for the absence of autocorrelation in the returns and squared returns, but we found no evidence of it. Based on the Bera–Jarque test, we rejected for 19 of the 50 stocks the null hypothesis of the normal distribution at the 5% significance level, for 21 of the 50 at the 10% significance level, and for 23 of the 50 at the 15% significance level. From the results reported in Table 17.1 we can observe that about half of the 50 stocks could be well described by a normal distribution. In Table 17.2 we report the average of the estimated parameters α, β, γ for the α-stable distribution and the average of the degree of freedom for the t-student computed as the mean of 20 estimations over the rolling window. A similar analysis is done for a rolling window of increasing size, with no significant changes in the results. Therefore, we do not report those results. From Table 17.2 we note that only

TABLE 17.1 Statistics on the Single Time Series

Ticker	Company	% of S&P 500	% of S&P 50	Mean	Volatility	Skewness	Kurtosis	BJ (p-value)
XOM	Exxon Mobil Corp.	3.33	6.86	11.73%	0.90	−0.20	3.90	0.361
GE	General Electric	3.31	6.82	10.96%	0.60	0.12	2.93	0.870
MSFT	Microsoft Corp.	2.40	4.94	13.41%	1.24	−0.19	5.58	0.000
C	Citigroup Inc.	2.16	4.45	17.07%	1.08	−0.02	4.23	0.246
WMT	Wal-Mart Stores	1.81	3.73	12.15%	0.84	−0.39	3.28	0.128
PFE	Pfizer Inc.	1.81	3.73	11.46%	0.89	0	2.68	0.811
JNJ	Johnson & Johnson	1.73	3.56	11.97%	0.67	−0.08	3.16	−0.930
BAC	Bank of America Corp.	1.62	3.34	10.79%	1.06	−0.41	4.46	0.057
INTC	Intel Corp.	1.45	2.99	10.85%	1.50	−0.88	5.74	0.000
AIG	American Intl Group	1.37	2.82	11.44%	0.87	0.01	3.40	0.600
MO	Altria Group Inc.	1.21	2.49	8.92%	1.11	−0.33	3.85	0.216
PG	Procter & Gamble	1.18	2.43	10.61%	0.85	−1.69	11.18	0.000
JPM	JPMorgan Chase & Co.	1.11	2.29	6.62%	1.25	−0.18	5.41	0.001
CSCO	Cisco Systems	1.09	2.24	16.87%	1.62	−0.63	4.60	0.000
IBM	Int. Business Machines	1.08	2.22	10.11%	1.17	0.37	4.86	0.003
CVX	Chevron Corp.	1.07	2.20	7.87%	0.63	0.38	3.31	0.140
WFC	Wells Fargo	0.93	1.92	13.53%	0.80	−0.05	3.82	0.737
KO	Coca Cola Co.	0.91	1.87	2.36%	0.84	−0.29	3.02	0.376
VZ	Verizon Communications	0.86	1.77	1.56%	0.83	0.00	3.32	0.990
DELL	Dell Inc.	0.85	1.75	33.87%	1.72	−0.66	4.89	0.000
PEP	Pepsico Inc.	0.80	1.65	8.67%	0.73	−0.02	4.14	0.183
HD	Home Depot	0.76	1.57	13.12%	1.08	−0.67	4.38	0.001
COP	Conoco Philips	0.74	1.52	11.47%	0.76	0.18	3.32	0.839
SBC	SBC Communications Inc.	0.71	1.46	−0.09%	0.93	0.08	4.91	0.017
TWX	Time Warner Inc.	0.70	1.44	20.97%	1.96	0.43	3.72	0.340
UPS	United Parcel Service	0.69	1.42	6.86%	0.95	−0.60	4.39	0.008
ABT	Abbott Labs	0.68	1.40	8.89%	0.76	−0.32	3.70	0.162
AMGN	Amgen	0.68	1.40	17.35%	1.05	0.58	4.64	0.002
MRK	Merck & Co.	0.61	1.26	2.68%	1.00	−0.69	5.34	0.691
ORCL	Oracle Corp.	0.61	1.26	13.80%	1.68	−0.13	3.41	0.752
HPQ	Hewlett-Packard	0.61	1.26	4.03%	1.46	0.25	4.76	0.021
CMCSA	Comcast Corp.	0.60	1.24	10.99%	1.22	−0.06	5.52	0.000
UNH	United Health Group Inc.	0.60	1.24	20.27%	1.27	−2.31	14.11	0.000
AXP	Amercian Express	0.60	1.24	13.44%	0.94	−0.34	4.26	0.059
LLY	Lilly(Eli) & Co.	0.56	1.15	9.80%	0.97	0.03	3.48	0.654
MDT	Medtronic Inc.	0.56	1.15	15.04%	0.97	0.07	3.89	0.698
WYE	Wyeth	0.54	1.11	7.69%	1.13	−1.69	12.61	0.000
TYC	Tyco International	0.53	1.09	14.57%	1.43	−1.00	6.17	0.000
MWD	Morgan Stanley	0.52	1.07	13.50%	1.35	−0.33	3.85	0.174
FNM	Fannie Mae	0.51	1.05	8.61%	0.79	−0.36	2.98	0.280
MMM	3M Company	0.50	1.03	9.47%	0.71	0.20	3.17	0.559
QCOM	Qualcomm Inc.	0.49	1.01	23.80%	1.97	0.16	4.11	0.353
BA	Boeing Company	0.48	0.99	6.28%	1.04	−0.35	3.87	0.198
UTX	United Technologies	0.47	0.97	14.82%	0.89	−0.13	5.50	0.000
MER	Merrill Lynch	0.47	0.97	13.20%	1.30	−0.36	4.80	0.007
VIA.B	Viacom Inc.	0.47	0.97	2.23%	1.11	0.08	3.10	0.870
DIS	Walt Disney Co.	0.46	0.95	2.81%	1.05	−0.01	3.40	0.940
G	Gillette Co.	0.45	0.93	3.94%	0.94	−0.50	4.05	0.004
BMY	Bristol-Myers Squibb	0.44	0.91	7.94%	−0.41	−0.36	3.67	0.264
BLS	Bell South	0.44	0.91	4.32%	0.80	0.13	3.54	0.769

Notes: From the table we observe the skewness and kurtosis of the single stocks on the complete sample. We perform Bera–Jarque test and we cannot reject the null hypothesis of normal distribution for 19 of the 50 at 5% significance level, for 21 of the 50 at 10% significance level, and for 23 of the 50 at 15% significance level.

TABLE 17.2 Estimated Parameters with α-stable and t-student Distributions

Numbering	Ticker	Company	% of S&P 500	% of S&P 50	α	β	γ	Degree of freedom
1	XOM	Exxon Mobil Corp.	3.33	6.86	1.95	−0.91	0.64	18
2	GE	General Electric	3.31	6.82	2.00	0.96	0.42	100
3	MSFT	Microsoft Corp.	2.40	4.94	1.71	0.49	0.79	7
4	C	Citigroup Inc.	2.16	4.45	1.88	−0.18	0.74	10
5	WMT	Wal-Mart Stores	1.81	3.73	1.85	−1.00	0.58	22
6	PFE	Pfizer Inc.	1.81	3.73	2.00	−0.85	0.64	100
7	JNJ	Johnson & Johnson	1.73	3.56	2.00	−0.14	0.48	54
8	BAC	Bank of America Corp.	1.62	3.34	1.48	−0.44	0.58	4
9	INTC	Intel Corp.	1.45	2.99	1.81	−0.86	0.97	6
10	AIG	American Intl Group	1.37	2.82	2.00	−0.20	0.61	17
11	MO	Altria Group Inc.	1.21	2.49	1.90	−0.84	0.78	10
12	PG	Procter & Gamble	1.18	2.43	1.82	−1.00	0.51	4
13	JPM	JPMorgan Chase & Co.	1.11	2.29	1.69	−0.14	0.76	5
14	CSCO	Cisco Systems	1.09	2.24	1.77	−0.69	1.05	5
15	IBM	Int. Business Machines	1.08	2.22	1.77	0.39	0.73	6
16	CVX	Chevron Corp.	1.07	2.20	1.86	1.00	0.42	11
17	WFC	Wells Fargo	0.93	1.92	1.87	−0.01	0.56	8
18	KO	Coca Cola Co.	0.91	1.87	1.95	−1.00	0.59	92
19	VZ	Verizon Communications	0.86	1.77	2.00	−0.19	0.61	37
20	DELL	Dell Inc.	0.85	1.75	1.76	−0.69	1.08	5
21	PEP	Pepsico Inc.	0.80	1.65	1.80	−0.02	0.47	6
22	HD	Home Depot	0.76	1.57	1.83	−1.00	0.72	9
23	COP	Conoco Philips	0.74	1.52	2.00	0.57	0.53	10
24	SBC	SBC Communications Inc.	0.71	1.46	1.81	−0.36	0.61	7
25	TWX	Time Warner Inc.	0.70	1.44	1.82	0.91	1.30	14
26	UPS	United Parcel Service	0.69	1.42	1.46	−0.45	0.51	4
27	ABT	Abbott Labs	0.68	1.40	1.94	−0.93	0.53	18
28	AMGN	Amgen	0.68	1.40	1.84	0.88	0.69	4
29	MRK	Merck & Co.	0.61	1.26	1.93	−1.00	0.67	16
30	ORCL	Oracle Corp.	0.61	1.26	2.00	−0.40	1.26	36
31	HPQ	Hewlett-Packard	0.61	1.26	1.74	0.05	0.94	5
32	CMCSA	Comcast Corp.	0.60	1.24	1.73	−0.16	0.75	4
33	UNH	United Health Group Inc.	0.60	1.24	1.54	0.15	0.54	3
34	AXP	Amercian Express	0.60	1.24	1.53	−0.84	0.54	8
35	LLY	Lilly(Eli) & Co.	0.56	1.15	1.94	−0.22	0.68	14
36	MDT	Medtronic Inc.	0.56	1.15	1.81	0.27	0.62	6
37	WYE	Wyeth	0.54	1.11	1.65	−0.56	0.62	4
38	TYC	Tyco International	0.53	1.09	1.71	−0.59	0.87	4
39	MWD	Morgan Stanley	0.52	1.07	1.82	−0.75	0.92	10
40	FNM	Fannie Mae	0.51	1.05	1.91	−1.00	0.54	100
41	MMM	3MCompany	0.50	1.03	1.96	1.00	0.50	57
42	QCOM	Qualcomm Inc.	0.49	1.01	1.76	0.19	1.28	6
43	BA	Boeing Company	0.48	0.99	1.72	−0.42	0.67	6
44	UTX	United Technologies	0.47	0.97	1.67	−0.43	0.53	5
45	MER	Merrill Lynch	0.47	0.97	1.81	−0.35	0.86	6
46	VIA.B	Viacom Inc.	0.47	0.97	2.00	−0.55	0.83	100
47	DIS	Walt Disney Co.	0.46	0.95	2.00	0.89	0.78	61
48	G	Gillette Co.	0.45	0.93	1.84	−0.87	0.65	7
49	BMY	Bristol-Myers Squibb	0.44	0.91	1.87	−0.97	0.60	7
50	BLS	Bell South	0.44	0.91	1.97	0.69	0.57	19

Notes: In this table we report the average estimate of the α-stable computed on a rolling window.

9 stocks show a value of the α parameter equal to 2. The α-stable distribution looks more appropriate in describing the behaviour of the remaining stock returns.

In order to assess the hypothesis of non-normal behaviour of the stocks and the statistical significance of the α-stable parameters, we estimated an autoregressive model on each estimated parameter of the α-stable distribution along the 20 consecutive months. Our reason for doing so is to have a statistical model that describes the evolution of the α-stable parameters over time. We then used the estimated statistical model to construct a confidence interval for the parameters. The following AR(1,1) model was estimated in order to construct a 90% confidence level for α and β

$$y_t = a_1 + a_2 y_{t-1} + \epsilon_t \sqrt{a_3{}^2}, \tag{17.22}$$

where y_t is the α (or β) series, and a_1, a_2, and a_3 the parameters to be estimated.

The estimated coefficients, together with their statistical significance, are reported in Table 3 for α and in Table 17.4 for β. We do not consider stocks with $\alpha = 2$ and $\beta = 1$ or $\beta = -1$ for all the out-of-sample period. So we exclude 9 stocks from Table 17.3 and 17 stocks from Table 17.4. Each table contains the ticker, the estimated coefficients of the AR(1,1) process, the ratio between the value of the coefficients and the standard error, the value of the likelihood function, the 5th, the 50th and the 95th percentiles for the 40 stocks. We observe that the autoregressive coefficient is significant for 80% of the stocks for α and for 70% of the stocks for β (see columns T_{a_2}). We used the model given by (17.22) to create 5000 scenarios for each of the parameters α and β and each stock in the benchmark, thus obtaining the related distributions: we report in Tables 17.3 and 17.4 the median, the 5th and 95th percentiles of those ones which construct the 90% confidence level.

The analysis of the 90% confidence interval confirms that for 82% of the stocks considered, the true value of α is less than 2, suggesting the presence of leptokurtic behaviour. Only for 9 stocks is the normal distribution suitable. Moreover, the upper value of the confidence level of β is less than 0 for 19 stocks and β is equal to -1 for 6 stocks, suggesting a left fat tail distribution for 50% of the stocks. The lower value of the confidence level of β exceeds 0 for 7 stocks and β is equal to 1 for 2 stocks of the 50 stocks considered, suggesting a right fat tail for 18% of the stocks. Only for 7 stocks does the confidence interval include the null value, suggesting that at most 16 stocks can show a symmetric behaviour. This explains the poor behaviour of the symmetric t-student distributional model which indeed seems to give the same result as the normal distribution. This is a further confirmation that the α-stable distribution is suitable for describing the returns of our data.

17.5   COMPUTATIONAL RESULTS

In general under the normal and t-student distributions we get very similar equilibrium returns and portfolio composition under all the different risk measures while the α-stable hypothesis implies different equilibrium returns and portfolio composition, see Giacometti *et al.* (2006) for the detailed analysis.

TABLE 17.3 Estimated Parameters of the AR(1,1) Process on α Values

Numbering	Ticker	a_1	a_2	a_3	T_{a_1}	T_{a_2}	T_{a_3}	LLF	Perc. 5%	Perc. 50%	Perc. 95%
1	XOM	0.724	0.629	–	1.997	3.376	2.136	79.249	1.9357	1.9433	1.9507
2	GE	–	–	–	–	–	–	–	–	–	–
3	MSFT	0.349	0.793	0.005	0.831	3.497	3.406	25.658	1.4907	1.5982	1.7095
4	C	0.072	0.961	–	0.345	8.717	3.338	82.661	1.8551	1.8614	1.8680
5	WMT	0.361	0.805	–	1.130	4.677	1.543	70.703	1.8483	1.8598	1.8711
6	PFE	–	–	–	–	–	–	–	–	–	–
7	JNJ	–	–	–	–	–	–	–	–	–	–
8	BAC	0.232	0.842	–	0.962	5.265	2.959	47.765	1.4178	1.4538	1.4905
9	INTC	0.600	0.668	–	1.192	2.403	2.805	62.988	1.7805	1.7978	1.8144
10	AIG	–	–	–	–	–	–	–	–	–	–
11	MO	0.122	0.934	–	0.297	4.306	3.432	49.386	1.7882	1.8219	1.8567
12	PG	0.015	0.993	–	0.062	7.664	1.876	78.473	1.8378	1.8457	1.8534
13	JPM	0.187	0.889	–	0.612	4.949	3.775	59.791	1.6460	1.6667	1.6873
14	CSCO	0.827	0.532	–	2.981	3.400	2.789	48.111	1.7286	1.7640	1.7991
15	IBM	0.145	0.917	–	0.366	4.103	3.847	60.213	1.7101	1.7295	1.7491
16	CVX	0.991	0.468	–	3.667	3.203	3.582	64.867	1.8459	1.8610	1.8771
17	WFC	0.096	0.946	0.001	0.262	4.979	3.766	37.453	1.7114	1.7738	1.8348
18	KO	0.674	0.655	–	1.560	2.967	2.501	82.666	1.9445	1.9510	1.9575
19	VZ	–	–	–	–	–	–	–	–	–	–
20	DELL	0.025	0.980	0.003	0.031	2.158	2.347	29.273	1.4427	1.5345	1.6267
21	PEP	0.296	0.834	–	1.214	6.148	1.750	58.839	1.7261	1.7466	1.7680
22	HD	0.505	0.724	–	1.671	4.380	1.710	73.083	1.8217	1.8320	1.8421
23	COP	–	–	–	–	–	–	–	–	–	–
24	SBC	0.275	0.848	–	1.880	4.920	3.815	77.941	1.8063	1.8145	1.8228
25	TWX	0.191	0.894	0.001	0.434	3.700	3.648	46.213	1.7440	1.7825	1.8206
26	UPS	0.181	0.877	0.001	0.900	6.435	2.987	45.839	1.4429	1.4828	1.5227
27	ABT	1.853	0.045	–	0.620	0.029	1.511	57.538	1.9183	1.9411	1.9630
28	AMGN	0.695	0.622	0.002	1.501	2.358	2.351	33.662	1.7800	1.8543	1.9298
29	MRK	0.258	0.866	–	0.176	1.149	1.944	51.033	1.8728	1.9033	1.9341
30	ORCL	–	–	–	–	–	–	–	–	–	–
31	HPQ	0.133	0.922	–	0.519	6.250	3.758	57.519	1.6797	1.7025	1.7248
32	CMCSA	1.107	0.358	–	2.150	1.203	3.041	57.611	1.7001	1.7224	1.7452
33	UNH	0.412	0.733	0.001	0.791	2.153	2.913	43.604	1.4799	1.5252	1.5713
34	AXP	0.258	0.832	–	1.178	5.811	2.236	53.223	1.5128	1.5392	1.5675
35	LLY	0.168	0.912	0.001	0.411	4.361	3.423	44.314	1.8213	1.8654	1.9085
36	MDT	0.210	0.880	0.003	0.477	3.724	3.118	29.431	1.5760	1.6674	1.7598
37	WYE	0.484	0.706	–	0.806	1.943	3.730	51.040	1.6047	1.6354	1.6659
38	TYC	0.166	0.903	–	0.622	5.778	3.323	56.193	1.6705	1.6939	1.7178
39	MWD	0.675	0.630	0.001	1.041	1.808	1.784	40.896	1.7651	1.8172	1.8702
40	FNM	0.545	0.716	–	1.616	4.065	2.147	65.505	1.9010	1.9161	1.9321
41	MMM	0.456	0.767	–	1.552	5.105	2.480	73.090	1.9474	1.9578	1.9683
42	QCOM	0.268	0.847	0.001	0.619	3.463	3.689	44.868	1.6753	1.7170	1.7593
43	BA	0.647	0.624	0.001	2.402	3.948	2.584	45.355	1.6726	1.7132	1.7538
44	UTX	0.464	0.723	–	1.491	3.890	2.847	58.561	1.6539	1.6762	1.6973
45	MER	0.243	0.865	–	0.546	3.524	4.650	69.746	1.7862	1.7982	1.8103
46	VIA.B	–	–	–	–	–	–	–	–	–	–
47	DIS	–	–	–	–	–	–	–	–	–	–
48	G	1.100	0.401	–	1.505	1.011	2.053	55.033	1.8080	1.8329	1.8578
49	BMY	1.618	0.135	0.001	0.256	0.040	1.787	37.088	1.7882	1.8497	1.9112
50	BLS	1.299	0.339	–	2.196	1.143	1.333	54.315	1.9363	1.9633	1.9895

Notes: The analysis of the 90% confidence interval confirms that for 82% of the stocks considered (i.e. those with α different from 2) the true value of α is smaller than 2, suggesting the presence of leptokurtic behaviour. Only for 9 stocks is the normal distribution suitable.

TABLE 17.4 Estimated Parameters of the AR(1,1) Process on β Values

Numbering	Ticker	a_1	a_2	a_3	T_{a_1}	T_{a_2}	T_{a_3}	LLF	Perc. 5%	Perc. 50%	Perc. 95%
1	XOM	0.016	0.990	0.014	0.060	2.859	2.032	14.274	−0.5791	−0.3036	−0.1924
2	GE	−	−	−	−	−	−	−	−	−	−
3	MSFT	0.261	0.472	0.192	1.036	1.658	1.895	−11.859	−0.1468	0.5880	1.3106
4	C	−0.066	0.623	0.002	−1.289	2.198	3.043	35.618	−0.2072	−0.1405	−0.0746
5	WMT	−	−	−	−	−	−	−	−	−	−
6	PFE	−	−	−	−	−	−	−	−	−	−
7	JNJ	−	−	−	−	−	−	−	−	−	−
8	BAC	−0.014	0.959	−	−0.151	4.331	3.426	48.559	−0.4008	−0.3660	−0.3313
9	INTC	−0.015	0.975	0.001	−0.091	5.404	2.399	37.926	−0.7761	−0.7170	−0.6570
10	AIG	−	−	−	−	−	−	−	−	−	−
11	MO	−0.192	0.757	0.017	−0.720	2.353	1.905	12.488	−0.7972	−0.5816	−0.3709
12	PG	−	−	−	−	−	−	−	−	−	−
13	JPM	−0.037	0.737	0.001	−1.552	4.105	3.456	47.167	−0.1706	−0.1342	−0.0964
14	CSCO	−0.195	0.712	0.005	−1.527	4.227	2.171	24.582	−0.7508	−0.6355	−0.5192
15	IBM	0.004	0.966	0.003	0.041	3.872	4.751	30.636	0.1086	0.1927	0.2794
16	CVX	−	−	−	−	−	−	−	−	−	−
17	WFC	0.096	0.946	0.001	0.262	4.979	3.766	37.453	1.7114	1.7738	1.8348
18	KO	−	−	−	−	−	−	−	−	−	−
19	VZ	−	−	−	−	−	−	−	−	−	−
20	DELL	−0.011	0.936	0.051	−0.011	0.703	1.182	1.441	−0.3535	0.0161	0.3791
21	PEP	−0.008	0.599	0.001	−0.884	1.510	3.668	38.309	−0.0925	−0.0346	0.0262
22	HD	−	−	−	−	−	−	−	−	−	−
23	COP	−	−	−	−	−	−	−	−	−	−
24	SBC	−0.120	0.668	−	−1.930	3.780	3.230	47.633	−0.3885	−0.3516	−0.3138
25	TWX	0.199	0.785	0.005	0.782	2.638	2.198	23.887	0.8665	0.9836	1.1046
26	UPS	−0.190	0.577	−	−2.036	2.759	3.232	54.503	−0.4741	−0.4477	−0.4219
27	ABT	−0.977	−0.055	0.104	−0.056	−0.003	0.390	−5.739	−1.4412	−0.9204	−0.3926
28	AMGN	0.333	0.624	0.055	1.732	1.612	1.853	0.586	0.5667	0.9582	1.3432
29	MRK	−	−	−	−	−	−	−	−	−	−
30	ORCL	−	−	−	−	−	−	−	−	−	−
31	HPQ	0.012	0.728	0.001	0.848	3.593	1.909	43.827	−0.0299	0.0151	0.0604
32	CMCSA	0.005	0.980	0.002	0.146	5.190	5.146	34.298	−0.0683	0.0044	0.0790
33	UNH	0.026	0.860	0.003	0.405	2.431	3.941	29.404	0.1545	0.2446	0.3351
34	AXP	−0.179	0.788	−	−0.600	2.217	2.260	63.458	−0.8579	−0.8406	−0.8242
35	LLY	0.007	0.897	0.110	0.095	5.689	4.415	−6.319	−0.1457	0.4035	0.9520
36	MDT	0.196	0.259	0.006	3.042	1.386	2.744	23.535	0.1365	0.2571	0.3789
37	WYE	−0.176	0.687	0.001	−1.140	2.481	3.373	47.454	−0.6063	−0.5695	−0.5312
38	TYC	−0.036	0.930	0.002	−0.241	3.828	4.387	35.015	−0.5459	−0.4764	−0.4081
39	MWD	−0.255	0.653	0.013	−0.392	0.910	0.878	15.149	−0.8484	−0.6553	−0.4647
40	FNM	−	−	−	−	−	−	−	−	−	−
41	MMM	−	−	−	−	−	−	−	−	−	−
42	QCOM	0.034	0.808	0.001	0.815	3.773	3.859	37.019	0.0862	0.1486	0.2110
43	BA	−0.321	0.244	−	−2.092	0.681	2.442	57.426	−0.4422	−0.4200	−0.3975
44	UTX	−0.150	0.648	0.001	−1.570	2.889	2.430	45.302	−0.4441	−0.4042	−0.3628
45	MER	−0.049	0.852	0.001	−0.520	3.123	3.812	45.335	−0.3261	−0.2847	−0.2452
46	VIA.B	−	−	−	−	−	−	−	−	−	−
47	DIS	−	−	−	−	−	−	−	−	−	−
48	G	−0.249	0.708	0.004	−0.955	2.390	2.783	25.784	−0.8737	−0.7651	−0.6516
49	BMY	−1.972	−1.000	0.014	−0.001	−0.001	0.109	14.410	−1.7079	−1.5127	−1.3212
50	BLS	0.496	0.287	0.359	0.918	0.378	0.503	−18.137	−0.2001	0.7659	1.7537

Notes: We can observe that, at 90% confidence level, 50% of the stocks show a left fat tail, 18% a right fat tail. Since for 9 stocks, that account for 18% of the total, $\beta = 0$ by definition since $\alpha = 2$, and only for 7 stocks the confidence interval includes the null value, we can suggest that at most 16 stocks can show a symmetric behaviour.

TABLE 17.5 Squared Errors among the Optimal Composition and the Unconditional Mean over the Rolling Window Horizon

	Dispersion		VaR		CVaR		
Date	Normal	Stable	Normal	Stable	Normal	Stable	Unc. mean
01/28/04	9.08	9.18	8.90	9.03	8.99	9.37	9.18
02/25/04	22.89	21.74	22.24	22.07	22.71	23.01	21.81
03/24/04	47.70	38.47	43.12	42.89	46.35	49.26	41.72
04/21/04	16.34	17.26	16.81	16.64	16.57	16.19	15.87
05/19/04	22.58	17.26	19.81	19.60	21.70	23.29	19.63
06/16/04	10.60	14.26	12.19	12.30	11.05	10.23	12.75
07/14/04	27.42	20.90	24.27	23.97	26.55	28.54	22.69
08/11/04	29.53	24.04	26.63	26.62	28.60	30.69	26.59
09/08/04	17.30	20.40	18.74	18.79	17.79	17.04	18.23
10/06/04	44.86	44.51	44.49	44.61	44.74	45.32	44.47
11/03/04	15.68	14.38	14.77	14.88	15.33	16.29	15.32
12/01/04	14.33	16.59	15.31	15.54	14.69	14.49	14.73
12/29/04	9.31	11.13	10.01	10.11	9.51	9.24	10.15
01/05/05	25.65	19.38	22.54	22.39	24.72	27.27	21.76
02/23/05	22.58	20.54	21.36	21.35	22.15	23.32	21.71
03/23/05	19.71	15.55	17.58	17.39	19.05	20.85	17.23
04/20/05	35.37	29.25	32.27	31.85	34.38	36.63	32.26
05/18/05	15.82	20.07	17.71	18.01	16.42	15.31	17.58
06/15/05	12.24	11.22	11.53	11.61	12.00	12.92	11.59
07/13/05	11.21	10.19	10.59	10.54	11.04	11.74	10.23

Notes: The hypothesis of stable distribution and the use of dispersion as the risk measure gives the best combination. For 13 of the 20 considered months, it is the combination that gives the best forecast (65% success rate). The second best combination is the α-stable distribution and the use of CVaR as the risk measure. For this combination 3 of the 20 months gives the best forecast (15% success rate). The third best combination is the unconditional mean resulting in the best forecast in 2 of the 20 months (10% success rate).

We consider the equilibrium returns as a forecast of the future returns. Of course, we assume that when we compare the forecast with the future realizations, the data that we observe in the future are the products of a market in equilibrium. In Table 17.5 we report the sum of squared errors for 20 months between the neutral view and realization of the month after using a rolling window of 110 months. Note that we re-estimate the parameters of the distribution as we move the rolling window. We observe that the hypothesis of a stable distribution and the use of dispersion as a risk measure gives the best combination.[5] For 13 months of the 20 months, it is the combination that gives the best forecast (65% success rate). The second best combination is the α-stable distribution and the use of CVaR as a risk measure. For this combination 3 of the 20 months give the best forecast (15% success rate). The third best combination is the unconditional mean which resulted in 2 of the 20 months (10% success rate). Finally,

---

[5] The use of a symmetric risk measure is not particularly surprising since we are dealing with a model of strategic allocation to find the optimal composition of our portfolio on a relatively long time horizon. Generally this phase is followed by a tactical allocation strategy where it is more likely that a relative *VaR* can be considered by the market, i.e. the additional tail risk that we accept when we move from the benchmark replication strategy assuming specific risk.

following Satchell and Scowcroft (2000), we compute the optimal composition for 28 January 2004 under a specific view for the different distributions and different risk measures with absolute certainty and with uncertainty.

We observe that the difference between the new returns and neutral view equilibrium are larger for the dispersion measure and normal returns, see Giacometti *et al.* (2006) for details. Indeed, this is the case with the highest variation in the portfolio composition. Once again the α-stable hypothesis with the same risk measure led to a more stable portfolio. If we consider the same view with uncertainty, we have similar effects but mitigated by the uncertainty that we put in our view.

17.6 CONCLUSIONS

The purpose of our work is twofold. The first is to improve the classical BL model by applying more realistic models for asset returns. We compare the BL model under the normal, *t*-student, and the stable distributions for asset returns. The second is to enhance the BL model by using alternative risk measures which are currently used in risk management and portfolio analysis. They include dispersion-based risk measures, value at risk, and conditional value at risk.

For the stocks in our sample, only a minority can be characterized as having a normal return distribution based on the statistical tests we performed. As a result of incorporating heavy-tailed distribution models for asset returns and alternative risk measures, we obtained the following results: (1) the appropriateness of the α-stable distributional hypothesis is more evident when we compute the equilibrium returns and (2) the combination of α-stable distribution and the choice of dispersion risk measure provides the best forecast.

ACKNOWLEDGEMENTS

The authors acknowledge the support given by research projects MIUR 60% 2003 'Simulation models for complex portfolio allocation' and MIUR 60% 2004 'Models for energy pricing,' by research grants from Division of Mathematical, Life and Physical Sciences, College of Letters and Science, University of California, Santa Barbara and the Deutschen Forschungsgemeinschaft. The authors thank the referees for helpful suggestions.

REFERENCES

Artzner, P., Delbaen, F., Eber, J.M. and Heath, D., Coherent measures of risk. *Math. Finance*, 1998, **9**(3), 203–228.

Bertocchi, M., Giacometti, R., Ortobelli, S. and Rachev, S.T., The impact of different distributional hypothesis on returns in asset allocation. *Financ. Lett.*, 2005, **3**(1), 17–27.

Best, M.J. and Grauer, R.R., On the sensitivity of mean-variance efficient portfolios to changes in asset means. Some analytical and computational results. *Rev. Financ. Stud.*, 1991, **January**, 315–342.

Black, F. and Litterman, R., Asset allocation: Combining investor views with market equilibrium. *J. Fixed Income*, 1991a, **September**, 7–18.

Black, F. and Litterman, R., Global asset allocation with equities, bonds and currencies. Unpublished paper, Goldman Sachs, 1991b, **October**.

Black, F. and Litterman, R., Global portfolio optimization. *Financ. Analysts J.*, 1992, **September/October**, 28–43.

Black, F. and Scholes, M., The pricing of options and corporate liabilities. *J. Polit. Econ.*, 1973, **81**, 637–659.

Eberlein, E. and Keller, U., Hyperbolic distributions in finance. *Bernoulli*, 1995, **1**, 281–299.

Embrechts, P., Lindskog, F. and McNeil, A., Modelling dependence with copulas and application to risk management. In *Handbook of Heavy Tailed Distributions in Finance*, edited by S.T. Rachev, pp. 329–385, 2003 (Elsevier/North Holland).

Fama, E.F. and French, K.R., The equity premium. *J. Finance*, 2002, **57**, 637–659.

Fusai, G. and Meucci, A., Assessing views. *Risk*, 2003, **March**, S18–S21.

Gaivoronski, A. and Pflug, G.C., Value-at-risk in portfolio optimization: Properties and computational approach. *J. Risk*, 2005, **7**(2), 1–31.

Giacometti, R., Bertocchi, M., Rachev, S. and Fabozzi, F., Stable distributions in the Black–Litterman approach to asset allocation, *Technical Report*, 2006 (PSTAT, University of California, Santa Barbara, University of Karlsruhe).

Green, R.C. and Hollifield, B., When will mean-variance efficient portfolios be well diversified? *J. Finance*, 1992, **47**(5), 1785–1810.

He, G. and Litterman, R., The intuition behind Black–Litterman model portfolios. *Invest. Manag. Res.*, 1999, **December**, 1–27.

Lamantia, F., Ortobelli, S. and Rachev, S.T., Value at risk, conditional value at risk and time rules with elliptical and asymmetric stable distributed returns: Theoretical advances, *Technical Report*, 2005 (Institute of Statistics and Mathematical Economic Theory, University of Karlsruhe).

Litterman, B. and The Quantitative Resources Group., *Modern Investment Management: An Equilibrium Approach*, 2003 (Wiley: New Jersey).

Kallberg, J.G. and Ziemba, W.T., Remarks on optimal portfolio selection, In *Methods of Operations Research*, edited by G. Bamberg and O. Optiz, pp. 507–520, 1981 (Oelgeschlager, Gunn & Hain: Cambridge, MA).

Kallberg, J.G. and Ziemba, W.T., Mis-specification in portfolio selection problems, In *Risk and Capital*, edited by G. Bamberg and O. Optiz, pp. 74–87, 1984 (Springer-Verlag: New York).

Markowitz, H., Portfolio selection. *J. Finance*, 1952, **March**, 77–91.

Mehra, R. and Prescott, E.C., The equity premium: A puzzle. *J. Monetary Econ.*, 1985, **15**, 145–161.

Mehra, R. and Prescott, E.C., The equity premium in retrospect, In *Handbook of the Economics of Finance*, edited by G.M. Constantinides, M. Harrius, and R. Stulz, pp. 887–936, 2003 (Elsevier North-Holland: Amsterdam).

Meucci, A., *Risk and Asset Allocation*, 2005 (Springer-Verlag: New York).

Meucci, A., Beyond Black–Litterman in practice: A five-step recipe to input views on non-normal markets *Risk*, 2006a, **19**, 114–119.

Meucci, A., Beyond Black–Litterman: Views on non-normal markets *Risk*, 2006b, **19**, 87–92.

Michaud, R.O., The Markowitz optimization enigma: Is "optimized" optimal? *Financ. Analysts J.*, 1989, **45**(1), 31–42.

Mittnik, S. and Paolella, M.S., Conditional density and value-at-risk prediction of Asian currency exchange rates. *J. Forecast.*, 2000, **19**, 313–333.

Ortobelli, S., Huber, I., Rachev, S.T. and Schwartz, E., Portfolio choice theory with non-Gaussian distributed returns, In *Handbook of Heavy Tailed Distributions in Finance*, edited by S.T. Rachev, pp. 547–594, 2002a (Elsevier Science: Amsterdam).

Ortobelli, S., Rachev, S.T. and Schwartz, E., The problem of optimal asset allocation with stable distributed returns, In *Stochastic Processes and Functional Analysis: A Volume of Recent Advances in Honor of Prof. M.M. Rao*, edited by A.C. Krinik and R.J. Swift, Vol. 238, pp. 295–347, 2002b (Marcel Dekker: New York).

Ortobelli, S., Rachev, S.T., Huber, I. and Biglova, A., Optimal portfolio selection and risk management: A comparison between the stable Paretian approach and the Gaussian one, In *Handbook of Computational and Numerical Methods in Finance*, edited by S.T. Rachev, pp. 197–252, 2004 (Springer-Verlag: Berlin).

Panorska, A., Mittnik, S. and Rachev, S.T., Stable GARCH models for financial time series. *Math. Comput. Modelling*, 1995, **29**, 263–274.

Rachev, S.T. and Mittnik, S., *Stable Paretian Models in Finance*, 2000 (Wiley: Chichester).

Rockafeller, T. and Uryasev, S., Optimization of conditional value-at-risk. *J. Risk*, 2000, **2**(3), 21–41.

Satchell, S. and Scowcroft, A., A demystification of Black–Litterman model: Managing quantitative and traditional portfolio construction. *J. Asset Manag.*, 2000, **1**(2), 138–150.

Stoyanov, S., Rachev, S.T., Samorodnitsky, G. and Ortobelli, S., Computing the portfolio Conditional Value-at-Risk in the α-stable case. *Probab. Math. Stat.*, 2006, **26**, 1–22.

Tokat, Y., Rachev, S.T. and Schwartz, E., The stable non-Gaussian asset allocation: A comparison with the classical Gaussian approach. *J. Econ. Dyn. Control*, 2003, **27**, 937–969.

Tokat, Y. and Schwartz, E., The impact of fat tailed returns on asset allocation. *Math. Methods Operat. Res.*, 2002, **55**, 165–185.

Zimmermann, H., Drobetz, W. and Oertmann, P., *Global Asset Allocation: New Methods and Applications*, 2002 (Wiley: Chichester).

CHAPTER 18

Ambiguity in Portfolio Selection

GEORG PFLUG and DAVID WOZABAL

CONTENTS

18.1   INTRODUCTION: THE AMBIGUITY PROBLEM

T HE DECISION ABOUT OPTIMAL COMPOSITION of a portfolio is a complex process, not just a single optimization task. It comprises of the selection of a statistical model, the collection of data, the estimation of the model in a parametric, semi-parametric or nonparametric way, the choice of an appropriate optimization criterion and finally the numerical solution of an optimization problem. It is well understood that the precision of the final decision depends on the quality of the described complete chain of subtasks.

In his 1921 book, the American economist Frank Knight (1921) made a famous distinction between 'risk' and 'uncertainty.' In Knight's view, 'risk' refers to situations where the decision-maker can assign mathematical probabilities to the randomness, which he is faced with. In contrast, Knight's 'uncertainty' refers to situations when this randomness cannot be expressed in terms of specific mathematical probabilities.

Since the days of Knight, the terms have changed. As introduced by Ellsberg (1961), we refer today to the *ambiguity problem* if the probability model is unknown and to the *uncertainty problem*, if the model is known, but the realizations of the random variables are unknown. While large classes of portfolio optimization problems under uncertainty have been successfully solved (see e.g. the surveys by Yu *et al.* (2003) or Ziemba and Vickson (1973) and references therein), there is a common observation that these solutions lack stability with respect to the chosen parameters (see e.g. Klein and Bawa 1976; Chopra and Ziemba 1993).

The aim of this chapter is to discuss an approach that explicitly takes into account the ambiguity in choosing the probability model and therefore is robust in the following sense: the selected portfolio is slightly suboptimal for the given basic probability model, but performs also well under neighbouring models. In contrast, non-robust portfolio decisions show a dramatic drop in performance, when deviating from the underlying model.

The organization of this chapter is as follows. In Section 18.2, the decision model is formulated. The successive convex program as a solution technique is presented in Section 18.3. In Sections 18.4 and 18.5, the choice of ambiguity sets is discussed. Section 18.6 illustrates the approach by an example.

18.2   PROBLEM FORMULATION

As introduced in the seminal work of Markowitz (1959), the basic portfolio selection problem in this chapter is to minimize the risk under a constraint for the expected return or—equivalently—maximize the expected return under a risk constraint. We follow the latter approach here.

Let $Y_x = \sum_{m=1}^{M} x_i \xi_i$ be the random return of a portfolio consisting of M assets with individual returns ξ_i and portfolio weights x_i. The chosen or estimated probability model P determines the distribution of the return vector $(\xi_1, \ldots, \xi_m)$ on $\mathbb{R}^M$.

Once a probability model is specified, the expected return is well defined. There are however several ways of quantifying the risk. In this chapter, risk is measured as negative acceptability. Acceptability in turn is measured by acceptability functionals A defined on the random asset returns.

Let $(\Omega, \mathcal{F})$ be a measure space, X a linear space of $\mathcal{F}$ measurable functions, $X: \to \mathbb{R}$, then a coherent acceptability functional A from X to $\bar{\mathbb{R}}$ is required to fulfill the following axioms (see Pflug (2006) for a discussion of coherent acceptability functionals)

[A1]   Concavity: $\forall X, Y \in X, \lambda \in [0,1]$

$$A(\lambda X + (1 - \lambda)Y) \geq \lambda A(X) + (1 - \lambda)A(Y).$$

[A2]   Monotonicity: if $X \leq Y$ then $A(X) \leq A(Y)$.
[A3]   Translation equivariance: $A(X + a) = A(X) + a$.
[A4]   Positive homogeneity: if $\lambda > 0$ then $A(\lambda X) = \lambda A(X)$.

We interpret the negative acceptability as risk, i.e. a risk functional is of the form $\mathcal{R} = -A$. The resulting risk functionals have properties coinciding with the convex risk functionals commonly used in the literature (see e.g. Delbaen 2002; Föllmer and Schied 2002). The goal in this chapter is to find decisions with high acceptability and thereby low risk. Note that since acceptability functionals and risk functionals differ only by their sign, they have the same level sets, which are called acceptance sets. For a given return Y, the acceptability value $A(Y)$ indicates the maximal shift of the distribution in the negative direction, which still keeps the return acceptable. Note that since the acceptability functions are concave, the level sets are convex sets. We will further assume that the acceptability functionals $\mathcal{A}$ are continuous, which assures that the respective level sets are closed.

Examples for such acceptability functionals $\mathcal{A}$ are for instance

- the average-value-at risk (conditional value-at-risk, expected shortfall)

$$\mathcal{A}(Y) = \mathbb{A}V@R_\alpha(Y) = \frac{1}{\alpha} \int_0^\alpha G_Y^{-1}(p)dp, \qquad (18.1)$$

where $G_Y^{-1}(p) = \inf\{v : P\{Y \le v\} \ge p\}$ is the quantile function of Y. An alternative representation is

$$\mathbb{A}V@R_\alpha(Y) = \max\left\{a - \frac{1}{\alpha}\mathbb{E}([Y - a]^-) : a \in \mathbb{R}\right\}.$$

(see Rockafellar and Uryasev 2000). A dual representation of $\mathbb{A}V@R_\alpha(Y)$ is

$$\mathbb{A}V@R_\alpha(Y) = \min\{\mathbb{E}(Y Z) : \mathbb{E}Z = 1, 0 \le Z \le 1/\alpha\}.$$

- also utility functionals of the following types
  1. $\mathcal{A}(Y) = \mathbb{E}(Y) - \lambda \, \mathbb{V}ar(Y)$,
  2. $\mathbb{E}(Y) - \lambda \mathbb{S}td(Y)$,
  3. $\mathcal{A}(Y) = \mathbb{E}(Y) - \mathbb{V}ar([Y - \mathbb{E}Y]^-)$,
  4. $\mathbb{E}(Y) - \lambda \mathbb{S}td([Y - \mathbb{E}Y]^-)$,

  with $0 < \lambda \le 1$ respectively, fall in the category of acceptability functionals (see Tokat *et al.* (2003) for a more detailed discussion of functionals of this form).

To translate acceptability back into risk, one may use $-\mathcal{A}(Y)$, or, as we will do in the numerical examples, $1 - \mathcal{A}(Y)$ as a risk functional.

Having identified the criteria for the optimization and the risk bound, we can now write the problem as an optimization problem under uncertainty using the basic probability model $\hat{P}$ which can be regarded as the 'best guess' for the real model. In most cases $\hat{P}$ will be determined by empirical data, but also other ways of obtaining $\hat{P}$ like the incorporation expert opinion may be considered (see e.g. Clemen and Winkler 1999).

As mentioned before the portfolio selection model we consider is a simple Markowitz type model without shortselling. The asset returns are modelled to be uncertain. Assuming there is no ambiguity about the statistical model of the returns the model reads

$$\left\| \begin{array}{l} \text{Maximize (in } x) : \mathbb{E}_{\hat{P}}(Y_x) \\ \text{subject to} \\ \mathcal{A}_{\hat{P}}(Y_x) \ge q \\ x^\top \mathbb{1} = 1 \\ x \ge 0. \end{array} \right. \qquad (18.2)$$

Note that the restriction to positive portfolio weights is chosen just for simplicity of exposition. It would pose neither theoretical nor computational problems to allow short selling in this setting as long as the feasible set of asset compositions remains bounded. In fact it might be rewarding to study the effects of ambiguity in the more risky setting where shortselling is permitted.

Let $\hat{x}^*$ be the optimal portfolio composition found by solving (18.2). The robustness of this solution is often checked by *stress testing*. A stress test consists in finding an alternative probability model P and calculating $\mathbb{E}_P(Y_{\hat{x}^*})$ as well as $\mathcal{A}_P(Y_{\hat{x}^*})$ to judge the change in the return dimension as well as in the risk dimension under model variation. While stress testing helps in assessing the robustness of a given portfolio selection, it does not help to find a robust portfolio.

For the latter goal, one has to replace the basic model (18.2) by its ambiguity extension. To this end, let $\mathcal{P}$ be an ambiguity set, i.e. the set of probability models, to which the modeller is indifferent. The portfolio selection model under $\mathcal{P}-ambiguity$ is of maximin type and reads

$$
\begin{Vmatrix}
\text{Maximize (in } x) : \ \min\{\mathbb{E}_P(Y_x) : P \in \mathcal{P}\} \\
\text{subject to} \\
\mathcal{A}_P(Y_x) \geq q \quad \text{for all } P \in \mathcal{P} \\
x^\top \mathbb{1} = 1 \\
x \geq 0.
\end{Vmatrix}
\tag{18.3}
$$

Let us comment on the structure of problem (18.3). It is a combination of a robust and a stochastic problem (see Zackova 1966 for the earliest occurrence of problems of the form (18.3)). Recall the definition of a robust optimization problem (see Ben-Tal and Nemirovski 2002):

If a deterministic optimization problem

$$
\begin{Vmatrix}
\text{Maximize (in } x) : f(x, \zeta) \\
\text{subject to} \\
f_i(x, \zeta) \leq 0, \quad i = 1, \ldots, k
\end{Vmatrix}
\tag{18.4}
$$

contains some parameters ζ and the decision maker only knows some range Z of these parameters, he/she may use the robust version of (18.4), namely

$$
\begin{Vmatrix}
\text{Maximize } \min\{f(x, \zeta) : \zeta \in Z\} \\
\text{subject to } f_i(x, \zeta) \leq 0, \quad i = 1, \ldots, k \quad \text{for all } \zeta \in Z.
\end{Vmatrix}
\tag{18.5}
$$

While in stochastic optimization one has at least a distributional information about the unknown parameters, the only information one has in robust optimization is a given set of parameters. Thus one may say that stochastic programs look at the *average* situation, while robust programs look at the *worst-case* situation.

The proposed portfolio selection under ambiguity contains both aspects: while assuming that the realizations of the return vectors come from some probability distribution, we allow on the other hand to vary this distribution within a certain set $\mathcal{P}$ without further structuring it. Should we impose a probability distribution on this set of probabilities (called a *prior distribution*), we would still solve an uncertainty problem, but of Bayesian type. In our approach we do not specify a prior and our problem (18.3) has the structure of a robust-stochastic problem.

18.3   SOLUTION TECHNIQUES

Introducing the set of constraints for the vector of portfolio weights x

$$\mathbb{X} = \{x : x^\top \mathbb{1} = 1, \ x \geq 0, \ \mathcal{A}_P(Y_x) \geq q \quad \text{for all } P \in \mathcal{P}\},$$

the ambiguity problem (18.3) reads

$$\max_{x \in \mathbb{X}} \min_{P \in \mathcal{P}} \mathbb{E}_P[Y_x]. \tag{18.6}$$

By continuity and concavity of $\mathcal{A}$, $\mathbb{X}$ is a compact convex set. Moreover, $(P, x) \mapsto \mathbb{E}_P[Y_x]$ is bilinear in P and x and hence convex-concave. Therefore $x^* \in \mathbb{X}$ is a solution of (18.6) if and only if there is a $P^* \in \mathcal{P}$ such that (P^*, x^*) is a saddle point, i.e.

$$\mathbb{E}_{P^*}[Y_x] \leq \mathbb{E}_{P^*}[Y_{x^*}] \leq \mathbb{E}_P[Y_{x^*}] \tag{18.7}$$

for all $(P, x) \in \mathcal{P} \times \mathbb{X}$ (see Rockafellar 1997).

Since there are infinitely many constraints present, problem (18.6) is a semi-infinite program. There are several solution methods for such problems. Direct methods use gradients to find the saddle point (Rockafellar 1976; Nemirovskii and Yudin 1978). In Shapiro and Ahmed (2004) it is proposed to dualize the inner minimization problem in order to get a pure optimization problem: in particular they suppose that the set $\mathcal{P}$ is of the form

$$\mathcal{P} = \left\{ P : \int \phi_i(u) dP(u) \leq b_i; \ i = 1, \ldots, k; P_1 \prec P \prec P_2 \right\},$$

where P_i are some measures, the ϕ_i are P integrable and $b_i \in \mathcal{R}$. Then using the dual representation

$$\min\{\mathbb{E}_P[Y_x] : P \in \mathcal{P}\} = \max \left\{ \int \left[\sum_i \lambda_i \phi_i(u) \right]^+ dP_1(u) + \int \left[\sum_i \lambda_i \phi_i(u) \right]^- dP_2(u) \right\},$$

the maximin problem is transformed into a pure maximization problem.

In this chapter, we propose a successive convex programming (SCP) solution method, which uses a finitely generated inner approximation of the ambiguity set $\mathcal{P}$. To be more precise, we approximate the infinitely many constraints $A_P(Y_x) \geq q$ for all $P \in \mathcal{P}$ by finitely many ones. One starts with no risk constraints. In every new step, two new probabilities enter the set of constraints. These new probabilities are chosen as current worst case probabilities and this makes the algorithm work.

In particular, the successive SCP algorithm proceeds as follows.

1. Set $n = 0$ and $\mathcal{P}_0 = \{\hat{P}\}$ with $\hat{P} \in \mathcal{P}$.
2. Solve the outer problem

$$
\left\|
\begin{aligned}
&\text{Maximize (in } x, t) : t \\
&\text{subject to} \\
&t \leq E_P(Y_x) \quad \text{for all } P \in \mathcal{P}_n \\
&A_P(Y_x) \geq q \quad \text{for all } P \in \mathcal{P}_n \\
&x^\top \mathbb{1} = 1 \\
&x \geq 0
\end{aligned}
\right.
\tag{18.8}
$$

and call the solution (x_n, t_n).
3. Solve the first inner problem

$$
\left\|
\begin{aligned}
&\text{Minimize (in } P) : \; E_P(Y_{x_n}) \\
&\text{subject to} \\
&P \in \mathcal{P}.
\end{aligned}
\right.
\tag{18.9}
$$

and call the solution $P_n^{(1)}$.
4. Solve the second inner problem

$$
\left\|
\begin{aligned}
&\text{Minimize (in } P) : A_P(Y_{x_n}) \\
&\text{subject to} \\
&P \in \mathcal{P}
\end{aligned}
\right.
\tag{18.10}
$$

call the solution $P_n^{(2)}$ and let $\mathcal{P}_{n+1} = \mathcal{P}_n \cup \{P_n^{(1)}\} \cup \{P_n^{(2)}\}$.
5. If
   a. $\mathcal{P}_{n+1} = \mathcal{P}_n$ or
   b. the optimal value of (18.9) equals t_n and the solution of (18.10) is equal to $\min_{P \in \mathcal{P}_n} A_P(Y_{x_n})$

then a saddle point is found and the algorithm stops. Otherwise set $n = n + 1$ and go to 2.

The two inner problems (18.9) and (18.10) may yield non-unique optimizers. In this case we simply choose one of the optimizing worst case measures and proceed with the algorithm.

To see that the second stopping criterion actually yields a saddle point, consider a situation where the condition 5(b) is fulfilled in the nth run. Note that since (18.10) does not yield a lower acceptability, then the minimum of the measures in $\mathcal{P}_n$ the point x_n is feasible for the original problem. Since (18.9) gives the optimal value $\mathbb{E}_{P'}(Y_{x_n})$, there is no measure P giving smaller expectation than P'. Furthermore it is clear from the optimality in (18.8), that there can be no x, such that $\mathbb{E}_{P'}(Y_x)$. This establishes that (x_n, P') is indeed a saddle point of the problem.

Note that if the measures $P_n^{(1)}$ or $P_n^{(2)}$ are in the convex hull of the measures in $\mathcal{P}_n$ they do not have to be added to the set $\mathcal{P}_{n+1}$ (since the functions E_p and $\mathcal{A}$ are concave and therefore take their minima on the extreme points of convex sets).[1] Whether or not a discrete probability measure on $\mathbb{R}^n$ is a convex combination of other such measures can be easily checked by linear programming techniques.

Notice that the set t_n is a decreasing sequence of numbers and $\mathcal{P}_n$ is an increasing sequence of sets. The convergence of this algorithm is stated below. Since one cannot exclude that there are several saddle points (in this case the set of saddle points is closed and convex), only a weak limit result is available in general.

Proposition: *Assume that $\mathcal{P}$ is compact and convex and that $(P, x) \mapsto \mathbb{E}_P[Y_x]$ as well as $(P, x) \mapsto \mathcal{A}_P[Y_x]$ are jointly continuous. Then every cluster point of (x_n) is a solution of (18.6). If the saddle point is unique, then the algorithm converges to the optimal solution.*

Proof Since both $\mathbb{X}$ and $\mathcal{P}$ are compact sets, the sequence x_n has a cluster point x^*. W.l.o.g. we may in fact assume that this is a limit point. Let

$$\mathbb{X}_n = \{x : x^\top \mathbb{1} = 1, x \geq 0, \mathcal{A}_P(Y_x) \geq q \text{ for all } P \in \mathcal{P}_n\}.$$

Then $\mathbb{X}_n$ is a decreasing sequence of compact convex sets, which all contain $\mathbb{X}$. Let $\mathbb{X}^+ = \bigcap_n \mathbb{X}_n$. On the other hand, $\mathcal{P}_n$ is an increasing sequence of sets with upper limit $\mathcal{P}^+$, which is the closure of $\bigcup_n \mathcal{P}_n$.

By construction, x_n is a solution of

$$\max_{x \in \mathbb{X}_n} \min_{P \in \mathcal{P}_n} \mathbb{E}_P[Y_x], \qquad (18.11)$$

i.e. there is a probability P_n^* such that (P_n^*, x_n) is a saddle point. W.l.o.g. we may assume that the sequence (P_n^*) has a limit P_*. Therefore, (P^*, x^*) is a saddle point for

$$\max_{x \in \mathbb{X}^+} \min_{P \in \mathcal{P}^+} \mathbb{E}_P[Y_x]. \qquad (18.12)$$

---

[1] We are indebted to a referee for pointing this out to us and also led us to discover the second condition for having found a saddle point in step 5 of the algorithm.

Since $\min_{P\in\mathcal{P}} \mathbb{E}_P[Y_{x_n}] = \min_{P\in\mathcal{P}_{n+1}} \mathbb{E}_P[Y_{x_n}]$ one sees by continuity that (x^*, P^*) is a saddle point for

$$\max_{x\in\mathbb{X}^+} \min_{P\in\mathcal{P}} \mathbb{E}_P[Y_x]. \tag{18.13}$$

Finally, notice that $x^* \in \mathbb{X}$. If not, $\inf_{P\in\mathcal{P}} \mathcal{A}_P(Y_{x^*}) < q$. But then there is an x_n such that $\inf_{P\in\mathcal{P}} \mathcal{A}_P(Y_{x_n}) < q$ and in the next step this x_n together with an open neighbourhood will be excluded from $\mathbb{X}_{n+1}$, a contradiction.

Therefore (P^*, x^*) is a saddle point for

$$\max_{x\in\mathbb{X}} \min_{P\in\mathcal{P}} \mathbb{E}_P[Y_x], \tag{18.14}$$

i.e. x^* is a solution of the original problem. □

18.4   AMBIGUITY SETS

Typically, ambiguity sets are in some sense neighbourhoods of basic models. Basic models are found by estimation from historic data, consistency considerations as the no-arbitrage rule, expert choice or a combination thereof. In all these cases, the found basic model is the most likely one, but model ambiguity is present. To express this ambiguity, one may allow some variations of the basic models in such a way that they differ in distance from the basic model not more than some ε.

In particular, consider ambiguity sets of the form

$$\mathcal{P} = \{P : d(P, \hat{P}) \le \epsilon\}, \tag{18.15}$$

where d is some distance for probability measures.

The choice of the distance d is crucial for the final result. In this chapter we use the Kantorovich distance (also called the $\mathcal{L}_1$ distance—see Vallander 1973 and Dall'Aglio 1972) defined as

$$d(P_1, P_2) = \sup\left\{ \int f(u)\mathrm{d}P_1(u) - \int f(u)\mathrm{d}P_2(u) : \text{where } f \text{ has Lipschitz} \right.$$
$$\left. \text{constant 1, i.e. } f(u) - f(v) \le \|u - v\|_1 \text{for all } u, v \right\}.$$

Here $\|u - v\|_1 = \sum_i |u_i - v_i|$. To ensure that the Kantorovich distance is finite, we restrict ourselves to the space of measures with finite first absolute moment, i.e. $\mathcal{P} = \left\{ P : \int_{\mathbb{R}^n} \|x\|\mathrm{d}P(x) < \infty \right\}$.

The choice is motivated by the fact that the expected return has Lipschitz constant 1 under this distance, i.e.

$$|\mathbb{E}_{P_1}(Y_x) - \mathbb{E}_{P_2}(Y_x)| \leq d(P_1, P_2)$$

for all portfolios x. Hence, the distance of probability models provides a bound for the difference in expectations and therefore a bound in the optimal values of the considered problems.

Furthermore if the chosen acceptability functional is the $\mathbb{AV@R}$, then one has also Lipschitz continuity of the risk functional with respect to the Kantorovich distance, i.e.

$$|\mathbb{AV@R}_{P_1,\alpha}(Y_x) - \mathbb{AV@R}_{P_2,\alpha}(Y_x)| \leq \frac{1}{\alpha} d(P_1, P_2)$$

for all α and all portfolio compositions x. If the acceptability functional involves higher moments, then the Fortet–Mourier (1953) extension of the Kantorovich distance appears more appropriate.

By the well-known theorem of Kantorovich–Rubinstein (see Rachev 1991), the Kantorovich ambiguity set (18.15) can be represented as

$$\{P : d(P, \hat{P}) \leq \epsilon\} = \left\{ P : \text{there is a bivariate probability } K(\cdot, \cdot) \text{ such that} \right.$$

$$\left. \int_v K(u, dv) = P(u); \int_u K(du, v) = \hat{P}(v); \int_u \int_v \|u - v\|_1 K(du, dv) \leq \epsilon \right\}.$$

If the probability space Ω is finite, i.e. if it consists of S scenarios $\xi^{(1)}, \ldots, \xi^{(S)}$, with $\xi^{(s)} \in \mathbb{R}^M$ then a probability model is just a S-vector $(P_1, \ldots P_S)$ and the ambiguity set is a polyhedral set

$$\{P : d(P, \hat{P}) \leq \epsilon\} = \left\{ P = (P_1, \ldots, P_S) : P_j = \sum_i K_{i,j}; \sum_j K_{i,j} = \hat{P}_i; \right.$$

$$\left. K_{i,j} \geq 0; \sum_{i,j} \|x^{(i)} - x^{(j)}\|_1 K_{i,j} \leq \epsilon \right\}.$$

The bivariate probability K has the interpretation as the solution of *Monge's mass transportation problem*. The Kantorovich distance describes the minimal effort (in terms of expected transportation distances), to change the mass distribution $\hat{P}$ into the new mass distribution P (see Rachev and Rüschendorf 1998).

In the case of a finite probability space Ω it is not difficult to find a solution for the inner problems, i.e. to determine $\inf\{\mathbb{E}_P(Y_x) : d(P, \hat{P}) \leq \epsilon\}$ and $\inf\{\mathcal{A}_P(Y_x) : d(P, \hat{P}) \leq \epsilon\}$.

For a given portfolio composition x, let $y = (y_i)$ with $y_i = x^\top \xi^{(i)}$; $i = 1, \ldots, S$. Denote by $\hat{P}_i$ the probability mass sitting on y_i under $\hat{P}$. To find the worst case probability $P \in \mathcal{P}$, one has to consider mass transportation from scenarios i to other scenarios j, which in total do need more than ε as expected transportation distance.

To this end, for minimizing $\mathbb{E}_P$, let for every pair (i,j), $w_{i,j} = y_i - y_j$ and $v_{i,j} = \|\xi^{(i)} - \xi^{(j)}\|_1$. The needed worst case is found by transferring masses from i to j in a stepwise manner: starting with the pair i,j for which $w_{i,j}/v_{i,j}$ is maximal, the new masses are assigned in descending order of $w_{i,j}/v_{i,j}$, but only if $w_{i,j} > 0$ until the maximal allowed distance ϵ is reached.

In a similar manner, for minimizing $\mathcal{A}_P$, one sets

$$w_{i,j} = \frac{\partial \mathcal{A}_P(Y_x)}{\partial P_i} - \frac{\partial \mathcal{A}_P(Y_x)}{\partial P_j}$$

and proceeds as before. Therefore the two inner problems are in fact directly solvable and do not need an optimization run.

If the acceptability functional is the average-value at risk, then constraint sets are polyhedral and the outer problems are linear. Therefore, in this case, the whole maximin algorithm is a sequential linear program, as studied in Byrd *et al.* (2005).

18.5   STATISTICAL CONFIDENCE SETS

Scenario models for asset returns are typically based on statistical data. If the portfolio decision follows a parametric model, as for instance the Markowitz model, then these parameters, as the mean return, the volatility and the correlations are estimated from the given data material and the estimation error may be quantified by estimating the standard errors or by determining confidence sets, see for instance Goldfarb and Iyengar (2003).

Since our approach uses a nonparametric setup, nonparametric confidence sets have to be found. Starting with a basic estimate $\hat{P}_n$ for the probability model for asset returns, a nonparametric confidence set has to be found: the basic estimate may be either the empirical distribution, i.e. the historical model or some variants of it, for instance models with parametric tail estimates to better accommodate extremal events. The second step is the choice of the size a such that a certain confidence level ϵ is reached, i.e.

$$P\{d(P, \hat{P}_n) \geq a\} \leq \epsilon \tag{18.16}$$

for a large class of models P. The parameter n refers to the number of observations.

As argued before, we use the Kantorovich distance here. Moreover, we consider only model variants which differ only in probabilities and not in values. The argument for doing so is that all model variants must be discrete for the numerical treatment and that variations in values would need some parametric modelling, which we want to avoid.

For getting a confidence set of the form (18.16), one has to assume that the true probability model P has no mass outside a ball in $\mathbb{R}^M$. For the empirical estimate, which puts mass $1/n$ on each of the n historical observations, one has that

$$\mathbb{E}_P[d(P, \hat{P}_n)] \leq Cn^{-1/M}$$

for some constant C as was shown by Dudley (1968). Consequently, using Markov's inequality, it follows that

$$P\left\{d(P, \hat{P}_n) \geq \frac{Cn^{-1/M}}{\epsilon}\right\} \leq \frac{\mathbb{E}[d(P, \hat{P}_n)]\epsilon}{n^{-1/M}C} \leq \epsilon.$$

If some smoothness properties of the unknown model P are known, the confidence sets may be improved, see for instance Kersting (1978). In any case, increasing the size n of the data set reduces the confidence set and leads to smaller costs for ambiguity. It should also be noted that in general the shrinking of the confidence sets can be arbitrary slow if the tails of the involved probability measures are heavy enough (see the results in Kersting 1978).

18.6   A NUMERICAL EXAMPLE

The following analysis is intended to demonstrate the impact of the size of the confidence sets (i.e. the robustness) on the optimal solutions of the outlined portfolio selection problem. The presented analysis furthermore makes it possible to asses the value of information in the model by comparing the optimal expected values for different levels of robustness.

To perform this analysis the maximin approach was implemented and applied to the following data set, downloaded from finance.yahoo.com: the data consists of monthly returns within the period 1 January 1990 to 31 December 2004 of stocks from six companies, namely,

- IBM—International Business Machines Corporation,
- PRG—Procter & Gamble Corporation,
- ATT—AT&T Corporation,
- VER—Verizon Communications Inc.,
- INT—Intel Corporation,
- EXX—Exxon Mobil Corporation.

The selection of these six assets was motivated as follows. Among all assets represented in the Dow–Jones index, IBM and PRG show the least correlation and ATT and VER show the highest correlation. INT is the asset with smallest and EXX with largest variance. Under the basic model $\hat{P}$, all observations have the same probability of $1/180$.

The 10% average value-at-risk $\mathbb{AV@R}_{0.1}$ was chosen as the acceptability functional. In order to translate the acceptance level into a risk level, we used $1 - \mathbb{AV@R}_{0.1}$ as the risk functional. The bounds were set to $\mathbb{AV@R}_{0.1} \geq 0.9$, i.e. the risk was bounded by 0.1. The ambiguity sets were determined as

$$\mathcal{P} = \{P : d(P, \hat{P}) \leq a\}.$$

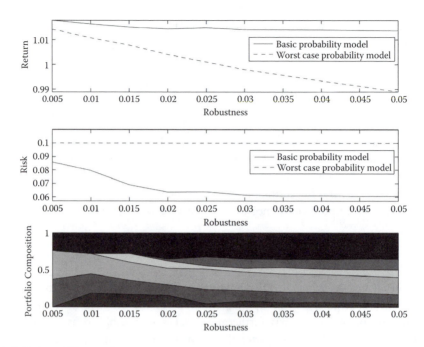

FIGURE 18.1 For different robustness parameters γ, the upper figure shows the expected returns, the middle figure shows the risks and the lower figure shows the portfolio composition. The assets are ordered from top to bottom as: EXX, VER, ATT, PRG, INT, IBM.

Since the a in the above formula is hard to interpret, for our analysis we vary the a with dependence on a robustness parameter $\gamma, \gamma \mapsto a(\gamma)$ and $\mathcal{P} = \{P : d(P, \hat{P}) \leq a(\gamma)\}$ whereby

$$a(\eta) = \max\{\eta : \sup_{d(P,\hat{P}) \leq \eta} \mathbb{E}_P(\xi^{(i)}) \leq (1+\gamma)\mathbb{E}_{\hat{P}}(\xi^{(i)}) : \text{for all } i\}.$$

The parameter as robustness parameter γ is displayed in Figures 18.1–18.3. The maximin problem was solved by successive linear programming as described above.

Figure 18.1 shows the change in return, risk and portfolio composition as a function of the robustness parameter γ. As one can see, more robust portfolios are more diversified. The worst case expected return decreases with robustness, but the basic model expected return $\mathbb{E}_{\hat{P}}$ only drops slightly. On the other hand, the worst case risk stays at the bound, because this bound is binding, while the basic model risk $1 - \mathbb{A}V@R_{0.1,\hat{P}}$ drops significantly. Thus, for the given data set, the price for model robustness is very small while the portfolio composition changes dramatically with the increase in robustness.

The robustness parameter influences the efficient frontier. We have plotted the efficient surface: the optimal expected return as a function of risk and robustness. Figure 18.2 shows this surface, when risk and return are calculated under the basic probability, while Figure 18.3 shows the surface when risk and return are calculated under the pertaining worst case. Notice that for each point on the surface, this might be another worst case

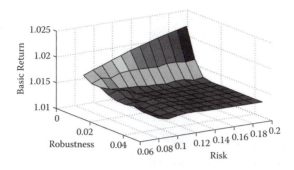

FIGURE 18.2 Efficient frontiers with dependence on the robustness parameter γ. Risk and return are calculated with respect to the basic model $\hat{P}$.

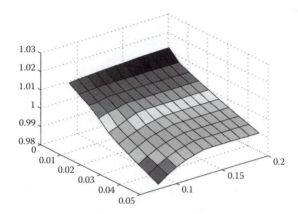

FIGURE 18.3 Efficient frontiers with dependence on the robustness parameter γ. Risk and return are calculated with respect to the worst case model.

model. While under $\hat{P}$, although the efficient frontier is greatly deformed by increasing robustness, there is a much smaller influence on the worst cases.

18.7   CONCLUSIONS

We have presented a maximin approach for portfolio selection, which accommodates scenario uncertainty (aleatoric uncertainty) and probability ambiguity (epistemic uncertainty). Ambiguity is modelled by Kantorovich neighbourhoods of a basic probability model.

The size of the ambiguity neighbourhood may be chosen to correspond to a probabilistic confidence region for the estimated basic model. The more information is collected about the model, the smaller is the ambiguity set and the smaller is the loss in expected return one has to sacrifice for the sake of robustness. Therefore, the value of better statistical information expressed in possible shrinks of the ambiguity set can be assessed by looking at the pertaining losses in the expected return for the basic model.

The chosen example showed a relative small drop in expected return, but a considerable decrease of risk in exchange for a reasonable gain in robustness. It suggests that looking at portfolios which are robust against model ambiguities is very advisable for portfolio managers.

REFERENCES

Ben-Tal, A. and Nemirovski, A., Robust optimization—methodology and applications. *Math. Program. Ser. B*, 2002, **92**(3), 453–480.

Byrd, R.H., Gould, N.I.M., Nocedal, J. and Waltz, R.A., On the convergence of successive linear-quadratic programming algorithms. *SIAM J. Optim.*, 2005, **16**(2), 471–489; (electronic).

Chopra, V.K. and Ziemba, W.T., The effect of errors in means, variances and covariances on optimal portfolio choice. *J. Portfolio Manag.*, 1993, **19**, 6–11.

Clemen, R.T. and Winkler, R.T., Combining probability distributions from experts in risk analysis. *Risk Anal.*, 1999, **19**(2), 187–203.

Dall'Aglio, G., Fréchet classes and compatibility of distribution functions. *Symposia Mathematica, Vol IX (Convegno di Calcolo delle Probabilità, INDAM, Rome, 1971)*, pp. 131–150, 1972 (Academic Press: London).

Delbaen, F., Coherent measures of risk on general probability spaces. In *Advances in Finance and Stochastics. Essays in Honour of Dieter Sondermann*, edited by K. Sandmann and P.J. Schönbucher, pp. 1–38, 2002 (Springer Verlag: New York).

Dudley, R.M., The speed of mean Glivenko-Cantelli convergence. *Ann. Math. Statist.*, 1968, **40**, 40–50.

Ellsberg, D., Risk, ambiguity, and the savage axioms. *Q. J. Econ.*, 1961, **75**(4), 643–669.

Föllmer, H. and Schied, A., Convex measures of risk and trading constraints. *Finance Stoch.*, 2002, **6**, 429–447.

Fortet, R. and Mourier, E., Convergence de la répartition empirique vers la répartition théorique. *Ann. Sci. Ecole Norm.*, 1953, **70**(Sup. 3), 267–285.

Goldfarb, D. and Iyengar, G., Robust portfolio selection problems. *Math. Oper. Res.*, 2003, **28**(1), 1–38.

Kersting, G-D., Die Geschwindigkeit der Glivenko-Cantelli-Konvergenz gemessen in der Prohorov-Metrik. *Math. Z.*, 1978, **163**(1), 65–102.

Klein, R.W. and Bawa, V.S., The effect of estimation risk on optimal portfolio choice. *J. Financ. Econ.*, 1976, **3**(3), 215–231.

Knight, F.H., *Risk, Uncertainty and Profit*, 1921 (Houghton Mifflin Company: Boston, MA).

Markowitz, H.M., Portfolio selection: Efficient diversification of investments, *Cowles Foundation for Research in Economics at Yale University, Monograph 16*, 1959 (John Wiley & Sons Inc.: New York).

Nemirovskii, A. and Yudin, D.B., Cesaro convergence of the gradient method for approximation of saddle points of convex-concave functions. *Dokl. AN SSSR*, 1978, **239**, 1056–1059.

Pflug, G.Ch., Subdifferential representations of risk measures. *Math. Program. Ser. B*, 2006, **108** (2–3), 339–354.

Rachev, S.T., Probability metrics and the stability of stochastic models. *Wiley Series in Probability and Mathematical Statistics: Applied Probability and Statistics*, 1991 (John Wiley & Sons Ltd.: Chichester).

Rachev, S.T. and Rüschendorf, L., *Mass Transportation Problems, Vol I, Probability and its Applications*, 1998 (Springer-Verlag: New York).

Rockafellar, R.T., Monotone operators and the proximal point algorithm. *SIAM J. Contr. Opt.*, 1976, **14**(5), 877–898.

Rockafellar, R.T., Convex Analysis, in the series, *Princeton Landmarks in Mathematics*, 1997 (Princeton University Press: Princeton, NJ); [reprint of the 1970 original, Princeton Paperbacks].

Rockafellar, R.T. and Uryasev, S., Optimization of conditional value-at-risk. *J. Risk*, 2000, **2**(3), 21–41.

Shapiro, A. and Ahmed, S., On a class of minimax stochastic programs. *SIAM J. Optim.*, 2004, **14**(4), 1237–1249; (electronic).

Tokat, Y., Rachev, S. and Schwartz, E., The stable non-gaussian asset allocation: A comparison with the classical gaussian approach. *J. Econ. Dyn. Control*, 2003, **27**, 937–969.

Vallander, S.S., Calculation of the Wasserstein distance between probability distributions on the line. *Theor. Prob. Appl.*, 1973, **18**, 784–786.

Yu, L.Y., Ji, X.D. and Wang, S.Y., Stochastic programming models in financial optimization: A survey. *Adv. Modeling Opt.*, 2003, **5**(1), 1–26.

Zackova, J., On minimax solutions of stochastic linear programming problems. *Casopis Pěst Math.*, 1966, **91**, 423–430.

Ziemba, W.T. and Vickson, R.G., *Stochastic Optimization Models in Finance*, 1973 (Academic Press: London).

CHAPTER 19

Mean-Risk Models Using Two Risk Measures: A Multi-Objective Approach

DIANA ROMAN, KENNETH DARBY-DOWMAN and GAUTAM MITRA

CONTENTS

19.1   INTRODUCTION AND MOTIVATION

MEAN-RISK MODELS ARE STILL THE MOST WIDELY USED APPROACH in the practice of portfolio selection. With mean-risk models, return distributions are characterized and compared using two statistics: the expected value and the value of a risk measure. Thus, mean-risk models have a ready interpretation of results and in most cases are convenient from a computational point of view. Sceptics on the other hand may question these advantages since the practice of describing a distribution by just two parameters involves great loss of information.

It is evident that the risk measure used plays an important role in making the decisions. Variance was the first risk measure used in mean-risk models (Markowitz 1952) and, in spite of criticism and many proposals of new risk measures (see e.g. Fishburn 1977; Yitzhaki 1982; Konno and Yamazaki 1991; Ogryczak and Ruszczynski 1999, 2001, Rockafellar and Uryasev 2000, 2002), variance is still the most widely used measure of risk in the practice of portfolio selection. For regulatory and reporting purposes, risk measures concerned with the left tails of distributions (extremely unfavourable outcomes) are used. The most widely used risk measure for such purposes is Value-at-Risk (VaR). However, it is known that VaR has undesirable theoretical properties (it is not subadditive, as shown, for example, in Tasche (2002) and thus fails to reward diversification). In addition, optimization of VaR leads to a non-convex NP-hard problem which is computationally intractable. In spite of a considerable amount of research, optimizing VaR is still an open problem (see e.g. Larsen *et al.* 2002, Leyffer *et al.* 2005 and references therein). For these reasons, another risk measure concerned with the left tail, the Conditional Value-at-Risk (CVaR), is gaining more popularity. CVaR has attractive theoretical properties: it controls the magnitude of losses beyond VaR and it is coherent (see e.g. Artzner *et al.* 1999; Pflug 2000; Acerbi and Tasche 2002; Rockafellar and Uryasev 2002; Tasche 2002). In addition, CVaR is easy to optimize. Optimizing CVaR is a convex programming problem. In the case when the random variables under consideration are discrete, with a finite number of outcomes, represented by various outcomes under different scenarios, optimizing CVaR leads to a linear programming model of finite dimension (Rockafellar and Uryasev 2000, 2002).

Variance and CVaR quantify risk from different perspectives. Variance measures the spread around the expected value of a random variable, while CVaR measures the expected loss corresponding to a number of worst cases, depending on the chosen confidence level. Thus, the mean-variance and the mean-CVaR models may lead to very different solutions. A portfolio obtained as a solution in the mean-variance model may be considered unacceptable by a regulator, since it may have an excessively large CVaR, leading to big losses under unfavourable scenarios. On the other hand, traditional fund managers may consider a portfolio obtained with the mean-CVaR model unacceptable since it may have an excessively large variance and thus an excessively small Sharpe index (see Luenberger 1998).

In this chapter, we seek to address the requirements of the traditional fund manager and the regime imposed by the regulator. We propose a model for portfolio selection that uses both variance and CVaR in order to make decisions. We call this model the mean-variance-CVaR model. Random variables are described and compared using three statistics: the expected value, variance and CVaR. Thus, the model may be considered as belonging to the family of mean-risk models.

We formally define the preference relation for random variables in this model. The efficient solutions with respect to this preference relation are such that we cannot improve on one statistic (of the three: expected value, variance and CVaR) without worsening another. Mathematically, the problem is multi-objective (maximize expected return, minimize variance, minimize CVaR) and the efficient solutions of the mean-variance-CVaR model are the Pareto optimal solutions of the multi-objective problem.

We prove that the efficient solutions of this model may be found by solving a single objective optimization problem in which variance is minimized while constraints are imposed on the expected return and the CVaR level. The practical importance of this approach is twofold. Firstly, a solution obtained in this way has an intuitive appeal. For example, if the CVaR of a mean-variance efficient portfolio is considered as unacceptably large, a constraint could be imposed on the CVaR level and a new portfolio obtained, which has a minimal variance under these conditions. Secondly, the problem is tractable from a computational point of view. In the case where the random variables under consideration are discrete and described by their realizations under various scenarios, the problem is one of quadratic programming.

Generally, the mean-variance and mean-CVaR efficient portfolios are particular efficient solutions of the proposed model.[1] However, most of the efficient portfolios in the mean-variance-CVaR model are dominated in both mean-variance and mean-CVaR models, although they may represent improved distributions: a compromise between the classical fund managers' and the regulators' points of view.

The rest of this chapter is structured as follows. In Section 19.2 the portfolio selection problem is described. Section 19.3 is concerned with mean-risk models, in particular with the mean-variance and the mean-CVaR models. In Section 19.4 we present the mean-variance-CVaR model. Firstly, the preference relation among random variables is defined. The efficient solutions of the proposed model are Pareto non-dominated solutions of a multi-objective problem. Secondly, an optimization approach for solving the multi-objective problem is proposed. With this approach, the efficient solutions of the proposed model are found by solving a single optimization problem, in which variance is minimized and constraints are imposed on the expected value and the CVaR level. Thirdly, we describe how all the efficient solutions of the model may be obtained. Finally, the algebraic form of the mean-variance-CVaR model for the case of scenario models is presented. Section 19.5 presents the computational results. A dataset, drawn from the FTSE 100 index is used to evaluate the performance of the proposed model. For several fixed levels of expected return, we consider the mean-variance and the mean-CVaR efficient portfolios together with other portfolios, efficient only in the mean-variance-CVaR model. We

---

[1] There may be a situation when several mean-CVaR efficient portfolios have the same mean return and the same (optimal) CVaR, but different variances. Only the portfolio with the minimal variance is efficient in the proposed model. The same discussion applies for mean-variance efficient portfolios. We reconsider the issue in Section 19.4.4.

evaluate their performances using both in-sample and out-of-sample analysis. Section 19.6 presents the conclusions.

19.2   THE PORTFOLIO SELECTION PROBLEM

The problem of portfolio selection with one investment period is an example of the general problem of deciding between random variables when larger outcomes are preferred. Decisions are required on the amount (proportion) of capital to be invested in each of a number of available assets such that at the end of the investment period the return is as high as possible. Consider a set of n assets, with asset j in $\{1, \ldots, n\}$ giving a return R_j at the end of the investment period. R_j is a random variable, since the future price of the asset is not known. Let x_j be the proportion of capital invested in asset j ($x_j = w_j/w$ where w_j is the capital invested in asset j and w is the total amount of capital to be invested), and let $x = (x_1, \ldots, x_n)$ represent the portfolio resulting from this choice. This portfolio's return is the random variable: $R_x = x_1 R_1 + \cdots + x_n R_n$, with distribution function $F(r) = P(R_x \le r)$ that depends on the choice $x = (x_1, \ldots, x_n)$.

   To represent a portfolio, the weights $(x_1, \ldots, x_n)$ must satisfy a set of constraints that forms a feasible set $\mathcal{A}$ of decision vectors. The simplest way to define a feasible set is by the requirement that the weights must sum to 1 and short selling is not allowed. For this basic version of the problem, the set of feasible decision vectors is

$$\mathcal{A} = \left\{ (x_1, \ldots, x_n) \middle| \sum_{j=1}^{n} x_j = 1, \ x_j \ge 0, \ \forall j \in \{1, \ldots, n\} \right\}. \tag{19.1}$$

   Consider a different portfolio defined by the decision vector $y = (y_1, \ldots, y_n) \in \mathcal{A}$, where y_j is the proportion of capital invested in asset j. The return of this portfolio is given by the random variable $R_y = y_1 R_1 + \cdots + y_n R_n$.

   The problem of choosing between portfolio $x = (x_1, \ldots, x_n)$ and portfolio $y = (y_1, \ldots, y_n)$ becomes the problem of choosing between random variables R_x and R_y. The criteria by which one random variable is considered 'better' than another random variable need to be specified and models for choosing between random variables (models for preference) are required. The purpose of such models is firstly, to define a preference relation among random variables and secondly, to identify random variables that are non-dominated with respect to that preference relation.

   The next issue is to consider a practical representation for the random variables that describe asset and portfolio returns. We treat these random variables as discrete and described by realizations under T states of the world, generated using scenario generation or finite sampling of historical data. For any $i \in 1, \ldots, T$, let state ω_i occur with probability p_i, $\sum_{i=1}^{T} p_i = 1$. Thus, the random returns are defined on a discrete probability space $\{\Omega, \mathcal{F}, P\}$ with $\Omega = \{\omega_1, \ldots, \omega_T\}$, $\mathcal{F}$ a σ-field and $P(\omega_i) = p_i$.

   Let r_{ij} be the return of asset j under scenario i, $i \in \{1, \ldots, T\}$, $j \in \{1, \ldots, n\}$. Thus, the random variable R_j representing the return of asset j is finitely distributed over $\{r_{1j}, \ldots, r_{Tj}\}$ with probabilities $p_1, \ldots, p_T$. The random variable R_x representing the return of portfolio

$x = (x_1, \ldots, x_n)$ is finitely distributed over $\{R_{x1}, \ldots, R_{xT}\}$, where $R_{xi} = x_1 r_{i1} + \cdots + x_n r_{in}$, $\forall i \in \{1, \ldots, T\}$.

19.3   MEAN-RISK MODELS

19.3.1   The General Case

Mean-risk models were developed in the early 1950s for the portfolio selection problem. In his seminal work 'Portfolio selection,' Markowitz (1952) proposed variance as a risk measure. Since then, many alternative risk measures have been proposed. The question of which risk measure is most appropriate is still the subject of much debate.

In mean-risk models, two scalars are attached to each random variable: the expected value (mean) and the value of a risk measure. Preference is then defined using a trade-off between the mean where a larger value is desirable and risk where a smaller value is desirable:

> In the mean-risk approach with the risk measure denoted by ρ, random variable R_x dominates (is preferred to) random variable R_y if and only if: $E(R_x) \geq E(R_y)$ and $\rho(R_x) \leq \rho(R_y)$ with at least one strict inequality. Alternatively, we can say that portfolio x dominates portfolio y.

In this approach, the choice x (or the random variable R_x) is efficient (non-dominated) if and only if there is no other choice y such that R_y has higher expected value and less risk than R_x. This means that, for a given level of minimum expected return, R_x has the lowest possible risk, and, for a given level of risk, it has the highest possible expected return. Plotting the efficient portfolios in a mean-risk space gives *the efficient frontier*.

Thus, the efficient solutions in a mean-risk model are Pareto efficient solutions of a multi-objective problem, in which the expected return is maximized and the risk is minimized:

$$\max\{(E(R_x), -\rho(R_x)) : x \in A\}.$$

Generally, for a multi-objective problem:

$$\max\{f(x) = (f_1(x), \ldots, f_T(x)) : x \in A\}, \tag{19.2}$$

the Pareto preference relation is defined as follows:

> A feasible solution $x^1 \in A$ *Pareto dominates* another feasible solution $x^2 \in A$ if $f_i(x^1) \geq f_i(x^2)$ for all i with at least one strict inequality.

x_0 is a Pareto efficient (non-dominated) solution of (19.2) if and only if there does not exist a feasible x such that x Pareto dominates x_0. In other words, a Pareto efficient solution is a feasible solution such that, in order to improve upon one objective function, at least one other objective function must assume a worse value.

In order to find an efficient portfolio, we solve an optimization problem with decision variables $x_1, \ldots x_n$:

$$\text{Minimize } \rho(R_x)$$
$$\text{Subject to} : E(R_x) \geq d \quad \text{and} \quad (x_1, \ldots x_n) \in A,$$

where d represents the desired level of expected return for the portfolio.

Varying d and repeatedly solving the corresponding optimization problem identifies the minimum risk portfolio for each value of d. These are the efficient portfolios that compose the efficient set. By plotting the corresponding values of the objective function and of the expected return respectively in a return-risk space, we trace out *the efficient frontier.*

An alternative formulation, which explicitly trades risk against return in the objective function, is

$$\text{Maximize } E(R_x) - \tau \rho(R_X)(\tau \geq 0)$$
$$\text{Subject to} : (x_1, \ldots x_n) \in A.$$

Varying the trade-off coefficient τ and repeatedly solving the corresponding optimization problems traces out the efficient frontier.

19.3.2   The Mean-Variance Model

The variance of a random variable R_x is defined as its second central moment:

$$\sigma^2(R_x) = E[(R_x - E(R_x))^2].$$

An important property is that the variance of the portfolio return $R_x = x_1 R_1 + \cdots + x_n R_n$, resulting from choice $(x_1, \ldots, x_n)$, can be expressed as:

$$\sigma^2(R_x) = \sum_{k=1}^{n} \sum_{j=1}^{n} x_k x_j \sigma_{kj},$$

where σ_{kj} is the covariance of R_k and R_j, and thus variance is expressed as a quadratic function of $x_1, \ldots, x_n$.

The mean-variance model can be formulated for the portfolio selection problem as follows:

$$\text{Minimize} \sum_{k=1}^{n} \sum_{j=1}^{n} x_k x_j \sigma_{kj}$$

$$\text{Subject to} \sum_{j=1}^{n} \mu_j x_j \geq d$$

$$\sum_{i=1}^{n} x_j = 1$$

$$x_j \geq 0, \quad \forall j = 1, \ldots, n,$$

where μ_j is the expected rate of return of asset j, $j \in \{1, \ldots, n\}$; σ_{kj} is the covariance between returns of asset k and asset j, with $k, j \in \{1, \ldots, n\}$; and d is the desired expected value of the portfolio return.

19.3.3 The Mean-CVaR Model

Let R_x be a random variable representing the return of a portfolio x over a given holding period and $A\% = \alpha \in (0,1)$ a percentage which represents a sample of 'worst cases' for the outcomes of R_x (usually, $\alpha = 0.01$ or $\alpha = 0.05$).

The definition of CVaR at the specified level α is the mathematical transcription of the concept 'average of losses in the worst $A\%$ of cases'[2] (Acerbi and Tasche 2002), where a 'loss' is a negative outcome of R_x (thus the loss associated with R_x is described by the random variable $-R_x$).

Formally, the Conditional Value-at-Risk at level α of R_x is defined as minus the mean of the α-tail distribution of R_x, where the α-tail distribution is obtained by taking the lower α part of the distribution of R_x (corresponding to extreme unfavourable outcomes) and rescaling its distribution function to span $[0,1]$:

$$\text{CVaR}_\alpha(R_x) = -\frac{1}{\alpha}\left\{E(R_x 1_{\{R_x \leq q^\alpha(R_x)\}}) - q^\alpha(R_x)[P(R_x \leq q^\alpha(R_x)) - \alpha]\right\}, \qquad (19.3)$$

where q^α is an α-quantile of R_x, meaning a real number r such that $P(R_x < r) \leq \alpha \leq P(R_x \leq r)$ (see Laurent (2003) for more details on α-quantiles), and

$$1_{\{\text{Relation}\}} = 1, \text{ if Relation is true}$$
$$0, \text{ if Relation is false}$$

(see Rockafellar and Uryasev (2000, 2002) for more details).

An important result, proved by Rockafellar and Uryasev (2000, 2002), and independently by Ogryczak and Ruszczynski (2002), is that the CVaR of a random variable R_x can be calculated by solving a convex optimization problem. Moreover, CVaR can be minimized over the set of feasible decision vectors. These results are summarized below:

Proposition 19.1: (CVaR calculation and optimization). *Let R_x be a random variable depending on a decision vector x that belongs to a feasible set $\mathcal{A}$, and $\alpha \in (0,1)$. Consider the function:*

---

[2] This is not necessarily the same as 'the expected value of losses exceeding VaR at confidence level α,' as it is defined in earlier papers on CVaR. The two definitions lead to the same results when the distribution of the random variable under consideration is continuous, but differences may appear when the considered distribution has discontinuities—see Acerbi and Tasche (2002), and Rockafellar and Uryasev (2002) for more details.

$$F_\alpha(x, v) = \frac{1}{\alpha} E\{[-R_x + v]^+\} - v,$$

where

$$[u]^+ = u \text{ for } u \geq 0,$$
$$[u]^+ = 0 \text{ for } u < 0.$$

Then:

(a) As a function of v, F_α is finite and continuous (hence convex) and

$$\text{CVaR}_\alpha(R_x) = \min_{v \in R} F_\alpha(x, v).$$

In addition, the set consisting of the values of v for which the minimum is attained, denoted by $A_\alpha(x)$, is a non-empty, closed and bounded interval (possibly formed by just one point).

(b) Minimizing CVaR_α with respect to $x \in A$ is equivalent to minimizing F_α with respect to $(x, v) \in A x R$:

$$\min_{x \in A} \text{CVaR}_\alpha(R_x) = \min_{(x,v) \in A x R} F_\alpha(x, v).$$

In addition, a pair (x^*, v^*) minimizes the right-hand side if and only if x^* minimizes the left-hand side and $v^* \in A_\alpha(x^*)$.

(c) $\text{CVaR}_\alpha(R_x)$ is convex with respect to x and $F_\alpha(x, v)$ is convex with respect to (x, v).

Thus, if the set A of feasible decision vectors is convex (which is the case for the basic version of the portfolio selection problem), and even if we impose a further lower limit on the expected return, minimizing CVaR is a convex optimization problem.

In the case when R_x is a discrete random variable (as described in Section 19.2), calculating and optimizing CVaR are linear programming problems. Suppose that R_x has T possible outcomes $R_{x1}, \ldots, R_{xT}$ with probabilities $p_1, \ldots, p_T$. Then:

$$F_\alpha(x, v) = \frac{1}{\alpha} \sum_{i=1}^{T} p_i [v - R_{xi}]^+ - v.$$

For the portfolio selection problem, as presented in Section 19.2, where $R_{xi} = \sum_{j=1}^{n} x_j r_{ij}$ with r_{ij} the return of asset j under scenario i,

$$F_\alpha(x, v) = \frac{1}{\alpha} \sum_{i=1}^{T} p_i \left[v - \sum_{j=1}^{n} x_j r_{ij} \right]^+ - v.$$

Thus, the mean-CVaR_α model can be formulated for the portfolio selection problem as follows:

$$\text{Minimize} - v + \frac{1}{\alpha} \sum_{i=1}^{T} p_i y_i$$

$$\text{Subject to} \sum_{j=1}^{n} -r_{ij}x_j + v \leq y_i, \quad \forall i = 1, \ldots, T \quad y_i \geq 0, \quad \forall i = 1, \ldots, T$$

$$\sum_{j=1}^{n} \mu_j x_j \geq d$$

$$\sum_{j=1}^{n} x_j = 1$$

$$x_j \geq 0, \quad \forall j = 1, \ldots, n.$$

19.4   THE MEAN-VARIANCE-CVaR MODEL

19.4.1   The Theoretical Background

In this section, a model for portfolio selection is proposed, in which random variables are described by three statistics: the expected value, the variance and the CVaR at a specified confidence level $\alpha \in (0,1)$. We claim that taking three parameters into consideration, instead of two, gives a better modelling power. The proposed model may bring an improvement in the solution, in the case where a mean-variance efficient portfolio has an excessively large CVaR, or a mean-CVaR efficient portfolio has an excessively large variance.

The idea of restricting the risk of a distribution from two different perspectives has been used before in various contexts.

Konno *et al.* (1993) proposed a 'mean-absolute deviation skewness portfolio optimization model,' in which the lower semi-third moment of the portfolio return is maximized subject to constraints on the mean and on the absolute deviation of the portfolio return. A 'mean-variance-skewness portfolio optimization model' was proposed by Konno *et al.* (1995): they maximized the third moment of the portfolio return subject to constraints on the mean and on the variance of the portfolio return. Optimization approaches are provided, in which the corresponding cubic and quadratic functions are approximated by linear functions.

Wang (2000) proposed a model in which the portfolio return has constraints on both variance and Value-at-Risk (VaR), and a maximum expected return under these conditions. However, no practical optimization approach is provided.

Harvey *et al.* (2003) proposed a model in which random variables are chosen with respect to their expected value, variance and skewness. Thus, it may be considered that they use two risk measures in order to control the selection of a solution: the variance and the negative of skewness. Their model has a distributional assumption for portfolio returns and uses an expected utility maximization approach, with the utility function depending on the expected value, variance and skewness.

Jorion (2003) proposed that a portfolio return distribution should have constraints on both variance and 'tracking error volatility,' which is 'the volatility of the deviation of the active portfolio from the benchmark,' with a maximum expected return under these conditions. Thus, this approach may also fall into the category of index-tracking models.

There have been various formulations of portfolio selection problems as multiple criteria models (see e.g. Ogryczak 2000, 2002). However, to the best of our knowledge, the use of CVaR together with variance within a multi-attribute model is novel. A categorized bibliography on the applications of multiple criteria decision-making techniques in finance is provided in Steuer and Na (2003).

The model proposed in this chapter does not assume a particular distribution for the returns and, in addition, is convenient from a computational point of view. We define a preference relation for random variables and provide an optimization approach for finding the efficient solutions with respect to this preference relation.

Consider again the portfolio selection problem described in Section 19.2, with the random variable R_x and R_y describing the returns of portfolios x and y respectively, with $x, y \in \mathcal{A}$.

We consider a model for choice under risk that we refer to as *the mean-variance-CVaR model*, in which the preference relation among random variables is defined as follows:

> In the mean-variance-CVaR model, a random variable R_x is preferred to a random variable R_y (or, similarly, the portfolio x is preferred to portfolio y) if and only if $E(R_x) \geq E(R_y)$, $\sigma^2(R_x) \leq \sigma^2(R_y)$ and $\text{CVaR}_\alpha(R_x) \leq \text{CVaR}_\alpha(R_y)$, with at least one strict inequality.

Thus, the non-dominated (efficient) solutions in the mean-variance-CVaR model are the Pareto efficient solutions of a multi-objective problem in which the expected value is maximized while the variance and the CVaR are minimized:

$$(\text{MVC}): \ \max\big(E(R_x), -\sigma^2(R_x), -\text{CVaR}_\alpha(R_x)\big)$$
$$\text{Subject to} : x \in \mathcal{A}.$$

When plotting the efficient solutions in a mean-variance-CVaR space, a surface is obtained; we refer to this surface as 'the efficient frontier' of the mean-variance-CVaR model.

19.4.2 An Optimization Approach

The next issue to address is how to obtain the efficient solutions of the mean-variance-CVaR model.

Firstly, the multi-objective problem (MVC) is transformed into a single objective problem in which one objective function is optimized while lower limits are imposed on the remaining objective functions and transformed into constraints. This method, known in multi-objective optimization as the 'ε-constraint method' (Haimes *et al.* 1971, see also Steuer 1986) generally requires some regularization in order to guarantee that an optimal solution of the single-objective problem obtained is a Pareto optimal solution of the original multi-objective problem.

We choose to minimize variance for two reasons. Firstly, it is more intuitively appealing to impose limits on the expected value and CVaR, rather than on variance. Secondly, we show that minimizing variance is more convenient from a computational point of view. In either case, a convex optimization problem would be obtained[3] irrespective of which statistic we choose for the objective function, but, when optimizing variance, a quadratic programming problem is obtained, as shown below.

In what follows, for a random variable R_x that depends on the decision vector x, the variance of R_x is denoted alternatively by $\sigma^2(x)$ or $\sigma^2(R_x)$. Similarly, the Conditional Value-at-Risk at level α of R_x is denoted by $\text{CVaR}_\alpha(x)$ or $\text{CVaR}_\alpha(R_x)$, and the expected value of R_x by $E(x)$ or $E(R_x)$.

We consider the following optimization problem:

(P1): $\min \sigma^2(x)$

        Subject to: $\text{CVaR}_\alpha(x) \leq z; \quad E(x) \geq d; \quad x \in \mathcal{A}$

where z and d are real numbers.

It is easy to prove that: if x^* is a Pareto optimal solution of (MVC) then x^* is also an optimal solution of (P1) with $z = \text{CVaR}_\alpha(x^*)$ and $d = E(x^*)$.

Indeed, assume that x^* is not an optimal solution of (P1). Obviously x^* is a feasible solution of (P1). Denote by x' an optimal solution of (P1). It follows that $\sigma^2(x') \leq \sigma^2(x^*)$, $\text{CVaR}_\alpha(x') \leq \text{CVaR}_\alpha(x^*)$ and $E(x') \geq E(x^*)$, which means that x' Pareto dominates x^* and we have a contradiction.

The converse is also true, with the additional assumption of uniqueness of the optimal solution:

If x^* is the unique optimal solution of (P1), then x^* is also a Pareto optimal solution of (MVC).

Indeed, assume that x^* is Pareto dominated in (MVC) and denote by x' a point that Pareto dominates x^*. This means that $\sigma^2(x') \leq \sigma^2(x^*)$, $\text{CVaR}_\alpha(x') \leq \text{CVaR}_\alpha(x^*) \leq z$ and $E(x') \geq E(x^*) \geq d$ with at least one strict inequality. Thus x' is another feasible solution of (P1) such that $\sigma^2(x') \leq \sigma^2(x^*)$, which is a contradiction.

Remark 1: If the covariance matrix of returns is positive definite, then variance is a strictly convex function of x. In this case, minimizing variance over a convex set has at most one optimal solution; thus, the possibility of multiple optimal solutions for (P1) is eliminated. This is usually the case; if there are no redundant assets (ones that can be replicated by the remaining of the assets) or risk-free assets in the collection of assets considered, then the covariance matrix is positive definite.

We summarize these results below:

---

[3] As stated in Proposition 1, CVaR is a convex function of x. Variance is also a convex function of x, since the variance-covariance matrix is positive semi-definite. The expected value is a linear function of x.

Proposition 19.2: *If the covariance matrix is positive definite, a point* x^* *is a Pareto efficient solution of* (MVC) *if and only if* x^* *is an optimal solution of* (P1) *with* $z = CVaR_\alpha(x^*)$ *and* $d = E(x^*)$.

Thus, in the case of a positive definite covariance matrix of returns, the Pareto efficient solutions of (MVC) can be fully characterized as optimal solutions in (P1) with active constraints on mean and on CVaR. In Appendix 19.A we treat the general case of a positive semi-definite covariance matrix.

The next issue that arises is how to represent the CVaR constraint in (P1). As presented in Proposition (19.1), the function $F_\alpha(x, v) = (1/\alpha)E\{[v - R_x]^+\} - v$ may be used both for calculating the CVaR of a given random variable and for optimizing CVaR with respect to all feasible decisions vectors.

Furthermore, Krokhmal *et al.* (2002) proved that the same function $F_\alpha(x, v)$ may be used for imposing an upper limit on the CVaR of a random variable, while maximizing its expected value. Their result may be extended to a much more general case. In fact, the constraint 'CVaR$_\alpha(x) \leq z$' can be replaced with the constraint '$F_\alpha(x, v) \leq z$' in all optimization problems, irrespective of the form of the objective function or the feasible set.

Proposition 19.3: *Consider two optimization problems* (P) *and* (P′) *with* $A \subset R^n$ *a feasible set of decision vectors and the objective function* $f: R^n \rightarrow R$ *of any form:*

$$(P) : \min f(x)$$
$$\text{Subject to}: \quad CVaR_\alpha(x) \leq z \quad x \in A$$

$$(P') : \min f(x)$$
$$\text{Subject to}: \quad F_\alpha(x, v) \leq z \quad x \in A, \quad v \in R.$$

In (P), *the variables are* $x_1, \ldots, x_n$ *while in* (P′), *the variables are* $x_1, \ldots, x_n$ *and* v.

Then: (P) *and* (P′) *achieve the same optimal value. Moreover, a point* $x^* \in A$ *is an optimal solution for* (P) *iff there exists* $v^* \in R$ *such that* (x^*, v^*) *is an optimal solution for* (P′). *If, in addition, the constraint* $CVaR_\alpha(x) \leq z$ *in* (P) *is active, then* $v^* \in A_\alpha(x^*)$ *(meaning that* $F(x^*, v^*) = \min_{v \in R} F_\alpha(x^*, v)$*).*

Proof: As stated in Proposition 1, $CVaR_\alpha(x) = \min_{v \in R} F_\alpha(x, v)$. Thus, the problem (P) may be written as:

$$(P) : \min f(x)$$
$$\text{Subject to}: \quad \min_{v \in R} F_\alpha(x, v) \leq z \quad x \in A.$$

Suppose now that x^* is an optimal solution for (P). Obviously (x^*, v^*) is a feasible solution for (P′), where v^* is such that $F(x^*, v^*) = \min_{v \in R} F_\alpha(x^*, v)$. Assume that there exists (x', v') another feasible solution for (P′) such that $f(x') < f(x^*)$. Since $F_\alpha(x', v') \leq z$ it follows that $\min_{v \in R} F_\alpha(x', v) \leq z$; thus, x' is a feasible solution of (P1) which improves the objective function as compared to x^*, which is a contradiction.

Similarly, in a straightforward way, the converse may be proven; the last part of the proposition is obvious. □

Thus, we consider another optimization problem, with variables $x = (x_1, \ldots, x_n) \in A \subset R^n$ and $v \in R$:

$$(P2) : \min \sigma^2(x)$$
$$\text{Subject to: } F_\alpha(x, v) \leq z$$
$$E(x) \geq d$$
$$x \in A,$$
$$v \in R,$$

where A is the (convex) set of feasible decision vectors, as given, for example, by (19.1).

The result below follows from Propositions 19.2 and 19.3:

Proposition 19.4: *If the covariance matrix of returns is positive definite, the Pareto efficient solutions of (MVC) are fully characterized as optimal solutions of (P2) with active constraints on mean and on CVaR.*

In other words, x^* is a Pareto efficient solution of (MVC) if and only if there exists $v^* \in R$ such that (x^*, v^*) is an optimal solution to (P2) with $z = F_\alpha(x^*, v^*)$ and $d = E(x^*)$.

Therefore, varying d and z in the problem (P2) such that the constraints on CVaR and on the expected value are active produces all the efficient solutions of the mean-variance-CVaR model. As shown in Section 19.4.4, this means varying d and z between some finite limits that can be easily determined.

19.4.3 Alternative Optimization Approaches

The optimization approach described in the previous subsection is not unique. A commonly used method of obtaining a Pareto efficient solution of a multi-objective optimization problem is to use a scalarizing function, meaning a real-valued function that is a composite of all objective functions. When optimized, the scalarizing function produces a Pareto efficient solution of the multi-objective optimization problem. Thus, the problem is reduced to a single objective optimization problem. We give below two examples of scalarizing functions, leading to two alternative optimization approaches for the mean-variance-CVaR model.

The most common scalarizing function is a weighted sum of the objective functions in the original multi-objective optimization problem. The general requirement on weights is that they should be strictly positive but usually they are normalized such they sum to 1. In our case, the single objective optimization problem that results is:

$$\max w_1 E(x) - w_2 \sigma^2(x) - w_3 \text{CVaR}_\alpha(x) \qquad (P3)$$
$$\text{Subject to} : x \in A$$

where w_1, w_2, w_3 are strictly positive.[4]

It is clear that every optimal solution of (P3) is a Pareto efficient solution of (MVC). The converse is not always true, in the sense that there may be Pareto optimal solutions of (MVC) that cannot be obtained as optimal solutions of a problem (P3) with strictly positive w_1, w_2 and w_3 (for example, the Pareto optimal solution of (MVC) that globally minimizes variance).

However, due to the convexity of all objective functions on (MVC), every Pareto optimal solution of (MVC) can be obtained as an optimal solution of (P3) with non-negative weights (see Jahn 1985). For example, the Pareto optimal solution of (MVC) that globally minimizes variance is obtained as an optimal solution of (P3) with $w_1 = w_3 = 0$, $w_2 = 1$.

This approach has several disadvantages (see Das and Dennis 1997), one of them being the fact that the weights w_1, w_2, w_3 are rather difficult to interpret. It is more meaningful to set desired levels of expected return and of CVaR and solve (P2).

Another example of a scalarizing function is obtained by considering target values (called reference points or aspiration points) for the values of the objective functions. This technique for multi-objective optimization, named The Reference Point Method is fully described in Wierzbicki (1998). Consider the general multi-objective problem

$$(\text{MO}') : \quad \max(f_1(x), f_2(x), \ldots, f_T(x))$$
$$\text{Subject to:} \quad x \in X,$$

and let $w_1^*, w_2^*, \ldots, w_T^*$ be the user-defined aspiration points for the objective functions. The simplest form of scalarizing function is:

$$\gamma_{w^*}(x) = \min_{1 \leq k \leq T} (f_k(x) - w_k^*) + \varepsilon \sum_{k=1}^{T} (f_k(x) - w_k^*), \qquad (19.4)$$

where $\varepsilon > 0$ is an arbitrary small parameter.

The terms $f_k(x) - w_k^*$ in (4) are usually replaced by more complicated functions of x and w_k^*, $\gamma_k(x, w_k^*)$, which must satisfy certain properties (see e.g. Wierzbicki 1998; Makowski and Wierzbicki 2003). These functions are called partial achievement functions since they measure the actual achievement of the kth objective function with respect to its corresponding aspiration level w_k^*.

Various functions $\gamma_k(x, w_k^*)$ provide a wide modelling environment for measuring individual achievements. Other examples of such functions may be found in Wierzbicki (1998), and Makowski and Wierzbicki (2003).

Provided that all the reference points lie between the lower and the upper bound of the corresponding objective function, the maximization of (19.4) provides a Pareto efficient solution of (MO'). The converse is true, in the sense that for every Pareto efficient solution

---

[4] If additionally there is the assumption of unique optimal solutions of (P3) when some of the weights are zero, then only the non-negativity condition is required for w_1, w_2 and w_3.

of (MO′), there exist aspiration levels such that this efficient solution maximizes the corresponding achievement function (see Wierzbicki 1998). In our case, the scalarizing achievement function to maximize is:

$$\gamma_{w^*}(x) = \min\{E(x) - w_1^*, w_2^* - \sigma^2(x), w_3^* - \mathrm{CVaR}_\alpha(x)\}$$
$$+ \varepsilon[E(x) - w_1^* + w_2^* - \sigma^2(x) + w_3^* - \mathrm{CVaR}_\alpha(x)\}],$$

where $\varepsilon > 0$ is an arbitrary small parameter.

The Reference Point Method is primarily designed for obtaining a specific solution of a multi-objective problem rather than the whole set of efficient solutions. Although all the efficient solutions may be obtained with this method by choosing appropriate reference points, care must be taken in choosing the reference points between the lower and upper bound of each objective function. The lower bounds for the objective functions are difficult to find and often approximations are used.

In contrast, the optimization method described in Section 19.4.2 produces the entire set of efficient solutions of the mean-variance-CVaR model with no difficulty, as described in the next section.

19.4.4 The Efficient Frontier of the Mean-Variance-CVaR Model

We consider the case when the covariance matrix of returns is positive definite; the general case of a positive semi-definite covariance matrix is treated in Appendix 19.A.

As presented in Section 19.4.2, varying the right-hand sides d and z in (P2) such that the corresponding constraints on mean and CVaR are active produces all the efficient solutions of (MVC).

Thus, the level d for the expected value must lie in the interval $[d_{min}, d_{max}]$. We define $d_{min} = \max\{d_{minvar}, d_{minCVaR}\}$, where d_{minvar} and $d_{minCVaR}$ are the expected returns of the minimum variance portfolio (mean-variance efficient) and minimum CVaR portfolio (mean-CVaR efficient) respectively. d_{minvar} may be found as the optimal value of the variable d_0 in the problem:

$$\min \sigma^2(x)$$
$$\text{Subject to: } E(x) \geq d_0$$
$$x \in \mathcal{A},$$
$$d_0 \in R.$$

$d_{minCVaR}$ may be found as the optimal value of the variable d_1 in the problem:

$$\min F_\alpha(x, v)$$
$$\text{Subject to: } E(x) \geq d_1 \quad x \in \mathcal{A}, \quad v \in R, \quad d_1 \in R.$$

To be more precise, $d_{minCVaR}$ may be found as above only when the minimization of $F_\alpha(x, v)$ with respect to (x, v) over $\mathcal{A}xR$ provides a unique optimal solution. In the case of non-unique optimal solutions, we can obtain portfolios having the same minimal CVaR but different expected returns; among these, we are interested in the portfolio with the maximum expected return. To obtain this portfolio, we denote by $\text{CVaR}_{\min}$ the optimal value of the above problem and solve another optimization problem:

$$\max E(x)$$
$$\text{Subject to: } F_\alpha(x, v) \leq \text{CVaR}_{\min} \quad x \in \mathcal{A}, \quad v \in R.$$

We define $d_{\max}$ as the maximum possible expected return:[5] the optimal value of the objective function in the problem:

$$\max E(x)$$
$$\text{Subject to: } x \in \mathcal{A}.$$

Furthermore, for a specific $d^* \in [d_{\min}, d_{\max}]$, the level z of CVaR_α must lie in the interval $[z_{d^*,\min}, z_{d^*,\max}]$, where $z_{d^*,\min}$ is the best (minimum) CVaR_α level for the expected return d^* and $z_{d^*,\max}$ is the CVaR_α level of the (unique) portfolio that minimizes variance for the expected return d^*. $z_{d^*,\min}$ is the optimal value of the objective function in the problem:

$$\min F_\alpha(x, v)$$
$$\text{Subject to: } E(x) \geq d^*$$
$$x \in \mathcal{A},$$
$$v \in R.$$

$z_{d^*,\max}$ may be found as the optimal value of the objective function in the problem:

$$\min F_\alpha(x^*, v)$$
$$\text{Subject to: } v \in R,$$

where $x^* = (x_1^*, \ldots, x_n^*)$ is the (unique) portfolio that minimizes variance for the mean return d^*.

The fact that the imposed limit z on CVaR_α is greater than or equal to $z_{d^*,\min}$ ensures that the problem (P2) is not infeasible, while z being less than or equal to $z_{d^*,\max}$ ensures that the constraint on CVaR in (P2) is active. When solving problem (P2) for a level of expected return equal to d^* and a CVaR level equal to $z_{d^*,\min}$, we obtain a mean-CVaR efficient portfolio; more precisely, the mean-CVaR efficient portfolio with the lowest variance for expected return d^*.

When solving problem (P2) for a level of expected return equal to d^* and a CVaR level equal to $z_{d^*,\max}$, we obtain the mean-variance efficient portfolio with expected return d^*.

For a fixed level of expected return, the efficient solutions in the mean-variance-CVaR model form a curve when plotted in a variance-CVaR space, where the lower end of this

---

[5] $d_{\max}$ is also equal to the highest expected return of the component assets in the portfolio selection problem.

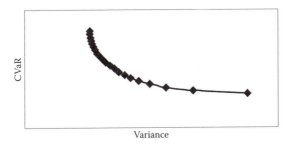

FIGURE 19.1 The efficient solutions of the mean-variance-CVaR model, for a fixed level of expected value, plotted in a variance-CVaR space.

curve is represented by the mean-CVaR efficient solution (with the lowest variance) and the upper end is represented by the mean-variance efficient solution (see Figure 19.1). The other points of this curve are not efficient in either the mean-variance or the mean-CVaR model.

For the maximum level of expected return d_{max}, this curve degenerates into just one point, with the coordinates equal to the variance and CVaR of the (only) efficient portfolio obtained for d_{max}, consisting of the asset with the highest expected return.

19.4.5 The Formulation of the Mean-Variance-CVaR Model for Scenario Models

For the portfolio selection problem, as presented in Section 19.2, consider T scenarios and n assets with

r_{ij} = the return of asset j under scenario i, for $i = 1, \ldots, T$ and $j = 1, \ldots, n$;
p_i = the probability of scenario i occurring, for $i = 1 \ldots T$;
μ_j = the expected return of asset j, $j = 1, \ldots, n$;
σ_{jk} = the covariance between the returns of assets j and k, for $j, k = 1, \ldots, n$.

As presented in Section 19.3.3, the function F_α can be written as:

$$F_\alpha(x, v) = \frac{1}{\alpha} \sum_{i=1}^{T} p_i \left[v - \sum_{j=1}^{n} x_j r_{ij} \right]^+ - v.$$

Thus, we write the mean-variance-CVaR model as:

$$\min \sum_{j,k=1}^{n} x_j x_k \sigma_{jk}$$

$$\text{Subject to:} \sum_{j=1}^{n} x_j \mu_j \geq d \qquad \frac{1}{\alpha} \sum_{i=1}^{T} p_i y_i - v \leq z$$

$$y_i \geq v - \sum_{j=1}^{n} x_j r_{ij}, \quad \forall i \in 1, \ldots, T$$

$$y_i \geq 0, \quad \forall i \in 1, \ldots, T$$

$$\sum_{j=1}^{n} x_j = 1$$

$$x_j \geq 0 \quad \forall j \in 1, \ldots, n.$$

The minimization is over $v, x_1, \ldots, x_n, y_1, \ldots, y_T$.

19.5   COMPUTATIONAL RESULTS

19.5.1   The Data Set and Methodology

The purpose of this section is to investigate the practical performance of the mean-variance-CVaR model as compared to that of the mean-variance or mean-CVaR model. Precisely, for several levels of expected return, we select portfolios that are efficient in the mean-variance-CVaR model, but dominated in the mean-variance or mean-CVaR model, and we also consider the corresponding mean-variance efficient portfolio and the mean-CVaR efficient portfolio. We compare their in-sample and out-of-sample performances.

We use CVaR at 0.01 confidence level. A dataset, drawn from the FTSE 100 index, was used for this analysis. The returns of the 76 stocks that belonged to the index throughout the period January 1993–December 2003 were considered (for each of the remaining 24 stocks data there is at least one missing data item in the specified period). The dataset consists of monthly returns and has 132 time periods, considered as equally probable scenarios ($n = 76$, $T = 132$). For the out-of sample analysis, the behaviour of the portfolios obtained was examined over the eighteen months following the date of selection (January 2004–June 2005). The models were written in the MPL modelling language (Maximal Software Inc. 2000) and processed using CPLEX 9.0 optimization solver (ILOG 2003). The matrix of covariances of the returns is computed from historical data.

19.5.2   In-Sample Analysis

We consider six levels of expected return, which divide the interval $[d_{\min}, d_{\max}]$ (see Section 19.4.4) into 5 equal parts: $d_1 = d_{\min} = 0.009268$, $d_2 = 0.014034$, $d_3 = 0.018801$, $d_4 = 0.023567$, $d_5 = 0.028334$, $d_6 = d_{\max} = 0.0331$. For each level of expected return d_i, with $i = 1, \ldots, 5$, we determine $z_{d_i, \min}$: the minimum level of CVaR (corresponding to the mean-CVaR efficient portfolio) and $z_{d_i, \max}$: the maximum level of CVaR (the lowest CVaR of a mean-variance efficient portfolio with expected return d_i) and, between them, another 3 equally spaced levels of CVaR. Thus, the interval $[z_{d_i, \min}, z_{d_i, \max}]$ for CVaR is divided into 4 equal parts. For a specific level of expected return, when solving the mean-variance-CVaR model with these CVaR levels, we obtain 5 portfolios, denoted by: P_{CVaR}, $P_{1/4\text{CVaR}}$, $P_{1/2\text{CVaR}}$, $P_{3/4\text{CVaR}}$ and P_{var} respectively. Thus, P_{CVaR} is the mean-CVaR efficient portfolio (with the lowest variance, for the specified expected return) and P_{var} is the

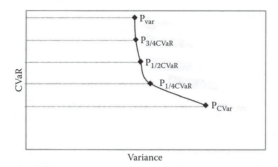

FIGURE 19.2 The efficient frontier for a fixed level of expected return, in a variance-CVaR space. The interval for CVaR is divided into 4 equal parts.

TABLE 19.1 The Number of Assets in the Composition of Mean-Variance-CVaR Efficient Portfolios

	P_{CVaR}	$P_{1/4\text{CVaR}}$	$P_{1/2\text{CVaR}}$	$P_{3/4\text{CVaR}}$	P_{var}
$d_1 = 0.00927$	10	18	21	21	23
$d_2 = 0.01403$	7	17	20	22	21
$d_3 = 0.01880$	7	12	14	13	13
$d_4 = 0.02357$	6	9	9	8	7
$d_5 = 0.02833$	4	5	5	6	6

(unique) mean-variance efficient portfolio for the specified expected return (see Figure 19.2).[6]

We first investigate the composition of the considered portfolios. For all levels of expected return, the mean-variance efficient portfolios have considerably more assets in their composition than the mean-CVaR efficient portfolios. This was expected, since the 'diversification effect' is the basis of the mean-variance theory.

The other three portfolios $P_{1/4\text{CVaR}}$, $P_{1/2\text{CVaR}}$, $P_{3/4\text{CVaR}}$ have usually a number of assets in composition significantly higher than mean-CVaR efficient portfolios, but usually smaller than mean-variance efficient portfolios. There are cases in which these portfolios are as well as or even more diversified than the mean-variance efficient portfolios (see Table 19.1) we notice that this happens when the expected return of the portfolio is high, thus, at high levels of risk. However, in most cases, the number of assets in the composition increases while the level of variance decreases (and the level of CVaR increases). Generally, the assets there are in the composition of mean-CVaR efficient portfolios are also in the composition of portfolios with a higher CVaR level. However, there are assets in the composition of the mean-CVaR portfolios but not in the composition of portfolios with a higher CVaR level. This aspect happens for small portfolio expected returns, thus, at low levels of risk. It may be noticed that, while the expected portfolio return (and thus the risk) increases, those assets are no longer in the composition of any efficient portfolio.

---

[6] The CVaR level of $P_{1/2\text{CVaR}}$ is the arithmetic mean of the CVaR levels of P_{CVaR} and P_{var}. Similarly, the CVaR level of $P_{1/4\text{CVaR}}$ is the arithmetic mean of the CVaR levels of P_{CVaR} and $P_{1/2\text{CVaR}}$.

The portfolio weights of the efficient portfolios considered are presented in Appendix 19.B. We next investigate the in-sample performances of $P_{1/4\text{CVaR}}$, $P_{1/2\text{CVaR}}$, $P_{3/4\text{CVaR}}$, as compared with those of P_{CVaR} and P_{var}. We analyse their return distributions using common in sample parameters. Obviously, the CVaR levels of $P_{1/4\text{CVaR}}$, $P_{1/2\text{CVaR}}$, $P_{3/4\text{CVaR}}$ are better than the CVaR of P_{var}. On the other hand, their variance is generally significantly smaller than that of P_{CVaR}. All the other in-sample parameters are between those of P_{CVaR} and P_{var}. In most cases, P_{CVaR} has the return distribution with the best skewness, kurtosis and minimum of returns but also with the worst variance.

In contrast, P_{var} has the return distribution with the best variance but usually the worst skewness, kurtosis and minimum of returns. This is in line with the modelling paradigm since minimization of CVaR leads to reduction in the (weighted) tail of the resulting portfolio return distribution. The other portfolios $P_{1/4\text{CVaR}}$, $P_{1/2\text{CVaR}}$, $P_{3/4\text{CVaR}}$ represent a compromise in between these two 'extremes.' Their return distribution improves in the left tail, as compared with P_{var} and also has a significantly smaller spread around the mean, as compared with P_{CVaR}. In particular, $P_{1/4\text{CVaR}}$ has return distributions with the variance significantly smaller than that of P_{CVaR} at the expense of a relatively small increase in CVaR. This aspect can be seen from Tables 19.C1–19.C5 (with the best values in italic bold and the worst values enclosed by rectangles) in Appendix 19.C and is also illustrated in Figure 19.2.

In Figure 19.3 the histogram of the return distribution of P_{CVaR} for expected return $d_1 = 0.009\,27$ is presented. This distribution is positively skewed, with a short left tail, a long right tail and a large probability of outcomes below the expected value. Therefore, the probability of large losses is very small, but there is a large probability of small losses. In addition, this distribution is particularly 'flat', that is, not concentrated around the expected value.

In Figure 19.4 the histogram of the return distribution of P_{var} for the same expected return $d_1 = 0.00927$ is presented. This distribution is negatively skewed, with a long left tail, a short right tail and also a large probability of outcomes above the expected value; thus, there is a large probability of small gains. This distribution is concentrated around the expected value.

In Figure 19.5 the histogram of the return distribution of $P_{1/4\text{CVaR}}$ for the same expected return $d_1 = 0.00927$ is presented. This distribution has approximately the same shape as the return distribution of P_{var}: concentrated around the expected value and with a large

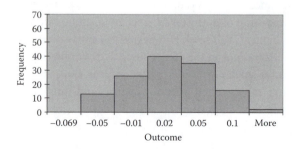

FIGURE 19.3 The histogram of the return distribution of P_{CVaR} for expected return $d_1 = 0.00927$.

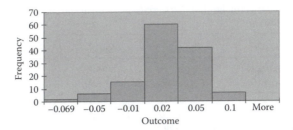

FIGURE 19.4 The histogram of the return distribution of P_{var} for expected return $d_1 = 0.00927$.

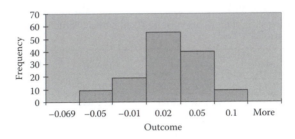

FIGURE 19.5 The histogram of the return distribution of $P_{1/4CVaR}$ for expected return $d_1 = 0.00927$.

probability of outcomes just above the expected value. However, its left tail is shorter, due to the constraint imposed on the CVaR level, and thus the probability of large losses is reduced.

19.5.3 Out-of-Sample Analysis

We analyse the performance of the portfolios described in the previous section over the next 18 time periods following the date of selection (January 2004–June 2005).

The portfolios that are non-efficient in either the mean-variance or the mean-CVaR model, denoted by $P_{1/4CVaR}$, $P_{1/2CVaR}$ and $P_{3/4CVaR}$, have an out-of-sample performance comparable to that of the mean-variance and the mean-CVaR efficient portfolios. It may be noted the generally good out-of-sample performance of the mean-CVaR portfolios and the somewhat poorer performance of the mean-variance portfolios, although the differences were not significant.

In general, the best out-of-sample parameters correspond to mean-CVaR portfolios, but for some levels of expected return, $P_{1/4CVaR}$ had equally good or even better out-of-sample parameters (see Tables 19.2, 19.3, with the best values in italic bold and the worse values enclosed by rectangles).

Figure 19.6 presents the compound out-of-sample returns of the mean-variance-CVaR efficient portfolios with in-sample mean return $d_1 = 0.009268$. $P_{1/4CVaR}$ had a better out-of-sample performance than P_{CVaR} in the first eight out-of-sample periods (January–August 2004) (moreover, P_{CVaR} had a compound return less than one in February 2004, which means that its value fell below the amount invested). At the same time, $P_{1/4CVaR}$ had a better out-of-sample performance than P_{var} in the last ten out-of-sample periods (September 2004–June 2005).

TABLE 19.2 Ex-Post Parameters of the Mean-Variance-CVaR Efficient Portfolios with In-Sample Mean Return $d_1 = 0.009\,268$

	P_{CVaR}	$P_{1/4\text{CVaR}}$	$P_{1/2\text{CVaR}}$	$P_{3/4\text{CVaR}}$	P_{var}
Mean	*0.016294*	0.01472	0.013835	0.013556	0.01345
Median	0.013106	*0.015918*	0.01456	0.012515	0.011549
Standard Deviation	0.029173	0.026514	0.025082	0.023893	*0.022882*
Minimum	−0.034941	−0.03156	−0.03316	−0.02945	−0.02491
Maximum	0.052624	*0.07282*	0.071134	0.068515	0.066001

TABLE 19.3 Ex-Post Parameters of the Mean-Variance-CVaR Efficient Portfolios with In-Sample Mean Return $d_3 = 0.01880$

	P_{CVaR}	$P_{1/4\text{CVaR}}$	$P_{1/2\text{CVaR}}$	$P_{3/4\text{CVaR}}$	P_{var}
Mean	0.01133	*0.012532*	0.012342	0.012352	0.012342
Median	0.010171	*0.013783*	0.013118	0.012365	0.01231
Standard Deviation	*0.028682*	0.031943	0.03221	0.032159	0.032581
Minimum	−0.04247	*−0.03263*	−0.03614	−0.03817	−0.04004
Maximum	0.081765	*0.08752*	0.082737	0.078024	0.072908

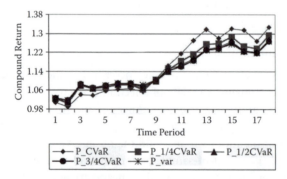

FIGURE 19.6 Ex-post compounded returns of the mean-variance-CVaR efficient portfolios with in-sample mean return $d_1 = 0.009268$.

Figure 19.7 presents the compounded out-of-sample returns of the mean-variance-CVaR efficient portfolios with in-sample mean return $d_3 = 0.018\,80$. $P_{1/4\text{CVaR}}$ had a better out-of-sample performance than both P_{CVaR} and P_{var}, although the differences are small.

19.6 SUMMARY AND CONCLUSIONS

In this chapter, we presented a model for portfolio selection, which selects a solution (distribution) on the basis of three parameters: the expected value, the variance and the CVaR at a specified confidence level. We called this model the mean-variance-CVaR model. The problem of selecting an efficient solution of this model is multi-objective:

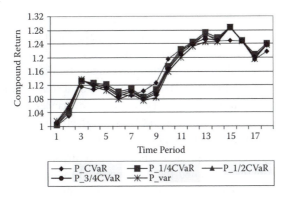

FIGURE 19.7 Ex-post compounded returns of the mean-variance-CVaR efficient portfolios with in-sample mean return $d_3 = 0.01880$.

the expected value is maximized, while the variance and CVaR are minimized. We chose variance and CVaR mainly because they are well established risk measures that quantify risk from different perspectives: variance measures the deviation around the expected value while CVaR measures the average loss over a specified number of worst cases.

Computationally, the problem reduces to solving a single objective problem in which variance is minimized, while constraints are imposed on the expected value and CVaR. In the practice of portfolio selection, the random variables under consideration are usually represented as discrete and described by realizations under various scenarios. In this case, the problem is one of quadratic programming, thus routinely solved by standard available software. Having a constraint on CVaR rather than on the variance has advantages not only from a computational point of view. It is more natural to impose a maximum CVaR level than a maximum variance level, since CVaR represents the mean of the worst outcomes of a distribution.

Varying the right-hand side of the constraints on the expected value and on CVaR such that these constraints are active produces all the efficient solutions of the mean-variance-CVaR model.

When solving the model for a fixed level of expected return, there is a range of efficient solutions. Plotted in a variance-CVaR space, they form a curve, with one end represented by the minimum variance portfolio (with the lowest CVaR), the other represented by the minimum CVaR portfolio (with the lowest variance).

The model was tested on a dataset drawn from the FTSE 100 index. Several levels of expected return were considered, and, for each level of expected return, five portfolios that were efficient in the mean-variance-CVaR model, were analysed: the minimum variance portfolio, the minimum CVaR portfolio and other three portfolios that were dominated in both mean-variance and mean-CVaR models. As expected, the best in-sample parameters concerning the left tail of distributions corresponded to mean-CVaR efficient portfolios: highest skewness, lowest kurtosis and highest maximum. However, the return distributions of mean-CVaR efficient portfolios have also the highest variances. In contrast, the mean-variance efficient portfolios have the return distributions with the lowest variance, but also with the 'worst' left tail (as described by skewness, kurtosis, minimum and CVaR). The other portfolios, efficient only in the mean-variance-CVaR model, improve on the left tail of the mean-variance efficient distributions: they have higher skewness, lower kurtosis higher maximum and higher CVaR. In some cases, this improvement comes at the expense of only a marginal increase in variance. The out-of-sample performances of these portfolios are comparable to those of the mean-variance and mean-CVaR efficient portfolios. In two out of five cases, such a portfolio achieved the highest mean of out-of-sample returns and in almost all cases led to the highest maximum of out-of-sample returns.

As a final remark, it may be noted that the proposed model does not dismiss mean-variance or mean-CVaR models, but on the contrary, it 'embeds' them. Most of the mean-variance and the mean-CVaR efficient solutions are particular solutions of the proposed model. For example, a mean-variance efficient solution is not a solution of the proposed model only if there is another mean-variance efficient solution with the same mean and variance but with lower CVaR. Likewise, from the set of mean-CVaR efficient solutions with a specified mean return, only the one(s) with the lowest variance is solution of the proposed model. Thus, the proposed model makes a 'positive' discrimination between mean-variance and mean-CVaR efficient solutions. In addition, the mean-variance-CVaR model has a range of solutions that are normally discarded by both mean-variance and mean-CVaR model. These solutions may bring an improvement in the distribution, in the case when the CVaR of a mean-variance efficient portfolio is considered to be unacceptably large. They represent a compromise between regulators' requirements for short tails and classical fund managers' requirements for small variance. In making the final choice, the personal preference of the decision-maker plays a key role.

REFERENCES

Acerbi, C. and Tasche, D., On the coherence of expected shortfall. *J. Bank. Finance*, 2002, **26**, 1487–1503.

Artzner, P., Delbaen, F., Eber, J.M. and Heath, D., Coherent measures of risk. *Math. Finance*, 1999, **9**(3), 203–228.

Das, I. and Dennis, J., A closer look at the drawbacks of minimising weighted sums of objectives for Pareto set generation in multicriteria optimisation problems. *Struct. Optimisation*, 1997, **14**(1), 63–69.

Fishburn, P.C., Mean-risk analysis with risk associated with below target returns. *Am. Econ. Rev.*, 1977, **67**, 116–126.

Haimes, Y.Y., Lasdon, L.S. and Wismer, D.A., On a bicriterion formulation of the problems of integrated system identification and system optimization. *IEEE Trans. Syst. Man Cybern.*, 1971, **1**, 296–297.

Harvey, C., Liechty, R.J., Liechty, M.W. and Mueller, P., *Portfolio selection with higher moments*, 2003 (Duke University); Working Paper (available online at: http://ssrn.com/abstract=634141).

ILOG., *CPLEX 9.0 User's Manual*, 2003 (ILOG SA: GentillyFrance).

Jahn, J., Some characterizations of the optimal solutions of a vector optimization problem. *OR Spektrum*, 1985, **7**, 7–17.

Jorion, P., Portfolio optimisation with tracking-error constraints. *Financ. Analysts J.*, 2003, **59**(5), 70–82.

Konno, H., Shirakawa, H. and Yamazaki, H., A mean-absolute deviation-skewness portfolio optimisation model. *Ann. Oper. Res.*, 1993, **45**, 205–220.

Konno, H. and Suzuki, K., A mean-variance-skewness portfolio optimisation model. *J. Oper. Res. Soc. Jpn*, 1995, **38**(2), 173–187.

Konno, H. and Yamazaki, H., Mean absolute deviation portfolio optimization model and its applications to Tokyo stock market. *Manag. Sci.*, 1991, **37**, 519–531.

Krokhmal, P., Palmquist, J. and Uryasev, S., Portfolio optimisation with conditional value-at-risk objective and constraints. *J. Risk*, 2002, **4**(2), 43–68.

Larsen, N., Mausser, H. and Uryasev, S., Algorithms for optimisation of value-at-risk. In *Financial Engineering, E-Commerce and Supply Chain*, edited by P. Pardalos. and V.K. Tsitsiringos, pp. 129–157, 2002 (Kluwer Academic Publishers: Norwell).

Laurent, J.P., *Sensitivity analysis of risk measures for discrete distributions*, 2003; available online at: http://laurent.jeanpaul.free.fr/var_risk_measure_sensitivity.pdf.

Leyffer, S. and Pang, J.S., *On the global minimization of the value-at-risk*, 2005; available online at: http://www.citebase.org/cgi-bin/citations?id=oai:arXiv.org:math/0401063.

Luenberger, D.G., *Investment Science*, 1998 (Oxford University Press: New York).

Makowski, M. and Wierzbicki, A.P., Modelling knowledge, In *Model Based Decision Support and Soft Computations, Applied Decision Support with Soft Computing*, Vol. 124, pp. 3–30, 2003 (Springer-Verlag: Berlin, New York) also available at: www.iiasa.ac.at/marek/pubs.

Markowitz, H.M., Portfolio selection. *J. Finance*, 1952, **7**(1), 77–91.

Maximal Software Incorporation., *MPL Modelling System*, 2000, Arlington, Virginia, USA, Release 4.11.

Ogryczak, W., Multiple criteria linear programming model for portfolio selection. *Ann. Oper. Res.*, 2000, **97**, 143–162.

Ogryczak, W., Multiple criteria optimization and decisions under risk. *Control and Cybernetics*, 2002, **31**, 975–1003.

Ogryczak, W. and Ruszczynski, A., From stochastic dominance to mean-risk models: Semideviations as risk measures. *Eur. J. Oper. Res.*, 1999, **116**, 33–50.

Ogryczak, W. and Ruszczynski, A., On consistency of stochastic dominance and mean-semideviations models. *Math. Program.*, 2001, **89**, 217–232.

Ogryczak, W. and Ruszczynski, A., Dual stochastic dominance and related mean-risk models. *SIAM J. Optimiz.*, 2002, **13**, 60–78.

Pflug, G., Some remarks on the value-at-risk and the conditional value-at-risk. In *Probabilistic Constrained Optimisation: Methodology and Applications*, edited by S. Uryasev, 2000; available online at: www.gloriamundi.org.

Rockafeller, R.T.. and Uryasev, S., Optimization of conditional value-at-risk. *J. Risk*, 2000, **2**, 21–42.

Rockafeller, R.T. and Uryasev, S., Conditional value-at-risk for general loss distributions. *J. Bank. Finance*, 2002, **26**(7), 1443–1471.

Steuer, R., *Multiple Criteria Optimization—Theory, Computation and Application*, 1986 (John Wiley & Sons: New York).

Steuer, R. and Na, P., Multiple criteria decision making combined with finance: A categorized bibliographic study. *Eur. J. Oper. Res.*, 2003, **150**, 496–515.

Tasche, D., Expected shortfall and beyond. *J. Bank. Finance*, 2002, **26**(7), 1519–1533.

Wang, J., *Mean-variance-VaR based portfolio optimisation*, 2000; available online at: www.gloriamundi.org.

Wierzbicki, A., Reference point methods in vector optimization and decision support. *IIASA Interim Report IR-98-017*, 1998 (IIASA).

Yitzhaki, S., Stochastic dominance, mean variance and Gini's mean difference. *Am. Econ. Rev.*, 1982, **72**, 178–185.

APPENDIX 19.A: THE GENERAL CASE OF A POSITIVE SEMI-DEFINITE COVARIANCE MATRIX

In the general case, when the covariance matrix of returns is positive semi-definite, the minimization of variance over a convex set may not have a unique optimal solution. Thus, when using the optimization problem (P2) as described in Section 19.4.4, we may obtain solutions that are Pareto dominated in (MVC).[7] However, we can still use (P2) for obtaining efficient solutions of (MVC), provided the right-hand sides d and z for the mean and CVaR constraints are chosen as described below.

The level d for the expected value must lie in the interval $[d'_{\min}, d_{\max}]$, where $d_{\max}$ is the maximum possible expected return (as presented in Section 19.4.4). We define $d'_{\min} = \max\{d'_{\minvar}, d_{\minCVaR}\}$, where $d'_{\minvar}$ and $d_{\minCVaR}$ are the expected returns of the minimum variance portfolio (mean-variance efficient) and minimum CVaR portfolio (mean-CVaR efficient) respectively. $d_{\minCVaR}$ may be found as described in Section 19.4.4. The expected return of the minimum variance portfolio $d'_{\minvar}$ cannot be determined so straightforward as for the case of a positive definite covariance matrix. We cannot just minimize variance over the whole feasible set $\mathcal{A}$ (with no constraints on the mean) since there may be different optimal solutions to this problem, with the same (optimal) variance but with different expected returns. Among these solutions that globally minimize variance, we consider only the one with the maximum expected return. To obtain this solution, we first solve the problem:

$$\min \sigma^2(x)$$
$$\text{Subject to: } x \in \mathcal{A}.$$

Denote the optimum value of this problem by $\sigma_{\min}$. In order to find the specific optimal solution of this problem with the maximum possible expected return, we propose a convex program with quadratic constraint:

$$\max E(x)$$
$$\text{Subject to: } \sigma^2(x) \leq \sigma_{\min} \quad x \in \mathcal{A}.$$

The optimal value of the above optimization problem is $d'_{\minvar}$. Furthermore, for a specific $d \in [d'_{\min}, d_{\max}]$, the right-hand side for the CVaR constraint z must lie in the interval $[z_{d,\min}, z'_{d,\max}]$; $z_{d,\min}$ is the best (minimum) CVaR$_\alpha$ level for the expected return d and may be found as described in Section 19.4.4. $z'_{d,\max}$ is the *minimum* CVaR$_\alpha$ level of the mean-variance efficient portfolios with expected return d.[8]

In order to determine $z'_{d,\max}$, one may solve two optimization problems. Firstly, the optimal variance for the expected return d (denoted by σ_d^2) may be found as the optimal value of the objective function in the problem:

---

[7] For example, multiple optimal solutions of (P2) may have the same variance, the same expected return but different CVaRs; only the one with the lowest CVaR is Pareto efficient in (MVC).

[8] In case there are several mean-variance efficient portfolios with expected return d, with different CVaR levels, only the portfolio with the lowest CVaR is efficient in the (MVC) model; its CVaR level is denoted by $z_{d,\max}$.

$$\min \sigma^2(x)$$
$$\text{Subject to:} \quad E(x) \geq d$$
$$x \in \mathcal{A}.$$

Secondly, $z'_{d,\max}$ may be found as the optimal value of the objective function in the problem:

$$\min F_\alpha(x, v)$$
$$\text{Subject to:} \quad E(x) \geq d$$
$$\sigma^2(x) \leq \sigma_d^2$$
$$x \in \mathcal{A},$$
$$v \in R.$$

Proposition 19.5: *Consider the optimization problem*

$$(\text{P1}): \quad \min \sigma^2(x)$$
$$\text{Subject to}: \text{CVaR}_\alpha(x) \leq z \quad E(x) \geq d \quad x \in \mathcal{A}.$$

If x^*; is an optimal solution of (P1) *for $d \in [d'_{\min}, d_{\max}]$ and $z \in [z_{d,\min}, z'_{d,\max}]$ (as described above), then x^*; is Pareto efficient in* (MVC).

Proof: Assume that x^*; is not Pareto efficient in (MVC). Denote by x' a feasible solution of (MVC) that Pareto dominates x^*. This means that $\sigma^2(x') \leq \sigma^2(x^*)$, $\text{CVaR}_\alpha(x') \leq \text{CVaR}_\alpha(x^*) \leq z$ and $E(x') \geq E(x^*) \geq d$ with at least one strict inequality. Thus, x' is a feasible solution of (P1). The case $\sigma^2(x') < \sigma^2(x^*)$ is excluded since this contradicts the fact that x^*; is an optimal solution of (P1). It only remains the possibility that x' and x^* are both optimal solutions of (P1) and $\text{CVaR}_\alpha(x') < \text{CVaR}_\alpha(x^*) \leq z$ or $E(x') > E(x^*) \geq d$.

Consider first the case: $\text{CVaR}_\alpha(x') < \text{CVaR}_\alpha(x^*) \leq z$; thus, x' is an optimal solution of (P1) and the constraint $\text{CVaR}_\alpha(x) \leq z$ is not binding. Since (P1) is a convex optimization problem, it follows that x' is an optimal solution of the 'reduced' problem, obtained from (P1) by removing the constraint on CVaR:

$$(\text{P1}_{\text{red}}): \min \sigma^2(x)$$
$$\text{Subject to}: E(x) \geq d$$
$$x \in \mathcal{A}.$$

This means that both x' and x^*; are mean-variance efficient portfolios with expected return $d \in [d'_{\min}, d_{\max}]$. Thus, we have two mean-variance efficient solutions with the same variance, the same expected return d but different CVaRs.

$\text{CVaR}_\alpha(x') < \text{CVaR}_\alpha(x^*) \leq z \leq z'_{d,\max}$. However, $z'_{d,\max}$ is, by construction, the lowest possible CVaR of a mean-variance efficient portfolio with mean return d and we have a contradiction.

Obviously the constraint $E(x) \geq d$ in (P1) is binding for $d \in [d'_{\min}, d_{\max}]$; thus, the case $E(x') > E(x^*) \geq d$ is also impossible and this ends the proof. □

Thus, when the right-hand sides d and v are chosen as above, the constraints on CVaR and on mean are active.

It was shown in Section 19.4.2 that the constraint $\text{CVaR}_\alpha(x) \leq z$ can be replaced with the constraint $F_\alpha(x, v) \leq z$, $v \in R$ and thus the problem (P2), equivalent to (P1), is obtained:

$$(P2): \min \sigma^2(x)$$
$$\text{Subject to: } F_\alpha(x, v) \leq z$$
$$E(x) \geq d$$
$$x \in \mathcal{A}, \quad v \in R$$

Solving problem (P2) with d varying between d'_{min} and d_{max} and z varying between $z_{d,min}$ and $z'_{d,max}$ as described above, gives an efficient solution of the mean-variance-CVaR model.

APPENDIX 19.B: THE COMPOSITION OF EFFICIENT PORTFOLIOS

The composition of efficient portfolios are given in Tables 19.B1–19.B5.

For the highest level of expected return $d_6 = d_{max} = 0.0331$, the efficient portfolio consists of the asset no. 58.

TABLE 19.B1 The Portfolio Weights of the Efficient Portfolios for $d_1 = 0.009268$

Asset no	P_{CVaR}	$P_{1/4CVaR}$	$P_{1/2CVaR}$	$P_{3/4CVaR}$	P_{var}
4	0	0.050	0.050	0.048	0.042
5	0.182	0.060	0.043	0.028	0.021
11	0	0	0.023	0.047	0.068
13	0	0.002	0.028	0.052	0.068
16	0	0.063	0.052	0.037	0.019
17	0	0.071	0.064	0.054	0.046
21	0	0	0	0	0.001
24	0	0.011	0.016	0.019	0.023
25	0	0.000	0.020	0.034	0.044
27	0	0.029	0.054	0.070	0.077
29	0.026	0	0	0	0
40	0.222	0.097	0.081	0.075	0.07
42	0	0.075	0.075	0.065	0.055
43	0.036	0.066	0.062	0.062	0.061
44	0.016	0.088	0.083	0.066	0.05
45	0.053	0	0.010	0.022	0.033
48	0	0	0	0	0.004
52	0	0.001	0	0	0
63	0	0.017	0.002	0.006	0.006
64	0	0.050	0.045	0.028	0.011
65	0	0	0	0	0.04
66	0	0.039	0.060	0.065	0.064
69	0.186	0.050	0.021	0.008	0
70	0	0	0	0.035	0.059
71	0.019	0	0	0	0
72	0.045	0.104	0.115	0.107	0.081
73	0.215	0.129	0.096	0.072	0.057

TABLE 19.B2 The Portfolio Weights of the Efficient Portfolios for $d_2 = 0.01403$

Asset no	P_{CVaR}	$P_{1/4\text{CVaR}}$	$P_{1/2\text{CVaR}}$	$P_{3/4\text{CVaR}}$	P_{var}
4	0.008	0.058	0.069	0.075	0.079
5	0	0.073	0.062	0.058	0.057
13	0	0.037	0.063	0.078	0.086
16	0	0.004	0.027	0.021	0.010
17	0.289	0.096	0.081	0.075	0.070
20	0.086	0	0	0	0
21	0	0.070	0.064	0.052	0.043
24	0	0.014	0.022	0.019	0.018
25	0	0	0.009	0.019	0.026
27	0	0.001	0.011	0.022	0.026
29	0.04	0.030	0	0	0
40	0	0.081	0.092	0.066	0.043
42	0	0	0.005	0.012	0.013
43	0	0.020	0.006	0.004	0
44	0	0.121	0.118	0.097	0.077
45	0.258	0.106	0.093	0.093	0.097
48	0	0	0	0.005	0.019
56	0	0.041	0.042	0.037	0.031
58	0	0	0.002	0.014	0.027
63	0	0.098	0.068	0.059	0.045
65	0	0	0	0.002	0.031
66	0	0.009	0.034	0.047	0.053
69	0.05	0	0	0	0
70	0	0	0.013	0.045	0.066
73	0.269	0.139	0.120	0.100	0.085

TABLE 19.B3 The Portfolio Weights of the Efficient Portfolios for $d_3 = 0.0188$

Asset no	P_{CVaR}	$P_{1/4\text{CVaR}}$	$P_{1/2\text{CVaR}}$	$P_{3/4\text{CVaR}}$	P_{var}
4	0	0.041	0.065	0.088	0.116
5	0	8.1E-05	0.029	0.048	0.066
13	0	0	0.009	0.023	0.032
16	0	0	0.002	0.017	0.035
17	0	0.113	0.119	0.102	0.065
20	0.016	0	0	0	0
21	0.116	0.183	0.163	0.144	0.124
28	0.02	0	0	0	0
29	0.134	0.023	0	0	0
40	0	0.052	0.028	0	0
44	0	0.121	0.126	0.125	0.125
45	0.303	0.179	0.173	0.168	0.161
56	0.209	0.145	0.137	0.124	0.112
58	0	0.021	0.034	0.048	0.066
63	0	0	0.001	0.009	0.012
73	0.202	0.102	0.095	0.084	0.068
76	0	0.022	0.019	0.018	0.019

TABLE 19.B4 The Portfolio Weights of the Efficient Portfolios for $d_4 = 0.02357$

Asset no	P_{CVaR}	$P_{1/4CVaR}$	$P_{1/2CVaR}$	$P_{3/4CVaR}$	P_{var}
4	0	0	0.046	0.09	0.134
17	0	4E-05	0.007	0	0
21	0.49	0.298	0.278	0.247	0.214
28	0.072	0	0	0	0
29	0	0.0138	0	0	0
44	0	0.027	0.041	0.053	0.059
45	0.373	0.269	0.225	0.222	0.217
56	0	0.249	0.248	0.225	0.2
58	6E-04	0.057	0.087	0.112	0.139
73	9E-04	0.007	0.016	0.007	0
76	0.063	0.079	0.054	0.045	0.036

TABLE 19.B5 The Portfolio Weights of the Efficient Portfolios for $d_5 = 0.02833$

Asset no	P_{CVaR}	$P_{1/4CVaR}$	$P_{1/2CVaR}$	$P_{3/4CVaR}$	P_{var}
4	0	0	0	0.003	0.018
21	0.472	0.418	0.367	0.325	0.324
45	0.119	0.105	0.066	0.032	0.022
56	0	0.065	0.138	0.197	0.196
58	0.29	0.342	0.370	0.394	0.397
76	0.119	0.069	0.059	0.049	0.043

APPENDIX 19.C: THE IN-SAMPLE PARAMETERS FOR THE RETURN DISTRIBUTIONS OF EFFICIENT PORTFOLIOS

The in-sample parameters for the return distributions of efficient portfolios are given in Tables 19.C1–19.C5.

TABLE 19.C1 In-Sample Parameters for the Return Distributions of Efficient Portfolios in the Mean-Variance-0.01CVaR Model with Expected Return $d_1 = 0.009268$

	P_{CVaR}	$P_{1/4CVaR}$	$P_{1/2CVaR}$	$P_{3/4CVaR}$	P_{var}
Median	0.010905	0.009989	0.010678	*0.011774*	0.011348
Standard Deviation	0.039557	0.032288	0.030899	0.030186	*0.030006*
Skewness	*0.175763*	−0.43318	−0.59261	−0.75996	−0.89894
Kurtosis	*−0.16328*	0.214433	0.763715	1.35481	1.964419
Minimum	*−0.05813*	−0.06857	−0.08198	−0.09601	−0.10946
Maximum	*0.128209*	0.085995	0.084375	0.081927	0.077194

TABLE 19.C2 In-Sample Parameters for the Return Distributions of Efficient Portfolios in the Mean-Variance-0.01CVaR Model with Expected Return $d_2 = 0.014034$

	P_{CVaR}	$P_{1/4CVaR}$	$P_{1/2CVaR}$	$P_{3/4CVaR}$	P_{var}
Median	0.009982	0.016801	0.016398	0.017359	*0.0176*
Standard Deviation	0.043277	0.035516	0.034453	0.03398	*0.033852*
Skewness	*0.238317*	−0.5367	−0.64824	−0.75897	−0.87193
Kurtosis	*0.100689*	0.329636	0.799505	1.213484	1.633637
Minimum	*−0.07056*	−0.07906	−0.08756	−0.09606	−0.10498
Maximum	*0.149618*	0.095584	0.093019	0.090123	0.087926

TABLE 19.C3 In-Sample Parameters for the Return Distributions of Efficient Portfolios in the Mean-Variance-0.01CVaR Model with Expected Return $d_3 = 0.018801$

	P_{CVaR}	$P_{1/4CVaR}$	$P_{1/2CVaR}$	$P_{3/4CVaR}$	P_{var}
Median	0.019982	0.021909	0.021945	*0.022453*	0.02225
Standard Deviation	0.051467	0.045116	0.043917	0.043138	*0.042869*
Skewness	*0.105138*	−0.27928	−0.35782	−0.44374	−0.50531
Kurtosis	0.816632	*0.588582*	0.748811	1.016336	1.309189
Minimum	*−0.09186*	−0.10046	−0.11094	−0.12183	−0.13216
Maximum	*0.188287*	0.139995	0.132851	0.127387	0.12672

TABLE 19.C4 In-Sample Parameters for the Return Distributions of Efficient Portfolios in the Mean-Variance-0.01CVaR Model with the Expected Return $d_4 = 0.023567$

	P_{CVaR}	$P_{1/4CVaR}$	$P_{1/2CVaR}$	$P_{3/4CVaR}$	P_{var}
Median	*0.026665*	0.02185	0.023582	0.022484	0.023786
Standard Deviation	0.071333	0.061135	0.059382	0.058374	*0.058031*
Skewness	*0.595438*	−0.12047	−0.23122	−0.30692	−0.36555
Kurtosis	3.354617	0.816052	*0.797705*	0.808283	0.834841
Minimum	*−0.12247*	−0.13142	−0.14231	−0.1528	−0.16327
Maximum	*0.367729*	0.204922	0.181086	0.162425	0.159635

TABLE 19.C5 In-Sample Parameters for the Return Distributions of Efficient Portfolios in the Mean-Variance-0.01CVaR Model with Expected Return $d_5 = 0.028334$

	P_{CVaR}	$P_{1/4CVaR}$	$P_{1/2CVaR}$	$P_{3/4CVaR}$	P_{var}
Median	*0.035256*	0.032523	0.027021	0.023606	0.022036
Standard Deviation	0.091039	0.088892	0.087699	0.087357	*0.087337*
Skewness	*0.319572*	0.215952	0.112308	0.050204	0.041352
Kurtosis	1.470049	1.069079	0.885207	*0.817357*	0.841093
Minimum	*−0.19129*	−0.19541	−0.19749	−0.19974	−0.20228
Maximum	*0.358639*	0.308329	0.26884	0.266499	0.267819

CHAPTER 20

Implied Non-Recombining Trees and Calibration for the Volatility Smile

CHRIS CHARALAMBOUS, NICOS CHRISTOFIDES, ELENI D. CONSTANTINIDE and SPIROS H. MARTZOUKOS

CONTENTS

20.1 INTRODUCTION

C ALIBRATING A TREE, OTHERWISE KNOWN AS CONSTRUCTING an implied tree, means finding the stock price and/or associated probability at each node in such a way that the tree reproduces the current market prices for a set of benchmark instruments. The main benefit of calibrating a model to a set of observed option prices is that the calibrated model is consistent with today's market prices. The calibrated model can then be used to price other more complex or less liquid securities, such as Over The Counter (OTC) options whose prices may not be available in the market.

The binomial tree is the most widely used tool in the fnancial pricing industry. The classic Cox–Ross–Rubinstein (CRR, 1979) binomial tree is a discretization of the Black–Scholes (BS, 1973) model since it is based on the assumption of the BS model that the underlying asset evolves according to a geometric Brownian motion with a constant volatility factor. This, however, contradicts the observed implied volatility, which suggests that volatility depends on both the strike and maturity of an option, a relationship commonly known as the volatility smile. This problem has motivated the recent literature on '*smile consistent*' *no-arbitrage models*. Consistency is achieved by extracting an implied evolution for the stock price from market prices of liquid standard options on the underlying asset. There are two classes of methodologies within this approach. Smile consistent *deterministic volatility* models (Derman and Kani 1994; Dupire 1994; Rubinstein 1994; Barle and Kakici 1995; Jackwerth and Rubinstein 1996; Jackwerth 1997; etc.); and *stochastic volatility* smile consistent models which allow for smile-consistent option pricing under the no-arbitrage evolution of the volatility surface (Derman and Kani 1998; Ledoit and Santa-Clara 1998; Britten-Jones and Neuberger 2000; etc.). The latter class of models is more general and it nests the former class of models (Skiadopoulos 2001). There also exist *non-parametric* methods, like Stutzer (1996) who uses the maximum entropy concept to derive the risk neutral distribution from the historical distribution of the asset price and Ait-Sahalia and Lo (1998) who propose a non-parametric estimation procedure for state-price densities using observed option prices.

Smile consistent deterministic volatility models are based on the assumption that the local volatility of the underlying asset is a known function of time and of the path and level of the underlying asset price. However, they do not specify local volatility in advance, but derive it endogenously from the European option prices. Therefore, they preserve the 'pricing by no-arbitrage' property of the BS model, and the markets are complete since the option's pay-off can be synthesized from existing assets.

Rubinstein (1994) finds the implied risk-neutral terminal-node probability distribution which is in the least-squares sense, closest to the lognormal subject to some constraints. The probabilities must add up to one and be non-negative. Moreover, they are calculated so that the present value of the underlying assets and all the European options calculated with these probabilities, fall between their respective bid–ask prices. This methodology allows for an arbitrary terminal-node probability distribution, but assumes that path probabilities leading to the same ending node are equal. Rubinstein's (1994) methodology suffers from the fact that options expiring at early time steps cannot be used for the construction of the tree. Thus, options with maturity other than the maturity of the options used during the construction of the tree are not consistent with market prices.

Jackwerth (1997) introduced generalized binomial trees as an extension of Rubinstein (1994). His model allows for an arbitrary terminal-node probability distribution, but also allows path probabilities leading to the same node to take different values.

Derman and Kani (1994) and Dupire (1994) constructed recombining binomial trees using a large set of option prices. For each node they need a corresponding option price with strike price equal to the previous node's stock price and expiring at the time associated with that node. Since they have fewer option prices than required, they need to interpolate and extrapolate from given option prices. Their trees are sensitive to the interpolation and extrapolation method and require adjustments to avoid arbitrage violations.

Barle and Cakici (1995) introduced a number of modifications which aimed to eliminate negative probabilities and improve the general stability of Derman's and Kani's (1994) model. Although their modified method fits the smile accurately, negative probabilities may still occur with increases in the volatility smile and interest rate. As they state, this is because of their '... strict requirement that continuous diffusion be modelled as a binomial process and on a recombining tree.' This problem can be referred to as a problem of *interdependencies* between nodes.

Possible methods that can be used to reduce the problem of inter-dependencies are the calibration of trinomial (or multinomial) trees or non-recombining trees. These extra degrees of freedom allow for more flexibility in the estimation of the distribution of the underlying asset.

Trinomial trees provide a much better approximation to the continuous time process than the binomial trees for the same number of steps. However, the extra degrees of freedom (additional number of nodes) require a larger number of simultaneous equations to be solved. Derman *et al.* (1996) proposed implied trinomial trees. In their model they use the additional parameters to conveniently choose the 'state space'of all node prices in the tree, and let only the transition probabilities be constrained by market options prices. Chriss (1996) generalized their method for American style options.

In this chapter we propose a method for calibrating a non-recombining (binary) tree, based on optimization. Specifically, we minimize the discrepancy between the observed market prices and the theoretical values with respect to the underlying asset at each node, subject to constraints that maintain risk neutrality and prevent arbitrage opportunities. Our model is built on a non-recombining tree[1] so as to allow the local volatility to be a function of the underlying asset and of time and to enable each node of the tree to act as an independent variable. Effectively, the problem under consideration is a non-convex optimization problem with linear constraints. We elaborate on the initial guess for the volatility term structure, and using methods from nonlinear constrained optimization we minimize the least-squares error function. Specifically, we adopt a penalty method and for the optimization we use a quasi-Newton algorithm. Because of the combinatorial nature of the tree and the large number of constraints, the search for an optimum solution as well as the choice of an algorithm that performs well becomes a very challenging problem.

Our model was created as a response for the need of a non-recombining implied tree. The main benefit of the model is its *analytical structure* which enables us to use efficient methods for nonlinear optimization. Although the method uses a large number of variables, due to the fact that we use efficient methods for optimization the model is not computationally intensive. Also, the proposed methodology can be easily modified to capture the observed bid/ask spreads in the market. This is very useful since the reported

---

[1] Other work we are aware of that uses a non-recombining tree is of Talias (2005) where for the calibration he uses genetic algorithms.

closing prices may not always be accurate, or may be inaccurate due to various market frictions. In addition, calibration of the non-recombining tree can be used for option replication with transaction costs as in Edirisinghe *et al.* (1993) and other related methodologies that require non-recombining trees.

In contrast to Rubinstein (1994), the proposed methodology can be easily modified to account for European contracts with different maturities. Our method does not need any interpolation or extrapolation across strikes and time to find hypothetical options as opposed to Derman and Kani (1994). Finally, the extra degrees of freedom and the analytical structure of the model would allow us to impose smoothness constraints on the distribution of the underlying asset if required.

We test our model using options data on the FTSE 100 index, for the year 2003 obtained from LIFFE. The results strongly support our modelling approach. Pricing results are smooth without the presence of an over-fitting problem and the derived implied distributions are realistic. Also, the computational burden is not a major issue.

The chapter continues as follows: In Section 20.2 we describe the proposed methodology and the initialization of the non-recombining tree. In Section 20.3 we discuss the imposed risk neutrality and no-arbitrage constraints. In Section 20.4 we describe the optimization algorithm. In Section 20.5 we test the model using FTSE 100 options data. Conclusions are in Section 20.6. In Appendix 20.A we prove the feasibility of the initialized tree, in Appendix 20.B we prove the feasibility of the initialized tree taking into account that the risk-free rate, dividend yield and time step are time dependent and in Appendix 20.C we adjust the formulas for time dependent risk free rate, dividend yield and step size.

20.2 THE PROPOSED METHODOLOGY AND INITIALIZATION OF THE NON-RECOMBINING TREE

Our goal is to develop an arbitrage-free risk neutral model that fits the smile, is preference-free, and can be used to value options form easily observable data. In order to allow more degrees of freedom, we use a non-recombining tree. In the following section we present the proposed methodology, and describe the initialization of the tree. Figure 20.1 shows a non-recombining tree with four steps.

The point (i, j) on the tree denotes:

i : the time dimension, $\quad i = 1, \ldots, n$

j : the asset (time specific) dimension, $\quad j = 1, \ldots, 2^{i-1}$

$S(i, j)$ is the value of the underlying asset at node (i, j).

Figure 20.2 shows a typical triplet in a non-recombining tree.

Let $C_{\text{Mkt}}(k)$, $k = 1, \ldots, N$ denote the market prices of N European calls, with strikes $K(k)$ and single maturities T. Also, let $C_{\text{Mod}}(x, k)$, $k = 1, \ldots, N$ denote the theoretical prices of the N calls obtained using the model. x denotes a vector containing the variables of the model which are the values of the underlying asset at each node of the tree, excluding its current value. The ideal solution is to find the values of the underlying asset (the model variables) at each node of the tree such that a perfect match is achieved

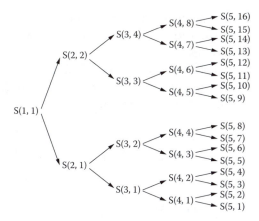

FIGURE 20.1 Non-recombining tree with 4 steps.

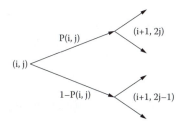

FIGURE 20.2 A typical triplet in a non-recombining tree.

between the option market prices and those predicted by the tree. However, due to market imperfections and other factors perfect matching may not always be possible. Therefore, we minimize the discrepancy between the observed market prices and the theoretical values produced by the model subject to constraints that prevent arbitrage opportunities.

We have to solve a non-convex constrained minimization problem with respect to the values of the underlying asset at each node:

$$\min_{x} \sum_{k=1}^{N} w_k f\left(C_{\mathrm{Mod}}(x, k), C_{\mathrm{Mkt}}(k)\right), \tag{20.1}$$

where f denotes a suitable objective function on the error between the observed and market prices. We can also allow for a weight factor, w_k.[2] In this chapter we use the least-squares error function which is defined as the sum of square differences between market prices and theoretical prices produced by the tree. The method can be adjusted easily for any other objective function.

---

[2] Weights can be related for example to the trading volume of the options.

The philosophy of the initialization of the non-recombining tree is the same as that of the construction of the standard CRR binomial tree, but we adjust the formulas so that the tree does not necessarily recombine.

We denote with $u(i,j)$ and $d(i,j)$ the up and down factors by which the underlying asset price can move in the single time step, Δt, given that we are at node (i,j). Δt, $u(i,j)$ and $d(i,j)$ factors are given by the following formulas:[3]

$$\Delta t = \frac{T}{n-1}, \tag{20.2}$$

$$u(i,j) = e^{\sigma(i)\sqrt{\Delta t}}, \quad i = 1,\ldots, n-1, \; j = 1,\ldots, 2^{i-1}, \tag{20.3a}$$

$$d(i,j) = e^{-\sigma(i)\sqrt{\Delta t}} = \frac{1}{u(i,j)}, \tag{20.3b}$$

where T is the option's time to maturity and $\sigma(i)$ is the volatility term structure at time step i.

We initialize the tree using the following volatility term structure:

$$\sigma(i) = \sigma(1)e^{\lambda(i-1)\Delta t}, \quad \lambda \in R, \quad i = 1,\ldots, n-1, \tag{20.4}$$

where λ is a constant parameter and $\sigma(1)$ is a properly chosen initial value for the volatility. If λ is positive, then volatility increases as we approach maturity and if λ is negative, then volatility decreases as we approach maturity.[4]

In order to preserve the risk neutrality at every time step and hence obtain a feasible initial tree, we choose λ to belong to the following interval (for proof see Appendix 20.A):

$$\lambda \in \left[\frac{1}{T}\log\left(\frac{|r_f - \delta|\sqrt{\Delta t}}{\sigma(1)}\right), +\infty\right). \tag{20.5}$$

By choosing λ from the above interval, we allow the initial volatility to increase or decrease across time. We make several consecutive draws from interval (20.5) until we find the value of λ that gives the 'optimal' tree.[5]

---

[3] For simplicity, we make the assumption that the risk free rate, the dividend yield and the step size do not change across time. Formulas adjusted for time dependence can be found in Appendices 20.B and 20.C.

[4] Other non-monotonic functions could also be used for σ (i) but what we have tried proved adequate for our purposes.

[5] Optimal tree is the one that gives the lowest-value objective function subject to the initial constraints.

We denote with $S(1,1)$ the current value of the underlying asset. The odd nodes of the tree $S(i,j)$, are initialized using the following equation:

$$S(i,j) = S\left(i-1, \frac{j+1}{2}\right) d\left(i-1, \frac{j+1}{2}\right), \quad i = 2, \ldots, n, \ j = 1, 3, \ldots, 2^{i-1} - 1.$$

(20.6a)

The even nodes of the tree $S(i,j)$, are initialized using the following equation:

$$S(i,j) = S\left(i-1, \frac{j}{2}\right) u\left(i-1, \frac{j}{2}\right), \quad i = 2, \ldots, n, \ j = 2, 4, \ldots, 2^{i-1}.$$ (20.6b)

We want to point out that Equations (20.3) to (20.6) are used only for initialization. Once the optimization process starts, each value of the underlying asset (except from $S(1,1)$) acts as an independent variable in the system.

Upward transition probabilities give the probability of moving from node (i,j) to node $(i+1, 2j)$ whereas downward transition probabilities give the probability of moving from node (i,j) to node $(i+1, 2j-1)$ for $i = 1, \ldots, n-1$ and $j = 1, \ldots, 2^{i-1}$. For the upward transition probabilities $p(i,j)$ between the various nodes of the tree we use the risk-neutral probability formula:[6]

$$p(i,j) = \frac{S(i,j)e^{(r_f - \delta)\Delta t} - S(i+1, 2j-1)}{S(i+1, 2j) - S(i+1, 2j-1)}, \quad i = 1, \ldots, n-1, \ j = 1, \ldots, 2^{i-1}, \quad (20.7)$$

where r_f denotes the annually continuously compounded riskless rate of interest and δ denotes the annually continuously compounded dividend yield. Their respective downward probability is equal to one minus the upward probability.

The call option value at the last time step is given by:

$$C(n,j) = \max\{S(n,j) - K, 0\}, \quad j = 1, \ldots, 2^{n-1}.$$ (20.8)

However, the function max is non-differentiable at $S(n, j) = K$. To overcome this problem, we propose the following smoothing approximation to $C(n, j)$:

$$\frac{C_\alpha(n,j)}{K} = \begin{cases} 0, & \text{for } S(n,j)/K \leq 1 - z/2, \\ \dfrac{S(n,j)}{K} - 1, & \text{for } S(n,j)/K \geq 1 + z/2, \\ \dfrac{1}{2z}\left[\left(\dfrac{S(n,j)}{K} - 1\right) + \dfrac{z}{2}\right]^2, & \text{for } 1 - z/2 < S(n,j)/K < 1 + z/2, \\ & j = 1, \ldots, 2^{n-1}, \end{cases}$$ (20.9a)

---

[6] Probability Equation (20.7) is effectively a martingale restriction (see Equation (6) and relevant discussion in Longstaff (1995)). Thus the numerical implementation of the model with this probability equation is restricted to a Markovian stochastic process.

where z is a small positive constant, for example 0.01 (see Figure 20.3).

The value of the call at intermediate nodes is given by the following equation:

$$C(i,j) = (p(i,j)C(i+1,2j) + (1 - p(i,j))C(i+1, 2j-1))e^{-r_f \Delta t},$$

$$i = n-1,\ldots,1, \quad j = 1,\ldots,2^{i-1}. \tag{20.9b}$$

20.3   RISK NEUTRALITY AND NO-ARBITRAGE CONSTRAINTS

In this section we describe the risk neutrality and no-arbitrage constraints. In order for the transition probabilities $p(i,j)$ defined in Equation (20.7) to be well specified, they should take values between zero and one. This implies the following *risk-neutrality* constraints: For $i = 1,\ldots,n-1, \quad j = 1,\ldots,2^{i-1}$,

$$S(i,j)e^{(r_f-\delta)\Delta t} \le S(i+1, 2j), \tag{20.10a}$$

and

$$S(i,j)e^{(r_f-\delta)\Delta t} \ge S(i+1, 2j-1). \tag{20.10b}$$

Risk neutrality constraints in the non-recombining tree prevent nodes $2j - 1$ and $2j$ to cross, for $i = 1,\ldots,n$ and $j = 1,\ldots,2^{i-1}$ (see Figure 20.1).

Options (puts and calls) have upper and lower bounds that do not depend on any particular assumptions on the factors that affect option prices. If the option price is above the upper bound or below the lower bound, there are profitable opportunities for arbitrageurs. To avoid such opportunities, we include the *no-arbitrage* constraints. Specifically, a European call with dividends should lie between the following bounds:

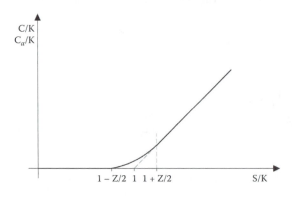

FIGURE 20.3 Smoothing of the option pay-off function at maturity.

$$\max\left(S(1,1)e^{-\delta T} - Ke^{-r_f T}, 0\right) \le C_{\text{Mod}} \le S(1,1).\tag{20.11}$$

Also, every value of the underlying asset on the tree should be greater or equal to zero. Thus, we also impose the following constraint:

$$S(i,j) \ge 0, \quad i = 2, \ldots, n, \quad j = 1, \ldots, 2^{i-1}.\tag{20.12}$$

20.4   THE OPTIMIZATION ALGORITHM

The objective of the problem is to minimize the least-squares error function of the discrepancy between the observed market prices and the theoretical values produced by the model. Thus, we have the following optimization problem:

$$\min_x \frac{1}{2} \sum_{k=1}^{N} (C_{\text{Mod}}(x, k) - C_{\text{Mkt}}(k))^2,\tag{20.13}$$

where $C_{\text{Mod}}(k)$ and $C_{\text{Mkt}}(k)$ denote the model and market price respectively of the kth call, $k = 1, \ldots, N$, subject to the constraints:

(i) $g_1(i,j) = S(i,j)e^{(r_f-\delta)\Delta t} - S(i+1, 2j-1) \ge 0,$
$$i = 1, \ldots, n-1, \quad j = 1, \ldots, 2^{i-1}.\tag{20.14a}$$

(ii) $g_2(i,j) = S(i+1, 2j) - S(i,j)e^{(r_f-\delta)\Delta t} \ge 0,$
$$i = 1, \ldots, n-1, \quad j = 1, \ldots, 2^{i-1}.\tag{20.14b}$$

(iii) $g_3(k) = S(1,1) - C_{\text{Mod}}(k) \ge 0, \quad k = 1, \ldots, N.\tag{20.14c}$

(iv) $g_4(k) = C_{\text{Mod}}(k) - \max(S(1,1)e^{-\delta T} - K(k)e^{-r_f T}, 0) \ge 0,$
$$k = 1, \ldots, N.\tag{20.14d}$$

(v) $g_5(i,j) = S(i,j) \ge 0, \quad i = 2, \ldots, n, \quad j = 1, \ldots, 2^{i-1}.\tag{20.14e}$

Since the problem under consideration is a non-convex optimization problem with linear constraints we adopt an exterior penalty method (Fiacco and McCormick 1968) to convert the nonlinear constrained problem into a nonlinear unconstrained problem. The Exterior Penalty Objective function that we use is the following:

$$P(x, \alpha) = \frac{1}{2} \sum_{k=1}^{N} (C_{\text{Mod}}(x, k) - C_{\text{Mkt}}(k))^2$$

$$+ \frac{\alpha}{2} \sum_{i=1}^{n-1} \sum_{j=1}^{2^{i-1}} ([\min(g_1(i,j), 0)]^2 + [\min(g_2(i,j), 0)]^2)$$

$$+ \frac{\alpha}{2} \sum_{k=1}^{N} ([\min(g_3(k), 0)]^2 + [\min(g_4(k), 0)]^2)$$

$$+ \frac{\alpha}{2} \sum_{i=2}^{n} \sum_{j=1}^{2^{i-1}} ([\min(g_5(i,j), 0)]^2). \tag{20.15}$$

The second, third and fourth terms in $P(x, \alpha)$ give a positive contribution if and only if x infeasible. Under mild conditions it can be proved that minimizing the above penalty function for strictly increasing sequence α tending to infinity, the optimum point $x(\alpha)$ of P tends to x^*, a solution of the constrained problem.

For the optimization we use a quasi-Newton algorithm. Specifically we use the BFGS formula[7] (Fletcher 1987). For the procedure of *Line Search* in the algorithm we use the Charalambous (1992) method. To achieve the best feasible solution, i.e. the solution that gives us a feasible tree with the smallest error function we force the algorithm to draw consecutively values of λ from the specified interval (20.5) until the objective function is smaller than 1.E-4 and also the penalty term equals zero, i.e. we have a feasible solution.

20.4.1   Implementation

For the implementation of the optimization method, we need to calculate the partial derivatives of $C_{\text{Mod}}(k)$[8] with respect to the value of the underlying asset at *each node*, for $k = 1, \ldots, N$, i.e. we want to find $\partial C(1, 1, k)/\partial S(i, j)$, $i = 2, \ldots, n$, $j = 1, \ldots, 2^{i-1}$[9] and $k = 1, \ldots, N$. For notational simplicity in the following, we assume that we have only one call option. For the computation of $\partial C(1,1)/\partial S(i, j)$, $\forall\ i, j$ we implement the following steps. We define the triplet vector (see Figure 20.2):

$$S_{i,j}^{(l)} = [S(i, j)\ S(i+1, 2j)\ S(i+1, 2j-1)]. \tag{20.16}$$

1st step: Compute the partial derivatives of the risk neutral transition probabilities, $\partial p(i, j)/\partial S(i, j)$, $\partial p(i, j)/\partial S(i+1, 2j)$ and $\partial p(i, j)/\partial S(i+1, 2j-1)$ for $i = 1, \ldots, n-1$, and $j = 1, \ldots, 2^{i-1}$. We summarize the derivatives in vector form (20.17):

---

[7] The BFGS formula was discovered in 1970 independently by Broyden, Fletcher, Goldfarb and Shanno.
[8] From now on we will use $C(1,1)$ instead of C_{Mod}.
[9] We do not calculate $\partial C(1, 1, k)/\partial S(1, 1)$ since $S(1, 1)$ is a known, fixed parameter, and thus does not take part in the optimization.

$$
\nabla_{S_{i,j}^{(l)}} p(i,j) \equiv \begin{bmatrix} \partial p(i,j)/\partial S(i,j) \\ \partial p(i,j)/\partial S(i+1,2j) \\ \partial p(i,j)/\partial S(i+1,2j-1) \end{bmatrix}
$$

$$
= \frac{1}{S(i+1,2j) - S(i+1,2j-1)} \begin{bmatrix} e^{(r_f-\delta)\Delta t} \\ -p(i,j) \\ -(1-p(i,j)) \end{bmatrix}. \tag{20.17}
$$

2nd step: Compute the partial derivatives $\partial C(i,j)/\partial S(i,j)$, for $i = 2,\ldots,n-1$ and $j = 1,\ldots,2^{i-1}$, $\partial C(i,j)/\partial S(i+1,\ 2j)$ and $\partial C(i,j)/\partial S(i+1,\ 2j-1)$ for $i=1,\ldots,n-1$, $j=1,\ldots 2^{i-1}$. We summarize the derivatives in vector form (20.18):

$$
\nabla_{S_{i,j}^{(l)}} C(i,j) \equiv \begin{bmatrix} \partial C(i,j)/\partial S(i,j) \\ \partial C(i,j)/\partial S(i+1,2j) \\ \partial C(i,j)/\partial S(i+1,2j-1) \end{bmatrix}
$$

$$
= \begin{bmatrix} \Delta(i,j) \\ p(i,j)\big(\Delta(i+1,2j) - \Delta(i,j)e^{\delta\Delta t}\big)e^{-r_f\Delta t} \\ (1-p(i,j))\big(\Delta(i+1,2j-1) - \Delta(i,j)e^{\delta\Delta t}\big)e^{-r_f\Delta t} \end{bmatrix}, \tag{20.18}
$$

where

$$
\Delta(i,j) = \frac{C(i+1,2j) - C(i+1,2j-1)}{S(i+1,2j) - S(i+1,2j-1)} e^{-\delta\Delta t}
$$

$$
= \frac{\partial C(i,j)}{\partial S(i,j)} \equiv \text{Delta Ratio.} \tag{20.19}
$$

3rd step: Compute the partial derivatives $\partial C_\alpha(n,j)/\partial S(n,j)$ for $j = 1,\ldots,2^{n-1}$. They are given by the following formula:

$$
\frac{\partial C_\alpha(n,j)}{\partial S(n,j)} = \begin{cases} 0, & \text{for } S(n,j) \le K(1-z/2), \\ 1, & \text{for } S(n,j) \ge k(1+z/2), \\ \dfrac{1}{z}\left[\left(\dfrac{S(n,j)}{K}-1\right)+\dfrac{z}{2}\right], & \text{for } k(1-z/2) \le S(n,j) \\ & \qquad < K(1+z/2). \end{cases} \tag{20.20}
$$

4th step: Compute the partial derivatives $\partial C(1, 1)/\partial S(i, j)$ for $i \geq 3$.

$$\frac{\partial C(1, 1)}{\partial S(i, j)} = \prod \{\text{of the probabilities on the path that take}$$

$$\text{us from node (1,1) to node } (i - 1, k)\}$$

$$\times \frac{\partial C(i - 1, k)}{\partial S(i, j)} e^{-(i-2)r_f \Delta t} \qquad (20.21)$$

$$k = \begin{cases} j/2, & \text{for even } j, \\ (j+1)/2, & \text{for odd } j. \end{cases}$$

For example,

$$\frac{\partial C(1, 1)}{\partial S(4, 6)} = p(1, 1)(1 - p(2, 2))\frac{\partial C(3, 3)}{\partial S(4, 6)} e^{-2r_f \Delta t},$$

$$\frac{\partial C(1, 1)}{\partial S(5, 3)} = (1 - p(1, 1))(1 - p(2, 1))p(3, 1)\frac{\partial C(4, 2)}{\partial S(5, 3)} e^{-3r_f \Delta t}.$$

20.5 APPLICATION USING FTSE 100 OPTIONS DATA

We use the daily closing prices of FTSE 100 call options of January 2003 to December 2003 as reported by LIFFE.[10] For the risk-free rate r_f, we use nonlinear cubic spline interpolation for matching each option contract with a continuous interest rate that corresponds to the option's maturity, by utilizing the 1-month to 12-month LIBOR offer rates, collected from Datastream.

Our initial sample (for the 12-month period) consists of 99051 observations. We adopt the following filtering rules.

i. Eliminate calls for which the call price is greater than the value of the underlying asset, i.e. $C_{Mkt} > S(1, 1)$. No observations are eliminated from this rule.
ii. Eliminate calls if the call price is less than its lower bound, i.e.
$C_{Mkt} < S(1, 1) e^{-\delta T} - Ke^{-r_f T}$. This rule eliminates 3206 observations.
iii. Eliminate calls with time to maturity less than 6 calendar days, i.e. $T < 6$. This rule eliminates 3109 observations.

---

[10] FTSE 100 options are traded with expiries in March, June, September and December. Additional serial contracts are introduced so that options trade with expiries in each of the nearest four months. FTSE 100 options expire on the third Friday of the expiry month. FTSE 100 option positions are marked-to-market daily based on the daily settlement price, which is determined by LIFFE and confirmed by the Clearing House.

iv. Eliminate calls if their closing price is less than 0.5 index points. This rule eliminates 11,373 observations.
v. Eliminate calls for which the trading volume is zero (since we want highly liquid options for calibration). This rule eliminates 66,826 observations.

The final sample consists of 14,537 observations.

In the implementation, for $\sigma(1)$ we use the at-the-money implied volatility given by LIFFE and for time to maturity, T, we use the calendar days to maturity. Also, since the underlying asset of the options on FTSE 100 is a futures contract, we make the standard assumption that the dividend yield equals the risk free rate. The model is applied every day, with $n = 6$ and also with $n = 7$. For each implementation, the options used have the same underlying asset and the same time to maturity.

The evidence for the behaviour of the futures volatility in the literature is not clear. According to Samuelson (1965) the volatility of futures price changes should increase as the delivery date nears. However, Bessembinder et al. (1996) find that the Samuelson hypothesis is not supported for options on financial futures. In order to choose the value of λ that gives the best feasible solution we make consecutive draws from interval (20.5), which allows for both, positive and negative values of λ. The first value of λ is that of its lower bound. However, since dividend yield equals risk free rate, instead of $|r_f - \delta|$ we set 1.E-8. The next value of λ equals the old plus an appropriately chosen step size.

For brevity, we present results only for the first trading day of each month of the year 2003 and only for $n = 6$ (Table 20.1). Trading Day is the trading day of each contract, Expiry is the expiration month of each contract, Asset is the value of the underlying asset at the specified trading day, N is the number of contracts used for the calibration (the contracts that on the same trading day, have the same underlying asset and the same expiration day), Error is the value of the objective function, Penalty is the value of the penalty term. Ideally we want the error function and the penalty term to tend to zero. Maturity is the calendar days until the maturity of the contract, and lambda is the value of λ that gives the best feasible solution. Also, we present results only when the number of option contracts is greater than 3, since with fewer options the distribution of the underlying asset taken will not be reliable.[11]

The results obtained support our modelling approach. As we can see in Table 20.1, in all cases the solution strictly satisfies the constraints since the penalty term equals zero. Also, we see that in 67 out of 69 cases, i.e. in 97.1% of the cases the error function tends to zero with an average value of 2.34E-08. In the other 2 cases, where the error function is greater than 1.E-4, the average error is 0.01. Similar results were found for $n = 7$.

Even though the problem requires a constrained non-convex optimization in $2(2^{n-1} - 1)$ variables, the use of efficient optimization algorithms prevents the calibration of the model from becoming computationally too intensive. On average, the computational time in minutes required for each calibration had a mean (median) 1.10 (0.03) for

---

[11] In Table 20.1 we note that for the same contract (same underlying asset, same expiration) the number of contracts used in the model changes across months. That is because some contracts were removed because of the filtering rules.

TABLE 20.1 Results for the Application of the Model on the 1st Trading Day of Each Month of the Year 2003: *Trading Day* is the Trading Day of Each Contract, *Expiry* is the Expiration Month of Each Contract, *Asset* is the Value of the Underlying Asset at the Specified Trading Day, *N* is the Number of Contracts Used for the Calibration, *Error* is the Value of the Objective Function, *Penalty* is the Value of the Penalty Term, *Maturity* is the Calendar Days until the Maturity of the Contract and *lambda* is the Value of λ that Gives the Best Feasible Solution

Trading day	Expiry	Asset	N	Error	Penalty	Maturity	Lambda
01/02/2003	Jan-03	4014	17	7.933E-11	0	15	4.3429
01/02/2003	Feb-03	4019	19	4.2E-05	0	50	−1.3851
01/02/2003	Mar-03	3991	12	5.855E-12	0	78	−6.6823
01/02/2003	Jun-03	3995	11	2.721E-13	0	169	−0.3963
01/02/2003	Dec-03	3999	6	0.0208333	0	351	0.4096
02/03/2003	Feb-03	3675.5	16	7.254E-08	0	18	−3.9616
02/03/2003	Mar-03	3646	14	8.611E-11	0	46	−1.5180
02/03/2003	Apr-03	3644.5	15	2.722E-12	0	73	−6.4346
02/03/2003	May-03	3645	6	2.355E-14	0	102	−6.6918
02/03/2003	Jun-03	3647	7	2.196E-12	0	137	−4.9425
02/03/2003	Sep-03	3640	5	1.337E-14	0	228	−1.1252
02/03/2003	Dec-03	3653.5	7	5.859E-11	0	319	−0.9666
03/03/2003	Mar-03	3657	16	6.572E-13	0	18	−3.9616
03/03/2003	Apr-03	3655	13	2.573E-11	0	45	−1.5455
03/03/2003	May-03	3655	9	2.466E-12	0	74	−8.3274
03/03/2003	Jun-03	3655.5	7	9.825E-14	0	109	−6.2312
03/03/2003	Sep-03	3645	9	1.735E-12	0	200	−3.3359
04/01/2003	Apr-03	3684.5	16	4.548E-11	0	16	−5.0218
04/01/2003	May-03	3683.5	16	4.396E-12	0	45	−1.5510
04/01/2003	Jun-03	3686.5	10	2.096E-11	0	80	−5.0920
04/01/2003	Jul-03	3693	7	1.222E-11	0	108	−6.1873
04/01/2003	Sep-03	3676.5	5	4.563E-11	0	171	−3.9295
04/01/2003	Mar-04	3667	5	1.301E-11	0	352	−1.8670
05/01/2003	May-03	3874	15	8.178E-08	0	15	−4.1763
05/01/2003	Jun-03	3879	14	1.581E-11	0	50	−1.3588
05/01/2003	Jul-03	3885.5	10	2.535E-11	0	78	−8.6239
05/01/2003	Sep-03	3870.5	4	2.657E-12	0	141	−2.1062
05/01/2003	Mar-04	3869	5	2.448E-13	0	322	−1.8172
06/02/2003	Jun-03	4132	16	3.535E-09	0	18	−25.1320
06/02/2003	Jul-03	4138.5	9	6.685E-12	0	46	−1.4751
06/02/2003	Aug-03	4128.5	9	9.668E-13	0	74	−9.0523
06/02/2003	Sep-03	4124	11	5.437E-11	0	109	−6.0923
06/02/2003	Dec-03	4136.5	9	3.798E-14	0	200	−3.2689
06/02/2003	Jun-04	4124	5	1.825E-11	0	381	−1.6875
07/01/2003	Jul-03	3967	13	0.0007343	0	17	13.5874
07/01/2003	Aug-03	3958	12	4.422E-12	0	45	0.1871
07/01/2003	Sep-03	3955	12	3.807E-11	0	80	−1.5851
07/01/2003	Oct-03	3959	4	5.92E-14	0	108	−6.1578
07/01/2003	Dec-03	3964	11	1.544E-12	0	171	−1.8451
07/01/2003	Mar-04	3956	7	1.265E-13	0	261	−2.4906
08/01/2003	Aug-03	4091.5	11	7.597E-12	0	14	1.1396
08/01/2003	Sep-03	4088.5	14	7.022E-11	0	49	5.6305
08/01/2003	Oct-03	4094.5	4	5.318E-11	0	77	−8.6243
08/01/2003	Nov-03	4096.5	5	2.43E-13	0	112	−5.8741
08/01/2003	Dec-03	4100.5	5	1.804E-14	0	140	−0.4681
08/01/2003	Mar-04	4097	4	1.013E-12	0	230	−1.7231

Table 20.1   (*Continued*)

Trading day	Expiry	Asset	N	Error	Penalty	Maturity	Lambda
08/01/2003	Jun-04	4098.5	6	7.7014E-12	0	321	0.9996
09/01/2003	Sep-03	4215	16	9.231E-11	0	18	−3.8161
09/01/2003	Oct-03	4222	9	4.986E-13	0	46	−1.4569
09/01/2003	Nov-03	4225	9	4.4323E-12	0	81	−8.1673
09/01/2003	Dec-03	4229	12	1.0014E-12	0	109	−0.6026
09/01/2003	Mar-04	4224.5	4	1.1111E-11	0	199	−32.5850
10/01/2003	Oct-03	4162.5	12	1.1715E-06	0	16	−4.3573
10/01/2003	Nov-03	4167	12	1.7641E-13	0	51	−1.3282
10/01/2003	Dec-03	4169.5	19	2.796E-12	0	79	−8.4832
10/01/2003	Jan-04	4173.5	5	5.4217E-13	0	107	−62.1642
10/01/2003	Mar-04	4162	8	1.1045E-11	0	169	−3.8885
10/01/2003	Jun-04	4171.5	5	1.0698E-13	0	260	−2.4977
11/03/2003	Nov-03	4330	12	9.9737E-12	0	18	26.2023
11/03/2003	Dec-03	4333	15	4.1511E-11	0	46	0.5538
11/03/2003	Jan-04	4344	9	3.2388E-12	0	74	−8.7229
11/03/2003	Feb-04	4354	7	2.0997E-12	0	109	−2.1552
11/03/2003	Mar-04	4329	7	2.6622E-13	0	136	−4.7931
11/03/2003	Jun-04	4343.5	7	4.4587E-13	0	227	−28.4200
12/01/2003	Dec-03	4415.5	13	6.9791E-11	0	18	15.2128
12/01/2003	Jan-04	4426	10	2.5172E-13	0	46	−1.4475
12/01/2003	Feb-04	4433.5	13	3.7691E-12	0	81	−8.1270
12/01/2003	Mar-04	4410.5	10	5.3369E-12	0	108	−6.0614
12/01/2003	Jun-04	4423.5	4	9.9101E-16	0	199	−3.2410

$n = 6$ and 2.27 (0.08) for $n = 7$. The computer used for the calibration of the model had the following specifications: a Pentium 4 (3.2 GHz) CPU, Memory 1GB (RAM) and Windows XP Professional operating system. The codes were written in Matlab R2006a. The computational time needed would have decreased if the codes were written in the C/C++ language.

When models provide an exact fit there is always the concern of over-fitting. We checked the model for over-fitting by pricing options with strikes in-between those used for the optimization (calibration). Then we made plots of the call prices (market prices and estimated from the model) versus moneyness. Over-fitting was also checked using a restricted sample consisting only of options with moneyness between 0.8 and 1.1, since these options are expected to be more liquid and more accurately priced.[12] For brevity, we exhibit only the plots for optimizations done in the first trading day of June (middle of the year) for the two samples using a tree with $n = 6$. As we see, for both samples the estimated call values increase smoothly with increasing moneyness without any evidence of over-fitting (see Figure 20.4). Similar results were obtained when a tree with $n = 7$ was used for the calibration procedure.

As a further check for over-fitting we use only part of the information to calibrate the tree and the other part to check the model using $n = 6, 7, 8$. Specifically, we leave out

---

[12] This sub-sample has a total of 13 696 observations for the year 2003.

consecutively one of the N options at each time and we calibrate our model with the remaining options. In order to preserve the options' moneyness range stability and avoid problems of extrapolation, we do not remove the options with the highest and lowest moneyness. Over-fitting is checked like before using the full and the restricted sample of options. For the calibration only cases consisting of $N > 8$ were used. Results for the mean and median absolute errors are given in Table 20.2. We see that the error (given an average contract size of 90 for the full and 74.4 for the restricted sample) is small and rather stable.[13]

Since implied volatility changes with strike and time to maturity (volatility smile) the index should have a *non-lognormal* distribution which implies that the log-returns will deviate from normality. In order to see how realistic is the distribution obtained from our model for year 2003, we calculate the statistics of the 1-month log-returns obtained from our model and compare them with the historical 1-month log-returns for the year 2003 and the years 2001–2005. Specifically, for each calibration (with $n = 6$ and $n = 7$) for which the options maturity was between 28 and 32 calendar days, we calculate the first four moments (mean, variance, skewness and kurtosis). Then, in order to get a feeling for the representative statistics of 1-month log-returns we provide for each of those moments the mean and the median. The statistics for $n = 6$ are summarized in Table 20.3. Similar statistics were found for $n = 7$. Liu *et al.* (2005) discuss the derivations of historical, and implied real and risk-neutral distributions for the FTSE 100 index. They demonstrate that the needed adjustments to get the implied real variance, skewness and kurtosis from the implied risk-neutral ones are minimal. Thus, knowing that our implied risk-neutral moments (beyond the mean) are very close to the implied real ones, we can then compare them with the historical ones (without expecting the two distributions to be identical). As we would expect, the mean of the implied risk-neutral distribution of log-returns differs from that of the historical distribution. Also, as we see, both the implied risk-neutral and the historical distribution deviate from normality since they exhibit negative skewness and (mostly) excess kurtosis. This is an indication that the implied distribution is realistic.

In order to give further evidence for the implied distributions obtained by our model, representative implied distributions (histograms) for the 1-month log-returns in June 2003 are shown in Figure 20.5a (full sample) and Figure 20.5b (restricted sample) for $n = 6$ and $n = 7$. To make the histograms of the implied distributions we make use of the Pearson system of distributions[14] as applied in Matlab.[15] Using the first four moments of the data it is easy to find in the Pearson system the distribution that matches these moments and to generate a random sample in order to produce a histogram

---

[13] Also, we compare our model (with respect to over-fitting) with the Black–Scholes model using the Whaley (1982) approach. According to this approach we find the volatility that minimizes the sum of square differences of the Black–Scholes option prices with their corresponding market prices using nonlinear minimization. Results show that the mean (median) absolute error using this approach is 7.36 (5.94) for the full sample and 6.61 (5.60) for the restricted sample which are much higher than the errors obtained using our model for $n = 6, 7, 8$.

[14] In the Pearson system there is a family of distributions that includes a unique distribution corresponding to every valid combination of mean, standard deviation, skewness and kurtosis.

[15] Copyright 2005 The MathWorks, Inc.

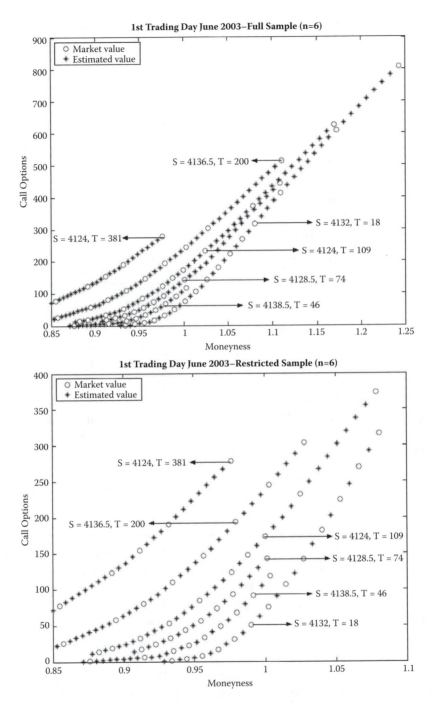

FIGURE 20.4 Plots of the call prices (market and estimated) for the FTSE 100 index, for the 1st trading day of June 2003. S denotes the value of the underlying asset and T the calendar days to maturity.

TABLE 20.2 Mean and Median Absolute Errors Using Our Model for $n = 6, 7, 8$ and Data from the Full and the Restricted Sample

Model		Absolute errors	
		Full sample	Restricted sample
$n = 6$	Mean	1.2458	1.1065
	Median	0.9163	0.8709
$n = 7$	Mean	1.1375	0.9792
	Median	0.8005	0.6929
$n = 8$	Mean	1.1286	0.9350
	Median	0.7928	0.6886
	Observations	446	405

TABLE 20.3 Implied Risk-Netural and Historical Statistics of the Distribution of the FTSE 100 1-Month Log-Returns

	Mean	Variance	Skewness	Kurtosis	Observations
Implied(2003, $n = 6$)					
Mean	−0.0024	0.0048	−0.6938	4.5075	58
Median	−0.0013	0.0027	−0.6653	3.6405	58
Historical					
2003	0.0106	0.0014	−0.6572	2.7689	12
2001−2005	−0.0021	0.0018	−1.1177	4.4749	59

corresponding to the implied distribution. From the figures, it is obvious that the implied distributions have negative skewness and positive kurtosis which is consistent with historical data. These figures are representative of the vast majority of cases.[16] Another interesting thing we observe is that distributions for $n = 6$ and $n = 7$ are practically indistinguishable for both samples.

20.6 CONCLUSIONS

In most options markets, the implied Black–Scholes volatilities vary with both strike and expiration, a relationship commonly known as the volatility smile. In this chapter we capture the implied distribution from option market data using a non-recombining (binary) tree allowing the local volatility to be a function of the underlying asset and of time. The problem under consideration is a non-convex optimization problem with linear constraints. We elaborate on the initial guess for the volatility term structure, and use nonlinear constrained optimization to minimize the least-squares error function on market prices. Specifically we adopt a penalty method and the optimization is implemented using a quasi-Newton algorithm. Appropriate constraints allow us to maintain risk neutrality and to prevent arbitrage opportunities. The proposed model can accommodate European options with single maturities and, with minor modifications,

---

[16] In rare exceptions only we have implied distributions close to normal or even leptokurtic.

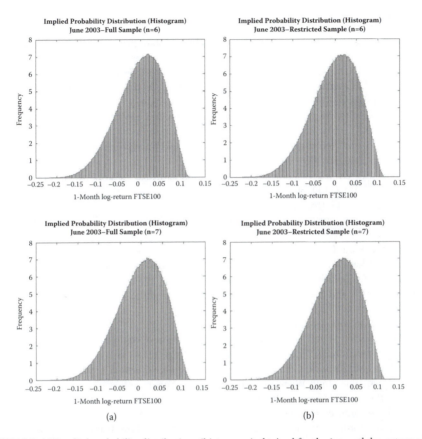

FIGURE 20.5 (a) Implied probability distributions (histograms) obtained for the 1-month log-return of June 2003 using the full sample for $n = 6$ and $n = 7$. (b) Implied probability distributions (histograms) obtained for the 1-month log-retrun of June 2003 using the restricted sample for $n = 6$ and $n = 7$.

options with multiple maturities. Also, this method is flexible since it applies to arbitrary underlying asset distributions, which implies arbitrary local volatility distributions. Market implied information embodied in the constructed tree can help the pricing and hedging of exotic options and of OTC options on the same underlying process. We test our model using FTSE 100 options data. The results obtained strongly support our modelling approach. Pricing results are smooth without the presence of an over-fitting problem, and the derived implied distributions are realistic. Also, the computational burden is not a major issue.

ACKNOWLEDGEMENTS

The authors are grateful for financial support from a research grant on Contingent Claims from the University of Cyprus.

REFERENCES

Ait-Sahalia, Y. and Lo, A.W., Nonparametric estimation of state-price densities implicit in financial asset prices. *J. Finan.*, 1998, **53**, 499–547.

Barle, S. and Cakici, N., Growing a smiling tree. *Risk*, 1995, **8**, 76–81.

Bessembinder, H., Coughenour, J.F. and Seguin, P.J., Is there a term structure of futures volatilities? Re-evaluating the Samuelson hypothesis. *J. Deriv.*, 1996, **4**, 45–48.

Black, F. and Scholes, M., Pricing of options and corporate liabilities. *J. Polit. Econ.*, 1973, **81**, 637–659.

Britten-Jones, M. and Neuberger, A., Option prices, implied price processes and stochastic volatility. *J. Finance*, 2000, **55**, 839–866.

Charalambous, C., Conjugate gradient algorithm for efficient training of artificial neural networks. *IEE Proc. G*, 1992, **139**, 301–310.

Chriss, N., Transatlantic trees. *Risk*, 1996, **9**, 143–159.

Cox, J., Ross, S. and Rubinstein, M., Option pricing: A simplified approach. *J. Financ. Econ.*, 1979, 7, 229–263.

Derman, E. and Kani, I., Riding on a smile. *Risk*, 1994, **7**, 32–39.

Derman, E. and Kani, I., Stochastic implied trees: Arbitrage pricing with stochastic term and strike structure of volatility. *Int. J. Theor. Appl. Finan.*, 1998, **1**, 61–110.

Derman, E., Kani, I. and Chriss, N., Implied trinomial trees of the volatility smile. *J. Deriv.*, 1996, **3**, 7–22.

Dupire, B., Pricing with a smile. *Risk*, 1994, **7**, 18–20.

Edirisinghe, C., Naik, V. and Uppal, R., Optimal replication of options with transaction costs and trading restrictions. *J. Financ. Quant. Anal.*, 1993, **28**, 117–138.

Fiacco, A.V. and McCormick, G.P., *Nonlinear Programming: Sequential Unconstrained Minimization Techniques*, 1968 (John Wiley and Sons, Inc.: New York).

Fletcher, R., *Practical Methods of Optimization*, 1987 (Wiley: Chichester).

Jackwerth, J., Generalized binomial trees. *J. Deriv.*, 1997, **5**, 7–17.

Jackwerth, J. and Rubinstein, M., Recovering probability distributions from option prices. *J. Finance*, 1996, **51**, 1611–1631.

Ledoit, O. and Santa-Clara, P., Relative pricing of options with stochastic volatility, Working Paper, 1998 (University of California: Los Angeles).

Liu, X., Shackleton, B.M., Taylor, J.S. and Xu, X., Closed transformations from risk-neutral to real-world distributions. *Working Paper*, 2005 (University of Lancaster).

Longstaff, F.A., Option pricing and the martingale restriction. *Rev. Financ. Stud.*, 1995, **8**, 1091–1124.

Rubinstein, M., Implied binomial trees. *J. Finance*, 1994, **49**, 771–818.

Samuelson, P., Proof that properly anticipated prices fluctuate randomly. *Ind. Manag. Rev.*, 1965, **6**, 41–49.

Skiadopoulos, G., Volatility smile consistent option models: A survey. *Int. J. Theor. Appl. Finance*, 2001, **4**, 403–437.

Stutzer, M., A simple nonparametric approach to derivative security valuation. *J. Finance*, 1996, **51**, 1633–1652.

Talias, K., Implied binomial trees and genetic algorithms, Doctoral dissertation, 2005, Imperial College, London.

Whaley, R., Valuation of American call options on dividend-paying stocks. *J. Financ. Econ.*, 1982, **10**, 29–58.

APPENDIX 20.A: FEASIBILITY OF THE INITIALIZED NON-RECOMBINING TREE

We initialize the tree using the following volatility term structure:

$$\sigma(i) = \sigma(1)e^{\lambda(i-1)\Delta t}, \quad \lambda \in R,$$

where $i = 1, \ldots, n$. The feasibility of the initial tree depends on the right choice of the local volatility term structure; hence to obtain a feasible initial tree we must find an interval with the appropriate values of λ. In order to preserve the risk neutrality at every time step, the following constraints must be satisfied:

$$S(i, j)e^{(r_f - \delta)\Delta t} \leq S(i + 1, 2j), \tag{20.A1a}$$

$$S(i, j)e^{(r_f - \delta)\Delta t} \geq S(i + 1, 2j - 1). \tag{20.A1b}$$

Also,

$$S(i + 1, 2j) = S(i, j)u(i, j) = S(i, j)e^{\sigma(i)\sqrt{\Delta t}}, \tag{20.A2a}$$

$$S(i + 1, 2j - 1) = S(i, j)d(i, j) = S(i, j)e^{-\sigma(i)\sqrt{\Delta t}}. \tag{20.A2b}$$

Substituting (20.A2a) and (20.A2b) to (20.A1a) and (20.A1b) respectively, we get the following inequalities:

$$\sigma(i) \geq (r_f - \delta)\sqrt{\Delta t}, \tag{20.A3a}$$

$$\sigma(i) \geq -(r_f - \delta)\sqrt{\Delta t}. \tag{20.A3b}$$

Thus we have that

$$\sigma(i) \geq |r_f - \delta|\sqrt{\Delta t} \quad \forall i. \tag{20.A4}$$

For $\lambda \geq 0$, $\sigma(i) = \sigma(1)e^{\lambda(i-1)\Delta t}$ is strictly increasing. Since (20.A4) holds for every i this means that

$$\min \sigma(i) \geq |r_f - \delta|\sqrt{\Delta t} \quad \text{or} \quad \sigma(1) \geq |r_f - \delta|\sqrt{\Delta t}. \tag{20.A5}$$

The minimum value of $\sigma(i)$ is for $i = 1$ ($\sigma(1)$) thus (20.A5) is independent of λ. Therefore, if λ is positive there is no upper bound for λ.

For $\lambda < 0$, $\sigma\,(i) = \sigma(1)e^{\lambda(i-1)\Delta t}$ is strictly decreasing. Since (20.A4) holds for every i this means that

$$\min \sigma(i) \geq \left| r_f - \delta \right| \sqrt{\Delta t} \Rightarrow$$

$$\sigma(n) \geq \left| r_f - \delta \right| \sqrt{\Delta t} \Rightarrow$$

$$e^{\lambda(n-1)\Delta t} \geq \frac{\left| r_f - \delta \right| \sqrt{\Delta t}}{\sigma(1)}.$$

But $(n - 1)\Delta t = T$, thus,

$$\lambda \geq \frac{1}{T} \log\left(\frac{\left| r_f - \delta \right| \sqrt{\Delta t}}{\sigma(1)} \right). \tag{20.A6}$$

If we allow λ to take both negative and positive values, then λ should belong in the interval,

$$\lambda \in \left[\frac{1}{T} \log\left(\frac{\left| r_f - \delta \right| \sqrt{\Delta t}}{\sigma(1)} \right), +\infty \right). \tag{20.A7}$$

APPENDIX 20.B: FEASIBILITY OF THE INITIALIZED NON-RECOMBINING TREE ASSUMING TIME DEPENDENT r_f, δ AND Δt

We denote with $r_f\,(i)$ and $\delta(i)$ the risk free rate and dividend yield respectively between two consecutive time steps, i.e. between time step i and $i + 1$, $i = 1, \ldots, n - 1$. (See Figure 20.B1.).

We initialize the tree using the following volatility term structure:

$$\sigma(i) = \sigma(1)e^{\lambda \sum_{j=1}^{i-1} \Delta t(j)}, \quad \lambda \in R,$$

where $i = 1, \ldots, n$. The feasibility of the initial tree depends on the right choice of the local volatility term structure; hence to obtain a feasible initial tree we must find an interval with the appropriate values of λ. In order to preserve the risk neutrality at every time step, the following constraints must be satisfied:

$$S(i, j)e^{(r_f(i) - \delta(i))\Delta t(i)} \leq S(i + 1, 2j), \tag{20.B1a}$$

$$S(i, j)e^{(r_f(i) - \delta(i))\Delta t(i)} \geq S(i + 1, 2j - 1). \tag{20.B1b}$$

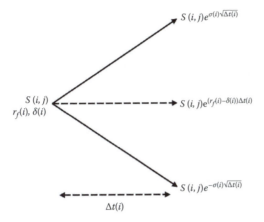

FIGURE 20.B1 A typical triplet in the initialization of the non-recombining tree assuming r_f, δ and Δt to be time dependent.

Also,

$$S(i+1, 2j) = S(i,j)u(i,j) = S(i,j)e^{\sigma(i)\sqrt{\Delta t(i)}}, \qquad (20.\text{B2a})$$

$$S(i+1, 2j-1) = S(i,j)d(i,j) = S(i,j)e^{-\sigma(i)\sqrt{\Delta t(i)}} \qquad (20.\text{B2b})$$

Substituting (20.B2a) and (20.B2b) to (20.B1a) and (20.B1b) respectively we get the following inequalities:

$$\sigma(i) \geq (r_f(i) - \delta(i))\sqrt{\Delta t(i)}, \qquad (20.\text{B3a})$$

$$\sigma(i) \geq -(r_f(i) - \delta(i))\sqrt{\Delta t(i)}. \qquad (20.\text{B3b})$$

Thus we have that

$$\sigma(i) \geq \left| r_f(i) - \delta(i) \right| \sqrt{\Delta t(i)} \quad \forall i. \qquad (20.\text{B4})$$

For $\lambda \geq 0$, $\sigma(i) = \sigma(1)\exp[\lambda \sum_{j=1}^{i-1} \Delta t(j)]$ is strictly increasing. Let $\xi_M = \max_i \left| r_f(i) - \delta(i) \right| \sqrt{\Delta t(i)}$. Then (20.B4) holds for every i if

$$\min_i \sigma(i) \geq \xi_M \quad \text{or} \quad \sigma(1) \geq \xi_M. \qquad (20.\text{B5})$$

The minimum value of $\sigma(i)$ is for $i=1$ ($\sigma(1)$), thus (20.B5) is independent of λ. Therefore, if λ is positive there is no upper bound for λ.

For $\lambda < 0$, $\sigma(i) = \sigma(1)\exp[\lambda\sum_{j=1}^{i-1}\Delta t(j)]$ is strictly decreasing. Let $\xi_m = \min_i |r_f(i) - \delta(i)|\sqrt{\Delta t(i)}$. Then (20.B4) holds for every i if

$$\min_i \sigma(i) \geq \xi_m,$$

$$\sigma(n) \geq \xi_m,$$

$$\sigma(1)e^{\lambda\sum_{j=1}^{n-1}\Delta t(j)} \geq \xi_m.$$

But, $\sum_{j=1}^{n-1}\Delta t(j) = T$, thus,

$$\lambda \geq \frac{1}{T}\log\left(\frac{\xi_m}{\sigma(1)}\right). \tag{20.B6}$$

If we allow λ to take both negative and positive values, then λ should belong to the interval,

$$\lambda \in \left[\frac{1}{T}\log\left(\frac{\xi_m}{\sigma(1)}\right), +\infty\right). \tag{20.B7}$$

APPENDIX 20.C: FORMULAS ADJUSTED FOR TIME DEPENDENT r_f, δ AND Δt

We denote with $r_f(i)$ and $\delta(i)$ the risk free rate and dividend yield respectively between two consecutive time steps, i.e. between time step i and $i + 1$, $i = 1, \ldots, n - 1$ and with r'_f and δ' we denote the risk free rate and dividend yield respectively from today until the maturity of the option, i.e. from $i = 1$ to $i = n$.

If we allow r_f, δ and Δt to be time dependent the equations of the main text are replaced with the following:

$$u(i, j) = e^{\sigma(i)\sqrt{\Delta t(i)}} \tag{20.3a'}$$

$$d(i, j) = e^{-\sigma(i)\sqrt{\Delta t(i)}} = \frac{1}{u(i, j)}, \quad i = 1, \ldots, n - 1, \ j = 1, \ldots, 2^{i-1} \tag{20.3b'}$$

$$\lambda \in \left[\frac{1}{T}\log\left(\frac{\xi_m}{\sigma(1)}\right), +\infty\right), \tag{20.5'}$$

where

$$\xi_m = \min_i \left| r_f(i) - \delta(i) \right| \sqrt{\Delta t(i)}.$$

$$p(i,j) = \frac{S(i,j)e^{(r_f(i)-\delta(i))\Delta t(i)} - S(i+1,2j-1)}{S(i+1,2j) - S(i+1,2j-1)},$$
$$i = 1,\ldots,n-1, \; j = 1,\ldots,2^{i-1}, \tag{20.7'}$$

$$C(i,j) = (p(i,j)C(i+1,2j) + (1-p(i,j))C(i+1,2j-1))e^{-r_f(i)\Delta t(i)}$$
$$i = n-1,\ldots,1, \; j = 1,\ldots,2^{i-1} \tag{20.9b'}$$

$$S(i,j)e^{(r_f(i)-\delta(i))\Delta t(i)} \le S(i+1,2j), \quad i = 1,\ldots,n-1,$$
$$j = 1,\ldots,2^{i-1} \tag{20.10a'}$$

$$S(i,j)e^{(r_f(i)-\delta(i))\Delta t(i)} \ge S(i+1,2j-1) \quad i = 1,\ldots,n-1,$$
$$j = 1,\ldots,2^{i-1} \tag{20.10b'}$$

$$\max\left(S(1,1)e^{-\delta'T} - Ke^{-r_f'T}, 0\right) \le C_{\text{Mod}} \le S(1,1), \tag{20.11'}$$

$$g_1(i,j) = S(i,j)e^{(r_f(i)-\delta(i))\Delta t(i)} - S(i+1,2j-1) \ge 0$$
$$i = 1,\ldots,n-1, \; j = 1,\ldots,2^{i-1}, \tag{20.14a'}$$

$$g_2(i,j) = S(i+1,2j) - S(i,j)e^{(r_f(i)-\delta(i))\Delta t(i)} \ge 0$$
$$i = 1,\ldots,n-1, \; j = 1,\ldots,2^{i-1}, \tag{20.14b'}$$

$$g_3(k) = S(1,1) - C_{\text{Mod}}(k) \ge 0, \quad k = 1,\ldots,N, \tag{20.14c'}$$

$$g_4(k) = C_{\text{Mod}}(k) - \max(S(1,1)e^{-\delta'T} - K(k)e^{-r_f'T}, 0) \ge 0,$$
$$k = 1,\ldots,N \tag{20.14d'}$$

$$\nabla_{S_{i,j}^{(l)}} p(i,j) \equiv \begin{bmatrix} \partial p(i,j)/\partial S(i,j) \\ \partial p(i,j)/\partial S(i+1,2j) \\ \partial p(i,j)/\partial S(i+1,2j-1) \end{bmatrix}$$

$$= \frac{1}{S(i+1,2j)-S(i+1,2j-1)} \begin{bmatrix} e^{(r_f(i)-\delta(i))\Delta t(i)} \\ -p(i,j) \\ -(1-p(i,j)) \end{bmatrix}, \qquad (20.17')$$

$$\nabla_{S_{i,j}^{(l)}} C(i,j) \equiv \begin{bmatrix} \partial C(i,j)/\partial S(i,j) \\ \partial C(i,j)/\partial S(i+1,2j) \\ \partial C(i,j)/\partial S(i+1,2j-1) \end{bmatrix}$$

$$= \begin{bmatrix} \Delta(i,j) \\ p(i,j)\left(\Delta(i+1,2j)-\Delta(i,j)e^{\delta(i)\Delta t(i)}\right)e^{-r_f(i)\Delta t(i)} \\ \left\{ \begin{matrix} (1-p(i,j))(\Delta(i+1,2j-1)) \\ -\Delta(i,j)e^{\delta(i)\Delta t(i)} \end{matrix} \right\} e^{-r_f(i)\Delta t(i)} \end{bmatrix} \qquad (20.18')$$

$$\Delta(i,j) = \frac{C(i+1,2j)-C(i+1,2j-1)}{S(i+1,2j)-S(i+1,2j-1)} e^{-\delta(i)\Delta t(i)}$$

$$= \frac{\partial C(i,j)}{\partial S(i,j)} \equiv \text{Delta Ratio}, \qquad (20.19')$$

$$\frac{\partial C(1,1)}{\partial S(i,j)} = \prod \{\text{of the probabilities on the path that take}$$

$$\text{us from node}(1,1)\text{to node}(i-1,k)\}$$

$$\times \frac{\partial C(i-1,k)}{\partial S(i,j)} e^{-\sum_{h=2}^{i-1} r_f(h)\Delta t(h)} \qquad (20.21')$$

$$k = \begin{cases} j/2, & \text{for even } j, \\ (j+1)/2, & \text{for odd } j. \end{cases}$$

Index

Z